CAMBRIDGE LIBRARY COLLECTION

Books of enduring scholarly value

Life Sciences

Until the nineteenth century, the various subjects now known as the life sciences were regarded either as arcane studies which had little impact on ordinary daily life, or as a genteel hobby for the leisured classes. The increasing academic rigour and systematisation brought to the study of botany, zoology and other disciplines, and their adoption in university curricula, are reflected in the books reissued in this series.

The Ferns (Filicales)

Frederick Orpen Bower (1855–1948) was a renowned botanist best known for his research on the origins and evolution of ferns. Appointed Regius Professor of Botany at the University of Glasgow in 1885, he became a leading figure in the development of modern botany and the emerging field of paleobotany, devising the interpolation theory of the life cycle in land plants. First published between 1923 and 1928 as part of the Cambridge Botanical Handbook series, *The Ferns* was the first systematic classification of ferns according to anatomical, morphological and developmental features. In this three-volume work Bower analyses the major areas of comparison between different species, describes primitive and fossil ferns and compares these species to present-day fern species, providing a comprehensive description of the order. Volume 1 describes and analyses the features of ferns which Bower uses in his system of classification.

Cambridge University Press has long been a pioneer in the reissuing of out-of-print titles from its own backlist, producing digital reprints of books that are still sought after by scholars and students but could not be reprinted economically using traditional technology. The Cambridge Library Collection extends this activity to a wider range of books which are still of importance to researchers and professionals, either for the source material they contain, or as landmarks in the history of their academic discipline.

Drawing from the world-renowned collections in the Cambridge University Library, and guided by the advice of experts in each subject area, Cambridge University Press is using state-of-the-art scanning machines in its own Printing House to capture the content of each book selected for inclusion. The files are processed to give a consistently clear, crisp image, and the books finished to the high quality standard for which the Press is recognised around the world. The latest print-on-demand technology ensures that the books will remain available indefinitely, and that orders for single or multiple copies can quickly be supplied.

The Cambridge Library Collection will bring back to life books of enduring scholarly value (including out-of-copyright works originally issued by other publishers) across a wide range of disciplines in the humanities and social sciences and in science and technology.

The Ferns (Filicales)

Treated Comparatively with a View to their Natural Classification

VOLUME 1:
ANALYTICAL EXAMINATION OF THE CRITERIA OF COMPARISON

F. O. BOWER

CAMBRIDGE UNIVERSITY PRESS

Cambridge, New York, Melbourne, Madrid, Cape Town, Singapore,
São Paolo, Delhi, Dubai, Tokyo, Mexico City

Published in the United States of America by Cambridge University Press, New York

www.cambridge.org
Information on this title: www.cambridge.org/9781108013161

This edition first published 1923
This digitally printed version 2010

ISBN 978-1-108-01316-1 Paperback

Cambridge Botanical Handbooks

Edited by A. C. Seward and A. G. Tansley

THE FERNS

VOLUME I

CAMBRIDGE UNIVERSITY PRESS
C. F. CLAY, MANAGER
LONDON : FETTER LANE, E.C. 4

LONDON : H. K. LEWIS & CO., LTD.,
136, Gower Street, W.C. 1
LONDON : WHELDON & WESLEY, LTD.,
2-4, Arthur Street, New Oxford Street, W.C. 2
NEW YORK : THE MACMILLAN CO.
BOMBAY, CALCUTTA, MADRAS : MACMILLAN AND CO., LTD.
TORONTO : THE MACMILLAN CO. OF CANADA, LTD.
TOKYO : MARUZEN-KABUSHIKI-KAISHA

View in the Kibble House, Botanic Garden, Glasgow, from a photograph by Mr Fullarton; showing the double-headed specimen of *Cyathea dealbata*, its trunk covered with *Trichomanes venosum*, with Dicksonias right and left, and with *Todea barbara* at its foot, together with other Ferns.

THE FERNS
(FILICALES)

TREATED COMPARATIVELY WITH A VIEW
TO THEIR NATURAL CLASSIFICATION

VOLUME I

ANALYTICAL EXAMINATION
OF THE CRITERIA OF COMPARISON

BY

F. O. BOWER, Sc.D., LL.D., F.R.S.

REGIUS PROFESSOR OF BOTANY
IN THE UNIVERSITY OF GLASGOW

CAMBRIDGE
AT THE UNIVERSITY PRESS
1923

PRINTED IN GREAT BRITAIN

PREFACE

"Little do ye know your own blessedness; for to travel hopefully is a better thing than to arrive, and the true success is to labour." R. L. STEVENSON, *Essay on "El Dorado."*

IN this passage Stevenson enunciates a truth that applies with singular force to those who enter on morphological enquiry. To travel hopefully is the chosen pursuit of all who study large groups of organisms with a view to reducing them to order, so as to throw light on their origin and evolution. In such quests no one need expect under present conditions to arrive at the final destination of complete and assured knowledge. If any one should indulge this hope his disappointment is certain. Even if he did so arrive, and found himself able fully to demonstrate the whole truth, how greatly would the quest lose in its interest. It is in the pursuit of his "El Dorado" of evolutionary history, not in the arrival there, that the true blessedness of the morphologist lies. It behoves then those who travel on this journey not to hurry unduly, but to consider with critical care the manner of their journeying, rather than to seek short cuts to an elusive goal.

Bacon in the *Novum Organum* laid it down that there are only two ways in which knowledge can be sought: viz. by anticipations of Nature, and by interpretations of Nature. In the former method men pass at once from particulars to the highest generalities, and thence deduce all intermediate propositions. In the latter they rise by gradual induction, and successively from particulars to axioms of the lowest generality, then to intermediate axioms, and so to the highest. He asserts that this is the true way. There will be good ground for hope, he says, when any one can be found content to begin at the beginning, and to apply himself to "experience and particulars."

It will depend in some degree upon the data available for study of the question in hand which of these two methods shall be used. Where the facts are few and disconnected the former may appear preferable. But the fewer the facts the less reliable will be the conclusions: till at last the results arrived at by the deductive method may be little better than speculations, liable to be modified or refuted by any positive discovery. The deductive study of evolution is in fact the refuge of those who, finding themselves destitute of the necessary data, are still determined to arrive at some conclusion. They accordingly use their imagination to make up the deficiency. The inductive method will be preferred in any case where the facts are many and cognate, so that they can be arranged in continuous sequences. It is true that the data may be read in divers ways, according as greater or less weight is assigned to one detail or to another. But against this criticism it may be urged that the several sequences of facts may be so linked into a coherent web that they will mutually support or check one another. It should be the constant practice of the morphologist to use them in this way. Stability will thus be given to the more general conclusions, and these will be based not upon preconceptions but upon the orderly use of "experience and particulars."

The Class of the Filicales stands pre-eminent as a field for the practice of this inductive method. It is represented by many thousands of living species, spread over all quarters of the globe. These plants have characters in common which seldom leave a doubt of the "Filical" nature of any species. Nevertheless the Ferns vary so greatly, and, as it is found, so consistently in their details that ample material is at hand for their comparison, and phyletic seriation. Besides this the Class is so well represented among the fossils that the sequences traced by comparison of living types may often be effectively checked by the order of occurrence of the several forms in the successive geological strata. But such a study of Ferns with a view to visualising their own inter-relationships need not be the sole aim before us here. If their comparison leads to a clear conception of some general type of primitive organisation from which in the long distant past the whole phylum may have sprung, this should serve to indicate probable relationships with other primitive phyla: and so the comparative study of their phylesis may contribute to still wider views on the Descent of Land-Living Plants. This has been in part the aim of the author in undertaking the present work. Primarily it is a treatise on the Filicales: but secondarily, it will touch broader questions of Morphology and of Evolution.

It has not been the author's intention to give an exhaustive summary of all knowledge relating to Ferns, nor will complete citations of the profuse literature of the subject be attempted. Instead of this, when works of other writers are quoted in which a special branch of the literature is fully cited, the fact will be noted. Readers will thus be given clues to the whole literature, which may readily be followed up by them; while the present volume will be relieved of much unnecessary print.

Where illustrations have been borrowed the debt is acknowledged in the legends, and here grateful thanks are accorded to their authors; some particular acknowledgments appear on p. viii. A large proportion of the figures in this book are original, and to them no author's name is attached. In the production of these, and in the general illustration of this volume, the author desires to acknowledge substantial assistance given by the Carnegie Trustees. By aiding the publication of results obtained during the present period of high prices, they promote the advancement of Science in a most practical way, and deserve not only the thanks of the immediate recipient, but the general regard of men of Science.

F. O. BOWER.

GLASGOW,
December, 1922.

CONTENTS

CHAP.		PAGE
I.	THE LIFE-HISTORY OF A FERN	1
II.	THE HABIT AND THE HABITAT OF FERNS	26
III.	A THEORETICAL BASIS FOR THE SYSTEMATIC TREATMENT OF FERNS	52
IV.	MORPHOLOGICAL ANALYSIS OF THE SHOOT-SYSTEM OF FERNS	66
V.	LEAF-ARCHITECTURE OF FERNS	81
VI.	CELLULAR CONSTRUCTION	105
VII.	THE VASCULAR SYSTEM OF THE AXIS	120
	(A) THE PROTOSTELE AND ITS IMMEDIATE DERIVATIVES	120
VIII.	THE VASCULAR SYSTEM OF THE AXIS	140
	(B) THE STELAR STRUCTURE OF THE LEPTOSPORANGIATE FERNS	140
IX.	THE VASCULAR SYSTEM OF THE LEAF	160
X.	SIZE A FACTOR IN STELAR MORPHOLOGY	177
XI.	DERMAL AND OTHER NON-VASCULAR TISSUES	195
XII.	THE SPORE-PRODUCING ORGANS	206
	(A) THE SORUS	206
XIII.	THE SPORE-PRODUCING ORGANS	242
	(B) THE SPORANGIUM	242
XIV.	THE GAMETOPHYTE, AND SEXUAL ORGANS	273
XV.	THE EMBRYO	298
XVI.	ABNORMALITIES OF THE LIFE-CYCLE	318
XVII.	ORGANOGRAPHIC COMPARISON OF THE FILICALES WITH OTHER PLANTS	336
	POSTSCRIPT TO VOLUME I	348
	INDEX	349

VIEW IN THE KIBBLE HOUSE, BOTANIC GARDEN, GLASGOW — *Frontispiece*

ACKNOWLEDGMENTS

Illustrations. Figures 1–23, 25–28, 37–39, 59, 66, 67, 74, 75, 83–85, 98, 99, 106, 107, 112, 128, 130, 140, 144, 148, 150, 163, 164, 171, 174, 197, 199, 200–202, 204, 207, 208, 212, 221, 238, 244–248, 252, 254, 255, 259, 265, 284, 285, 289–293, 295–302 are reproduced by arrangement with Messrs Macmillan & Co., Ltd.; Figs. 307, 308 by permission of the Royal Society, Edinburgh; Figs. 76–79, 81, 82, 91, 97, 114, 127, 134, 135, 141, 142, 149, 151, 161, 173, 175, 176, 178, 180–182 by permission from the *Proceedings* and *Transactions* of the Royal Society, Edinburgh; Figs. 35, 40, 41, 43, 44, 49, 51, 52, 54 are reproduced from H. Christ's *Geographie der Farne*, by arrangement with Herr Gustav Fischer; Figs. 24, 129 from G. F. Atkinson's *The Biology of Ferns*; Figs. 33, 34, 50, 53, 56, 66, 74, 75, 93, 94, 189, 195, 196, 236 by arrangement with Herr Wilhelm Engelmann. Numerous Figures by the author himself and by other Botanists have been drawn from the *Annals of Botany*, for the use of which due acknowledgment is made to the Clarendon Press, Oxford.

LIST OF WORKS OF GENERAL USE IN THE STUDY OF THE FILICALES

I. SWARTZ. Synopsis Filicum. 1806.
II. SCHKUHR. Die Farnkräuter. Wittemberg. 1809. 219 Plates.
III. PRESL. Tentamen Pteridographiae. Pragae. 1836.
IV. PRESL. Supplementum Tentaminis Pteridographiae. Pragae. 1845.
V. KUNZE. Die Farnkräuter. Suppl. zu Schkuhr. Leipzig. 1840–1854. 140 Plates.
VI. BAUER & HOOKER. Genera Filicum. London. 1842. 120 Plates.
VII. HOOKER & GREVILLE. Icones Filicum. 2 Vols. London. 1831. 240 Plates.
VIII. HOOKER. Species Filicum. 5 Vols. London. 1846–1864. 304 Plates.
IX. HOOKER & BAKER. Synopsis Filicum. 2nd Edn. London. 1883.
X. HOFMEISTER. Higher Cryptogamia. Engl. Edn. Ray Society. 1862.
XI. FÉE. Mémoires sur la Famille des Fougères. Strasbourg. 1844–1873. 289 Plates.
XII. METTENIUS. Filices Horti Lipsiensis. 1856. 30 Plates.
XIII. METTENIUS. Ueber einige Farrngattungen. I–VI. Frankfurt. 1857–1859.
XIV. BEDDOME. Ferns of British India. Madras. 1867–8. 345 Plates.
XV. LUERSSEN. Rabenhorst's Kryptogamen Flora. Vol. III. Leipzig. 1889. (The literature on Ferns of Central Europe is here fully quoted.)
XVI. BAKER. Summary of New Ferns. Oxford. 1892. New Ferns of 1892–3. London. 1896.
XVII. CHRIST. Die Farnkräuter der Erde. Jena. 1897.
XVIII. CHRIST. Geographie der Farne. Jena. 1910. (Fern-Floras are here fully quoted.)
XIX. RACIBORSKI. Flora von Buitenzorg. Die Pteridophyten. Leiden. 1898.
XX. ENGLER & PRANTL. Natürl. Pflanzenfam. I. 4. Leipzig. 1902. (The general literature on Ferns is here fully quoted.)
XXI. CAMPBELL. Mosses and Ferns. 2nd Edn. New York. 1905. (The literature on microscopic analysis is here fully quoted.)
XXII. CAMPBELL. The Eusporangiatae. Carnegie Institution. Washington. 1911.
XXIII. VAN ROSENBURGH. Malayan Ferns. Batavia. 1908.
XXIV. JENMAN. Ferns and Fern-Allies of the British West Indies, and Guiana. Bull. Bot. Dept. Trinidad.
XXV. CHRISTENSEN. Index Filicum. Copenhagen. 1906. Suppl. 1906–1912. Copenhagen. 1913. (An alphabetical list of all names of Ferns.)
XXVI. BOWER. The Origin of a Land Flora. London. 1908.
XXVII. BOWER. Studies on Spore-Producing Members. I–V. Phil. Trans. 1894–1903.
XXVIII. BOWER. Studies in the Phylogeny of Ferns. I–VII. Ann. of Bot. 1910–1918.
XXIX. LOTSY. Botanische Stammesgeschichte. Bd. II. Jena. 1909. (The general literature on Ferns is here very fully quoted.)
XXX. SEWARD. Fossil Plants. Vol. II. Cambridge. 1910. (The fossil literature is here fully quoted.)
XXXI. VON GOEBEL. Organographie. 2nd Edn. Part II. Jena. 1918.
XXXII. SCOTT. Fossil Botany. 3rd Edn. Part I. London. 1920.

Organic Nature is the better understood by consulting the fossils that illuminate its history. It is like an ancient and enduring document, whose diction may seem clear, but is found to be more intelligible and fuller of meaning when the text is studied historically, due thought being given to the conditions under which it was written.

CHAPTER I

THE LIFE-HISTORY OF A FERN

THE present volume will treat generally of the Plants included in the Class of the Filicales, or as they are commonly called, the Ferns. Any such treatise must take as its starting point a knowledge of the usual structure and successive phases of life of the plants in question. Experience shows that the cycle of life is uniform in its essentials for all normal Ferns, while its features correspond also to those seen in land-living plants at large. That life-cycle comprises two alternating phases, or "generations." The one is the relatively large leafy plant popularly recognised and designated as the "Fern-Plant": the other is a relatively small and simple green scale-like structure, called the "Prothallus." The relation of these to one another, and the constant alternation which they show, are best illustrated by a description of the successive incidents seen in the life-history of some common Fern. The familiar Male Shield Fern (*Dryopteris* (*Nephrodium*) *Filix-mas*, Rich.) will serve as a suitable example.

This Fern is known to everyone as growing in woods and hedgerows, and even in more exposed situations, such as open gills and hill-sides of higher-lying districts (Fig. 31, p. 26). It presents a robust appearance, and when fully developed it consists of a simple shoot with an oblique and massive stock, which is relatively short. This is entirely covered over by the bases of the leaves, of which the youngest form a closely packed terminal bud (Fig. 1, *A*, *D*). Those leaves which are seated further from the apex, and immediately below the terminal bud, may in summer be found to be of large size and complex structure (Fig. 2). Collectively they form a crown-like series surrounding the apex of the stem that bears them. Passing again further back from the apex of the stem, the surface is found to be closely invested by the bases of the numerous leaves of former seasons, the upper portions of which have rotted away. Clearly then the leaves are borne upon the stock in acropetal succession. While young they are densely covered by chaffy flattened scales, which take a rusty colour with age; but they mostly fall away from the mature leaf, though often persisting on its stalk (Fig. 2). If the plant be dug up, and the soil be carefully removed from it, an ample root-system will be seen, consisting of thin, wiry, and dark-coloured fibrils, which spring from the basal parts of the leaves.

All these parts of the Fern-Plant consist of tracts of tissue differentiated to subserve distinct functions. The most notable is the Vascular Skeleton, which appears in the Male Shield Fern as a cylindrical network of strands within the massive axis (Fig. 1, *E*, *F*). It throws off branches on the one hand

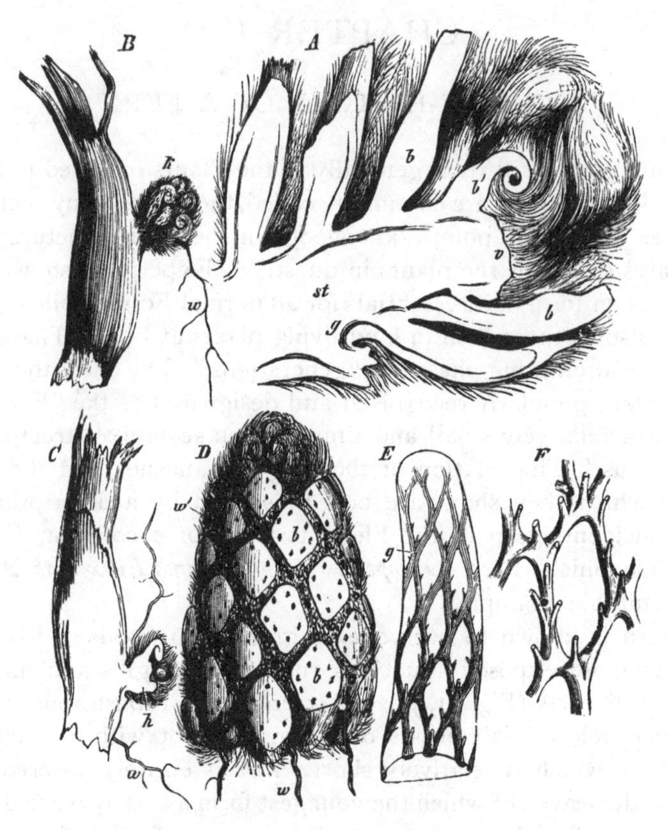

Fig. 1. *Dryopteris* (*Nephrodium*) *Filix-mas*, Rich. *A* = stock in longitudinal section: *v*=apex: *st*=stem: *b*=leaf-stalks: *b'*=one of the still folded leaves: *g*=vascular strands. *B*=leaf-stalk bearing at *k* a bud with root (*w*), and several leaves. *C*=a similar leaf-stalk cut longitudinally, bearing bud (*h*), with root (*w*). *D*=stock from which the leaves have been cut away to their bases, leaving only those of the terminal bud. The spaces between the leaves are filled with numerous roots, *w*, *w'*. *E*=stock from which the rind has been removed to show the vascular network (*g*). *F*=a single mesh of the network (foliar gap) enlarged, showing the insertion of the leaf-trace strands. (After Sachs.)

into the leaves, where they ramify and extend upwards to the extreme tips and margins. On the other hand strands of vascular tissue, springing from the leaf-bases, extend towards the tips of the roots, and laterally into their branchlets. The Vascular System is thus a connected conducting system throughout the plant. It is embedded in soft parenchymatous tissues, which serve various purposes in different parts. Thus in the root they may be absorbent, while on the other hand they transmit the fluids absorbed to the conducting system. In the stem they may serve the purpose of storage of reserve materials, while in the leaf the parenchyma carries on the function of photo-synthesis, together with the passing on of the supply thus acquired to the conducting system. The parts exposed to the air are covered by an

Fig. 2. *Dryopteris Filix-mas*, Rich. Fertile leaf about one-sixth natural size, the lower part with the under surface exposed. To the left a single fertile segment, bearing kidney-shaped sori, enlarged about seven times. (After Luerssen.)

epidermal layer with a cuticularised external wall, which prevents the indiscriminate loss of water by surface-evaporation. But the epidermis is perforated by numerous stomata the mobile guard-cells of which according to circumstances can control the width of the pores leading to the ventilating spaces within. Finally, there are also firm brown resistant tissues disposed sometimes near the outer surface, as in the stem and leaf-stalk: sometimes more deeply seated, as in the root, while in the leaf they follow the course of the vascular strands. These give to the several parts increased mechanical strength, while their thickened walls have also a special power of retaining water within their substance.

Thus constituted the Male Shield Fern is an organism which is capable of leading an independent life on exposed land-surfaces. It is in a position to nourish itself by taking up from the soil the water and salts which it requires, and to elaborate therefrom, and from the carbon-dioxide of the air, fresh supplies of organic food. Further, though it is frequently found

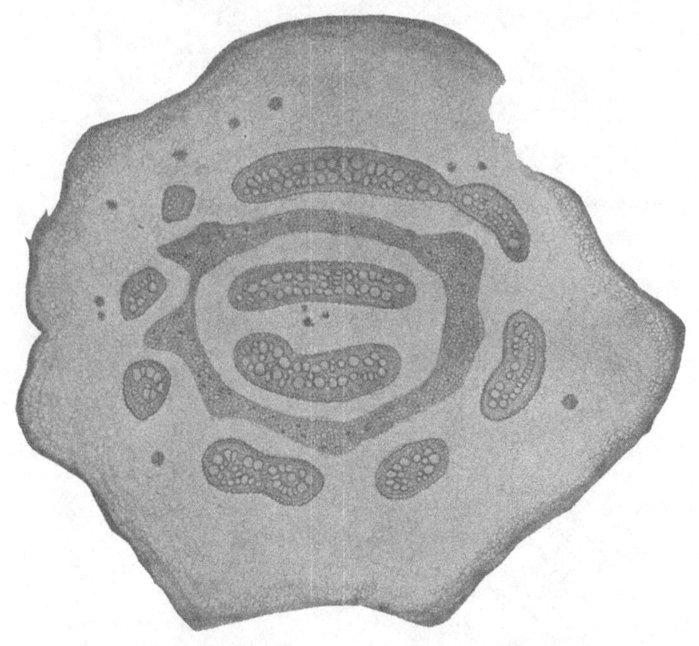

Fig. 3. Transverse section of rhizome of Bracken *Pteridium aquilinum*, showing the outer and inner series of meristeles, and the irregular bands of sclerenchyma between them. These are embedded in soft ground-parenchyma, with a hard sclerotic rind. (× 10.)

growing in situations where moisture is abundant and the air moist, still it can resist considerable drought, and is capable of living under as exacting conditions as any ordinary terrestrial plant.

For comparative purposes the vascular tissues, which are so marked a feature of land-living plants generally, are those which command the greatest attention. For the study of the tissues composing a vascular strand, or *meristele* as it is called, a rhizome with long internodes, such as the Bracken, gives the best results. In a transverse section of it taken between the leaf-insertions, an outer and an inner series of vascular strands are found, separated by an incomplete ring of sclerenchyma. The outer series correspond to the mesh-work of *Dryopteris* (*Nephrodium*), the inner are accessory or medullary meristeles (Fig. 3). Each one is a compact body of vascular tissue, and is circumscribed by a complete endodermal sheath, which delimits it sharply from the surrounding parenchyma. The details are shown respectively in transverse and in longitudinal section in Figs. 4 and 5. Each meristele consists of a central core of xylem, surrounded by phloem. A small part of one of them, examined either in transverse or in longitudinal section under a high power, gives the following succession of tissues (Figs. 4, 5). Passing inwards from the starchy ground-tissue, with intercellular spaces (*g*),

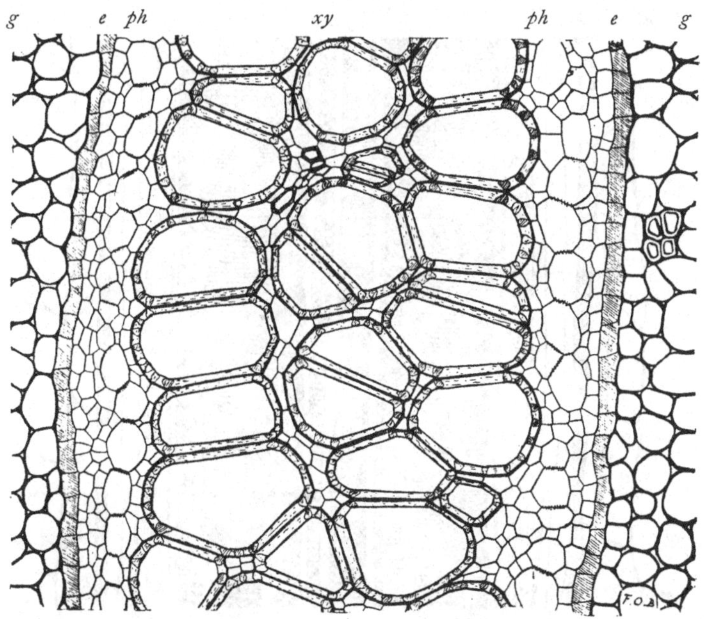

Fig. 4. Part of a transverse section of a meristele of Bracken. *g* = ground parenchyma. *e* = endodermis. *ph* = phloem with sieve-tubes. *xy* = xylem, with large scalariform tracheides. Some smaller tracheides lying centrally are the protoxylem. Note that no intercellular spaces are seen within the endodermis. (× 75.)

the layer of brownish cells of the *endodermis* (*e*) forms a continuous barrier, delimiting the strand sharply. Within it follows the *pericycle*, with its cells not very regularly disposed, but corresponding roughly to the cells of the endodermis, both having been derived by division from a single layer. Within this comes the *phloem* (*ph*), with large *sieve-tubes* as the characteristic elements. They are thin-walled, with watery contents. The lateral walls where two adjoin bear the sieve-plates, and are recognised by glistening globules that adhere to them. They are embedded in parenchyma, which extends inwards into the xylem, and may be called collectively *conjunctive parenchyma*. The chief features of the *xylem* (*xy*) are the *tracheides*, which are relatively large, with a very characteristic polygonal outline. They have woody walls, and no protoplasmic contents. Where two adjoin the walls are flattened, and of double thickness, showing that each has its own share of the thickening, which overarches the pit-membrane as in the pits of Conifers. But where the tracheide abuts on parenchyma-cells the pits are narrower. Internally, and usually about the foci of the elliptical meristele, smaller tracheides are found. These are the first-formed tracheides, or *protoxylem*. The meristele of a Fern is thus concentric in construction; it is strictly delimited, and has no provision for increase in size.

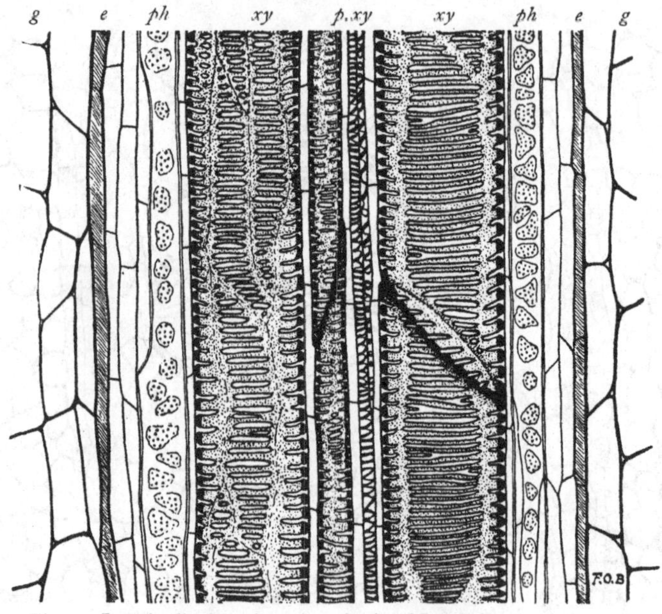

Fig. 5. Longitudinal section of meristele of Bracken. Lettering and magnification as in Fig. 4.

A transverse section (Fig. 4) gives only one aspect in which such complicated tissues can be studied. Its interpretation is aided by longitudinal sections (Fig. 5). It is then seen that the sieve-tubes, which are elongated and pointed, bear their numerous sieve-areas upon the lateral walls: and that the spindle-shaped tracheides bear also upon their lateral walls those transversely elongated pits which give them the so-called scalariform appearance. Interspersed between the tracheides and sieve-tubes are cells of the conjunctive parenchyma, and the meristele is delimited by endodermis (*e*), as seen in the transverse section.

The *tracheide* of the Fern resembles that of the Pine in being of spindle form, with its thickened lignified walls marked by bordered pits. But whereas the pits in the Pine are circular, those in the Fern are liable to be transversely elongated, as is natural in tracheides so wide as these are. Their features are well seen in longitudinal sections, but better if they are isolated by maceration (Fig. 6, *A*). The elongated pits lie parallel to one another, and this is specially well seen where two wide tracheides have faced one another. From the ladder-like appearance that results they have been called *scalariform tracheides*. Examined under a high power the double outline of the pits is seen, and when the pits are small and circular the similarity to those of the Pine is plain (Fig. 6, *B*). In most Ferns the pit-membranes persist, but in *Pteridium* they appear to be liable to be broken down, and the cavities of adjacent tracheids thrown together as they are seen to be in vessels. The tracheides of the protoxylem are seen in longitudinal section to be spiral or reticulate, as in other Vascular Plants (Fig. 5).

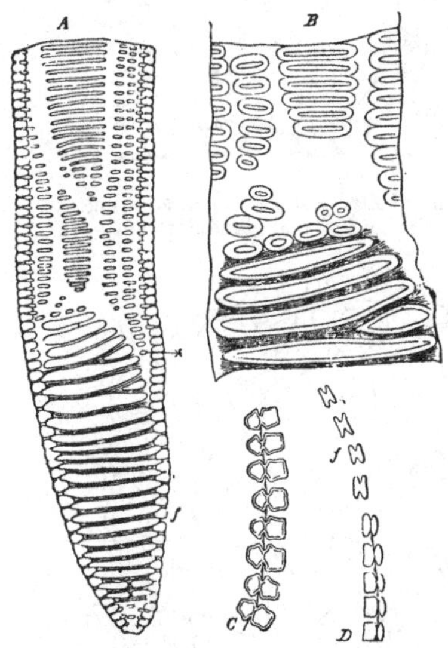

Fig. 6. Tracheides of *Pteridium*. *A* = the end and about one-third of the length of a tracheid, with part of the lateral wall in surface view, showing scalariform marking (× 100). *B* = part of *A* magnified 200. *C* = thin longitudinal section through a lateral wall where two tracheides adjoin (× 375). *D* = similar section through oblique wall at *f* (× 200). There the pit-membranes are not visible. (After De Bary.)

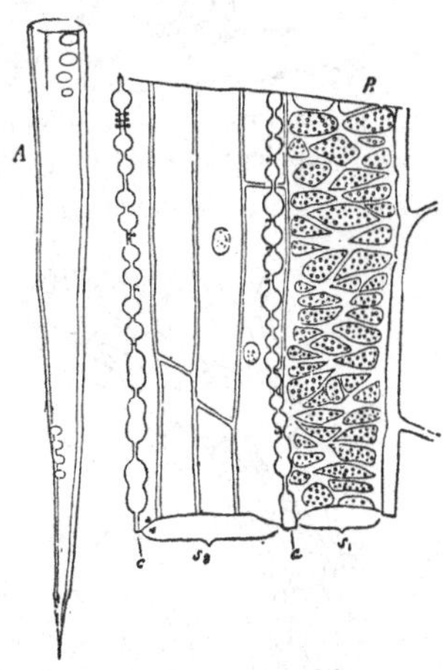

Fig. 7. Sieve-tubes of *Pteridium*. *A* = end of a tube separated by maceration (× 100). *B* = longitudinal section through phloem showing one sieve-tube with the sieve-plates (s_1) in surface view. *c*, *c* are walls shown in section, bearing sieve-pits (× 200).

The *sieve-tubes* are also spindle-shaped, and are without companion-cells. Their cellulose walls are swollen. Where two sieve-tubes adjoin, numerous thinner sieve-areas of irregular outline are borne. They are found to be perforated by very fine protoplasmic threads extending between highly refractive globules that adhere to the walls (Fig. 7). Such tracheides and sieve-tubes are characteristic of Ferns and, with differences of detail, of other Pteridophytes as well.

The anatomy of the *leaf* in Ferns resembles that of Seed-Plants down even to the collateral structure of the vascular strands. Being chiefly shade-loving plants chlorophyll is usually present in the cells of their epidermis, and the differentiation of the mesophyll into palisade and spongy parenchyma is not marked (Fig. 8). In these respects they resemble the leaves of Angiosperms of similar habit. In the *roots* of Ferns, as in those of Seed-Plants,

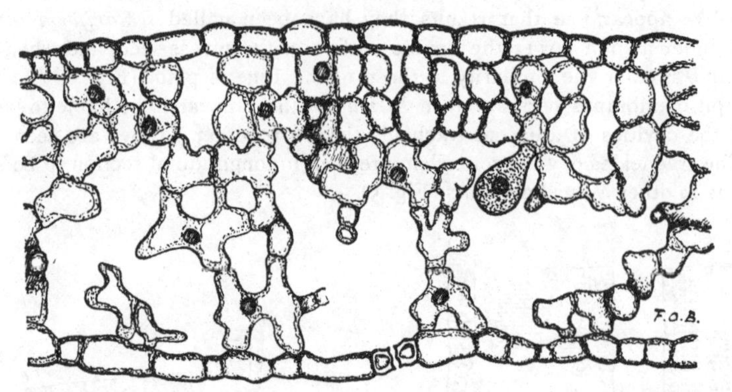

Fig. 8. Transverse section of part of pinnule of *Dryopteris* (× 150), showing epidermis, and the spongy mesophyll, with an internal glandular cell.

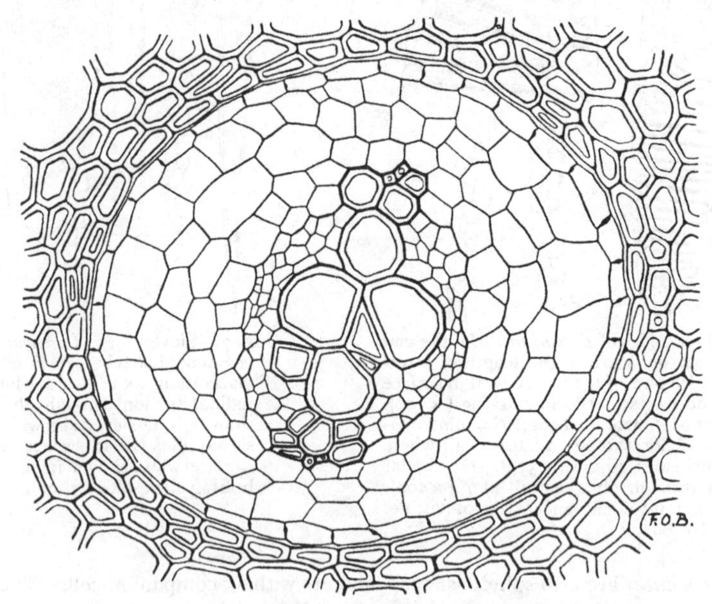

Fig. 9. Transverse section of a root of a Fern (*Pellaea*) (× 150). Outside lies the sclerotic cortex, limited internally by a definite endodermis. There are two groups of protoxylem; a very broad pericycle, of 3 or 4 layers, surrounds the vascular tissues.

there is a superficial piliferous layer, a broad cortex, and a contracted stele. But usually the inner cortex is very strongly lignified, up to the endodermis, which is thin-walled (Fig. 9). The pericycle which follows is variable, sometimes being greatly enlarged as a water-storage-tissue. The protoxylems are peripheral and two or sometimes more in number, the phloem-groups alternating with them. In fact the root of a Fern is constructed essentially on the plan of that in Seed-Plants. As there is no secondary thickening the roots of Ferns are all fibrous. The lateral roots arise opposite to the protoxylems, and there they originate from definite cells of the endodermis, which may often be recognised beforehand by their size and contents.

While we recognise the substantial similarity of Ferns and Seed-Plants in respect of form and structure of stem, leaf, and root, these plants differ in the construction of their *apical meristems*. In Seed-Plants these are small-celled tissues, and more or less definitely stratified. In Ferns such as *Osmunda*, *Dryopteris* or *Polypodium*, a single large cell, the *apical or initial cell*, occupies the tip of each growing part. It has a definite shape, and segments are cut off from its sides in definite succession. As the whole

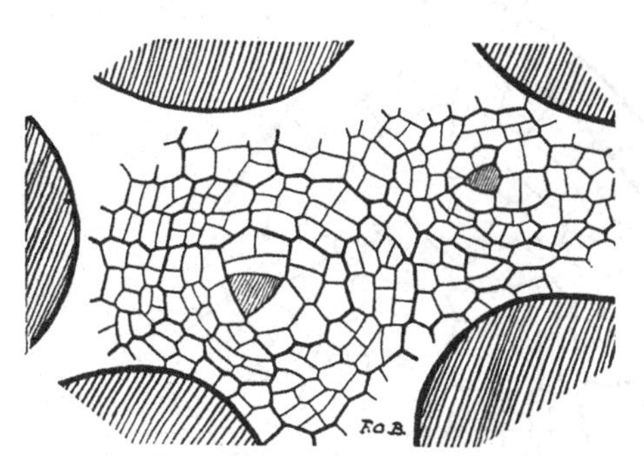

Fig. 10. Apex of stem of *Osmunda regalis*, seen from above, showing the three-sided apical cells of stem and of leaf shaded. The successive segments of the apical cell form the whole of the apical cone. (×83.)

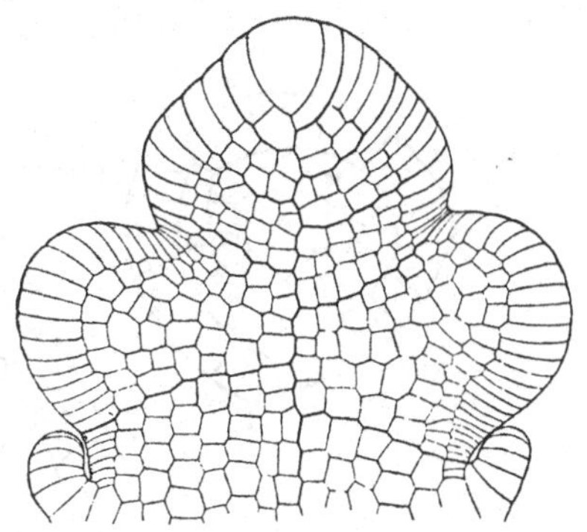

Fig. 11. Young leaf of *Ceratopteris*, in surface view, after Kny; showing two-sided apical cell; and the marginal series, continuous round the young pinnae. The latter do not correspond to the segments from the apical cell.

tissue of the stem, leaf, or root is derived from such segments, the whole of each part is referable in origin to its apical cell, which maintains its identity throughout. The form of the cell in roots, in most stems, and in some leaves (*Osmunda*) is that of a three-sided pyramid; but where the organ is flattened, as in some stems (*Pteridium*) and almost all leaves, it has two convex sides and is shaped like half of a biconvex lens. In the former case the segments are cut off in regular succession from the three sides (Fig. 10), in the latter alternately from the two sides (Fig. 11). The further subdivision of the segments to form the tissues is represented in surface-view for the case of *Osmunda* in Fig. 10: and Fig. 11 shows in the surface-view of a young leaf of *Ceratopteris* how the whole member may be built up from such segments. In roots the segmentation is complicated by the origin of the root-cap. This is provided by a segment cut off from the frontal face of the pyramid, after each cycle of three has been cut off from its sides (Fig. 12). Thus every fourth segment goes to form the protective cap, and renews it from within. Not only does the leaf show continued growth and apical segmentation from its two-sided apical cell, but the lateral wings or flaps originate by the activity of rows of marginal cells. There is also a definite segmentation seen in the origin of the sporangia. Thus Ferns have not stratified meristems like Seed-Plants. The tissues of all their parts originate from segmentation of superficial cells. This is a general character of the Pteridophyta, though the details of their segmentation and the number of the initial cells are open to variation.

Thus constituted the Fern-Plant carries out its Life on Land in essentially the same way as Seed-Plants. The structural differences are thosc of detail, the most important being the absence of secondary thickening in the

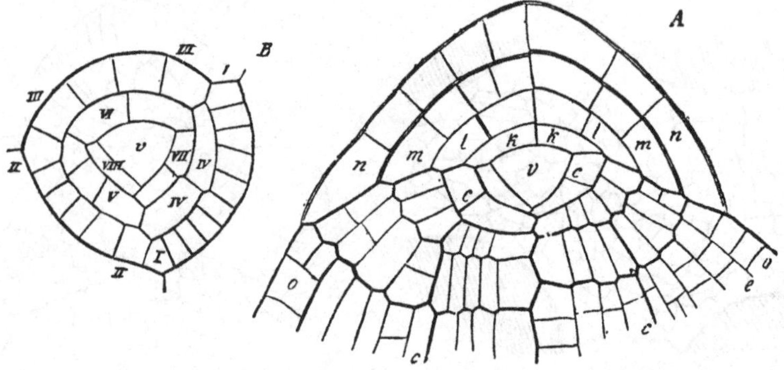

Fig. 12. (×250.) *A*=longitudinal section through apex of the root of *Pteridium*. *B*=transverse section through the apical cell of the root and neighbouring segments of *Athyrium*. (After Naegeli and Leitgeb.) *v*=apical cell. *k*, *l*, *m*, *n*= successive layers of root-cap. *o*=dermatogen. *c*=limit of stele. (From Sachs.)

stem. These plants have no automatic provision for increasing mechanical strength with size. In Tree Ferns this deficiency is partly made up by masses of hard brown sclerenchyma, which accompany and enclose the flattened meristeles; and their margins are usually curved outwards, thus securing increased mechanical resistance on the same principle as in corrugated columns. Their strength is further increased according to size and age by the development of masses of sclerotic, adventitious roots, matted together to form a thick investment to the original trunk, and adding to its stability by a method comparable mechanically to a cambial thickening, though quite different in origin (see Fig. 152, p. 158). But such mechanical provision for increase in size is only partially effective. There is no evidence that Ferns ever ranked among the largest of living Plants.

Many Ferns increase in number by *vegetative propagation*. This may follow simply on continued growth and branching, as in *Pteridium*, where the rhizome forks frequently. Whenever progressive rotting extends from the base beyond a branching, the two apices grow on as independent plants. In this way the Bracken multiplies habitually. In *Dryopteris* buds are formed near the bases of the leaves in old plants. Again, as rotting proceeds from the base, these buds become isolated, and root themselves as new individuals (Fig. 1, *B*, *C*). In other Ferns, as in the various species of *Asplenium* so commonly grown in dwelling rooms, buds or bulbils arise on the lamina. Being very lightly attached to the leaf they are readily shed, and root themselves independently in the soil. In some cases vegetative buds may replace the sori (Fig. 13). Such vegetative propagation of the Fern-Plant is a mere repetition of the sporophyte generation. But sooner or later the Fern-Plant bears the *spores*, which start the alternate generation.

Fig. 13. A pinna of a Fern (*Woodwardia*) showing many sporophytic buds on the upper surface. They correspond in position to sori on the lower surface, which are abortive, and so they may be held to be substitutionary growths.

The *spores* are produced on certain leaves of the mature plant which are therefore called *sporophylls*, to distinguish them from those which are only nutritive. In *Dryopteris* nutritive leaves and sporophylls are alike in outline (Fig. 2). The young plant only produces the former. But the leaves of older plants bear on their lower surface, and chiefly in the apical region, numerous groups of organs which are green or brown according to age. These are called *sori*, and consist of *sporangia* with certain protective structures. The sori vary greatly in size and form in different Ferns, which are classified according to their characters. In *Dryopteris*, as its old name *Nephrodium* implies, they are kidney-shaped, as is seen in Fig. 2. Each sorus is seated on a vein, which provides its necessary nourishment. It is protected by a covering called the indusium, of kidney-like outline, beneath which are numerous sporangia. If a leaf bearing mature sori be laid on a sheet of paper to dry, with its lower surface downwards, the indusia shrivel, and the bursting sporangia shed the spores in such numbers that they give a clear print of the outline of the sporophyll upon the paper. The spores are dark-coloured, very minute, and are produced in millions.

A vertical section through the sorus of *Dryopteris* shows an enlarged receptacle, traversed by the vascular strand. The indusium rising from it

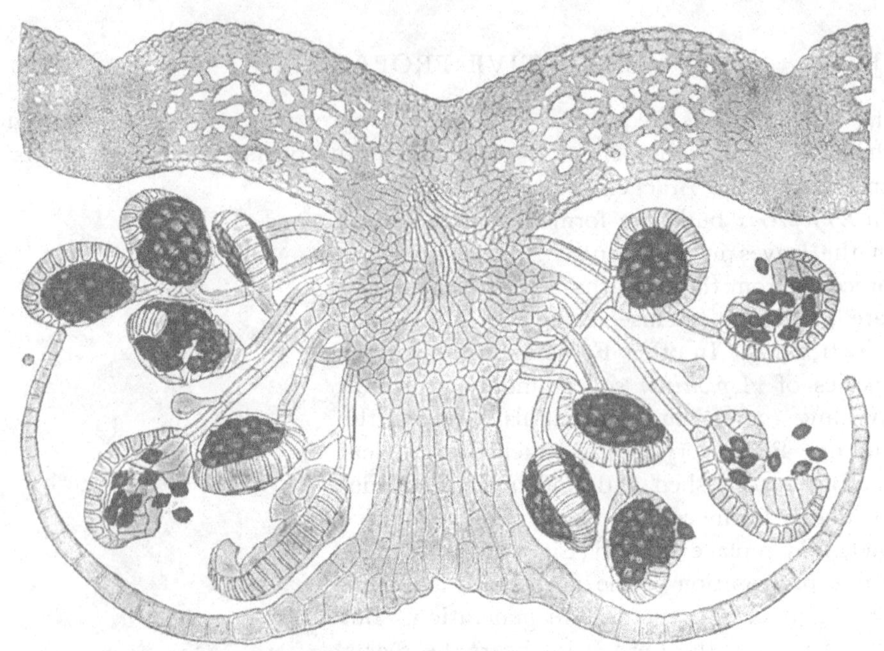

Fig. 14. Vertical section through the sorus of *Dryopteris Filix-mas*. (After Kny.) The adaxial surface is uppermost.

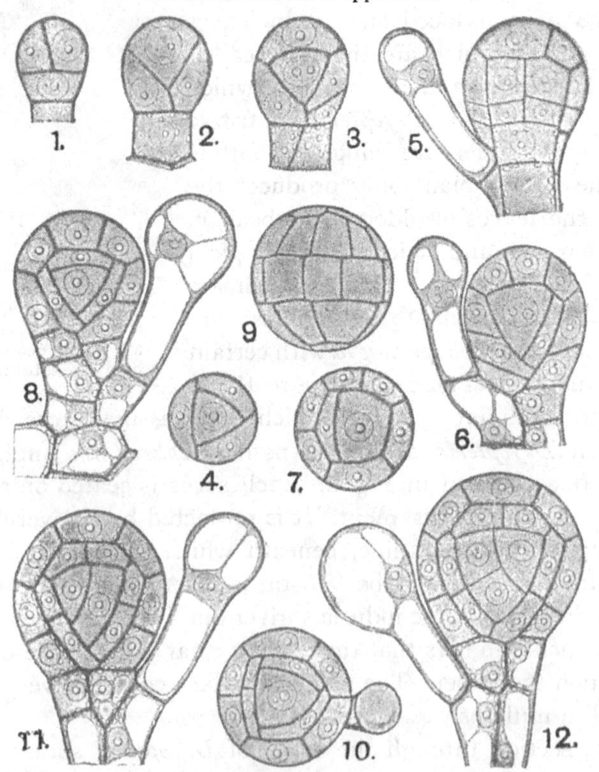

Fig. 15. Successive young stages in the segmentation of the sporangium of *Dryopteris Filix-mas*. (After Kny.)

overarches the numerous sporangia which are attached to it by long thin stalks (Fig. 14). The head of each sporangium is shaped like a biconvex lens; its margin is almost completely surrounded by a series of indurated cells, which form the mechanically effective *annulus*. This stops short on one side, where several thin-walled cells define the *stomium*, or point where dehiscence will take place (Figs. 14, 16, 4^a, 17). Within are the dark-coloured *spores*, which on opening a single sporangium carefully in a drop of glycerine may be counted to the number of 48. Normally the sporangia open in dry air, and the dry and dusty spores are forcibly thrown out.

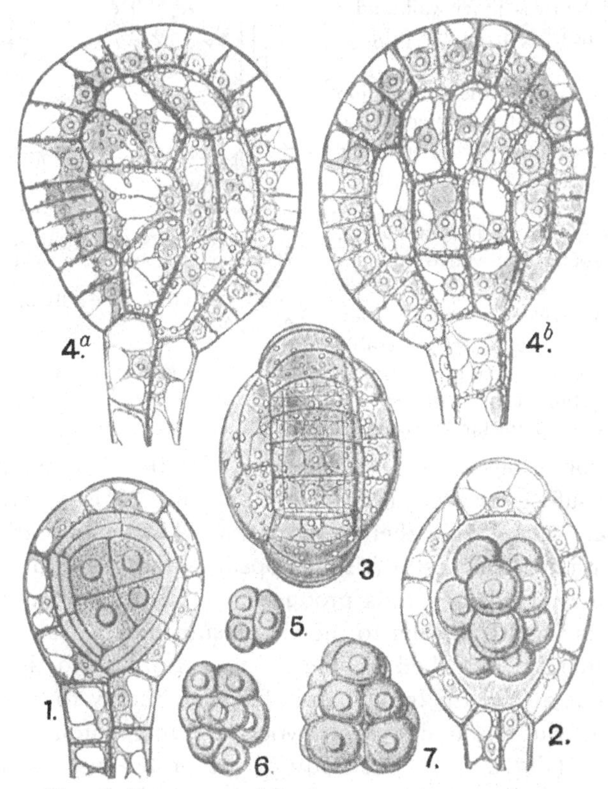

Fig. 16. Later stages of development of the sporangium of *Dryopteris Filix-mas*. (After Kny.)

The origin of a sporangium is by outgrowth of a single superficial cell of the receptacle, which undergoes successive segmentations as illustrated in Fig. 15, 1–3. A tetrahedral internal cell is thus completely segmented off from a single layer of superficial cells constituting the wall. The former undergoes further segmentation to form a second layer of transitory nutritive cells called the *tapetum* (Fig. 15, 6–12), subsequently doubled by tangential fission (Fig. 16, 1). The tetrahedral cell which still remains in the centre, having grown meanwhile, undergoes successive divisions till 12 *spore-mother-cells* are formed (Fig. 16, 2–7). These become spherical, and are suspended in a fluid which, together with the now disorganised tapetum, fills the enlarged cavity of the sporangium. Each spore-mother-cell then divides twice to form a *spore-tetrad*: in this process, just as in the formation of pollen-grains and other spores, the number of chromosomes is reduced to a half. Finally the resulting cells separate on ripening as individual *spores*, each covered by a protecting wall, rugged and dark brown at maturity. Owing to the absorption of the fluid contents of the sporangium the separate spores are dry and dusty, and are readily scattered. Since each of the 12 spore-mother-cells forms four spores, their number is 48 in each sporangium. Each mature spore consists of a nucleated protoplast, bounded by a colourless inner wall, and a brown epispore bearing irregular projecting folds.

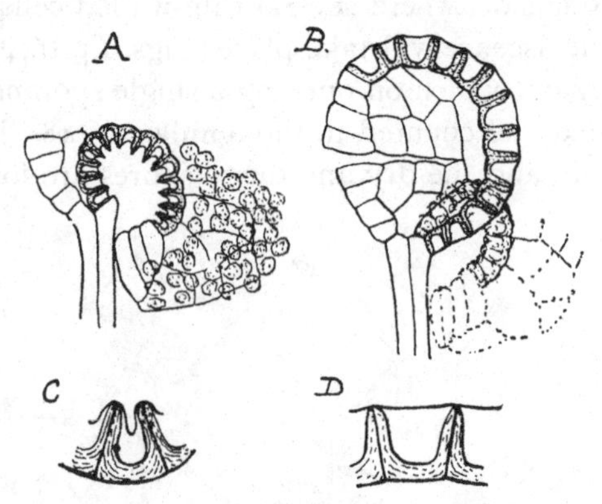

Fig. 17. *A*=sporangium with annulus everted. *B*=a similar sporangium after recovery by a sudden jerk. *C*=condition of cells of the everted annulus. *D*=cells of annulus before evertion.

Meanwhile the wall of the sporangium has differentiated into the thinner lateral walls of the lens-shaped head, and the annulus, which is a chain of about 16 indurated cells surrounding its margin (Fig. 16, 4^a, 4^b). These form a mechanical spring, which on rupture of the thin-walled stomium becomes slowly everted as its cells dry in the air, and then recovering with a sudden jerk throws out the spores to a considerable distance (Fig. 17). Dry conditions are necessary for this last phase of spore-production, viz. the dissemination of the numerous living germs. Each spore is a living cell, and may serve as the starting point for a new individual.

The dry conditions which are necessary for the dissemination of the spores do not suffice for their further development. Moisture and a suitable temperature are required for their germination. The outer coat then bursts, and the inner protrudes, cell-division appearing as the growth proceeds (Fig. 18). The body that is thus produced is called the *prothallus*, and it may vary in its form according to the circumstances. It usually grows first into a short filament attached by one or more rhizoids to the soil (4). It then widens out at the tip to a spatula-like and finally to a cordate form (Fig. 18, 5, 6). But when closely crowded the filamentous form may be retained longer (Fig. 20, 1). The body of the prothallus, exclusive of the downward-growing rhizoids, consists of cells which are essentially alike, arranged at first in a single-layered sheet. The peripheral parts retain this, but in the central region, below the emarginate apex, the cells divide by walls parallel to the flattened surfaces, and thus a massive central cushion is formed. The mature cells are thin-walled, with a peripheral film of protoplasm surrounding a central vacuole and embedding the nucleus and numerous chloroplasts: intercellular spaces are absent. The whole body is thus capable of an independent physiological existence, nourishing itself by absorption from the soil, and by photo-synthesis (Fig. 19). But there is a large proportion of surface to bulk, and no serious resistance is offered to the evaporation of water from it in dry air. Comparing the prothallus with the Fern-Plant as regards the water-relation, the former is plainly less adapted for life on land, and more immediately dependent on moisture.

The prothallus thus constituted is capable in some cases of vegetative propagation by "gemmae." But this *gametophytic budding* is less common here than in the Bryophytes.

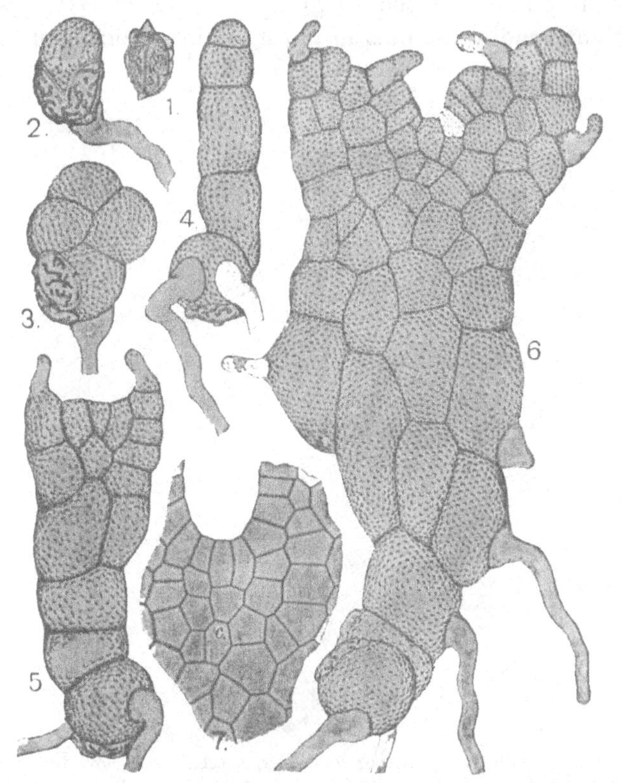

Fig. 18. Successive stages in germination of the spores of *Dryopteris Filix-mas*, to form the prothallus. (After Kny.)

The dependence on moisture is still more obvious in the behaviour of the *sexual organs* which the prothallus bears. These are male and female, and they may be found on the same prothallus or on different prothalli (Fig. 20, 1). In the former case the *antheridia, or male organs*, commonly appear first, and the *archegonia, or female organs*, later. There may thus be a separation of the sexes either in time or in space. The flattened prothallus of the ordinary cordate type usually bears both sex-organs (Fig. 19). When grown under normal circumstances on a horizontal substratum it produces them on its lower surface, the antheridia in the basal or lateral regions, the archegonia upon the massive cushion. The latter develop in acropetal order, the youngest being nearest to the incurved apex of the prothallus. The position of the sexual organs is evidently favourable to their continued exposure to moist air, or to fluid water which is necessary for carrying out their function.

The antheridium, which arises by outgrowth and segmentation of a single superficial cell (Fig. 20, 2, 3), consists when mature of a peripheral wall of tabular cells, surrounding a central group of *spermatocytes* (Fig. 20, 4, 5). The antheridium readily matures in moist air, but it does not open except

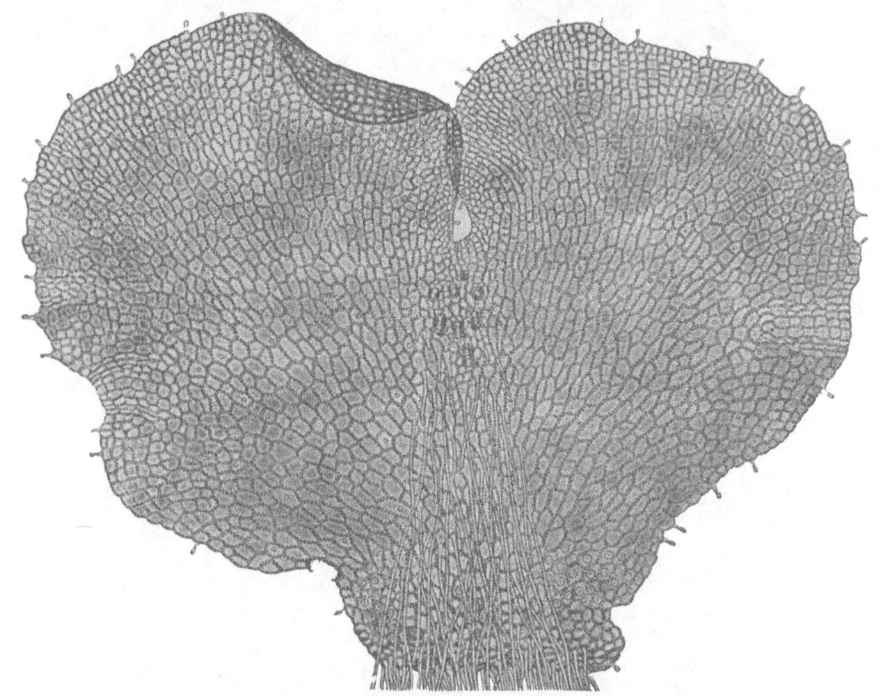

Fig. 19. Mature prothallus of *Dryopteris Filix-mas*, as seen from below, bearing antheridia among its rhizoids, and archegonia near to the apical indentation. (After Kny.)

in presence of external fluid water. This causes swelling of the mucilaginous walls of the spermatocytes and increased turgor of the cells of the wall. The tension is relieved by rupture of the cell covering the distal end, and the spermatocytes are extruded into the water; in this the cells of the wall assist by their swelling inwards, and consequent shortening (Fig. 20, 6). The spermatocytes thus extruded into the water which caused the rupture, soon show active movement, and the spermatozoid which had already been formed within each of them escapes from its mucilaginous sheath, and moves freely in the water by means of active cilia attached near one end of its spirally coiled body (Fig. 20, 8).

The archegonium also originates from a single superficial cell, and grows out so as to project from the downward surface of the thallus. It consists when mature of a peripheral wall of cells constituting the projecting neck, and a central group arranged serially. The deepest-seated of these is the large *ovum*, which is sunk in the tissue of the cushion; above this is a small *ventral-canal-cell*, and a longer *canal-cell* (Fig. 21, *A*). If prothalli be grown in moist air, and only watered by absorption from below, the archegonia

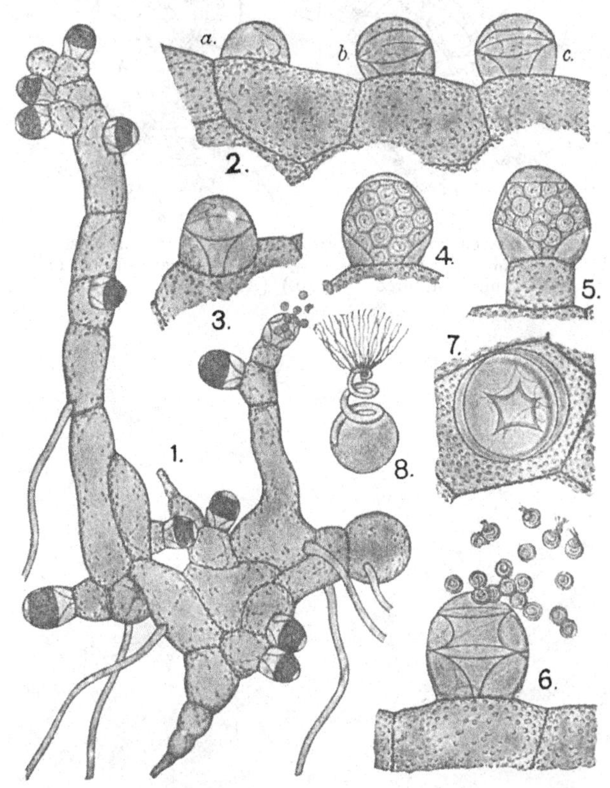

Fig. 20. 1, an attenuated male prothallus of *Dryopteris Filix-mas*. 2–5, stages of development of the antheridium. 6, 7, ruptured antheridia. 8, a spermatozoid highly magnified. (After Kny.)

will have no access to fluid water, and they will remain closed. Fertilisation is then impossible. But if they are watered from above, as they would be by rain in the ordinary course of nature, the external fluid water will bathe them, and rupture will result. This may be observed in living archegonia which have been kept relatively dry and then mounted in water. The neck bursts at the distal end, owing to internal mucilaginous swelling, and its cells diverge widely. The canal-cell and ventral-canal-cell are extruded, and the ovum remains as a deeply seated spherical protoplast, while access to it is gained through the open channel of the neck (Fig. 21, *B*). Thus the same condition leads to the rupture both of the male and female organs. In nature a shower of rain would supply the necessary water, which would serve also as the medium of transit of the spermatozoids to the ovum. But the move-

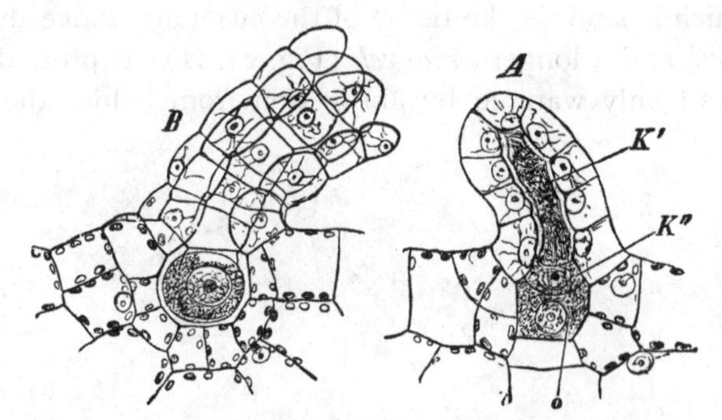

Fig. 21. Archegonia of *Polypodium vulgare*. *A*, still closed. *o*=ovum. *K'*=canal-cell. *K"*=ventral-canal-cell. *B*, an archegonium ruptured. (×240.) (After Strasburger.)

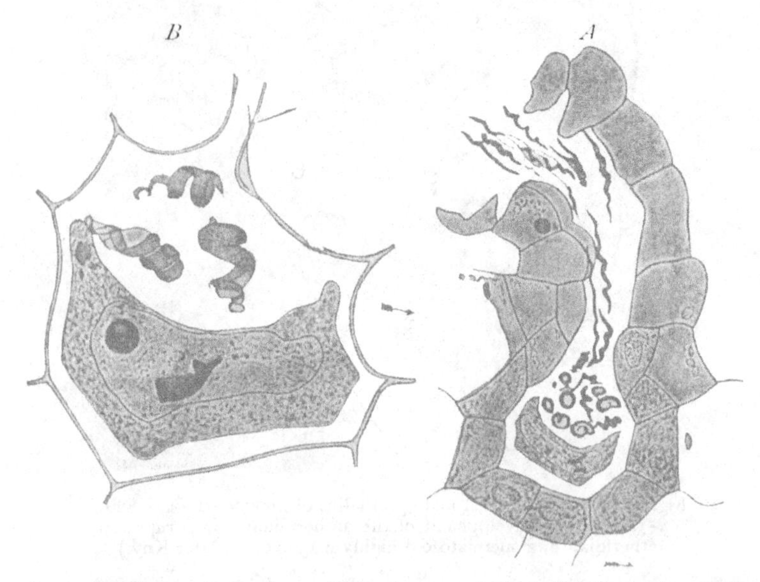

Fig. 22. Fertilisation in *Onoclea sensibilis*: the arrows indicate direction to the growing point. *A*=a vertical section through an archegonium probably within ten minutes after entrance of the first spermatozoid. (×500.) *B*=vertical section of the venter of an archegonium, containing spermatozoids, and the collapsed egg with a spermatozoid within the nucleus. Thirty minutes. (×1200.) (After Shaw.)

ments of the spermatozoids are not subject to blind chance. It has been shown that diffusion into water of a very dilute soluble substance, such as malic acid, serves as a guide, the spermatozoids moving towards the centre of diffusion. Probably it is in this way that they are attracted to the neck of the archegonium, which they may be seen to enter, and finally one spermatozoid coalesces with the ovum (Fig. 22). The male nucleus has been seen to enter into the female nucleus, and their complete fusion follows (Fig. 23). Thus the presence of external fluid water is essential for fertilisation in Ferns. Their normal life-cycle cannot be completed without it.

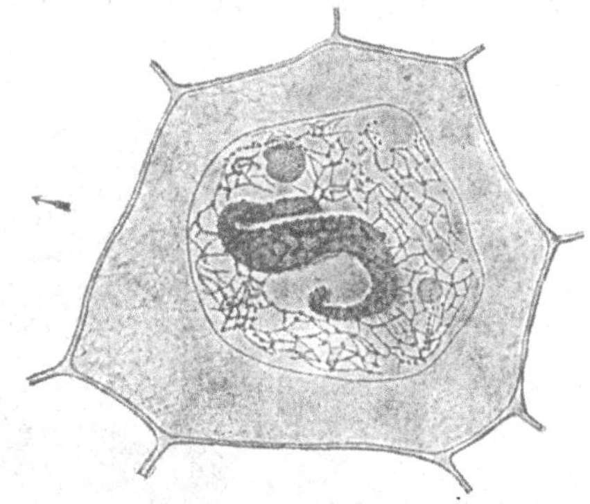

Fig. 23. Horizontal section of an egg showing coiled male nucleus within the female. Twelve hours. (× 1200.) (After Shaw.)

The immediate consequence of fertilisation is growth and segmentation of the *zygote*, which first secretes a cell-wall. It divides first into two, by a *basal wall*, the plane of which includes the axis of the archegonium: then into octants, four of which constitute an *epibasal hemisphere*, directed towards the apex of the parent thallus, giving rise to *axis* and *leaf* of the sporeling; four form a *hypobasal tier*, which gives rise to the first *root* and a suctorial organ called the *foot* (Fig. 24, *a*, *b*, *c*). These parts are soon distinguishable

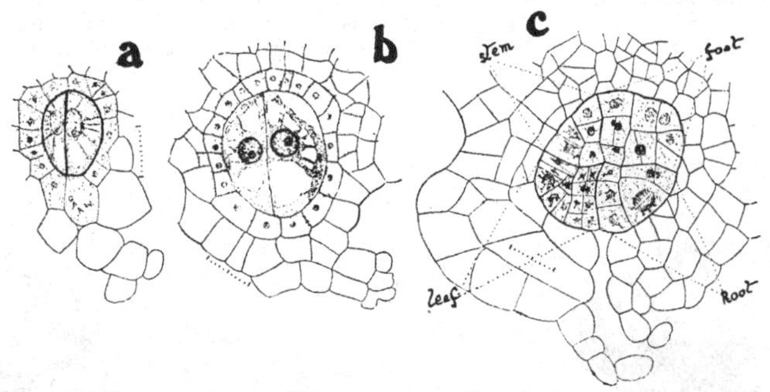

Fig. 24. Young embryos of Ferns, orientated with the archegonial neck downwards. The epibasal hemisphere is seen to the left, and the hypobasal to the right. *a*=two-celled embryo *Adiantum concinnum* (× 30 times scale). *b*=similar embryo of *Pteris serrulata* (× 30 times scale). *c*=more advanced embryo of *Adiantum concinnum*: the epibasal hemisphere has given rise to stem and leaf, and the hypobasal to root and foot. (After Atkinson.)

by their form and structure, and are seen in their relative positions, but still enclosed in the enlarged venter of the archegonium, in Fig. 25. Soon the cotyledon and first root burst their way out: the former expands as the first nutritive leaf, the latter buries itself in the soil (Figs. 26, 27). At first the young Fern-Plant is dependent upon the prothallus that encloses it, but by means of its cotyledon and its root it soon becomes self-dependent, and the prothallus rots away. It is then only a matter of time and opportunity for it to attain characters similar to those of the parent Fern-Plant.

These are the salient features in the life-cycle of a Fern as it is seen in its simplest form. They may be represented graphically to the eye in a diagram (Fig. 28). The two most notable points are those where the indi-

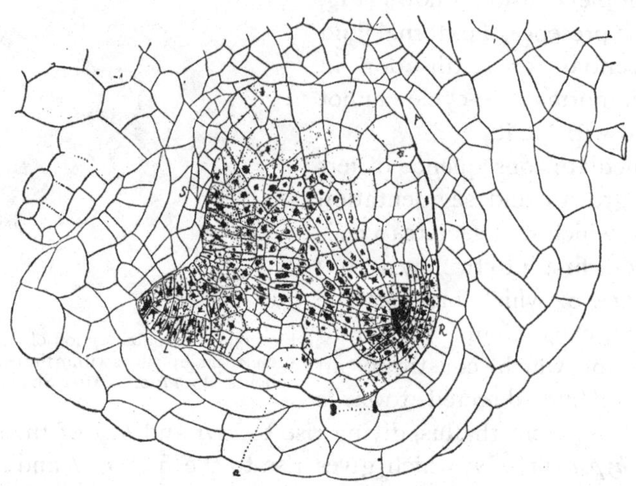

Fig. 25. Embryo of *Adiantum concinnum* in the enlarged venter of the archegonium, so far advanced as to show the parts of the embryo. The epibasal hemisphere is to the left, the hypobasal to the right. *L*=leaf or cotyledon. *R*=root. *S*=stem. *F*=foot. (After Atkinson.)

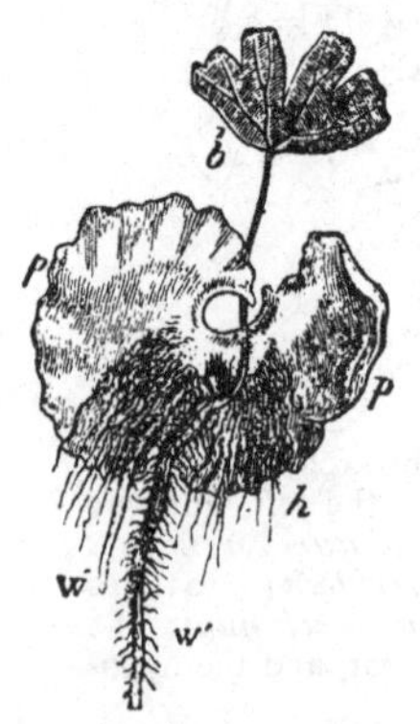

Fig. 26. *Adiantum Capillus Veneris.* The prothallus, *pp*, seen from below has a young Fern-Plant attached to it. *b*=first leaf. *w*, *w'*=first and second roots. *h*=rhizoids of the prothallus. (×about 30.) (After Sachs.)

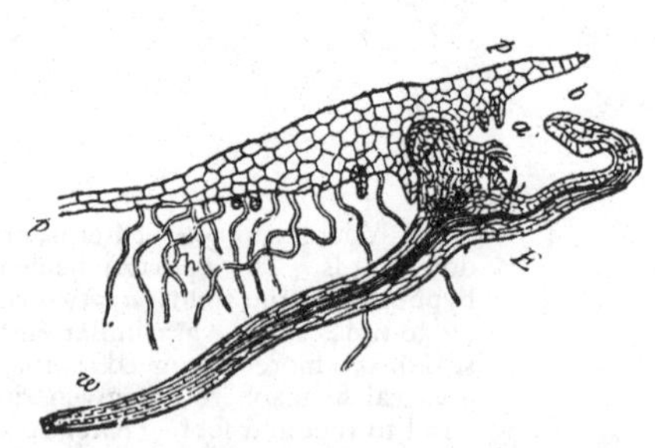

Fig. 27. *Adiantum Capillus Veneris.* Longitudinal section through the prothallus, *pp*, and young Fern-Plant *E*. *h*=root hairs of prothallus. *a*=archegonia. *b*=the first leaf. *w*=the first root of the embryo. (×10.) (After Sachs.)

vidual is represented only by a single cell, viz. the *spore*, and the *zygote*. These are two landmarks between which intervene two more extensive developments, on the one hand the sexual generation or prothallus, on the other the spore-bearing generation, or Fern-Plant. If the events above detailed recur in regular succession the two phases of life will alternate. Of these the one bears sexual organs, containing sexual cells or gametes, and it may accordingly be called the *gametophyte*; the other is non-sexual, but bears sporangia containing the spores, and is accordingly called the *sporophyte*. The study of Ferns at large leads to the conclusion that this regular alternation is typical for them all. These two alternating generations differ not only in form but also in their relation to external circumstances, and especially in the water-relation. The sporophyte is structurally a land-growing plant, with nutritive, mechanical, and conducting tissues, and a ventilating system. Not only is it capable of undergoing free exposure to the ordinary atmospheric conditions, but dryness of the air is essential for the final end of its existence, viz. the distribution of its spores. On the other hand, the gametophyte is structurally a plant ill-fitted for exposure, with undifferentiated and ill-protected tissues and no ventilating system, while the object of its existence, viz. fertilisation, can only be secured in the presence of external water. As regards the water-relation the whole life-cycle of a Fern might not inaptly be designated as *amphibious*, since the one phase is dependent on external liquid water for achieving its object of propagation, while the other is independent of it.

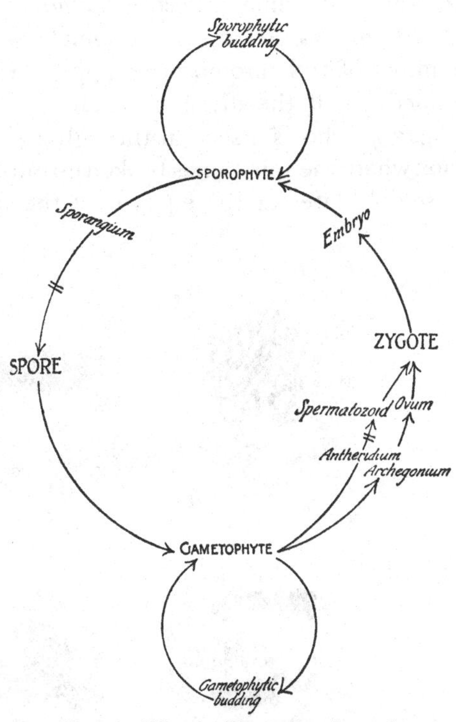

Fig. 28. Diagram illustrating the cycle of life of a Fern.

The normal cycle thus presented to the eye involves differences of nuclear condition of the alternating phases, those differences being established respectively by fertilisation and by the tetrad-division in the sporangium. The *sporophyte* or Fern-Plant is *diploid*, and the number of chromosomes is usually very large, being about 90 for *Athyrium*: it is 128 in *Dryopteris Filix-mas*, but only 32 for *Marsilia*. These numbers are reduced to one half in the tetrad-division of the spore-mother-cells, and the spores on germination produce the *gametophyte* which is *haploid*. But in fertilisation, when the gametes fuse, the diploid number is restored. This normal cycle corresponds to that seen in Vascular Plants at large, the substantive Plant being in all cases the diploid sporophyte.

The nuclear details of the normal chromosome-cycle have been observed for the Male Shield Fern (*Dryopteris*) by Yamanouchi, and delineated as shown in Figs. 29, 30. In the somatic divisions of the sporophyte the diploid number of 128 chromosomes appears (Fig. 29, *e*): and that number is maintained in all the divisions leading up to the definite spore-mother-cells. Fig. 29, *a* shows a section through a nucleus of a somatic cell, in the condition where the spirem has broken up into the constituent chromosomes. These assemble later at the equator of the nuclear spindle (*b*) and divide longi-

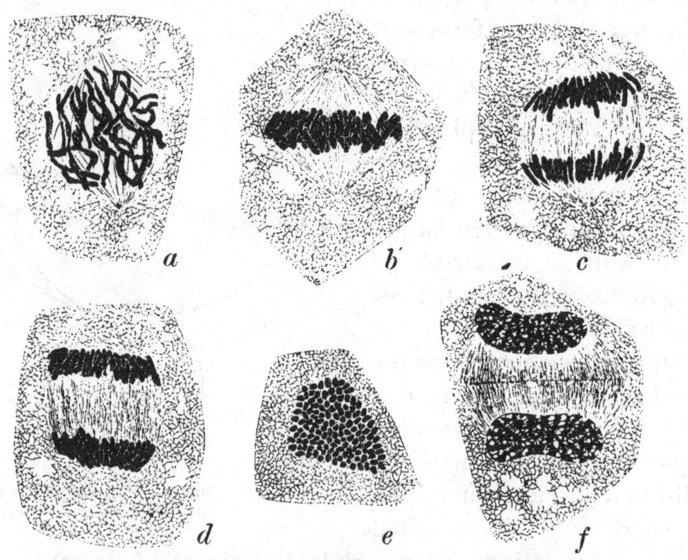

Fig. 29. Vegetative mitosis in sporogenous cell of *Dryopteris*, after Yamanouchi. *a* = spirem segmented into chromosomes; *b* = metaphase showing equatorial plate; *c* = anaphase, two sets of daughter chromosomes separated; *d* = late anaphase; *e* = polar view of stage seen in *d*, showing 128 chromosomes, which is the diploid number; *f* = daughter-nuclei after formation of membrane; the cell-plate has appeared equatorially between the two nuclei.

tudinally, the half-chromosomes separating (*c*) and collecting in the anaphase at its two poles (*d*). Seen from the pole, in this condition, it is possible to count the chromosomes, and their diploid number is seen to be 128 (*e*). They then coalesce to form the new nuclei (*f*), while the equatorial plate constitutes the partition wall between the cells.

Similar observations of the divisions leading to the formation of the spore-tetrad are illustrated in Fig. 30, *a*—*k*. The nucleus is at first clearly delimited, showing a chromatic reticulum, with a nucleolus (*a*). The nucleus enlarging enters the condition called "synapsis," from which it gradually recovers, loops of the chromatin extending towards the nuclear wall (*b*). The chromosomes then separate and contract, but their number cannot be exactly estimated in sections at this stage (*c*). In the metaphase which follows (*d*) the chromosomes, which are now bivalent by pairing, arrange themselves to form the equatorial plate, and when this is seen from the polar aspect, their number can be counted as 64, that is the haploid or reduced

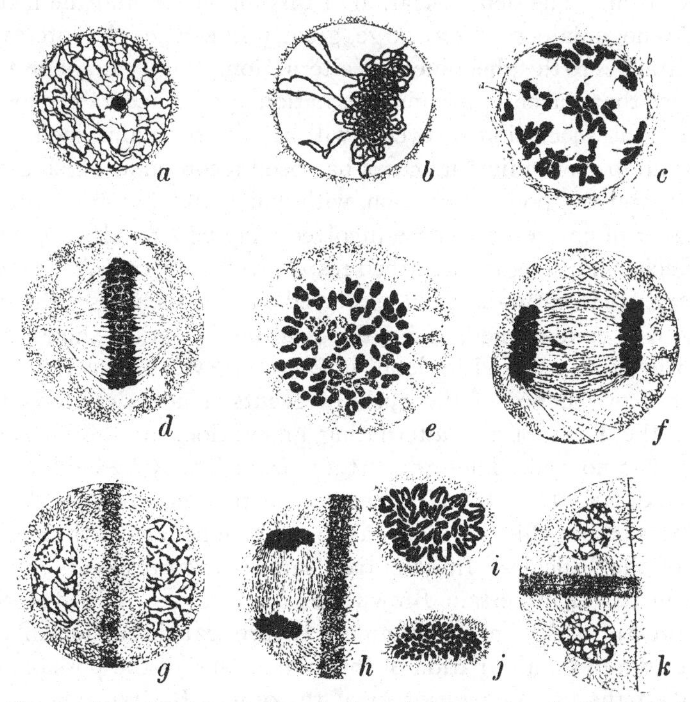

Fig. 30. Tetrad-division, with reduction of chromosomes in *Dryopteris*, after Yamanouchi. *a* = nucleus with chromatin-reticulum: *b* = nucleus emerging from synapsis: *c* = nucleus approaching the metaphase: *d* = early metaphase, the bivalent chromosomes arranged in an equatorial plate: *e* = polar view of early metaphase showing 64 chromosomes, that is the reduced number: *f* = early telophase, the daughter-chromosomes grouped at either pole: two are delayed: *g* = spore-mother-cell divided into two hemispheres by a granular zone: each contains a daughter-nucleus: *i* = polar view showing bivalent chromosomes: *j* = polar view of daughter-chromosomes of the second division, showing their number to be 64: *k* = the second division completed, the nuclei reconstituted, and the cell-plate dividing the two future spores of the half-tetrad from one another.

number (*e*). The paired chromosomes then separate and group themselves in the telophase at the poles of the spindle (*f*), constituting the new nuclei, each of which will accordingly contain the haploid or reduced number (*g*). Each of these nuclei then undergoes a further division (*h*), and views of this from the polar aspect show that the number of chromosomes involved is 64 (*i*, *j*). Accordingly the nuclei formed from them, which become the definitive nuclei of the spores (*k*), are seen to be haploid, and they start the gametophyte generation with nuclei containing 64 chromosomes. This is then continued in the gametophyte, till syngamy again doubles the number to 128, and establishes again the diploid sporophyte.

The alternating nuclear cycle thus demonstrated coincides in normal cases with the alternating phases of sporophyte and gametophyte, as defined by external form. The demonstration of this for Ferns, and the fact that it is so for Archegoniate Plants at large, greatly intensifies the interest which naturally attaches to the phenomena of alternation. But it cannot be assumed, from the fact that the chromosome-alternation applies accurately for normal cycles, that that succession of events will always be maintained. An ever increasing number of individual cases has been recorded in which the events of syngamy and of spore-production, with which the doubling and halving of the number of chromosomes are involved, may be excluded from the life-history. Such conditions are described respectively as *apomixis* (or *apogamy*), and *apospory*. It is naturally to be expected that in such cases the chromosome-cycle will be disturbed, and cytological investigation shows that it is so. They will be considered in detail in Chapter XVI. Meanwhile it may be held that the prevalence of the cycle of events as here described, whether as regards the form of the alternating generations or the chromosome-numbers, is the normal. These departures from it may be held as late and sporadic incidents. The natural inference will therefore be that the normal cycle, as described in this Chapter, was that which held good throughout the evolution of the Filicales. In essential features it corresponds to a similar cycle demonstrated in certain Brown and Red Algae. The materials are therefore present which might stimulate theoretical discussion of the evolutionary origin of an alternation of this nature, and of the possible priority in Descent of the one generation or of the other. But the time is not ripe for any decision of such questions: for it is quite possible that the alternations seen in the several distinct phyla of Algae, and that seen in Archegoniate Plants, may really be homoplastic in all these cases, and not truly homogenetic at all. Much more detailed comparative study will be necessary before any assured opinion on such points can be evolved.

The normal alternation appears clear-cut in the Archegoniatae at large, and evidences of nascent stages of it are deficient in them. Consequently the Archegoniatae are not likely to provide the necessary early steps which would illuminate the progression towards that alternation which is seen in them. The most promising field would appear to be the methodical study of the cytology of those phyla of Algae in which the most elaborate forms show a well-marked alternation, while in the simpler types the alternation is either rudimentary, or even absent. But this book deals with the Filicales, not with Algae: and the question must therefore be left over for the Algologists: always, however, with the willingness to apply any sound results which they may obtain in helping to elucidate the problem of alternation as it is seen in the Ferns and in other Archegoniate Plants.

It will be manifest from the foregoing sketch that the sporophyte generation, being larger and more elaborate in structure than the gametophyte, will be likely to yield more ample materials for comparison than the latter. This is actually the case, and accordingly its details will be described first. But none the less the features of the gametophyte must also be examined in detail: for conclusions as to relationship and descent must be based upon the whole sum of the characters which can be observed.

BIBLIOGRAPHY FOR CHAPTER I

1. MORRISON. Historia Plantarum. Oxonii. 1699. First raised Fern-Plants from spores[1].
2. EHRHART. Beyträge. Hannover. 1788. Described the formation of the prothallus.
3. KAULFUSS. Das Wesen der Farrenkräuter. Leipzig. 1827. First observed germination of the spores. He gives a full quotation, and extraction of the early literature.
4. BISCHOFF. Handb. botan. Terminologie. Nürnberg. 1842. Recognised the embryo attached to the prothallus.
5. NAEGELI. Zeit. f. wiss. Bot. Zurich. 1844. Discovered the antheridia and spermatozoids.
6. SUMINSKI. Z. Entwickelungsgesch. d. Farrenkräuter. Berlin. 1848. Ascertained the nature of the archegonium, and its relation to the embryo.
7. HOFMEISTER. Vergleichende Untersuchungen. Leipzig. 1851. Gave the first complete and comparative account of the life-history of a Fern. Historical review of discovery of propagation of Ferns. Engl. Edn. Ray Soc. p. 257.
8. KNY. Entw. d. Parkeriaceen. Nov. Act. Leopold. Bd. xxxvii, 4. Dresden. 1875. Also Bot. Wandtafeln. Texte. xciii. Berlin. 1894.
9. ENGLER & PRANTL. Natürl. Pflanzenfam. i, 4. Here the general literature is fully quoted.
10. CAMPBELL. Mosses and Ferns. 2nd. Edn. London. 1905. Here the general literature is fully quoted.
11. BOWER. Origin of a Land Flora. London. 1908.
12. YAMANOUCHI. Sporogenesis in *Nephrodium*. Bot. Gaz. Vol. xlv, 1908. Here the cytological papers up to date are fully quoted.

[1] Valerius Cordus (1515—1544) of Wittenberg asserted that all kinds of Ferns reproduce by means of dust that is developed on the backs of the leaves: but there appears to be no evidence how he arrived at this conclusion.

CHAPTER II

THE HABIT AND THE HABITAT OF FERNS

THE type of plant seen in *Dryopteris* (*Nephrodium*) *Filix-mas* is a familiar example of the megaphyllous habit in Ferns. Numerous large leaves are grouped closely round the apical bud, so as to form a basket-like tuft attached distally to the abbreviated and massive stock (Fig. 31). The like is seen in other common native Ferns belonging either to the same or to quite different genera. For instance in the Lady Fern (*Athyrium Filix-foemina*), the Hard

Fig. 31. Adult plant of *Dryopteris* (*Nephrodium*) *Filix-mas* grown in the open. Much reduced. (Photograph by Mr R. Whyte, Rothesay.)

Fern (*Blechnum spicant*), the Hart's Tongue (*Phyllitis Scolopendrium*), and many others. On the other hand there are native species of the genus *Dryopteris* which have a creeping habit with long internodes, and leaves isolated at a distance from one another, a condition which gives a quite different aspect to the plant as a whole. This is seen in *Dryopteris Thelypteris*, a marsh-growing species, and it appears also in the Oak Fern (*Dryopteris*

(*Polypodium*) *Linnaeana*, C. Chr.)(Fig. 32). Further, while in *Onoclea Struthiopteris* the basket-form is seen, in *O. sensibilis* there is an elongated creeping rhizome. In the former species, however, there are runners, which connect its habit of shoot with that of *O. sensibilis*. These facts illustrate what applies generally for Ferns, viz. that species which appear alike in general habit may not be really akin: and that those which are akin as indicated by other and more important characteristics, may differ in habit. In fact, the general form of the plant is no safe guide to affinity.

What has thus been noted for the general form of the shoot applies with even greater force to the outline of the leaf. For instance, the reniform blade borne on a long petiole is a very marked feature of some Ferns. It is seen in *Phyllitis* (*Scolopendrium*) *Delavayi*, in *Adiantum reniforme* (Fig. 33), in *Trichomanes reniforme* (Fig. 74, p. 83), and in *Pterozonium reniforme* (Fig. 34). But all of these Ferns, apparently so similar in their habit and particularly in

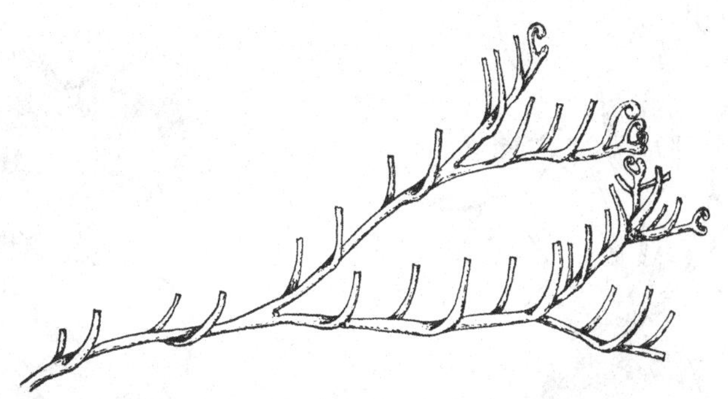

Fig. 32. Rhizome of *Dryopteris Linnaeana* C. Chr. (= *Polypodium Dryopteris*, L.) dichotomously branched, with alternating leaves. The roots are omitted. (After Velenovsky.) Reduced.

the outline of their leaves, belong to genera widely distinct in the character of their sori, as well as in their anatomy and other features. Again, the widely expanded lamina characteristic of *Christensenia* (*Kaulfussia*) *aesculifolia* appears to resemble that of *Hypoderris Brownii* in outline and venation; but the two are widely different when compared as regards other and more essential characters. These examples are mentioned in order to illustrate how similarity of some external feature, even when it is a marked one, does not in itself indicate affinity. Illustrations might be indefinitely extended, and might be taken not only from leaves of simple form such as those named, but from Ferns with various degrees of that pinnation which is so characteristic of their leaves. These will, however, suffice to show at the opening of this Chapter how limited is the value to be attached to habit in the Classification of Ferns. This fact greatly increases the difficulty of the systematic treatment of this large and varied Class. But by an exact analysis not only of form, but of structure and development, many features are brought to light which are more reliable than those of external form or general habit. These lead to decisions regarding affinity which are more trustworthy because they are founded on a more scientific method.

The study of the habit of Ferns may, however, be pursued profitably from the point of view of the surroundings in which they grow, and of their accommodation to them: in fact ecologically. It is found that their sporophyte has been remarkably plastic under varying conditions during their evolution, and that is the reason why their external habit is so unreliable as a guide to affinity. The basket-form already alluded to may be regarded

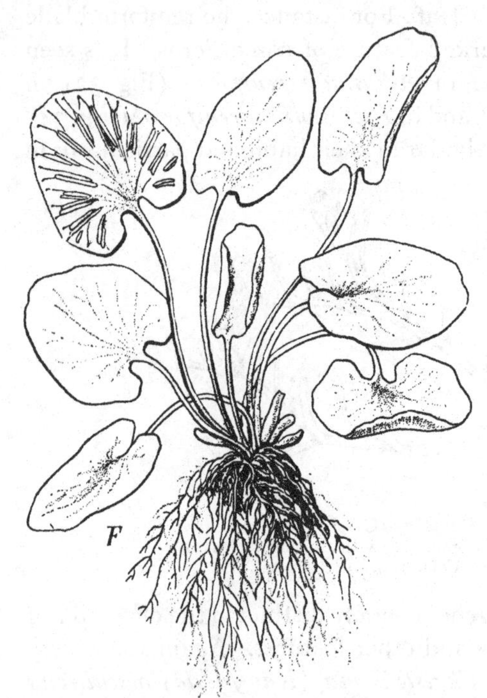

Fig. 33. Habit of *Phyllitis* (*Scolopendrium*) *Delavayi*, Franck. (After Engler and Prantl.)

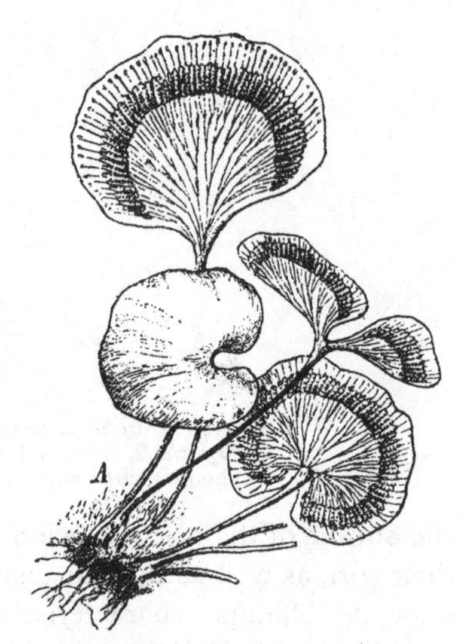

Fig. 34. Habit of *Pterozonium* (*Gymnogramme*) *reniforme* (Mart.) Fée. (After Engler and Prantl.)

as a central habit-type, and judging from their anatomy it appears to have existed frequently among the early fossils. It follows naturally from an upright position of a radially constructed axis, with slow but continuous apical growth, and an acropetal succession of numerous, closely set leaves of large size. A number of these developing almost simultaneously with overlapping margins produce the basket-form. The short stock is suitable for Ferns living as undergrowth among low vegetation, which is a very frequent habitat for those of temperate climates. These considerations may account for its prevalence among native Ferns. It is well represented in the

Osmundaceae, which link so naturally on to the early fossil Botryopterideae. The living Marattiaceae also provide good examples. In the latter it is the direct continuation of the growth of an embryo erect from the first: but in the Osmundaceae the embryo is prone, and its apex turns upwards later.

Fig. 35. Dwarf Tree Fern habit of *Blechnum tabulare*, S. Brazil. (From Christ, after Wacket.)

It is the same upright habit which leads to the *dendroid types*. Given continuous apical growth a vertical axis becomes an upright column. Dwarf examples of this are seen in the Blechnums (e.g. *B. tabulare*, Fig. 35). Continued further a "Tree Fern" is the result, whose trunk may bear 60 feet or more above the ground a huge basket-shaped group of leaves surrounding the apical bud. A practical limit of height is imposed by the mechanical strength of the stem, which does not increase by cambial activity, but is entirely primary in such Ferns. Nevertheless it derives additional support from the mass of adventitious roots which surround it, and often give it the appearance of thickening towards the base (Frontispiece). The dendroid

Fig. 36. Whole plant of *Ophioglossum palmatum* showing the distended storage-stock, from a drawing by Prof. Lawson. (½ nat. size.)

habit being thus the simple result of continued apical growth, it may be assumed by Ferns of various affinity. It is seen in a dwarf form, though sometimes attaining considerable height, in *Osmunda* and *Todea*; in *Blechnum*, *Sadleria*, and *Brainea*: also in *Thyrsopteris*. It attains full development of 60 or even of 70 feet in *Dicksonia* and in the genera closely allied with *Cyathea*. In all of these the sporeling is prone, and the apex assumes the upward position as it develops. This being so it cannot be assumed that the dendroid habit is in itself a sign of affinity of those Ferns, *inter se*, much less with the more distant Marattiaceae. It may well have been achieved independently by each type that shows it. The result is, however, the most striking of all the developments of Ferns. Tree Ferns flourish best under a high forest-canopy, though individuals often project upwards through low undergrowth, where the climate is sufficiently moist.

A curious modification of the upright habit is seen in the geophilous Ophioglossaceae, which are by its means able to adjust their mode of life

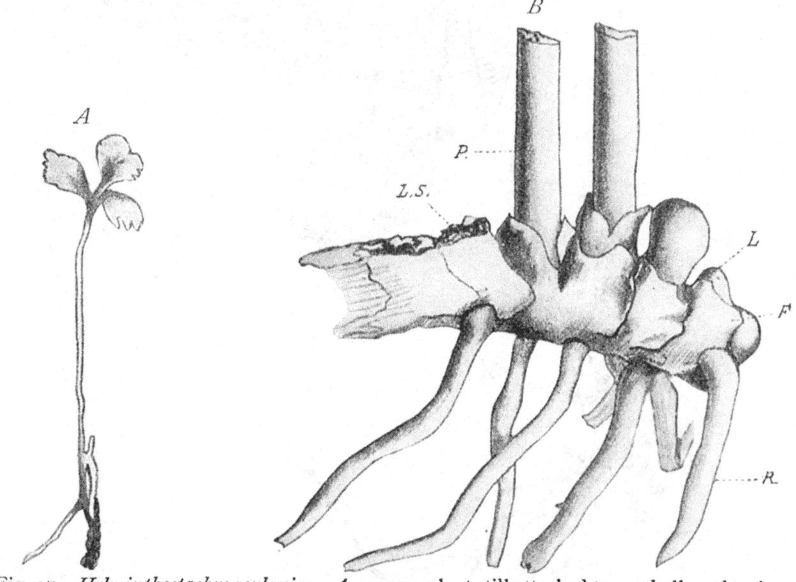

Fig. 37. *Helminthostachys zeylanica*. *A* = young plant still attached to prothallus, showing vertical position of axis. *B* = adult plant with rhizome horizontal. *A* after Lang, *B* after Farmer and Freeman. *F* = stipular flap; *L* = leaf; *P* = petiole; *LS* = leaf-scar; *R* = root.

to the requirements of marked seasonal change. The axis in *Botrychium* and *Ophioglossum* is short, upright, and deeply buried in the soil, and it is largely composed of storage-parenchyma. From it, in most of the large-leaved species, one leaf is expanded annually above ground, and it functionates for both nutrition and propagation. In some of the smaller species, such as *O. lusitanicum* and *O. bulbosum*, two or three, or as many as six leaves may be expanded in each year. At the end of the season the leaves rot away, and the product of their activity is stored in the stock for the succeeding season (Fig. 36). This monophyllous habit is suitable for meeting seasonal needs. During a period of drought or low temperature the plant perennates underground, but its expanded leaf or leaves carry on nutrition and propagation during the favourable season. It seems probable from the complex leaf-arrangement, and from the degrees of polyphylly seen in the smaller species, that the monophyllous habit is an adaptive modification of the usual basket-type, but with the successive leaves spread in their time of expansion over a prolonged period. Biologically it compares with what is seen in such Aroids as *Arum maculatum*, or *Amorphophallus*.

The creeping habit is in strong contrast to the upright. It results from an elongation of the internodes with a consequent greater or less isolation

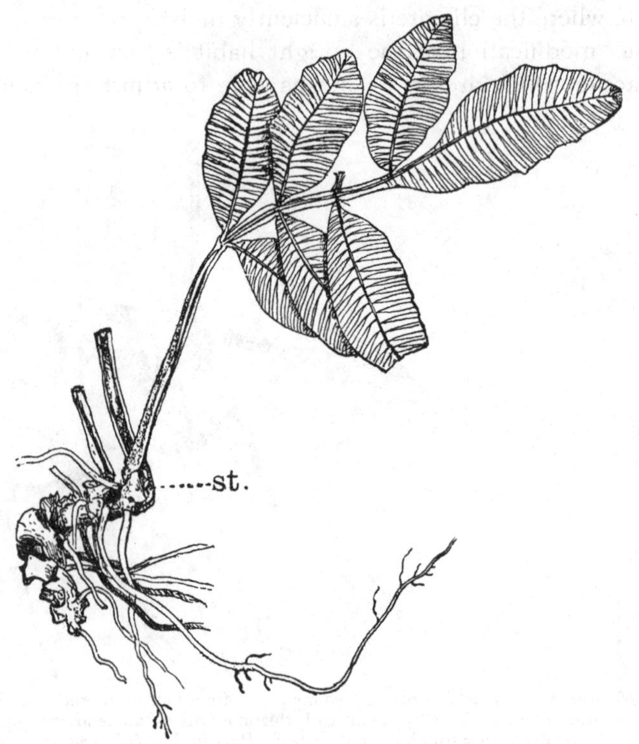

Fig. 38. A small plant of *Danaea alata*, showing the lower part of the axis vertical, the upper part obliquely horizontal. *st*=stipules. (½ size. After Campbell.)

of the leaves. Though the vertical and more compact shoot was probably the primitive type, the creeping habit was certainly acquired early, for it is seen in *Metaclepsydropsis duplex* from the Culm. Its secondary origin in the ontogeny of those primitive Ferns which have a vertical embryo may be traced in *Helminthostachys*, *Danaea*, and *Christensenia* (Fig. 38), in all of which the originally vertical sporeling falls over later to a creeping habit. Here it seems to be a natural consequence of the heavy leaves weighing down a weak stem. In the Leptosporangiate Ferns, however, where the embryo is prone from the first, the upright habit which so many of them show in the adult state would have been secondarily acquired (Fig. 39). But on the other hand the creeping habit is not necessarily a mere continuation of the prone axis of the sporeling. This is shown by the inversion of the axis in *Polypodium vulgare*, where by hyponastic growth the creeping axis arches backwards over the parent prothallus (Fig. 39). While the upright habit and basket-form are effective in growth in the open, or under forest-canopy, and in restricted space, the creeping habit is suitable for Ferns living as dense undergrowth, or where the space is not restricted. Moreover it brings many advantages. Rooting in the soil is not localised at a given centre, but adventitious roots spread over the whole area of the soil covered by the creeping stem. Each leaf can develop independently, and without competition for exposure to light. Branching is common and even prevalent in creeping Ferns (Fig. 32): while the original individual may be multiplied to many by the simple process of decay of the older parts progressing beyond a branching. This is conspicuous in the Bracken, and it accounts for a large proportion of the individuals of that species that we see. These advantages, together with the opportunity offered for a climbing habit, sufficiently account for the prevalence of a creeping form in Ferns at large.

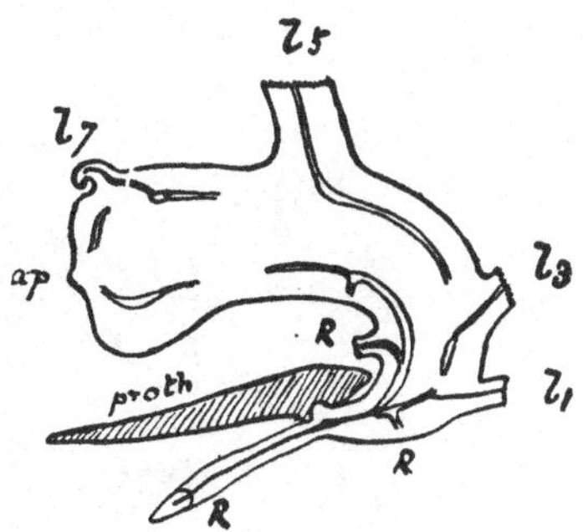

Fig. 39. *Polypodium vulgare*. (×6.) Median section through a prothallus and embryo, partly diagrammatic: showing one series only of the distichous leaves, l_1, l_3, etc.: R=roots: ap=apex of axis. The hyponastic shoot becomes completely inverted, growing backwards over the prothallus.

Like other Pteridophytes the Ferns are perennials, and have been so from the time of their early fossil progenitors. Occasional exceptions are seen, such as *Anogramme leptophylla*, which perennates by its storage-prothallus, while the sporophyte dies off each autumn (see Chapter XIV). *Ceratopteris thalictroides* is short-lived, but frequently perennates by means of its sporophytic buds. The Salviniaceae are also mostly annual plants, though they may perennate by their apical buds. Putting such special cases aside, it may be said of modern Ferns in general that they are perennials. Frequently their active vegetation may continue without any marked seasonal break, as evergreens. This is the case with most tropical Ferns, and especially with those of dendroid habit. Successive series of leaves may be added to the terminal tuft, and the older leaves may successively lose their firmness, shrivel, and fall, sometimes with a clean detachment as in *Cyathea Schansin* (Fig. 40); sometimes they leave ragged and persistent stumps, as in *Dicksonia squarrosa*. On the other hand, many of the Ferns of temperate climates show a regular seasonal leaf-fall, as in the Male Shield Fern, and the Bracken. Sometimes the shedding of the leaves is carried out piecemeal, the separate pinnae or pinnules falling away, as in *Nephrolepis*, *Photinopteris*, or *Didymochlaena*. Such cases make clear the necessity of storage of materials for the next season's leaves. The analogy with Flowering Plants in all this is too obvious to need any stressing.

Fig. 40. *Cyathea Schansin*. Crown of leaves, and stem with leaf-scars. S. Brazil. (From Christ, after Wacket.)

Though the storage-arrangements are a less prominent feature externally in Ferns than in Flowering Plants, they are often very efficient. The cortex and pith of the stem of Ferns are commonly packed with starch, and especially so in the deciduous Ferns such as those named. The quantity of material laid by in the rhizomes of the Bracken led to their being used as a staple food by the Maories. Beyond a slight distension of the stock the form is not altered, excepting in some few cases: for instance in *Ophioglossum bulbosum* and *palmatum*, and some other species, the stock is swollen, and stored with starch, so as to resemble the corm in Angiosperms (Fig. 36). The tubers borne distally on branches of the stolons of *Nephrolepis* provide an example still more marked externally. They are oval in form, and consist mainly of aqueous tissue stored with sugar. But such modifications of form are rare (17). The possession of such reserve materials is often coupled with a geophilous habit, but certain well-stored rhizomes are exposed above ground, as in *Davallia*, and many species of *Polypodium*. In *Pteridium*, however, they are deeply buried, a feature which together with its unlimited apical growth and frequent branching may have contributed to its world-wide success. The buried storage-stock of *Botrychium* and *Ophioglossum* may well have been a decisive condition of the survival of these ancient types.

The creeping habit is much more adaptable to the varied circumstances of life than the upright with its massive axis. The thinner creeping stem shows frequent distal branching, often with very perfect dichotomy (Fig. 32). It can thus adjust itself readily to the substratum. In particular this adjustment leads naturally to the scandent habit: and this finally to a full epiphytic mode of life. There is little doubt that the creeping and primitive *Dipteris*, the scandent *Cheiropleuria*, and the strikingly epiphytic genus *Platycerium* are themselves illustrations showing how this last state may have been acquired. Interesting parallels to this may again be found among the Aroids. The climbing, and finally the epiphytic habit bring the same advantages in Ferns as in Flowering Plants, and the analogies of method between the two are many. The weak axis fixed to the stem of some stronger plant by adventitious roots may climb to considerable heights, as is seen in *Trichomanes scandens*, and *auriculatum* (Fig. 41), or *venosum* (Frontispiece): in *Oleandra*, in *Stenochlaena aculeata* (Fig. 42), and *S. scandens*, in *Polybotrya osmundacea*, and many others. Still greater success in climbing follows from a prehensile function of the leaves, either in cooperation with a climbing stem or independent of it. The latter is the case in *Lygodium*, where a compact underground rhizome bears leaves endowed with unlimited apical growth (Fig. 43). The opposite pairs of divaricating pinnae are borne upon a thin wiry rachis which can twine round a support, and the plant thus climbs to a very considerable height in thickets and low forest. The habit is superficially like that of *Medeola*, and like this Flowering Plant it is sometimes grown by nursery-men on strings, and used as a table-decoration. *Blechnum* (*Salpichlaena*) *volubile* is also an example of the prehensile function of the rachis, and is a successful climber of the Western Tropics. Other Ferns have a straggling habit, chiefly due to unlimited apical growth of their leaves, together with divaricating pinnae. This is the case with *Gleichenia*, in which the leaf grows intermittently at its apex, the details varying greatly with the species. After forming one or more pairs of pinnae the

Fig. 41. *Trichomanes auriculatum*, a stem-climber from Northern India. (½ size. After Christ.)

circinate tip becomes dormant, and the pinnae enlarging give the misleading appearance of a dichotomy. In a fresh season its activity is renewed, and fresh pinnae are formed. This may be continued indefinitely. As the pinnae and often also the pinnules may repeat the process, the leaf with its widespread branching and wiry rachis forms involved thickets on exposed savannahs, or dense undergrowth in woods, often difficult to penetrate. But a still more formidable obstacle is presented by the "Bramble Ferns," the chief of which is *Odontosoria aculeata* (Fig. 44): these Ferns in addition to

Fig. 42. *Stenochlaena* (*Teratophyllum*) *aculeata* (Bl.), Kze., showing its climbing habit, and dimorphic leaves. (After Karsten.)

their scrambling habit, and the continued apical growth of their wiry leaves, bear reflexed prickles on the rachis and its branches, which assist the straggler like the prickles of the Bramble or Rose. Thus Ferns are occasionally as effective in climbing or straggling as Flowering Plants, and use means similar to theirs.

It may be shown experimentally that the leaves of Ferns, like those of Flowering Plants, react directly to the conditions under which they develop. The effect of exposure to full sun in relatively dry surroundings is seen in Fig. 45, (i), for *Dryopteris*, and the effect of growth under deep shade in

Fig. 43. Climbing leaf of *Lygodium palmatum*, from N. Carolina. (⅓ size. After Christ.)

Fig. 45, (ii). The average measurements for a number of cases showed that the area of the shade-form may be practically double that of the sun-form. But the sun-leaves are more robust in structure, and give a higher spore-output from their sori. Similar results were obtained by experiment from the Bracken and the Hart's Tongue. How direct the effect may be is shown by Fig. 46, where (i) represents a section of a pinna of *Pteridium* which had

Fig. 44. *Odontosoria aculeata*, a well known "Bramble Fern" from the West Indies. (½ size. After Christ.) (See text.)

been exposed to sun and wind during development, while (ii) shows a section of a pinna of the same leaf which developed in the shade. Such observations give the key to certain features in the general habit of Fern-leaves.

Comparison of these at large, when based upon average specimens from normal stations, shows that Ferns of exposed habit have leaves of relatively firm texture, with narrower and highly divided segments, while Ferns of shaded habit usually have broader and less cut leaves of thinner texture. Examples of shade-leaves are seen in *Christensenia aesculifolia*, and in *Hypo-*

Fig. 45. (i) Leaf of *Dryopteris* grown in the open with extreme exposure to sun and wind. (ii) Similar leaf from the same garden grown protected from the wind, and in heavy shade.

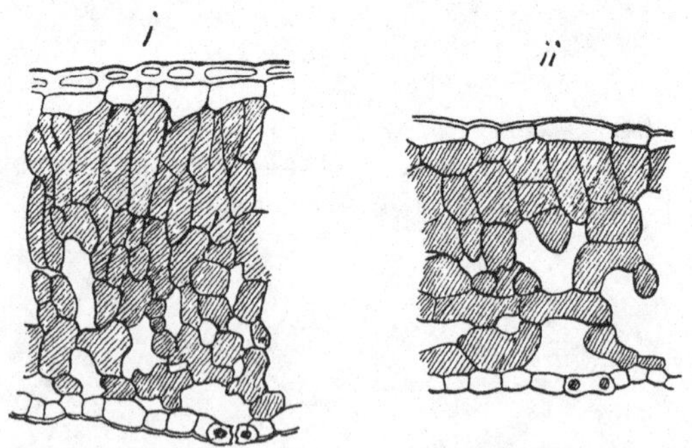

Fig. 46. Parts of the same leaf of the Bracken seen in section. (i) is from a pinna exposed to full sun during development. (ii) is from a pinna developed in shade. (After Boodle.) (× 200.)

derris Brownii. Ferns of shaded and moist habitat are often of still more delicate texture than these, and usually have naked surfaces. The extreme condition is that which is styled "filmy," in which the lateral wings of the expanded lamina may be reduced to a single layer of pellucid cells. The Hymenophyllaceae are the chief representatives of this habit, and they are often called "the Filmy Ferns." But there are also filmy species of *Asplenium* and of *Todea*, and even of so leathery a genus as *Danaea*. Specimens of *D. trichomanoides* were discovered by Spruce, growing in a wet habitat near Tarapoto in Peru (1856). This species, and *D. crispa*, and some other species in less degree, have pellucid leaves almost equalling the Hymenophyllaceae in their texture. The filmy character is in fact an adaptive hygrophytic condition, which has been assumed homoplastically in several quite distinct groups of shade-loving Ferns. It is secondary and derivative, and cannot rightly be accepted as any indication of a primitive and Moss-like nature, though this was once the accepted belief. A similar texture is approached in the leaves of some shade-loving species of *Selaginella*.

Notwithstanding the hygrophytic character of the Class as a whole, and their general preference for a shaded habit, as well as the fact that they may, like the Filmy Ferns, show sometimes extra adaptation to these conditions, many Ferns are able to endure exposure to strong insolation, and even to drought provided it be temporary. Others grow epiphytically, a habit which is congenial enough under shade in tropical rain-forests: but elsewhere the epiphyte, having no access to the soil, must needs have some means of economising its water-supply. The requirements are in either case met by such modifications of structure as are characteristic of other xerophytes, and are specially seen in Flowering Plants. There is a small Flora of British Ferns of rocks, and even of dry wall-tops, including such species as *Asplenium ruta-muraria*, and *A. Ceterach*: or of tree-trunks and rocks, such as *Polypodium vulgare*. These may sometimes be found with their firm leathery leaves dried to crispness in summer: and yet they tolerate those stations. In *A. Ceterach* the dense covering of scales is an efficient protection in addition to the texture of the leaf (Fig. 47). Abroad there are certain genera, such as *Notholaena*, *Cheilanthes*, and *Pellaea*, which are still more typically xerophytic, having stiff, attenuated, and often highly divided leaves of small area, with hard polished leaf-stalks, and a waxy glandular covering to their surface. Others, such as *Niphobolus* and *Elaphoglossum*, bear a covering of protective scales over the whole shoot: sometimes the protective armour is specially developed on the rhizome, as is seen in *Phlebodium aureum* (Fig. 48). Sometimes the scales are borne chiefly on the lower surface of the blade, the petiole, and the rhizome, while the upper surface is almost clear, as in *Polypodium* (*Lepicystis*) *incanum*. This rock-growing Fern is a most successful xerophyte. It may be seen shrivelled for weeks without rain under a tropical sun, but when moistened it swells again, and continues growing as Mosses do. In other cases isolated species even of peculiarly hygrophytic genera may be specially protected, as in the case of *Hymenophyllum sericeum*, whose long pendent leaves are densely covered with a felt of ferruginous hairs. Others may show a xerophytic folding of

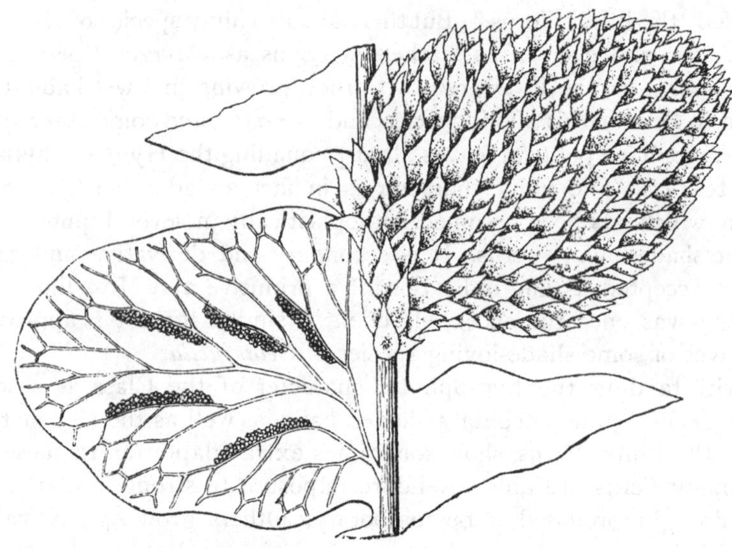

Fig. 47. *Asplenium Ceterach*, Willd. Two pinnae seen from below, showing on the right the covering of scales, on the left the sori and nervation with the scales removed. (×8.) (After Luerssen.)

Fig. 48. Vertical section through the scales covering the rhizome of *Phlebodium aureum*, showing their elaborate overlapping. (×35).

the pinnae, as in *Notholaena sinuata* or *ferruginea*, or in *Jamesonia*, with or without the addition of a felt of hairs such as occurs in *N. lanuginosa*. These xerophytic characters appear in Ferns exposed to drought either in actual dry areas, or on rocks, or where from their position as epiphytes they are without direct access to the soil.

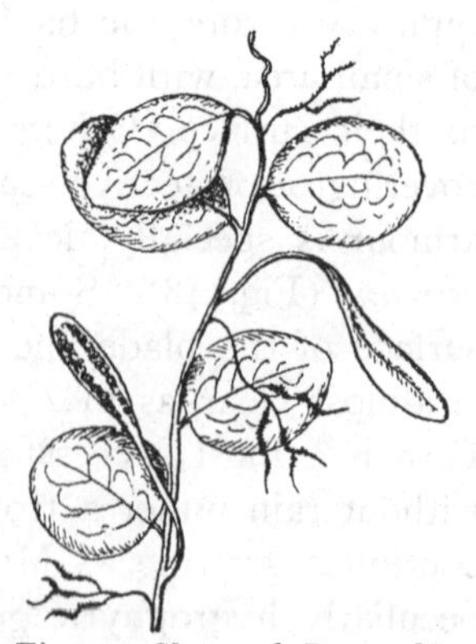

Fig. 49. Shoot of *Drymoglossum subcordatum* after Christ. (Nat. size.)

Another form which xerophytic specialisation may take is succulence of the stem or leaf, sometimes with a smooth surface and well-developed epidermis, but oftener with more or less numerous scaly hairs. This is seen in *Drymoglossum subcordatum*, where the barren leaves are elliptical and fleshy (Fig. 49). But in other cases the storage of water may be in the distended stem, while the leaves are leathery, as in *Photinopteris*. Peculiar cases of this are seen in *Polypodium sinuatum* and *Lecanopteris carnosa*, which grow on exposed rocks and tree-trunks (Fig. 50). Their stocks are distended while young by aqueous parenchyma. This dries up in the mature state forming hollow galleries in which ants habitually live. These galleries communicate with the outer air by passages excavated by the ants themselves at points corresponding to lateral buds, where the tissues are soft. The condition of these epiphytic Ferns is similar to that of *Myrmecodia*, *Hydnophytum*, or *Humboldtia*. As in these Flowering Plants, the ants by

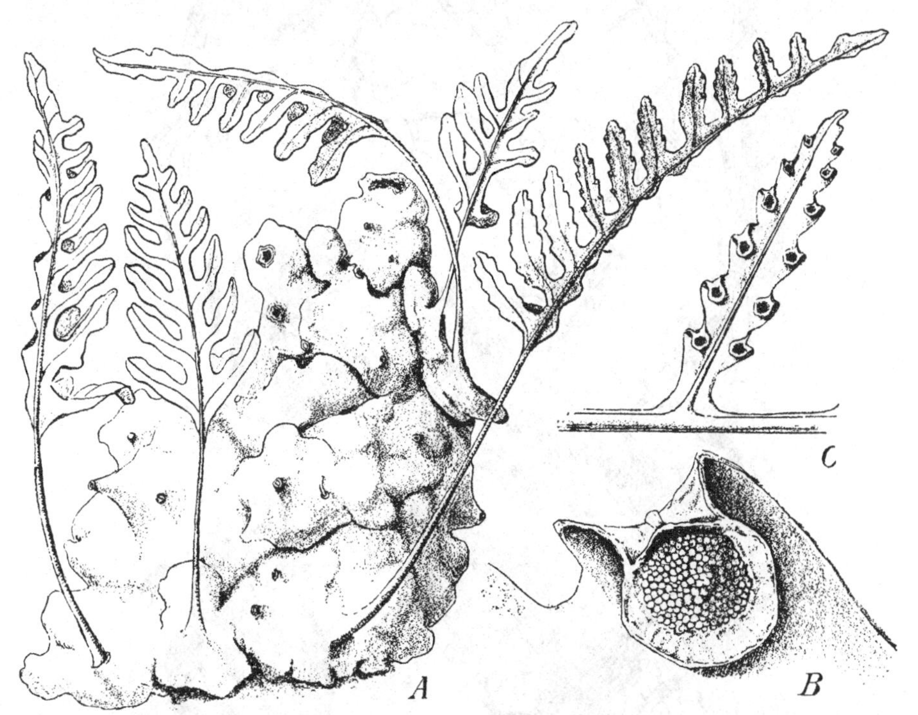

Fig. 50. *Lecanopteris carnosa*, Bl. *A* = habit of the plant; *B* = marginal flap, with sorus, enlarged; *C* = segment of a fertile leaf. (*A*, after Burck; *B*, *C*, after Diels.)

gnawing provide the entrance: they are the invaders, and there is no evidence of the adaptation of the plant directly to the convenience of its visitors.

The tubers of *Nephrolepis* have already been mentioned as containing a carbohydrate store. They are also reservoirs for water, and it has been noted that as they shrink under extreme drought the leaves wither, and the pinnae are shed from the leaf-tops downwards, thus gradually reducing their exposed surface. Still more remarkable provisions against drought are seen in *Polypodium Brunei* and *bifrons*, as described by Christ (Fig. 51). These epiphytic Ferns produce pouch-like urns in form not unlike those of the Asclepiadaceous plant *Dischidia*. Being hollow they naturally collect water when it is available, and as in *Dischidia* the cavities are penetrated by roots, which spring from neighbouring points on the same plant. Physiologically in both cases they appear to be a means of securing a supply of water during rains, to be absorbed at leisure into the tissues. But whereas the urns are specialised leaves in *Dischidia*, they appear in these Ferns to be highly modified branches of the rhizome.

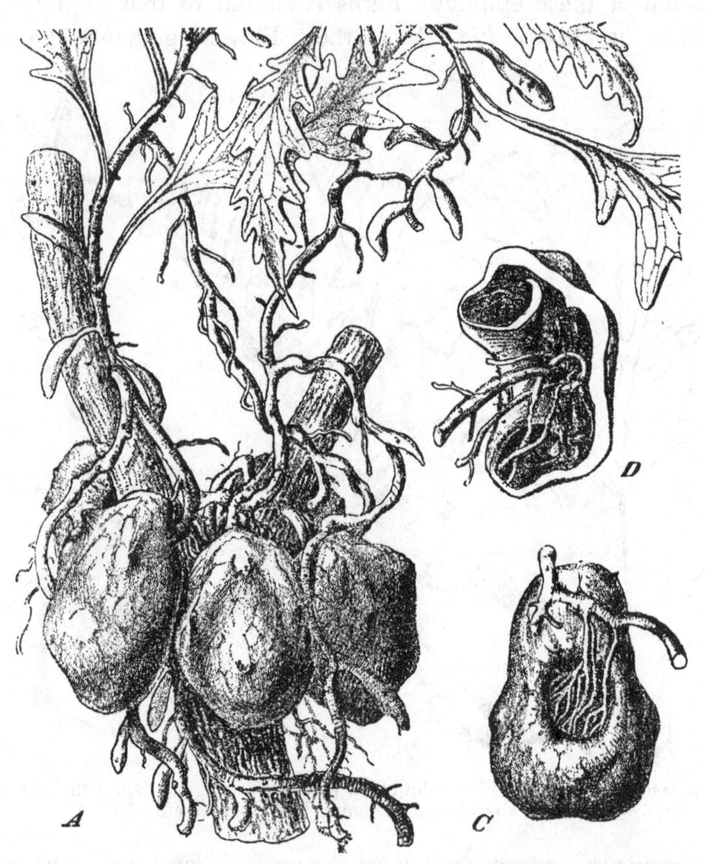

Fig. 51. *Polypodium bifrons*, after E. Ule, from Christ; showing the epiphytic habit, and sac-shaped urns, into which roots penetrate, as in *Dischidia*.

An alternative adaptation seen in epiphytic Ferns is the so-called "nest-habit," where the leaves are aggregated in a dense tuft, with roots below, the whole serving as a means of holding together a considerable mass of humus, which collects within the leaves, and retains water. A very successful example of this is seen in *Asplenium nidus*, a species whose leathery lanceolate leaves often attain large size, but are not specialised in form (Fig. 52). But the nest-habit is more effectively developed in *Platycerium* and in

Drynaria by the differentiation of the leaves into two types. In the latter genus the nest-leaves are widely expanded, growing appressed to the surface of the trunk or branch, and are sterile. They are pale coloured, and have a strong venation, which persists after the mesophyll decays, forming a basket-like receptacle for humus in which the adventitious roots are nourished.

Fig. 52. *Asplenium nidus*, an epiphytic Nest Fern. E. Borneo. (After M. Mühlberg, from Christ.)

Other leaves develop of a normal type, and are photo-synthetic and fertile (Fig. 53). In *Platycerium* the method is the same, though with different details (Fig. 54). The affinities of the two genera are widely distinct, and their resemblances are clearly homoplastic. The general method of these epiphytes is not unlike that seen in Bromeliads and Orchids.

Fig. 53. *Drynaria*, showing epiphytic habit, and dimorphic leaves. *A*, *B*=*D. quercifolia* (L.), Bory. In *B* the dimorphism is not yet fully developed, the plant being young. *C*=*D. Baronii* (Christ), Diels. (*A*, *B*, after von Goebel; *C*, after Christ; from Engler and Prantl.)

The climbing and humus habits are connected in Flowering Plants on the one hand with parasitism, and on the other with mycorhiza. The former is not seen in the Filicales, but a number of Ferns have adopted mycorhiza. It is found in its most complete form in the gametophyte of the Ophioglos-

Fig. 54. *Platycerium coronarium* (*biforme*), showing epiphytic habit. Buitenzorg. (After von Goebel, from Christ.)

saceae, where it has advanced to the state of complete saprophytic nutrition of the underground prothallus (see Chapter XIII). But the mycorhizic fungus invades the young sporophyte also. In *Botrychium Lunaria*, which has mycorhizic roots from the first, the eighth or ninth leaf of the sporeling is the first to appear above ground, and the young plant is thus holo-saprophytic up to its eighth or ninth year. In *Helminthostachys* the fungus is present in the first three or four roots of the young plant, but it is absent from the later roots. Fungal invasion has been observed in the adult plant of *Ophioglossum vulgatum*, though it is stated to be absent from the young plant. It is present in the adult plants of *O. pendulum*, and it is specially prevalent in the precocious mycorhizic root of its embryo. The most peculiar case of all is that of *Ophioglossum simplex*, in which a pronounced state of mycorhiza goes along with the apparently complete absence of a sterile lamina (Fig. 55). Here it would seem that the mycorhiza makes the nutrition of the large spike still possible in the dense wet forest in which the plant grows, notwithstanding that the usual photosynthetic organ is functionally absent. Reduction is not, however, apparent in the spike itself, for provided nutrition be kept up from whatever source it would still retain its character, being essentially a spore-bearing organ. Mycorhiza has also been found in the roots of the Marattiaceae, and in *Cyathea*, but without any consequent reduction. Strangely enough it is absent in *Asplenium nidus*. Thus its occurrence in Ferns is sporadic, and seemingly arbitrary: it may be associated in extreme cases with morphological reduction, but this is not a general feature, and no Fern is known in which the holo-saprophytic nutrition continues throughout the whole life-cycle. Irregular nutrition has never secured a complete hold among the Filicales.

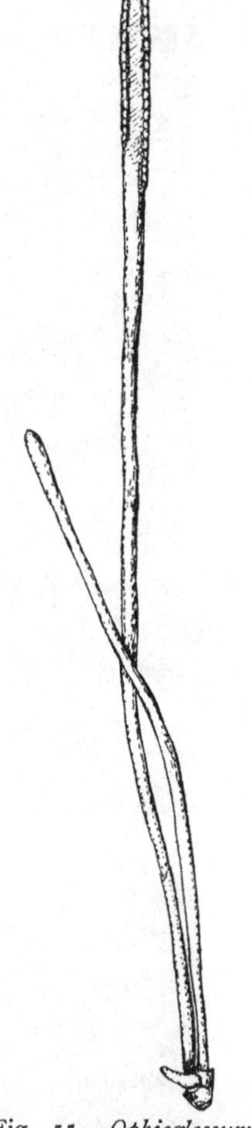

Fig. 55. *Ophioglossum simplex*, Ridley, slightly reduced. Three leaves are seen inserted on a short stock. But the leaves appear to consist each of a fertile spike, with no sterile lamina.

The hygrophilous habit so prevalent among Ferns has in some few been extended to a definite swamp-life, or even to a floating habit. *Dryopteris Thelypteris* grows habitually in fen-land: *Marsilia* and *Pilularia* live in swampy ground, and grow quite well with their rhizomes floating in water, but they do not appear to fruit except on firm ground. *Ceratopteris thalictroides* is a semi-aquatic Fern, and it is particularly at home among crops of rice, which are grown in artificially irrigated, muddy ground. *Azolla* floats with its roots pendent into the water, and *Salvinia* with a like habit has no proper roots, but root-like leaves. Thus various steps are seen leading to a completely aquatic life. Some few Ferns affect salt water. *Asplenium marinum* lives on maritime rocks within reach of the spray, and never far inland. Its leathery foliage may be held as halophytic. The same applies for *Acrostichum aureum*, which though it can be grown well without salt in greenhouses, is a constant companion of the Mangroves throughout the tropics, and roots downwards into the salt swamps, raising its thick leathery leaves six or eight feet into the full sun.

The differentiation of leaves has already been mentioned in the case of the "nest-habit." But it comes more prominently forward in relation to the propagative function. The primitive condition has doubtless been that of the "general-purposes" leaf, which is at once a nutritive organ and a base for the production of sporangia. Many Ferns have retained that undifferentiated state with all their leaves of one type to the present day. This is the condition seen in the Male Shield Fern, the Bracken, the Lady Fern, and the Common Polypody. But as in Flowering Plants, so here there may be a differentiation of leaves for distinct purposes. The recognised types for Flowering Plants are the scale-leaf, the foliage-leaf, and the sporophyll: and these may all be distinguished in some Ferns. In so primitive a type as *Osmunda regalis* certain leaves are found to have their apex arrested, while the lower sheathing region is fully developed. A like condition is seen in the leaves borne on the runners of the Ostrich Fern (*Matteuccia Struthiopteris*). These are in fact scale-leaves, comparable with those so often seen on the rhizomes of Flowering Plants, or covering their winter buds. The last-named Fern shows also a strong distinction of two other types of leaf, viz. the foliage-leaves which appear first in the seasonal growth, and are broadly expanded and quite sterile; and the sporophylls which appear later in the season, and are attenuated and fertile (Fig. 56). A like differentiation is also seen in *Onoclea sensibilis*, and in the Hard Fern. Sometimes the sterile and fertile regions are segregated as parts of the same leaf, a condition seen in *Thyrsopteris*, *Anemia*, and *Osmunda*, while it is a very marked feature in the Ophioglossaceae. Such examples, taken together with the differentiation seen in the "Nest Ferns," illustrate a considerable degree of foliar differentiation in Ferns of the same general nature as that seen in Flowering Plants. It is not carried out to the same degree as in these, but still the comparison is justified.

It thus appears that in the adaptation of their sporophyte to their habitat the Filicales have progressed along lines similar to those seen in Flowering Plants. But the specialisation of the shoot to the habitat is not so general or so exact as in them. In a very large proportion of living Ferns it appears only in a low degree. There are, however, occasional examples of a relatively high state of adaptation. Since such cases are isolated among various genera which as a whole do not appear specialised to meet peculiar circumstances, it follows that *the more extreme examples have probably resulted from relatively direct adaptation. In that case they will not form a suitable basis for*

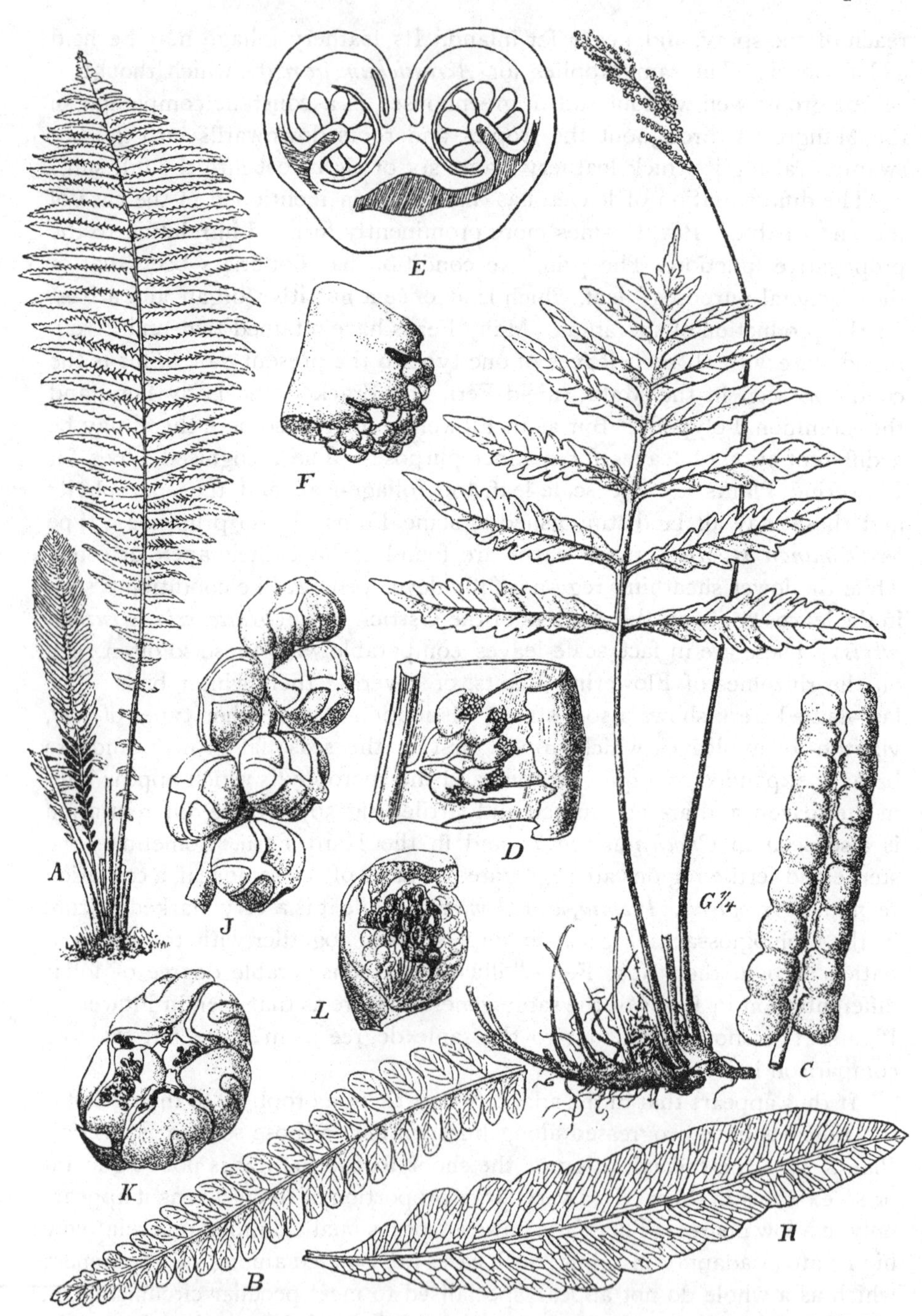

Fig. 56. *A—F = Matteuccia Struthiopteris* (L.), Todaro; *A* = a fertile and a sterile leaf, showing habit; *B* = a sterile pinna; *C* = a fertile pinna; *D* = part of a fertile leaf with venation and sori; *E* = transverse section through a fertile pinna; *F* = sorus with indusium. *G—L* = *Onoclea sensibilis*, L. *G* = habit; *H* = a sterile pinna; *J* = a fertile pinna; *K* = a fertile pinnule; *L* = sorus with indusium. (*A*, *D—F*, after Luerssen; *J—L*, after Bauer; *B*, *C*, *G*, *H*, after Diels. From Engler and Prantl.)

classification, and for that purpose therefore they naturally take a subsidiary place. There remain a large number of forms showing a general megaphyllous Fern-like habit, with a prevailing uniformity of the external character of the sporophyte. This makes their systematic grouping difficult. It will be necessary to examine them closely, and by exact observation of other details in all stages of their life-history to find features more permanent and therefore more reliable than those of external habit would be. The nature of those characters will be discussed in the following Chapters.

It may be thought that in this treatment of Habit and Habitat in Ferns the prothallus has been unduly ignored. In the comparisons which are to follow it will take its proper place. But here it must suffice to say that its relative inconspicuousness has led to its being entirely passed over in the ordinary Systematic Works. The fact that the details of the gametophyte are very imperfectly known in the large majority of the species has hitherto made its general use impossible for generic and specific comparison. Moreover, the very direct dependence of the prothallus, as regards its form, upon the external conditions ruling during its development is another serious obstacle to its use in comparison with a view to Classification. The comparative treatment of the prothallus and sexual organs will be reserved till Chapters XIII and XIV.

BIBLIOGRAPHY FOR CHAPTER II

13. ENGLER & PRANTL. Natürl. Pflanzenfam. i, 4. 1902. Here are many illustrations of general habit.
14. CHRIST. Geographie der Farne. 1910, where floristic literature is quoted.
15. VELENOVSKY. Vergleichende Morphologie der Pflanzen. Prag. 1905. Teil I.
16. VON GOEBEL. Organographie. 1918. Teil II, Heft II.
17. SAHNI. Tubers of *Nephrolepis*. New Phyt. xv, p. 73, where literature on storage is quoted.
18. McILROY. Proc. Roy. Phil. Soc. Glasgow. 1906.
19. BOODLE. Structure of the leaves of the Bracken. Linn. Journ. xxxv, p. 659.
20. KARSTEN. *Teratophyllum aculeatum*. Ann. Jard. Buit. xii, p. 143.
21. CHRIST. Biol. u. Syst. Bedeutung d. Dimorph. bei Epiphyt. Farnpfl. besonders *Stenochlaena*. Verh. d. Schw. Naturf. Ges. 89th Meeting. July 1906.
22. YAPP. Two Malayan myrmecophilous Ferns. Ann. of Bot. xvi, p. 185.
See also the Floristic works quoted for general use on p. ix.

CHAPTER III

A THEORETICAL BASIS FOR THE SYSTEMATIC TREATMENT OF FERNS

THE Filicales may be held to comprise all of the living Megaphyllous Pteridophytes, together with such fossils as show essentially similar characters. This is not a scientific definition, but is merely a provisional designation of the Plants that are here to be dealt with, all of which show, so far as investigation has extended, a life-history and essential features such as have been described in Chapter I. It is not possible to draw up a strict definition of the Filicales till their characters have been examined in detail and compared with those of other groups of Plants from which they are to be distinguished. The mere fact that their leaves are relatively large in proportion to the axis which bears them (megaphyllous) is not in itself a sufficient diagnosis. Some Filicales are actually microphyllous, for instance *Salvinia* and *Azolla*. On the other hand, the large and growing Flora of the Pteridosperms, known only as fossils, was certainly large-leaved. These plants were probably derivatives from some more or less Fern-like ancestry: but being actually Seed-Plants they are not to be included under the Filicales. Some Lycopodiales (*Sigillaria* and *Isoetes*) had relatively large leaves, and certain fossils related to the Equisetales or Sphenophyllales (*Pseudobornia*) shared that character. We may indeed hold it as possible that any phylum of Pteridophytes might have developed megaphyllous types, and in that case megaphylly could not be held in itself to be a feature indicating affinity or community of Descent. But as a matter of fact, excepting *Isoetes* which is plainly a Lycopod, no such Pteridophytes are living now upon the earth other than those which may naturally be included in the Filicales. Accordingly the provisional designation given above will serve to introduce the subject of this volume.

The Filicales are represented on the earth at the present time by about 150 genera and about 6000 species, according to Christensen's *Index Filicum* (23). Some are very minute, others attain a considerable size as Tree Ferns: but none can be reckoned among the largest of living plants, nor is there fossil evidence that Ferns ever attained extreme dimensions. Their geographical spread is very general. Some few are Arctic: but Ferns increase in numbers both of species and of individuals towards the Equator. They are mostly mesothermal hygrophytes: that is, they flourish under moist conditions with a moderate temperature, and the majority of them are shade-loving Plants. Hence their headquarters are in the mountains of the tropics, where they form a considerable part of the undergrowth below the forest-canopy. But their habitat is variable. Some specialised types are actually aquatic, while others are able to withstand conditions of moderate, some even of extreme drought.

The Ferns are much richer in genera, species, and individuals than any other living Pteridophytes. They probably present the climax of successful development in Homosporous Vascular Plants. They show also a high degree of variety both in their vegetative and in their propagative characters. These characters provide good diagnostic features upon which their Classification may be based. They have a full and long palaeontological history, which may be traced through successive horizons backwards to Palaeozoic times. The characters of the fossils have been proved to be so far comparable with those of certain of the living Ferns that their relationship cannot be held in doubt. The geological record can therefore be used as a valid check upon such conclusions as to relative antiquity as may be drawn from the comparison of living types of Ferns. There is in fact no group of living plants which offers so favourable an opportunity for phyletic classification: for none is there so long, rich, and consecutive a geological record, combined with so great a variety of living types, possessed of diagnostic characters so marked, and with progressions so susceptible of reasonable physiological interpretation. This gives the Filicales an interest peculiarly their own. But the classification of such a group need not have as its end merely the reduction of the group itself to phyletic order. It may be made a means of demonstrating methods of comparison applicable to other groups. This is the way in which it is proposed to treat them in the present work, which will thus illustrate phyletic method as seen in practice in a natural alliance of plants specially favourable for giving definite conclusions.

Current systematic arrangements of the Filicales

Before entering on the detailed examination of the facts which should thus supply the basis for a natural system of the Filicales it will be well to summarise the current classification of them, as set forth in standard works of the present time. Naturally it is to Christensen's *Index Filicum* (1906) that one looks first (23). In that great work,—which is primarily a catalogue of the accepted generic and specific names of Ferns, together with all the synonyms that have been used by various writers,—a general grouping of the genera and families is given at the outset. This corresponds in essentials with the scheme laid out in *Die Natürlichen Pflanzenfamilien*, by Engler and Prantl (1902) (24), which is the most comprehensive systematic treatise on Ferns of recent decades. The sequence of Families and their Orders as given by Christensen is as follows, a few explanatory and descriptive words being added after each:—

CLASS. **FILICALES**

SERIES I. **FILICALES LEPTOSPORANGIATAE**

SUB-SERIES A. EUFILICINEAE

Family I. Hymenophyllaceae. (2 genera, 459 species.)

These are the typical "Filmy Ferns," relatively small, sometimes minute, and mostly shade-loving. They were long regarded as the most primitive of all Ferns, and related to the Mosses on account of their delicate structure. Their sori are gradate, and the annulus oblique.

Family II. Cyatheaceae. (7 genera, 456 species.)

Order i. Dicksonieae.
Order ii. Thyrsopterideae.
Order iii. Cyatheae.

These include the typical "Tree Ferns," with upright stems: their sori are marginal or superficial, and gradate: the annulus is oblique.

Family III. Polypodiaceae. (114 genera, 4527 species.)

Order i. Woodsieae.
Order ii. Aspidieae.
Order iii. Oleandreae.
Order iv. Davallieae.
Order v. Asplenieae.
Order vi. Pterideae.
Order vii. Vittarieae.
Order viii. Polypodieae.
Order ix. Acrosticheae.

These include the main bulk of the species of living Ferns: they are in fact the dominant type of the present day: their sori are marginal or superficial, and mixed, and the annulus vertical.

Family IV. Parkeriaceae. (1 genus, 1 species.)

A monotypic tropical family of specialised, semi-aquatic Ferns, with superficial, solitary sporangia.

Family V. Matoniaceae. (1 genus, 2 species.)

A small tropical family, with only one living genus: probably a survival of a type prevalent in Mesozoic times. Simple, superficial sori, sporangia with oblique annulus.

Family VI. Gleicheniaceae. (2 genera, 80 species.)

Mostly straggling, tropical or sub-tropical Ferns, related to certain early fossils; with superficial simple sorus, and primitive anatomy, sporangia with oblique annulus.

Family VII. Schizaeaceae. (4 genera, 118 species.)
Ferns of varied habit and structure, but all with relatively simple sori, monangial, and typically marginal: related to certain early fossils. Annulus distal and almost transverse.

Family VIII. Osmundaceae. (3 genera, 17 species.)
This is the family of the Royal Fern. The habit, anatomy, and the large sporangia suggest a primitive type, related to certain early fossils. Some of them are "Filmy."

SUB-SERIES B. HYDROPTERIDINEAE

Family IX. Salviniaceae. (2 genera, 18 species.)
Floating Ferns of small size and simple structure, with sexually distinct sporangia.

Family X. Marsiliaceae. (3 genera, 63 species.)
Ferns of aquatic habit, whose relatively simple structure links on to more complex types. They also have sexually distinct sporangia.

SERIES 2. **MARATTIALES**

Family XI. Marattiaceae. (5 genera, 118 species.)
Sappy tropical Ferns often of large size, with leathery leaves and simple sori, and sporangia of large size. Their characters appear to be primitive, but Cycad-like. They are related to certain early fossils.

SERIES 3. **OPHIOGLOSSALES**

Family XII. Ophioglossaceae. (3 genera, 78 species.)
The Moonworts and Adder's Tongues, with very peculiar habit. The large sporangia are borne on a "fertile spike." Prothalli typically subterranean.

It may not have been the intention of Engler and Prantl, or of Christensen, to express in this arrangement any definite opinion as to the relationship of the several Families. Certainly the sequence as shown does not accord with any probable phyletic view as checked by palaeontological evidence. It will not be necessary here to enter on a detailed criticism of it. It must suffice to note a few examples which show the sort of discrepancies that exist in it. The Hymenophyllaceae are placed first in accordance with a common practice of many of the older writers, though these Ferns cannot now be held to be either the most primitive or the most advanced types of Ferns. The Hydropterideae are spliced between the Osmundaceae and the Marattiaceae, though they show no near affinity with either. *Dipteris* is placed in the Aspidieae (III. ii) far away from its relative *Platycerium*, which is ranked with the Acrosticheae (III. ix), and *Acrostichum* in the Acrosticheae (III. ix) far from *Pteris* in the Pterideae (III. vi). The Schizaeaceae are not placed in relation to the Hymenophyllaceae, Pterideae, or Marsiliaceae: nor are the Gleicheniaceae placed near to the Cyatheaceae, though there are certain species which seem to hover in their characters between the two. Grounds will, however, be advanced in the course of this work which will tend to support all of these relationships, though they are not in any way suggested by the order of the families in the Engler-Prantl scheme.

The fact has probably been that up to the present those who have written on the systematic treatment of Ferns have not seriously attempted to trace the relations of the larger groups by descent, or to suggest these by the order of their arrangement. The interest of Pteridologists has been centred in the relations of genera, species, and varieties rather than of families. That this was so in the *Synopsis Filicum*(27) of Hooker and Baker appears evident from the fact that that book was based upon the *Species Filicum*(26), and followed the sequence of "sub-orders" and "tribes" as shown in it. But as the Osmundaceae, Schizaeaceae, Marattiaceae, and Ophioglossaceae were not included in the *Species Filicum*, they were added at the end of the book. Convenience in compilation probably dictated this position for them rather than design, though they took a similar place in the sequence of Mettenius. In point of fact the list of families as given in the later systematic works is little more than a catalogue, and so far as their phyletic relations are concerned the families might almost as well have been disposed in alphabetical order as in the order in which they stand. That Sir William Hooker took some such view as this of the succession of families is shown in the Introduction to the *Synopsis Filicum* where he speaks (p. xiv) of Presl's system as "the completest *Catalogue* that has yet appeared." The retention by Engler and Prantl, and by Christensen, of a succession of the families so nearly the same as that of Presl and of Hooker shows that, even up to the present, the results of comparison have not fully found their place in the systematic works on Ferns.

It may of course be argued that, as it is impossible to represent the phylogeny of any complicated group of organisms in a linear sequence of their names, it is best to take refuge in a mere catalogue. Nevertheless such a catalogue, while it may fall far short of the complicated truth, may at least be so constructed as not to violate those probable relations which comparison and palaeontology together demonstrate.

Methods of early Pteridologists

It will be useful to take a backward glance at the methods of the earlier Pteridologists, and to see how the bases of classification have gradually been expanded(28). They first attempted to define species and genera by their detailed characters, using chiefly those of external form: but they also grouped them by features of wider than generic application. For this purpose search was made for differential data, and attention soon fell upon the annulus of the sporangium. Bernhardi (1800) distinguished *Filices Gyratae*, and *Filices Agyratae*. Swartz adopted this distinction, while the further difference was recognised between Ferns with an oblique annulus (*Helicogyratae*) and those with a vertical annulus (*Cathetogyratae*). Robert Brown (*Prodromus*, 1810), who also adopted a classification according to the annulus, added the use of venation, and founded the Families of Polypodiaceae, Gleicheniaceae, Osmundaceae, and Ophioglossaceae. Kaulfuss added to these the Family of the Marattiaceae. From this period the real establishment of the Natural Families of the Class is dated by Bommer (*l.c.* p. 22). But the Polypodiaceae, which bulked the largest in genera and species, monopolised to a high degree the attention of the Pteridologists of the first half of the 19th century. On the other hand, the general relation of the families was sometimes struck in singular accord with modern views. For instance Mettenius (29) disposed them thus: I. Polypodiaceae: II. Cyatheaceae: III. Hymenophyllaceae: IV. Gleicheniaceae: V. Schizaeaceae: VI. Osmundaceae: VII. Marattiaceae: VIII. Ophioglossaceae. It is difficult to understand why later writers should ever have diverged from so reasonable a sequence as this. If its order were inverted it would accord in essentials, though not in detail, with current views as to affinity.

The external characters of the vegetative system, and the venation of the leaves: the "vestiture" by hairs and scales: the form and position of the sorus: the presence or absence of an indusium, and its outline and position, together with the details of the mature sporangium, were the chief characters used by systematists up to the time of the *Synopsis Filicum* (1868). Of those who followed that era the writer of all others who brought a new method and fresh features into the classification of Ferns was Prantl. His views, already put in practice in his monographs of the Hymenophyllaceae (1875) and Schizaeaceae (1881), were summarised in the words "It is clear that the System of the Ferns must express the results of anatomical and developmental investigation (30)." But as he wisely adds, "the question is less clear which state of organisation is the more primitive, and which the derivative." Into these sentences are condensed much of the method and spirit of the later systematic investigation of Ferns,—and indeed of plants at large. That method applied in respect of the development of the sporangium by von Goebel led him to distinguish and to designate the Leptosporangiate as distinct from the Eusporangiate Ferns (35). He also proposed to use more fully the features of the gametophyte for purposes of comparison. On the other hand adult Fern-anatomy, already pursued by Mettenius, was placed upon a fresh theoretical basis by Van Tieghem in his theory of the "stele," while detailed observations were extended by Bertrand, Poirault, Gwynne-Vaughan, Jeffrey, Boodle, and others. The study of apical segmentation initiated by Naegeli (1845) has led to the general view that segmentation of stem, leaf, and root in Ferns runs parallel with that of their sporangia, and that the details seen in them connote a general difference of more or of less robust cell-constitution(34). The tendency has arisen to seriate the various types of Ferns according to such varied characters as these, recognising most clearly those types which appear to be extreme. The series would then extend roughly between the delicate Leptosporangiatae and the more robust Eusporangiatae. Thus the sequence already apparent in the arrangement of Mettenius seemed to be confirmed.

But here came in the insistent phyletic question which Prantl felt to be so difficult: viz. which extreme type is the more primitive, and which the derivative? It was first taken up on purely comparative grounds by Campbell (1890). He recognised the massive Eusporangiatae as the more primitive, and the relatively delicate Leptosporangiatae as the derivative types of Ferns(33). If this conclusion be true it should stand the test of palaeontological enquiry. It was shown in the following year that at least the great majority of the Palaeozoic Ferns were Eusporangiate, while the Leptosporangiate types become frequent in later periods, and that they are the characteristic Ferns of the present day(34). Such considerations, supported as they now are by greatly extended observations and comparisons, confirm that general seriation of the main Families of the Filicales already laid down by Mettenius. Subject to test by all the characters that can be used comparatively, that sequence may be upheld as probably illustrating in a quite general sense a progressive evolution. Inverting the order of the Families as given by Mettenius it would run as follows:—

I. Ophioglossaceae
II. Marattiaceae
III. Osmundaceae
IV. Schizaeaceae
V. Gleicheniaceae
VI. Hymenophyllaceae
VII. Cyatheaceae
VIII. Polypodiaceae

Relatively primitive.

Relatively advanced.

This is then the general position from which we may start upon a more searching comparison of the Filicales, with a view to the arrangement of the Class as nearly as may be according to their probable phylesis. *The general principle will be to widen to the utmost the range of the characters and comparisons that are to be used in attaining that end, and thus to base phylesis upon the sum of the available data.*

The Criteria of Comparison

The Sporophyte, or Fern-Plant, is the most prominent phase of the life-cycle, and it provides the most marked features which are used in comparison. But the alternate sexual phase, or Prothallus, must also be considered. Certain characters which it shows will be brought into the comparative treatment, though these are less reliable than those of the sporophyte. Theoretically all possible features should be regarded as criteria of comparison, and used in drawing conclusions as to relationship and descent. But the relative importance to be attached to each will vary according to its constancy, and to the directness of its bearing on the physiological success of the organism. Such weighing of evidence from the widest possible sources forms the proper basis for classification according to a Natural System. If due value be allowed to the various lines of evidence, the result should be the disposition of the members of a family or group according to their relation by descent. Convenience of arrangement should only find its place in such a method in default of sufficient evidence from comparison. Where the knowledge of detail is sufficient it should be possible to place not only groups and families, but genera and even species according to their natural affinities. This is the theoretical end of a Natural Classification.

The first step in the present case will be to ascertain for the Filicales what are the criteria upon which comparison may be based. It is desirable to use those available from both of the alternating generations. In the main the progressions traced should run substantially parallel in respect of the several criteria if their recognition be valid and true. The more nearly this proves to be the case the more solid and reliable will be the conclusions drawn from them. Any apparent discrepancy may have arisen from one or another of various sources. Of these the most important are (i) that the facts may have been erroneously observed: (ii) that they may have been wrongly interpreted: (iii) that the advance may not have been synchronous in the evolution of the plants compared in respect of the various criteria: (iv) that the suggested relationship may have been too wide to constitute part of a real sequence: or (v) that the facts are themselves deficient. Such difficulties are liable to arise in all sequences, but most frequently in the case of the fossils. It cannot then be anticipated that the phyletic conclusions arrived at by such comparison will in all cases be definite or final. To attain finality presumes a completeness of knowledge greater than we can reasonably expect at present. The limitation of fact and the uncertainty of interpretation make it impossible to advance such conclusions beyond the point of reasonable probability. But though a final grouping based on comparison, which is the theoretical end, may not be actually attained at the moment, the method employed to approach that end may be none the less sound in itself, and the attempt may find its justification in a partial success.

A knowledge of the whole life-cycle, such as is given in Chapter I, is necessary in order to select those criteria which are to be specially used in comparison. Since that cycle is essentially similar for them all, the Male Shield Fern (*Dryopteris Filix-mas*, Rich.) will serve as a fair example. Moreover comparison indicates that it is phyletically neither one of the most primitive nor one of the most advanced of Ferns, and is on that account all the more suitable. Its life as detailed in Chapter I will therefore give a mean example of those several features which will be taken as criteria for comparison of Ferns at large. They will be severally named in the following paragraphs.

I. The shoot of Ferns though constructed on the general plan seen in *Dryopteris* is liable to great differences of proportion and of pose. The stem may be erect, ascending, or prone. The number of the leaves and the distance between the individual leaves may vary. The shoot itself may branch in various ways, either at the tip, or by the formation of buds in divers positions remote from the tip. These and other features show that for the Ferns at large *the form and proportions of the shoot* will provide material suitable for comparative treatment.

II. An examination of the apical cone in Ferns shows that in most cases it is occupied by a single cell of definite form, with other cells regularly arranged around it in such a way as to show that they are really segments which have been successively cut off from that cell. It is accordingly styled an *initial cell* (Fig. 10). A similar examination of the tip of the leaf and of the root gives a like result, while somewhat similar cleavages are found to occur with like regularity in the formation of sporangia (Figs. 11, 12). Such segmentations from initial cells are general for all embryonic regions in Ferns: but there are differences of detail in the segmentations, and even in the number and the form of the initial cells. These differences again provide material for comparison, and it will be found that *the constitution of the plant-body as indicated by its initial segmentation* will serve as a further basis for comparative treatment.

III. The contours, texture, and venation of the expanded leaves in Ferns show great differences in detail. Usually, as in *Dryopteris* (Fig. 2), there is a central rachis which bears rows of pinnae right and left, diminishing in size towards the apex. These again are cut into pinnules or secondary branches, while their marginal toothing suggests an imperfect branching of a third order. But on the other hand in some Ferns, as in the Adder's Tongue and Hart's Tongue, the leaves are entire, not being cut into pinnae. Others again may show branching of a higher order than in *Dryopteris*. There may also be great diversity in outline of the segments, and in their relation to one another, and to the main rachis. Such features offer an easily observed criterion of comparison. They are closely connected with the venation. Sometimes, as in *Dryopteris* (Fig. 2), the veins end blindly near to the margin, the venation being then described as open. But in other Ferns they are connected so as to form a coarse network of large meshes (Fig. 16): or others again may show a finer network of smaller meshes (Figs. 17, 18). It is thus apparent that *the architecture and venation of the leaf* should provide another valuable criterion of comparison, and one easy of observation.

IV. If the veins of the leaf of *Dryopteris* be followed inwards from the margin of the pinnule, they connect downwards into the midrib of the pinna, and pass on through the rachis to the leaf-base. There they enter the stem as a number of isolated strands, which insert themselves on the margin of one of the meshes of the cylindrical network of vascular tissue which traverses the massive stem (Fig. 1, *D*, *E*, *F*). These together constitute the stele which is here broken up into parts (meristeles). Thus the shoot is traversed by a connected system of conducting tissue. It happens that this tissue, so important physiologically, retains with great persistence its characters of form and distribution in individuals of the same affinity. But a comparison of different types of Ferns discloses features, both in the stele of the stem and in the leaf-traces which spring from it, that are different and distinctive for the several groups. Moreover, since these resistant tissues are preserved commonly in the fossil state, they provide a wide basis for comparison. Accordingly *the vascular tissue of stem and leaf* affords perhaps the most reliable structural criterion of comparison of all those which lead towards a phyletic grouping of the Ferns.

V. The leaf-stalk of the mature leaf of *Dryopteris*, the whole leaf while young, and the apical region of the stem are closely invested by a dense covering of broad protective scales, or ramenta. These are pale coloured while young, but are rusty brown when mature, after which time many of them dry up and fall away. These broad scales are of the nature of trichomes originating from superficial cells. But there is a good deal of variety in this investiture in Ferns. Many possess no flattened ramenta, but only hairs composed of a single row of cells. Frequently the hairs may bear glands on their ends, or be variously branched. They may even be borne on large outgrowths which sometimes mature as hard woody emergences covering the older parts. Such *dermal appendages*, simple or complex, will also be brought into the comparative treatment.

VI. The preceding characters all relate to the vegetative system, which in plants at large is found to be so directly adaptable to circumstances that its features are commonly accorded only a minor place as a basis for classification, priority being given to the propagative organs. Though there may be less reason for this difference of valuation in Ferns than in some other cases, the interest in their propagative organs certainly predominates. These appear in *Dryopteris* on the under surface of certain adult leaves, and are grouped in "sori" disposed in a single linear series on either side of the midrib of the pinna or pinnule, being seated on the secondary veins (Fig. 2). But this position of the sorus is not constant in Ferns. It is true that in many of them it is on the lower surface of the sporophyll, but in others, and especially in those which are of archaic type, it may be borne close to, or actually on the margin of the leaf. The extent, individuality, and relation of the sori to the venation and to the margin of the leaf are very variable. It will then be found that *the position and relations of the sorus* are features useful for comparison.

VII. The indusium, which is so marked a feature in *Dryopteris*, is not constant for Ferns at large (Fig. 14). There is reason to believe that all primitive sori were without it: also that indusial protections of several distinct sorts have originated in the course of evolution along distinct phyletic lines. Thus *indusial protections* will afford another useful criterion for comparison.

VIII. The structure of the sporangium above described for *Dryopteris* (Figs. 14, 17) represents the type common for relatively advanced Ferns, though not for those of the most primitive, nor yet for those of the most advanced types. All the details of the sporangium are liable to vary from one type to another. It will be shown on the one hand that the sporangium is more massive in relatively primitive Ferns, but more attenuated in those which are relatively advanced. It will be found that *the origin of the sporangium, the details of its segmentation, the length and thickness of the stalk, the form of the head, and the position and structure of the annulus all offer points of comparative value.*

IX. Closely related with these features is also the *numerical spore-output.* In archaic Ferns, where the sporangia are large, the numbers are as a rule high compared with those of more modern Ferns such as *Dryopteris.* Lastly, *the form of the spores and the character of their walls* also present features that have their value in comparison. The rugged surface and dark colour of the spores of *Dryopteris* are examples showing how pronounced such characters may sometimes be.

X. Each spore on germination may produce a prothallus. In *Dryopteris* it is flattened, with no distinction of axis and appendages, and with undifferentiated structure. But this shape is not constant in the Ferns: sometimes it is filamentous, sometimes massive, and in the latter case it may even develop underground, and contain little or no chlorophyll, being nourished saprophytically. Thus *the form and physiology of the prothallus* may provide material for comparison, though this is not very reliable since it varies readily and often directly with the external conditions. The prothalli of *Dryopteris* itself are found to differ greatly in size and form according to the conditions under which they are grown. When crowded together they appear filamentous, a form which is normal for some Ferns.

XI. The prothallus bears the sexual organs: and again *Dryopteris* offers a type common for relatively advanced Ferns (Fig. 19). Speaking generally the sexual organs are more massive and more deeply sunk in the prothallial tissues in relatively primitive forms, and are less massive in relatively advanced types. Thus *the structure and position of the sexual organs will offer details of value for phyletic comparison.*

XII. The embryo-sporophyte of the Fern, which arises as the consequence of fertilisation and is nursed for a time by the prothallus, is very uniform in its structure and position in most of the Ferns of relatively recent type (Fig. 25). Its parts, resulting from segmentations of the zygote which correspond closely in sequence and relation to one another in the different types, show striking uniformity. But this uniformity is not found among the more primitive types. In some of these a suspensor is present, a feature not seen in any of the most recent types of Ferns, though it is commonly present in the Lycopods and in Seed-Plants. Further, the position of the first shoot, which is prone in the later Ferns, is upright in some of the more primitive types. Thus *the embryology of the sporophyte offers certain features of comparative value.*

We are thus in possession of a number of characteristics of Ferns which will serve as criteria for their comparison with a view to their phyletic seriation. They are these:—

(1) The external morphology of the shoot.
(2) The initial constitution of the plant-body as indicated by segmentation.
(3) The architecture and venation of the leaf.
(4) The vascular system of the shoot.
(5) The dermal appendages.
(6) The position and structure of the sorus.
(7) The indusial protections.
(8) The characters of the sporangium, and the form and markings of the spores.
(9) The spore-output.
(10) The morphology of the prothallus.
(11) The position and structure of the sexual organs.
(12) The embryology of the sporophyte.

Before these criteria can be properly used as a basis for phyletic seriation each must be considered comparatively as seen in a large number of distinct instances. The scope of variation in respect of it must be examined, and extreme types recognised. It will then be a question in relation to each criterion how far the extreme types can be held as relatively primitive or as advanced. Such questions in the case of each criterion can be resolved with some degree of probability by comparison in respect of other characters. But more material assistance to a conclusion is afforded by reference to the geological record. For instance, some living Ferns are grouped as Eusporangiate, that is, they have relatively bulky sporangia: others as Leptosporangiate, being characterised by relatively attenuated sporangia. It is now known that the Ferns of the Palaeozoic Period had relatively bulky sporangia, corresponding to the Eusporangiate Ferns of the present day. It is even a question whether true Leptosporangiate Ferns existed in the Primary Rocks, though they constitute the bulk of the present-day species. The natural conclusion is that the Eusporangiate type is relatively primitive and the Leptosporangiate derivative. Again, an open venation, without fusion of veins, is characteristic of the earliest fossil Ferns. Fusion of veins is first seen in the Middle Coal Period, and a small-meshed reticulum appears only in Ferns of the Mesozoic Era. The natural conclusion is that open venation is a primitive, and a small-meshed reticulum a derivative character. Such conclusions commonly running parallel to and supporting one another, produce a cumulative effect. This may even help towards decisions respecting yet other criteria. And thus by coordinated study of the whole series of criteria the web of evidence may be ever more closely drawn. The result in favourable cases will often be a confident opinion as to the seriation of distinct but related families, genera, and even species according to descent. In less favourable cases a reasonable probability can usually be established.

As a further test or corroboration of the conclusions thus acquired the method of observation of the successive steps in the individual life may be used. For instance, the outline and venation of the earliest leaves of the sporophyte are liable to differ from those formed later by the same plant, showing characters more primitive than those of the adult leaf. It is, however, in the study of the vascular tissues that this line of ontogenetic evidence finds its best opportunity, though owing to singular neglect in collecting the necessary facts by serial sections it has not hitherto been developed as it should have been. The ontogenetic development is in point of fact a natural key by which to interpret the stelar elaborations so characteristic of the Filicales. But it can only be applied subject to limitations, since the ontogeny is often abbreviated, and steps that might have been anticipated are frequently condensed, or even omitted from the individual life.

Following the lines thus briefly sketched the first duty will be to examine each of the criteria of comparison critically. A Chapter or more will be devoted to each. The nature of the variations within the Filicales will be considered, and in each case the relatively primitive will be distinguished from the relatively advanced. Thus the way will be prepared for using that criterion, in conjunction with others, for the purpose of phyletic seriation of the Ferns. If upon this wide basis, and subject to the palaeontological check, the comparisons and arguments are correctly pursued, the result should provide a trustworthy classification of this large and varied class of plants. It should in fact reflect the probable course which the evolutionary history of the Filicales has actually pursued. Incidentally a wider appreciation of the plants themselves will have been acquired, and the basis found for sane views as to the evolutionary origin of those parts which are seen highly elaborated in their more advanced types.

BIBLIOGRAPHY FOR CHAPTER III

23. CHRISTENSEN. Index Filicum. 1906.
24. ENGLER & PRANTL. Natürl. Pflanzenfam. i, 4. 1902.
25. CHRIST. Die Farnkräuter der Erde. 1897.
26. HOOKER. Species Filicum. 1846—1864.
27. HOOKER & BAKER. Synopsis Filicum. 1868.
28. BOMMER. Monographie de la Classe des Fougères. Bull. de la Soc. Roy. de Bot. de Belgique. Tome v, No. 3. A useful summary is here given of the Classifications of Ferns devised by various authors up to 1867.
29. METTENIUS. Filices Horti Lipsiensis. 1856.
30. PRANTL. System der Farne. Arbeiten a. d. Königl. Bot. Gart. zu Breslau. 1892. Here a complete list of the author's own writings on the Classification of Ferns up to that date is given.
31. NAEGELI. Zeitschr. f. wiss. Bot. iii. 1845.
32. BOWER. Ann. of Bot. iii, p. 305. 1889.
33. CAMPBELL. Bot. Gaz. xv, p. 1. 1890.
34. BOWER. Ann. of Bot. v, p. 109. 1891.
35. VON GOEBEL. Bot. Zeit. 1881, p. 681, etc.

CHAPTER IV

MORPHOLOGICAL ANALYSIS OF THE SHOOT-SYSTEM OF FERNS

THE sporophyte of modern Ferns, however complex, is built up on the basis of the simple shoot, constructed on essentially the same plan as that of the Higher Vascular Plants. The differences are those of detail rather than of principle. The Shoot consists of an axis terminating in a growing point endowed with unlimited powers of development, which bears leaves as lateral appendages produced exogenously upon it, and in acropetal order. This shoot is fixed in the soil by roots which are of adventitious origin, and variable in number. Though they may at times show some degree of regularity both in number and in position relatively to the leaves, this is by no means a general rule. Nevertheless a frequent relation is such that one root arises immediately at the base of each leaf. This is well seen in *Ophioglossum vulgatum*, and in *Blechnum Patersoni* (Fig. 57).

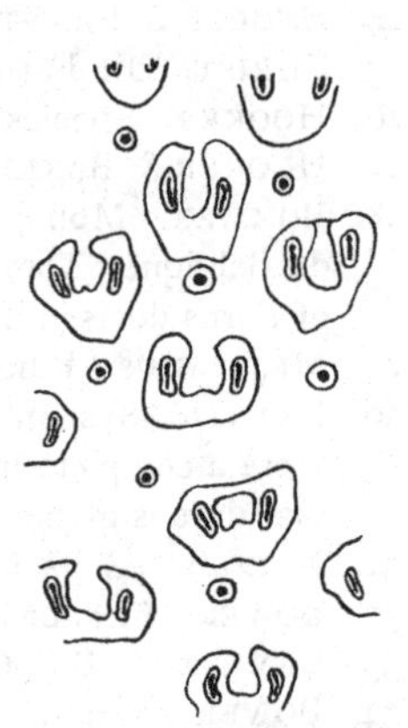

Fig. 57. *Blechnum Patersoni* (R. Br.), Mett. Tangential section of the stock, showing a constant relation of one root below each leaf-trace. (×6.)

As has been seen in Chapter II, the position of the axis may be erect, ascending, or prone. Closely connected with the pose of the shoot is its symmetry. In Ferns as in other plants a shoot directed vertically upwards is usually radial, while ascending or creeping shoots show varying degrees of dorsiventrality, the difference of symmetry being referred to the effect of gravity upon it. The same question of priority will then arise regarding the radial and dorsiventral symmetry as for the erect and prone positions. In Vascular Plants at large there is a strong probability of the erect position and radial symmetry having been primitive in the sporophyte, and the prone position and dorsiventral symmetry derivative. The analogies between Ferns and other Vascular Plants in this respect are so close as to justify a similar conclusion for their adult shoots, and it would apply to the juvenile shoots also in the most primitive Ferns, such as the Marattiaceae and Ophioglossaceae, where the embryo is erect from the first. But the prone position of the Leptosporangiate embryo presents a difficulty in accepting the conclusion as general for all Ferns. Moreover the relation of the symmetry of the shoot to gravity is not always so simple and direct as it usually appears to be. For instance, the young plant of *Polypodium vulgare* is from the first prone, but the adult rhizome assumes its creeping position not by continuous growth in the original direction, but by arching backwards over the prothallus in a strong hyponastic curve (Fig. 39). Again, in the case of *Polypodium heracleum* the dorsiventral, climbing rhizome is normally appressed to the support. But if it is grown on soil it takes a spiral curve, owing to stronger growth of the leaf-bearing side (von Goebel, *Organographie*, 1913, p. 221, Fig. 216). Such cases show that though gravity may have been a determining influence causing the dorsiventrality of the shoot in the first instance, it does not always account directly for the positions which such dorsiventral shoots may assume.

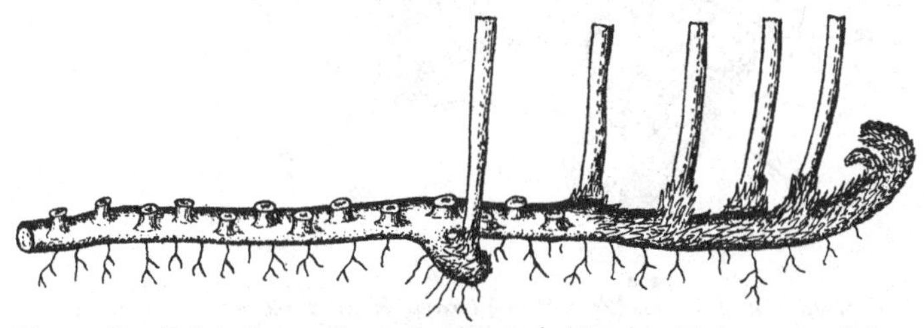

Fig. 57 *bis*. *Polypodium vulgare*, dorsiventral rhizome with two rows of alternating leaves. Those of the preceding year represented by scars; those of the current season have their petioles attached. A lateral branch of dichotomous origin shows the character of the dichopodial system. (After Velenovsky.)

The arrangement of the leaves on the axis in Ferns is alternate, a condition initiated by the first seedling leaf, which is always solitary, and may be followed by leaves arranged on a spiral plan. In creeping Ferns the leaves are usually inserted alternately right and left, forming two lax orthostichies upon the dorsiventral rhizome. This is well seen in *Polypodium vulgare* (Fig. 57 *bis*), and in the Bracken. But in certain cases the leaf-bases may appear as though forming a single row, owing to the close approximation of the two orthostichies on the upper surface of the rhizome, as in *Lygodium* (Fig. 58). In ascending or upright stems, on the other hand, the arrangement of the leaves is spiral, with divergences which approximate to those shown by Flowering Plants. A good instance is seen in *Ankyropteris Grayi*, a fossil Fern from the Palaeozoic

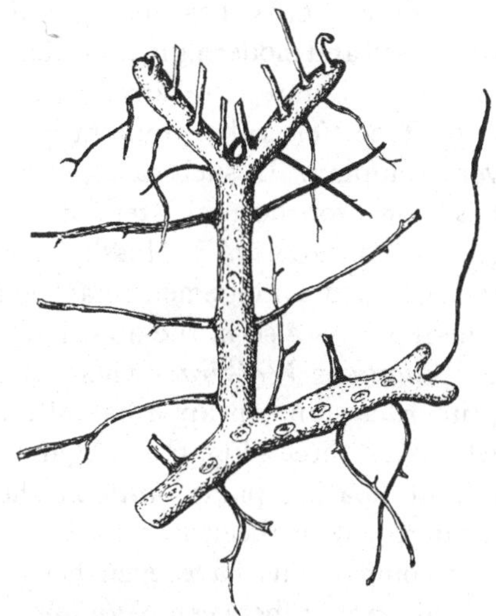

Fig. 58. The dorsiventral and dichotomous rhizome of *Lygodium scandens*, with leaves apparently in a single row. (After Velenovsky.)

Period. Here the two-fifths divergence is clearly indicated by the stelar anatomy (Fig. 59). In the adult plant of *Dryopteris Filix-mas* the leaf-

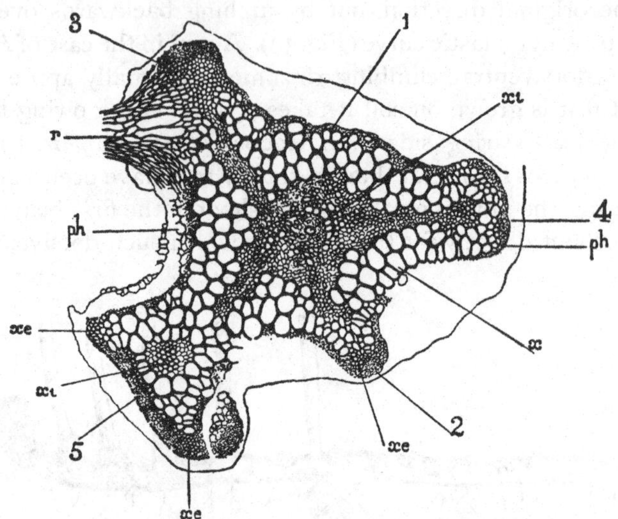

Fig. 59. *Ankyropteris (Zygopteris) Grayi.* Transverse section of stele, showing wood and remains of phloem. 1—5 = the five angles of the wood, from which leaf-traces are given off, in order of the phyllotaxis, no. 5 belonging to the lowest of the series. x = principal ring of xylem; xi = small tracheides of internal xylem; xe = small tracheides at periphery; ph = phloem; r = base of adventitious root. (× 14.) Will. Coll. 1919 B. (From Scott's *Studies in Fossil Botany*.)

arrangement is more complicated, giving the basket-like head of leaves so usual in the Ferns of ascending or upright habit (Fig. 31). The Osmundaceae, both fossil and modern, give good examples of more elaborate spiral arrangements. In large Tree Ferns the phyllotaxy may be very complex. In such cases, as Schoute has shown for the adult stem of *Alsophila glauca*, Sm. var. *setulosa*, Hassk, a tendency to a whorled arrangement may be found. This appears also in the apical region of *Amphicosmia Walkerae*, where the leaf-primordia stand in apparent alternating whorls of three (Fig. 60). Again, in the minute floating plant *Salvinia* the leaf-arrangement has been described as whorled and comparisons have even been drawn on this ground between *Salvinia* and the Sphenophylleae, where the leaves are typically whorled. Nevertheless, even in *Salvinia*, the ontogeny opens by

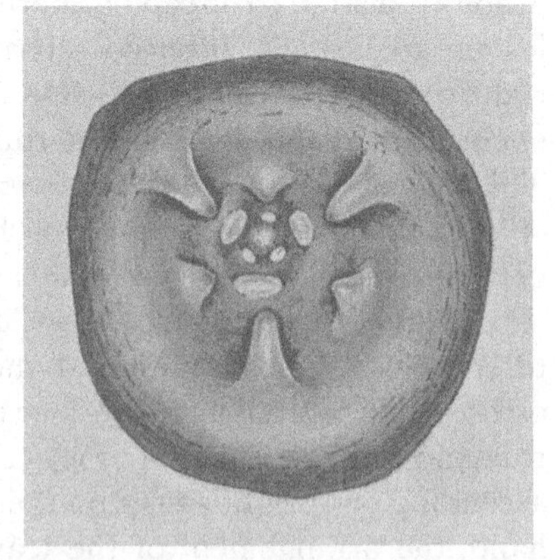

Fig. 60. Apex of stem of a large plant of *Amphicosmia Walkerae*, showing the arrangement of the leaves apparently in alternating tri-merous whorls.

the formation of a solitary leaf, and this may be held to suggest that there, as in other Ferns, the alternate arrangement is fundamental. That disposition is maintained throughout life by creeping Ferns, and by most upright Ferns as well. Where an apparently whorled arrangement of the leaves supervenes, it may be held to be in the race, as it certainly is in the individual, derivative from an alternate origin.

Buds and Branching

The shoot may remain simple, and unbranched, as it commonly is in the Marattiaceae, and many Tree Ferns. On the other hand it may branch with greater or less profusion. But the branches are disposed in such different ways in various cases that it appears difficult at first to recognise in them any common scheme. They fall, however, for the most part into certain types of branching, which will first be described and illustrated by definite examples:

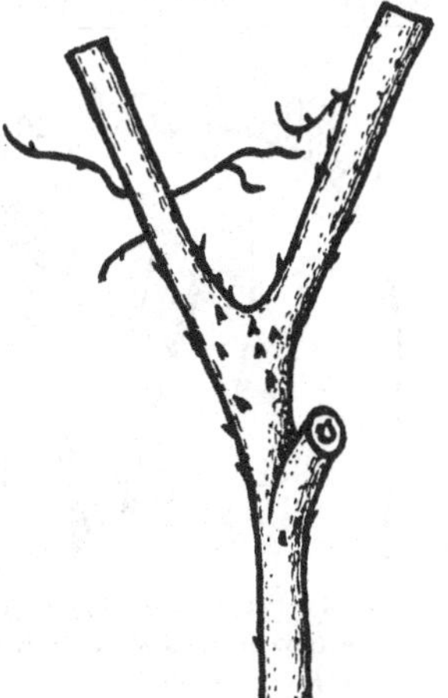

Fig. 61. *Gleichenia* (*Dicranopteris*) *fulva*, Underw., showing dichotomy of the rhizome. (Nat.? size.)

(I) Many Ferns show *dichotomy* of the apex of the shoot, often with very exact equivalence of development of the two limbs or shanks of the forking. This may be found occasionally in upright stems with crowded leaves, as in *Osmunda*, *Plagiogyria*, and *Cyathea* (Frontispiece). But it is in creeping axes, where the leaves are far apart, that bifurcation is most common, and is most readily seen. The branching may occur at some distance from any leaf-insertion (Fig. 61), though frequently a leaf may appear to be related to the forking (Fig. 62). Such forkings are frequently present in rhizomes of *Lygodium*, *Schizaea*, *Gleichenia*, *Lindsaya*, *Pellaea*,

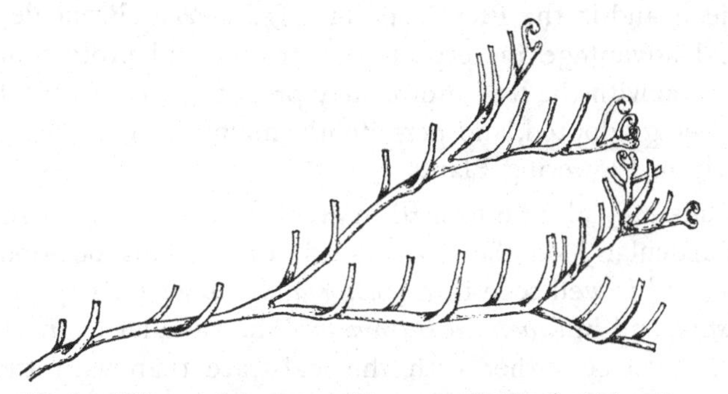

Fig. 62. Rhizome of *Dryopteris Linnaeana*, C. Chr. (= *Polypodium Dryopteris*, L.) dichotomously branched, with alternating leaves. The roots are omitted. (After Velenovsky.) Reduced.

and *Davallia*, and they have long been known in *Dryopteris*, *Polypodium*, *Pteridium*, and many others. Dichotomy has also been recorded among early fossils, such as *Botrychioxylon* and the Botryopterideae. These examples serve to show that *distal dichotomy of the axis* is widespread among Ferns. Moreover, while it appears in some advanced types, it is frequent among primitive types now living, and has been recorded among early fossils.

(II) Other Ferns show buds in a position seemingly very different from those produced by dichotomy. The *axillary position* is not uncommon, and it is found in some of the most archaic types. Among living Ferns it is best seen in the Hymenophyllaceae (Fig. 63). It has been shown to occur in the Ophioglossaceae, being first demonstrated in them by Gwynne-Vaughan (48) for *Helminthostachys* (Fig. 64), and later by Lang (47) for *Botrychium*. In them

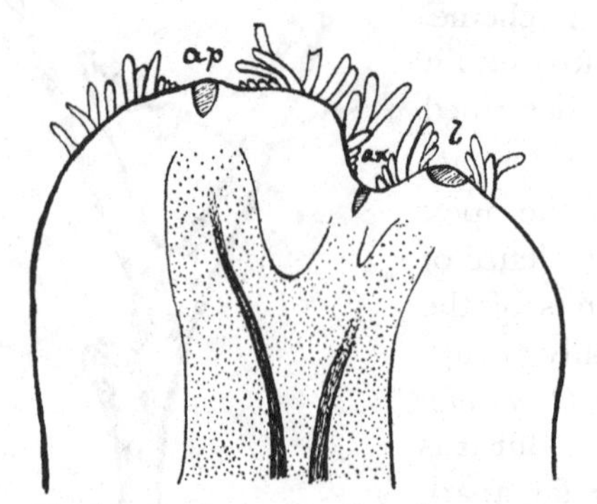

Fig. 63. *Trichomanes radicans*, longitudinal section of the apex (*ap*), with axillary bud (*ax*), and subtending leaf (*l*). (× 20.)

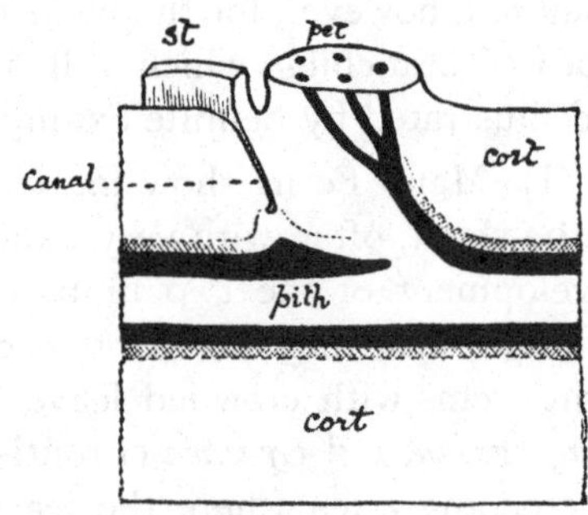

Fig. 64. Longitudinal section through a leaf-insertion of *Helminthostachys*, showing the petiole (*pet*) and stipule (*st*), in the axil of which a canal arises leading obliquely down to a dormant bud. (After Gwynne-Vaughan.)

the position of the bud is comparable to that which is prevalent in Flowering Plants. It has also been shown to exist in such early fossils as the Zygopterideae, and in the Pteridosperm, *Lyginopteris* (Brenchley (57)). The physiological advantage in respect of nutrition and protection which the axillary relation with the leaf affords may probably account for that position of the bud being adopted, and persistently maintained in certain groups, as it is regularly in Flowering Plants.

(III) Buds may also be formed in *extra-axillary positions* related to the leaf-base, particularly on the abaxial side of it. This position is often a definite one, and is well seen in *Lophosoria*, *Metaxya*, *Cheiropleuria* (Fig. 65), *Elaphoglossum*, and in *Cibotium Barometz*. The vascular connection appears in these cases to be rather with the leaf-trace than with the supply of the main axis, though it is difficult to draw any clear line between these. The physiological convenience of a bud which can thus tap the stream of nutriment from the leaf before it reaches the stem, and can at the same time grow directly outwards from the parent plant, may be held as conducing to the perpetuation of buds of this type. Other positions, either laterally upon the leaf-base or upon the axis slightly removed from the leaf-insertion, are also frequent, and especially upon dorsiventral rhizomes. But such positions are not so fixed or constant as are the buds on the abaxial side of the leaf-base.

Those buds which appear to be formed early and in definite positions upon the leaf-bases of *Angiopteris* may here also be noted (63). They arise at the corners where the margins of each of the stipules pass over into the leaf-base, and there are accordingly four of them on each leaf-base. They are seated, as in *Helminthostachys*, each at the base of a narrow canal. The buds remain dormant till the leaf-bases separate with age from the trunk: after which event they may grow on, each forming a new plant. The analogy with the buds of *Helminthostachys* is very striking except that their position is not axillary, and their number is greater.

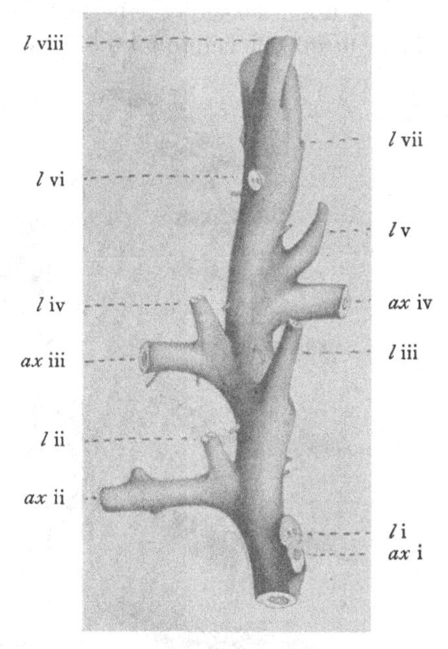

Fig. 65. Drawing by Dr J. M. Thompson of a rhizome of *Cheiropleuria bicuspis* (Bl.), Presl, with the superficial hairs removed, so as to expose the successive leaf-bases, which are numbered *l* i to *l* viii, and the lateral axes which spring from their bases, numbered *ax* i to *ax* iv. But the leaves iii, vi, vii, viii have no associated axes. The leaf-arrangement is alternate, and the climbing shoot is seen from the side facing away from the support. (× 2.)

(IV) Buds are also formed at various points on the expanded lamina, but naturally these are later in their origin than those above described. They may be found related apparently either to the lower or to the upper surface, though in some cases their origin appears to be in the first instance marginal. The former is the case in *Asplenium bulbiferum*, the latter in *Asplenium viviparum*, *Diplazium celtidifolium*, and *Dennstaedtia rubiginosa*: and this is the more common type. The buds frequently appear close to the bases of the pinnae, as in *Woodwardia radicans*, and *Cystopteris bulbifera* (Fig. 66): but often their position is less definite than this. In certain cases their origin has been traced to single superficial cells, which awake relatively late to renewed activity of growth and division. Being seated on a narrow base these buds are easily detached, and provide a ready means of vegetative propagation, as seen in various species of *Asplenium*. In certain cases the production of such sporophytic buds may be definitely related to the arrest of sori, as in the case of *Cystopteris bulbifera*. A good instance of this has been noted in *Woodwardia radicans* (Fig. 67), where buds are found on the upper surface of the leaf, at points immediately above arrested sori. Their formation is undoubtedly connected with the arrest, and the buds may

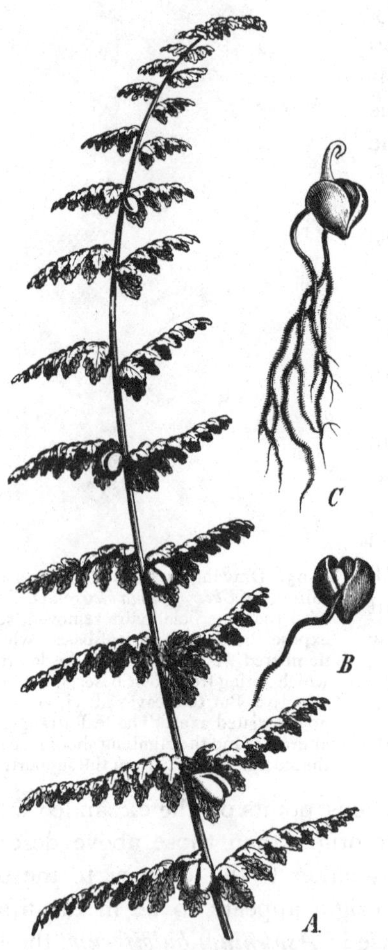

Fig. 66. *Cystopteris bulbifera* (L.), Bernh. *A* = part of a leaf with adventitious buds. Natural size. *B* = an adventitious bud which has fallen off, forming a root. *C* = an adventitious bud further developed. *B* and *C* somewhat enlarged. (After Matonschek.)

Fig. 67. A pinna of *Woodwardia radicans*, showing many sporophytic buds on the upper surface. They correspond in position to sori on the lower surface, which are imperfectly developed.

be held as an alternative method of using the material available for spore-formation. The buds borne thus upon the lamina cannot be related to any regular method of branching of the shoot, and their origin is held to be adventitious.

(V) In other cases buds may be formed upon the roots: but this is less common. In *Platycerium* and in *Asplenium esculentum* it has been shown that the apex of a root may be directly transformed into a leafy bud (64, p. 63, where the literature is quoted). Similar buds appear upon the roots of *Ophioglossum vulgatum*, and other species: but these have been traced in origin to one of the segments cut off from the apical cell itself (Rostowzew, (46)). By such means an effective vegetative propagation is secured.

Under these five headings diverse means of increase of the shoot in Ferns have been included. Those first named appear in definite positions, and are produced by primary development, distally. This would suggest that their origin was primitive. Those mentioned under (IV) are less definite in position, and are often late in their origin, so that they bear no direct relation to the apex of the shoot. Those under (V) have evidently no relation whatever to the terminal bud. There may, however, be a question in the case of (II) and (III) what relation, if any, the buds so produced bear to the apex of the main shoot: that is, how far they may be held to be of the nature of distal branchings of it, diverted into a lateral position and delayed in their origin. In the case of (I) the distal origin of the true dichotomies or equal forkings is clear. The genesis of the second apex has been traced in *Polypodium*, by Klein (41), to a segment of the apical cell. The original apical cone thus gives rise to two approximately equal cones, each of which acts as an equivalent apex. This may take place independently of any leaf. In the adult state the axis may be seen to fork at a point some distance above the insertion of the nearest leaf (Fig. 61). But frequently a leaf may appear to be related to the forking, and this has been designated by Velenovsky the "angular leaf" (50). He states that the leaf nearest to the dichotomy halves its angle, and is to some extent a constant character. That this relation is not close or constant is shown by Velenovsky's own drawings. Further, the orientation of the nearest leaf relatively to the shanks of the dichotomy is variable. A group of drawings from transverse sections through the forking axes of certain allied Ferns is represented in Fig. 68, i–vi. They show that sometimes there may be one leaf-trace related to the forking, sometimes two: and it has been seen that sometimes there is none. It appears also that the orientation of the leaf-traces varies relatively to the shanks of the dichotomy, and that this variation may be found even in the same species (Fig. 68, ii and v, are both from *D. pinnata*). These facts suggest that there is no definite relation of a leaf to the forking of the axis which would justify its recognition as an "angular leaf" in the sense of Velenovsky. It appears rather that the forking is purely an axial phenomenon, and that it may or may not be associated with the insertion of one or more leaves.

It is interesting to see that this inconstancy holds also in the case of an ancient fossil. For in *Rachiopteris* (*Botryopteris*) *cylindrica* it has been shown that the forking of the axis may occur either with or without the departure of an associated leaf-trace (57, p. 543, Text Fig. 7; also p. 544, Text Fig. 8).

The inconstancy of orientation of the leaf which may thus accompany forking provides facts material for an explanation of the positions of those buds which are ranked under (II) and (III), and their possible reference in origin to dichotomy. Almost any bifurcating Fern is apt to show unequal development of the two shanks of the fork. Occasionally these may be equal, as in the Frontispiece, and as shown by Velenovsky's drawings (50, Figs. 165—167). But various degrees of inequality may be recognised, and these

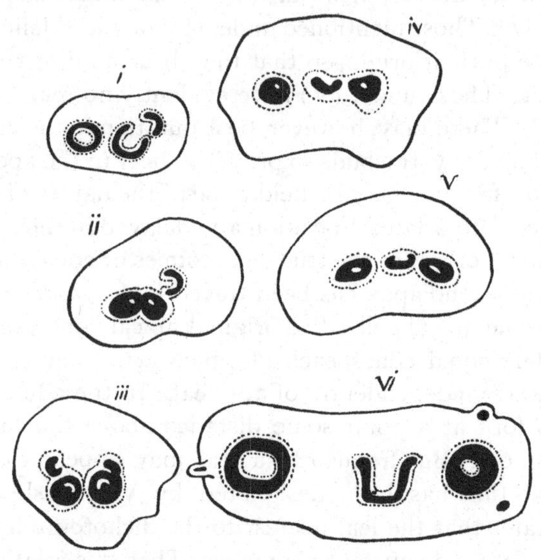

Fig. 68. Sections from point of dichotomy of the rhizome of various Ferns, showing the absence of uniformity of relation of the "angular" leaf to the branches of the dichotomy. i, ii, iii suggest that the minor branch is axillary; iv, v, vi suggest that it would be dorsal to the leaf. i = *Pellaea falcata* (G.-V. Coll. 1222); ii = *Davallia pinnata* (G.-V. Coll. 883, 892); iii = *Davallia tenuifolia* (G.-V. Coll. 927); iv = *Lindsaya reniformis* (G.-V. Coll. 1000); v = *Davallia pinnata* (G.-V. Coll. 888); vi = *Paesia viscosa*.

are often more pronounced as the parts become mature. The case of *Pteridium*, which has long been a subject of discussion, serves as a good example. The young plant starts with a single axis bearing 7 to 9 alternating leaves, spirally arranged, after which it undergoes distal and equal dichotomy, and on both of the elongated shanks a leaf may be borne near to its base (Fig. 69). The two branches burrow downwards into the soil, bearing leaves alternately right and left: in the later phases of their development they also show dichotomies, but with unequal shanks: the shorter of these commonly bears a leaf at once, so placed that when mature the shank that bears it appears as a small bud at the base of the strongly developed leaf. The arrested

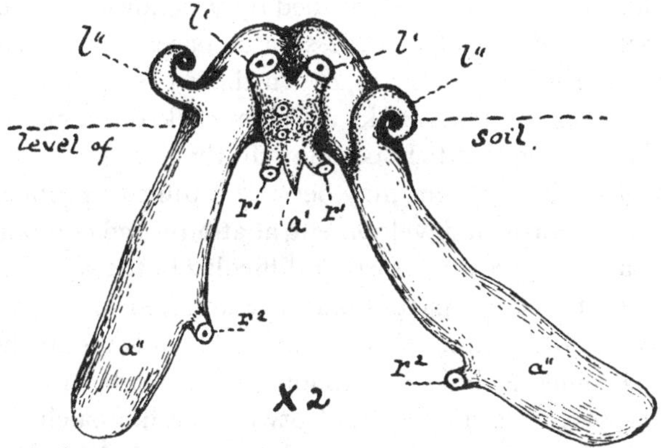

Fig. 69. Young plant of *Pteridium aquilinum*, seen from the convex side (obliquely downward directed) of the curved primary axis, and showing the first dichotomy, with downward directed shanks. a' = primary axis; a'' = shanks of first dichotomy; l' = leaves borne on the primary axis; l'' = leaves borne on shanks of dichotomy; r^1 = roots attached to the primary axis; r^2 = roots on shanks of dichotomy.

bud may take up active growth later (Fig. 70). That this is the probable evolutionary relation of the parts is indicated by the fact that the region between the insertion of the bud and the forking which gave rise to it shows characteristic stem-structure rather than that typical of the leaf. From this it may be concluded that here it is a leaf that is borne on an abbreviated axis, not a bud that is seated on a leaf-base. A similar interpretation will apply in the case of *Stromatopteris*, which has been anatomically described with like results by Thompson (59, p. 152). Though the point has not often been thus investigated, and though the anatomical test may not always be decisive in such cases, examples of varying degrees of inequality of distal forking can easily be multiplied, leading to conditions where one shank is represented only by a dormant bud associated with the base of its own first leaf.

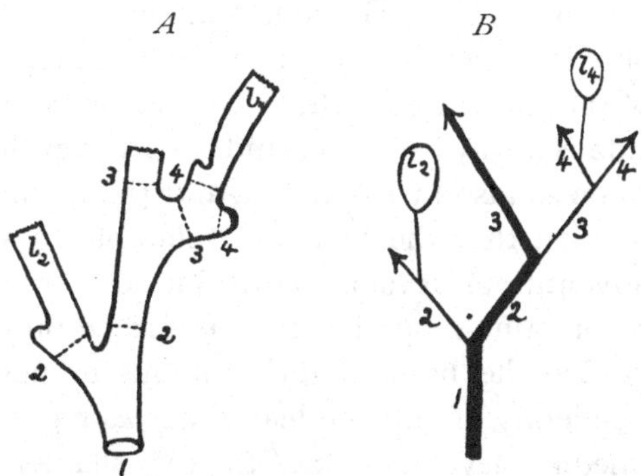

Fig. 70. Diagram constructed so as to show the branching in *Pteridium* on a hypothesis of unequal dichotomy modified from Velenovsky, Fig. 169. Sections cut below a bud, as at 2.2, show stem-structure, with slight bias towards leaf-characters on the side away from the dormant bud. Sections cut above the bud show characteristic structure of the petiole. Thus the anatomy supports the hypothesis of dichotomy with arrest of one of the branches, which forthwith produces a leaf. The numerals indicate the successive branchings on this hypothesis, actually on left, and diagrammatically on right.

Once the inequality of development of the shanks of dichotomy is established in any Fern, a sympodial succession of them becomes possible, so as to form what has been aptly described by Velenovsky as a *dichopodium*, that is, a false axis made up of a succession of bases of dichotomy, such as is shown diagrammatically in Fig. 70, *B*. In these cases, which are common in Ferns, the more the weaker shanks are side-tracked the more they appear as mere appendages upon the dichopodium, till the final interpretation of the branching that gave rise to them may become a difficult problem. This can only be solved on the basis of development, anatomy, and comparison studied in each several case. Moreover, a further difficulty in the solution arises from the general fact that where a part is partially arrested in its development it is apt to be delayed in its time of appearance. This leads to the result that the branching becomes technically monopodial, though still it may be held to have been dichotomous in origin. Notwithstanding such difficulties the branchings of Ferns may be very generally interpreted as derivative from distal dichotomy.

Occasionally more than a single bud may be found at the base of a leaf. This occurs in *Pteridium* (Fig. 70, *A*), and it has been noted by Gwynne-Vaughan for *Hypolepis repens*. In the former case an explanation based on a second dichotomy has been suggested by Velenovsky, and it is indicated diagrammatically by Fig. 70, *B*. In *Hypolepis* a similar explanation is also possible. Gwynne-Vaughan notes for *Hypolepis repens* that if one bud only is present its vascular supply always arises from the basiscopic margin of the leaf-trace: if there are two or more, then the lowest on the basiscopic side is always stronger and further developed than the others. This appears to be so also in *Pteridium*, and the regularity thus observed strengthens the opinion that the buds are referable to a regular branching rather than to adventitious development. But developmental observations will be necessary before a final opinion can be formed on these difficult cases (Fig. 71).

On the basis of dichotomous branching of the axis, carried out independently of any one leaf either as regards orientation or identity, and with unequal development of the two shanks of the fork, it is possible to account for certain other well-known branchings in Ferns. If an unequal dichotomy be associated with the regular formation of a leaf on the free side of the weaker shank, the result would be an *axillary bud* (II). This condition is seen in the Hymenophyllaceae and Ophioglossaceae, among living Ferns. An early condition of the bud is shown in Fig. 63, together with its vascular connections, which indicates that the axillary bud and the subtending leaf have a common vascular supply, a relation which agrees with the theory of dichotomy. In that case the subtending leaf would be the first borne upon the arrested shoot lying in its axil (compare Chambers (60), p. 1037). The fact of axillary branchings is also well authenticated in *Zygopteris* and

Lyginopteris, and it has been compared with that in the Hymenophyllaceae. These facts confer a special interest upon the unequal branching seen in *Botryopteris cylindrica*, which occurs with or without a subtending leaf (Bancroft(56), Text-Figs. 7, 8). Given a high degree of inequality of the shanks, and a constant subtending leaf, the condition of *Zygopteris* would follow from the comparison of these ancient fossils. The fact that axillary branching is the only regular method also for the Ophioglossaceae shows the hold which it has had among primitive types, in which the biological advantage may well have fixed it as a permanent feature (Hofmeister(40), p. 264; Schoute(55)).

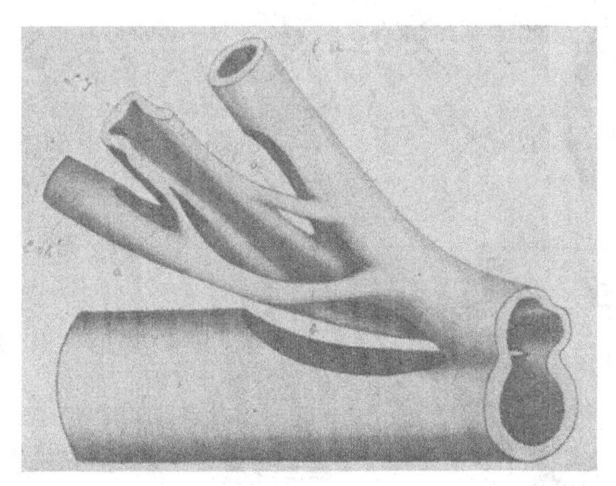

Fig. 71. Vascular system of *Hypolepis repens*, showing the departure of a leaf-trace (*L.T.*) from the solenostele, and the attachment to it of two lateral shoots, the one (*l.sh.*) arising from the basiscopic margin of the leaf-trace, the other (*l.sh.'*) from the acroscopic margin. (After Gwynne-Vaughan.)

If on the other hand a leaf intervene between the limbs of the dichotomy (as it is seen to do in Fig. 68, iv, v, vi), and if one of those limbs be arrested in its development, the effect would be a bud on the abaxial face of the leaf-base, attached either in the median plane, as it is seen to be in *Lophosoria*, *Metaxya*, *Cheiropleuria*, and others (Fig. 65): or it might be obliquely, as it frequently is in creeping rhizomes, such as *Dennstaedtia*, *Hypolepis*, *Marsilia*, or *Pilularia* (Fig. 72). Such differences of orientation as are suggested by comparison of Fig. 68, v, vi, if slightly more accentuated, would explain the difference between the conditions shown in the former and the latter groups of examples. It thus becomes possible to refer both the abaxial buds and those attached laterally to the leaf-base to an origin in dichotomy of the axis, with unequal development of the shanks, and with a close relation of a leaf to the base of the arrested shank. The vascular connections do not appear to be distinctive for or against this interpretation. In *Lophosoria*,

Metaxya, and *Cheiropleuria* the supply to the bud appears to arise from the base of the leaf-trace: in *Dennstaedtia* and *Hypolepis*, where the bud is lateral, the relation is with the margin of the leaf-trace. The question here is whether the leaf bears the bud, or the bud on its departure immediately gave rise to the leaf. Gwynne-Vaughan on anatomical grounds held the former view, but the latter seems the more probable interpretation on grounds of comparison. That the reference of the buds to an origin by dichotomy is correct as applied to *Marsilia* and *Pilularia* is indicated by reference to the related genus *Schizaea*, in which there is equal dichotomy.

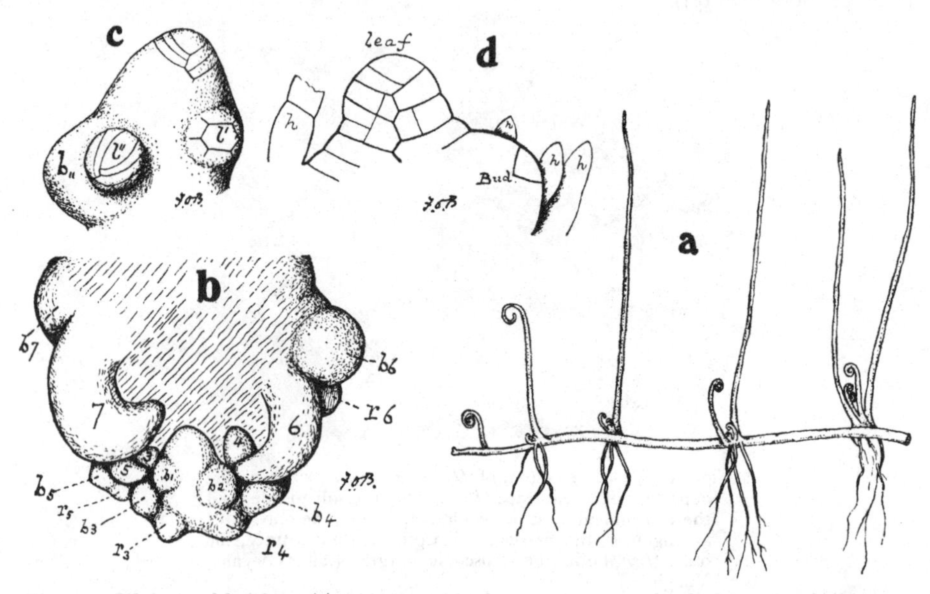

Fig. 72. *Pilularia globulifera*. (*a*) Rhizome with alternate leaves, and a bud associated with each of them; note circinate vernation (after Velenovsky). (*b*) Apical bud, with hairs removed, seen from above. 3—7 the successive leaves; b_1—b_7 the successive buds, one at the base of each leaf; r_3—r_6, the corresponding roots. (*c*) The extreme up-turned tip of a similar apex seen from above; $l'l''$ the youngest alternate leaves; $b_{,,}$ the bud corresponding to l''. (*d*) Section vertically through a corresponding leaf and bud, showing the apical cell of each; *h*, *h* = hairs. [(*a*) about natural size; (*b*) × 70; (*c*) × 175; (*d*) × 325.]

While it may thus be possible to interpret the buds about the leaf-base in Ferns in terms of distal branching unequally developed, there are cases which are obviously not of that nature. The buds borne on the upper regions of the leaf, whether substitutionary for soral development or not, are locally distinct in their origin from the axis. This and their late appearance mark them as adventitious. But there are other buds which present difficulty in interpretation. As examples of these the buds on the persistent leaf-bases of *Dryopteris Filix-mas* may be quoted (Fig. 1, *B, C*). Their position at some distance from the leaf-insertion and their relatively late appearance have been held to indicate an adventitious origin, though their constancy and similarity of position suggest comparison with certain buds of the leaf-base which result from dichotomy. Each such case presents its own problem, and it is only by the solution of such separate problems that the line can be drawn between those innovations which are referable to distal dichotomy and those which are really adventitious. Both methods of amplification of the vegetative system certainly exist in Ferns, but recent investigations show that the former may account for a larger proportion of the buds observed than had been anticipated by earlier writers. Consequently the position comes to be substantially that stated by Hofmeister, viz. that *in Ferns the branching may be referred either to dichotomy, with equal or unequal development of the resulting shanks: or to the formation of adventitious buds. But the line of distinction between these is not always clearly perceptible.*

The conclusion thus reached by comparative analysis of the shoot-system of the Filicales has important bearings upon the morphology of the Higher Plants on the one hand, and on the other upon that of the primitive sporophyte. In the light of what is seen in Ferns the prevalent axillary position of the buds in Flowering Plants is no longer a thing apart. The sharp distinction between the monopodial and the dichotomous branching is obliterated by the recognition of the gentle intermediate steps that exist between them: and the whole construction of the shoot-system of Phanerogams takes its natural place as a final derivative from a dichopodial state, already foreshadowed in the shoot-system of the Filicales and Pteridosperms. On the other hand, the recognition of the various branchings seen in Ferns as derivatives of dichotomy will suggest the possibilities that lay before a primitive sporophyte having a dichotomous system still undifferentiated: a state which is actually seen in the Psilophytales(62). By its sympodial development leading to strong inequality, and even to positive difference of form of the two shanks, the possibility of the evolution of a shoot with distinct axis and appendicular branches ranking as leaves is clearly present (see Chapter XVII).

As regards the phyletic treatment of the Filicales themselves, the equal dichotomy being held as the original type of branching, those shoots which dichotomise equally will be held as primitive in that feature, and any departure from equality will be held as derivative. Those Ferns in which the anatomical structure of the petiole most nearly approaches that of the axis, as it does in the early fossil *Botryopteris cylindrica*, and in *Stromatopteris* and the Hymenophyllaceae among living Ferns, will be held to be relatively primitive in respect of that feature in their morphology.

BIBLIOGRAPHY FOR CHAPTER IV

36. BRONGNIART. Histoire des Végétaux Fossiles, ii, p. 30.
37. KARSTEN. Vegetationsorgane der Palmen. Berlin. 1847.
38. METTENIUS. Ueber Seitenknospen bei den Farnen. Leipzig. 1860.
39. METTENIUS. Ueber den Bau von *Angiopteris*. Leipzig. 1863.
40. HOFMEISTER. Higher Cryptogamia. Ray Soc. 1862, p. 208.
41. KLEIN. Vegetationspunkt dorsiventrale Farne. Bot. Zeit. 1884.
42. SADEBECK. Natürl. Pflanzenfam. i, 4, p. 72: where a full list of earlier literature on adventitious buds is given.
43. BOWER. Comparative Morphology of the Leaf. Phil. Trans. 1884.
44. BOWER. Studies in the Phylogeny of the Filicales, ii, iii. Ann. of Bot. xxvi, xxvii.
45. ROSTOWZEW. Adventivknospen bei *Cystopteris bulbifera*. Ber. d. D. Bot. Ges. 1894.
46. ROSTOWZEW. Beitr. z. Kenntniss d. Ophioglosseen, i. Moscow. 1892.
47. LANG. Studies in the Ophioglossaceae, i, ii, iii. Ann. of Bot. xxvii, xxviii, xxix.
48. GWYNNE-VAUGHAN. Anatomy of Solenostelic Ferns, i, ii. Ann. of Bot. xvi, p. 170.
49. GIESENHAGEN. *Niphobolus*. Jena. 1901, p. 24.
50. VELENOVSKY. Vergl. Morphologie. 1905, p. 246.
51. SCOTT. On *Botrychioxylon paradoxum*. Trans. Linn. Soc. Botany. Vol. vii. 1912, p. 373.
52. SCOTT. Studies in Fossil Botany. Vol. i, Chapter ix.
53. SCOTT. Ann. of Bot. xxvi, p. 57.
54. VON GOEBEL. Organographie, 1 Auflage. 1913, p. 85, etc.
55. SCHOUTE. Ueber verästelte Baumfarne. Groningen. 1914.
56. BANCROFT. *Rachiopteris cylindrica*. Ann. of Bot. 1915, p. 543.
57. BRENCHLEY. *Lyginodendron*. Linn. Journ. 1913, p. 349.
58. CHRIST. Die Geographie der Farne. Jena. 1910.
59. THOMPSON. On *Stromatopteris*. Trans. Roy. Soc. Edin. lii, p. 152.
60. CHAMBERS. Ann. of Bot. xxv. 1911, p. 1037.
61. SAHNI. Anat. of *Nephrolepis*. New Phyt. xiv. 1915, p. 251.
62. KIDSTON & LANG. Fossils of the Rhynie Chert. Trans. Roy. Soc. Edin. li (1917), lii (1920). Parts I, II, III, IV.
63. VAN LEEWIN. Ueb. Veg. Vermehrung v. *Angiopteris*. Buit. Ann. 1912, p. 202.
64. SAHNI. New Phytologist. Vol. xvi. 1917, p. 1. Here useful references to the literature are given.

CHAPTER V

LEAF-ARCHITECTURE OF FERNS

ACCORDING to the definition given in Sachs' Textbook, the Leaf is a part which originates below the apex of the stem, as a lateral outgrowth. Leaves arise from the primary meristem, in acropetal order, and are always exogenous. The conceptions of leaf and stem are correlative, and their tissues are continuous: but the leaf usually grows more rapidly than the axis which bears it, and assumes a form different from it. In all these respects the leaves of living Ferns accord with the morphological definition: but they commonly show continued apical growth, sometimes to an indefinite degree (*Gleichenia*, *Lygodium*). This, together with the preponderant size and complexity of the leaf of most Ferns, has led to its designation as a "Frond." But the small size of the leaf in *Azolla* and *Salvinia* shows that large size is not a constant feature, nor should it in any case be held to override the more general grounds of morphological definition. The circinate vernation which is seen in Fern-leaves is a striking characteristic of their youth, but it is not a permanent feature. It is due to hyponastic growth in the young state, which is equalised as the leaf matures. A passing feature such as this, though interesting as a biological device giving protection to the vulnerable apex, is not to be held to characterise the Fern-leaf as a thing apart from other leaves: for many of these may also show temporary nastic differences, though in a less marked degree. The leaves of Ferns may then be ranked as parts of the same general category as the leaves of other Vascular Plants. Assenting to this does not imply that all such leaves were common in origin by Descent. It should always be contemplated as possible that organs of foliar nature may have been initiated along a plurality of phyletic lines.

The leaves of living Ferns are markedly bifacial, a condition shown also by many of the early fossils. But others belonging to the Zygopterideae present the unusual condition of bearing on organs, which are ranked as leaves, secondary appendages arranged in alternate pairs, so that they constitute four longitudinal rows (Fig. 73). At present it must suffice to mention these archaic fossils, and we turn from them to the architecture of the leaves of living Ferns, which conform to the dorsiventral type usual for other plants. They are very variable in outline. Sometimes they appear simple and unbranched: most frequently they are branched, and often in high degree. The branches are then arranged in two longitudinal rows, one on either side of a central stalk or rachis, which is continuous below into the stipe or petiole, often of considerable length. The whole of this, including petiole and rachis, may be styled the *Phyllopodium*[1]. The usual appearance of the foliar organ in Ferns is thus that of a pinnate, petiolate leaf, comparable with that of some leaves of Dicotyledons (see Fig. 2, p. 3). But others may be fan-like in construction, with lobes radiating out from a common point, or with veins traversing in similar fashion a webbed, that is a simple, blade or lamina (Fig. 74, *A*). Many intermediate conditions may also exist, both in the webbing and the cutting of the leaf, and in the disposal of the veins, though with persistent dorsiventral symmetry.

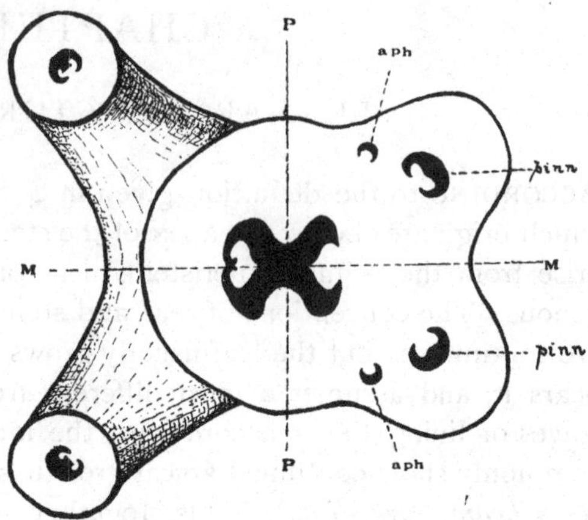

Fig. 73. *Diplolabis Römeri*, Solms. Diagrammatic drawing showing the ramification of the primary petiole, as it might be seen in a thick transverse section. *PP* = principal plane of symmetry; *MM* = median plane; *aph* = aphlebia-trace; *pinn* = pinna-trace. (From P. Bertrand, after Gordon.)

Comparison of these raises the whole question of the architecture of the leaves of the Filicales, and the recognition of the principles which underlie their diverse construction. It will be found that, however complex, they are all deducible from dichotomous branching of a primordial leaf. The fact that equal dichotomy of the leaf is also seen in the Sphenophyllales and Equisetales indicates that this has been a widespread feature among primitive Vascular Plants.

There are three chief avenues for the study of the leaves of Ferns, which may lead to a knowledge of the principles that have ruled in their evolution and resulted in their architecture as now seen in the adult state [Bower (77)]. The same method would apply equally for other plants in which elaborate leaf-architecture is seen, provided the necessary facts were available. They are:—

(I) Comparison of the juvenile with the adult leaves of the individual plant.

[1] This term was introduced in 1884 (71, p. 565) in the sense here applied in the text. It ranks in construction and in meaning with such words as "dichopodium," "sympodium," "monopodium." The name connotes the dominant axis or rachis of a branch-system of foliar character, bearing appendages the relation of which to the rachis or axis (phyllopodium) may be either that of monopodial branching or of sympodial dichotomy. The same word has since been used by Luerssen (xiii, p. 10) as connoting the base only of the leaf-stalk, which often persists as a scar upon the rhizome. It is not used in that sense here; and priority as well as preference of meaning and construction are claimed for its use in the sense above defined.

(II) Comparison of the details of the adult leaves in different species and genera.

(III) Comparison of the related fossils.

All these avenues should be pursued in order to arrive at a scientific knowledge of the principles on which leaf-architecture has progressed from simple beginnings: and if the reasoning from the facts be correct, the conclusions from them all should substantially coincide.

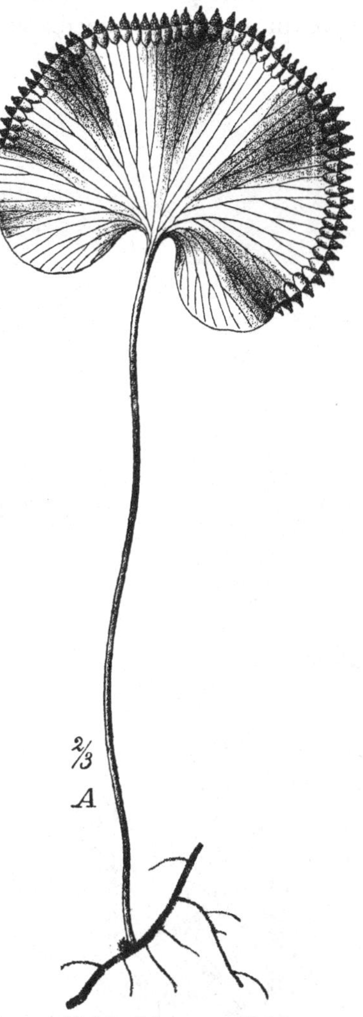

Fig. 74. Habit of the genus *Trichomanes*. *A* = *Tr. reniforme*, Forst. (After Sadebeck, from Engler and Prantl.)

WEBBING

We ought to distinguish at the outset two factors which do not necessarily advance parallel, though they may frequently be found to do so. They are, the *venation* and the *cutting* of the leaf-blade. In the simplest unbranched cotyledonary leaves there may be only a single vein: this is seen in *Hymenophyllum* and *Trichomanes*. In the branched adult leaves of these Ferns, as also in the cotyledons and adult leaves of other Ferns, each separate segment contains only a single vein (Fig. 75). This, which may be held as a primitive state, is seen in most of the species of *Hymenophyllum* and *Trichomanes*, and in *Todea superba* (Fig. 76). But in others of the Hymenophyllaceae and Osmundaceae, and very commonly elsewhere in Ferns, each segment may contain more than one vein. Various degrees of lateral fusion of narrow, one-veined segments to form a broad many-veined lobe may be found, which clearly demonstrate the close relation between the separate-lobed and the "webbed" conditions. An instructive parallel is found in the Sphenophylls, where there is great variety in the webbing of the leaves. All grades of it can be illustrated in the genus *Sphenophyllum*, from narrow linear dichotomising segments with a single vein each, as in *S. tenerrimum*, to webbed segments with a marginal tooth at the end of each vein, as in *S. cuneifolium*. The margin may even be entire, as in *S. verticillatum*. Potonié states (72, p. 176) that on passing from lower to later geological horizons the size of the leaves increases, but the degree of their cutting decreases. The ancient *S. tenerrimum* has narrow linear

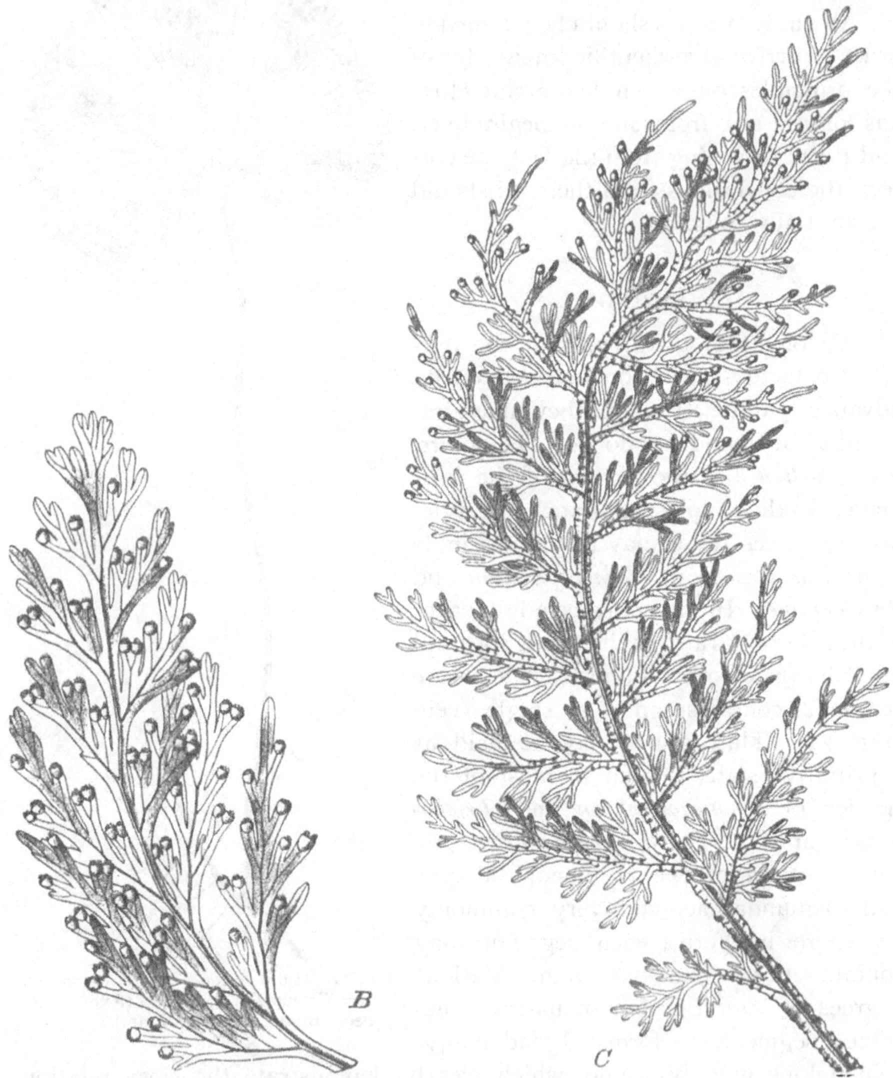

Fig. 75. Habit of leaves of *Hymenophyllum*. *B* = *H. dilatatum*, Sw. *C* = *H. australe*, Willd. In *B* and *C* the single-veined segments are separate one from another, which is held to be a more primitive condition. (After Sadebeck, from Engler and Prantl.)

radiating segments: the more modern *S. Thoni* has large undivided leaves (Fig. 99, p. 103). These facts are important for comparison with the Ferns. It seems probable that a primitive condition of the leaves in Ferns as well as in Sphenophylls was that in which each segment is separate laterally, and contains only one vein, as in the *Sphenopteris* type. It appears that later the segments became progressively webbed by lateral fusion. Moreover the margin often shows distal teeth, each containing a vein-ending, which marks the constituent segment (Fig. 77). If then progressive webbing had really taken place the venation as actually observed would be a more direct guide to the branching which underlies the structure of the leaf, lobe, or segment, than is the mere outline. In deeply cut leaves the two may coincide, but in fully webbed leaves the margin may be quite entire, though with highly branched venation, as in the Hart's Tongue, or in *Trichomanes* (Fig. 78). In that case venation and not the outline will be the safe guide in a morphological analysis of the blade whether of juvenile or of adult

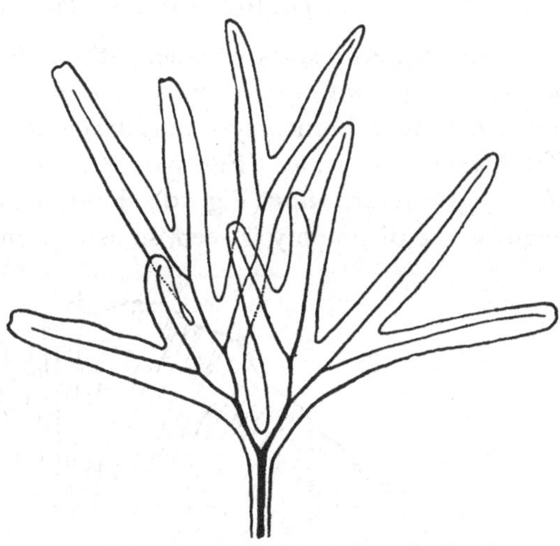

Fig. 76. Juvenile leaf of *Todea superba*, showing single-nerved segments all separate. (× 3.)

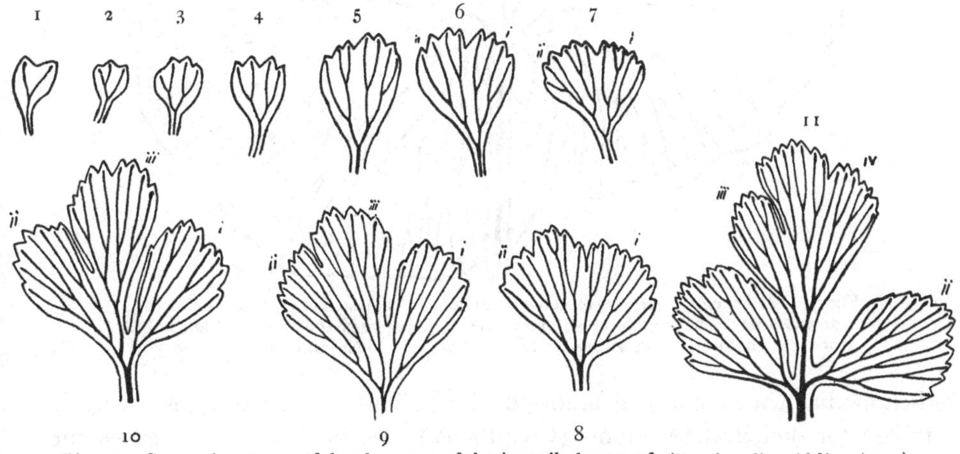

Fig. 77. Successive stages of development of the juvenile leaves of *Anemia adiantifolia*. (× 4.)

leaves. This line of comparison can be followed even in the fossils, for in them the course of the veins is often well preserved. The general conclusion following from such considerations is: that *in a primitive type of leaf-construction the segments resulting from distal branching were separate laterally one from another: but that by progressive webbing they became coherent laterally to form broader expanses in later and derivative types.*

DICHOTOMY AND ITS DERIVATIVES

In many Fern-leaves, whether with blades deeply cut, toothed at margin, or entire, *dichotomy* of the veins is a quite obvious feature. In some of them the dichotomous branch-system is developed with all its branches equal, as in *Trichomanes reniforme* (Fig. 78), *Actiniopteris radiata*, or *Elaphoglossum* (*Rhipidopteris*) *peltatum* (Fig. 79). But more frequently a sympodial development of the dichotomy is seen, so as to form a *dichopodium* or a number of

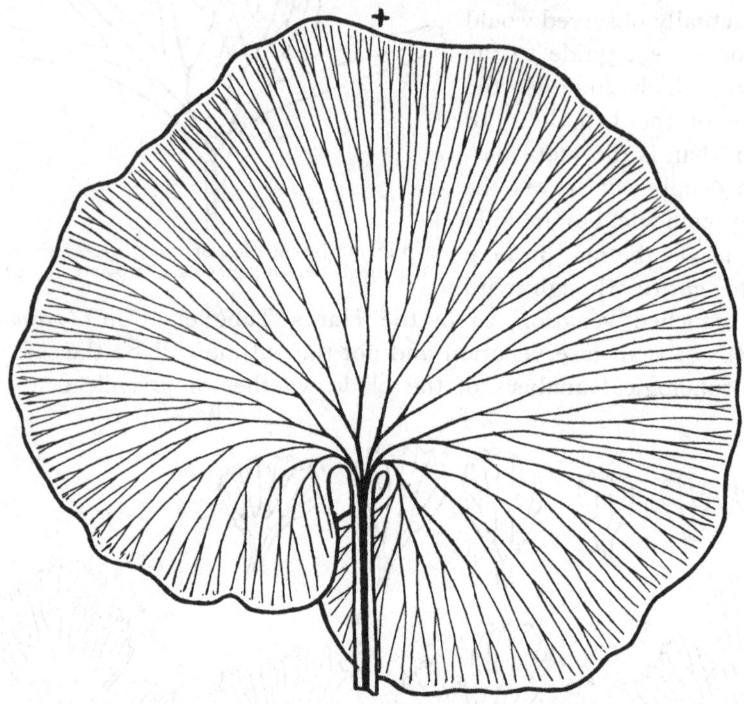

Fig. 78. *Trichomanes reniforme*, showing detail of venation and vascular connections downwards to the petiole. (+) marks the limit between the right and left halves of the dichotomy. (Drawn by Dr J. M. Thompson.) Natural size.

dichopodia, which may be followed along lines similar to those already traced for dichotomous stems (Chapter IV). As in them so in leaves the dichopodium is probably a secondary development as regards its evolution: it is produced in relation to rapid growth in length, in the one case of the whole shoot-system, in the other of the branching leaf. Biologically this is specially an advantage in leaves, for the longer the leaf the greater is the radius outwards from the central stem of the area functional for the capture of photo-synthetic rays of light. The ontogenetic facts frequently support this progressive history. The first leaf may be simple, with a single vein

(*Trichomanes*, *Ceratopteris*). But usually the cotyledons show equal dichotomy, while the later leaves show inequality of the forking, leading gradually

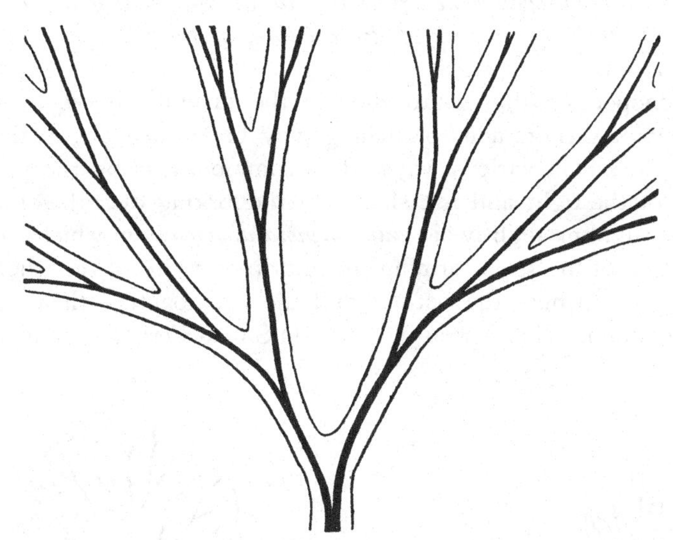

Fig. 79. Basal part of the lamina of an adult sterile leaf of *Elaphoglossum* (*Rhipidopteris*) *peltatum*, showing very perfect dichotomy. (× 4.)

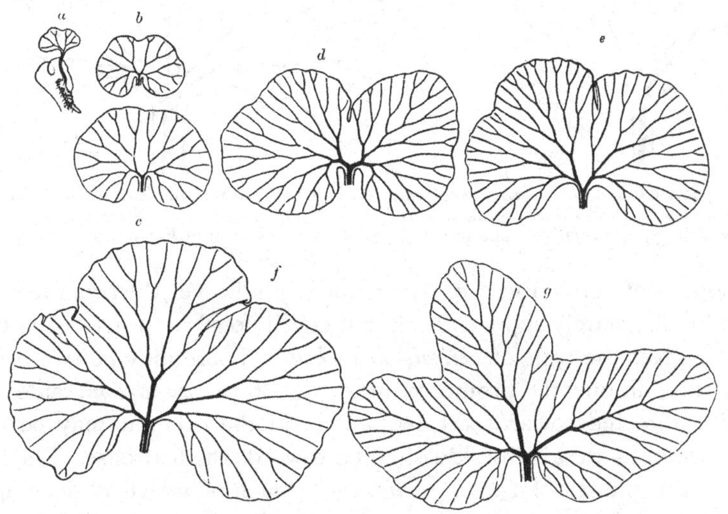

Fig. 80. Successive juvenile leaves of *Osmunda regalis*, showing successive steps of progression from equal dichotomous venation to sympodial branching, and the establishment of a terminal lobe. (× $2\frac{1}{2}$.)

to a dichopodium (*Cyathea*, *Anemia*, *Osmunda*, Fig. 80). Where, however, the first leaf is relatively large it may at once step into a more advanced position, and the initial state of equal dichotomy is apt to be omitted. This is seen in *Helminthostachys* (Fig. 81, *B*), or *Botrychium* (Fig. 81, *A*). Comparison of the juvenile leaves of other Ferns supports the view that *equal dichotomy was the prior, and probably the original state in the construction of the leaf, and that some form or another of dichopodium of the main veins is a state derivative from it.*

The dichopodium thus established in the juvenile leaves by unequal development of dichotomous branching may be worked out in the leaves of the adult Fern in various ways. The commonest is by the promotion alternately of the right and left shanks of the forking over the other. The result is a straight or slightly zig-zag *scorpioid dichopodium*, which appears as a continuation of the stipe, forming the midrib or rachis of the whole blade. It apparently continues the petiole, and the two together then constitute the phyllopodium. This is seen in the Male Shield Fern, and in most of the

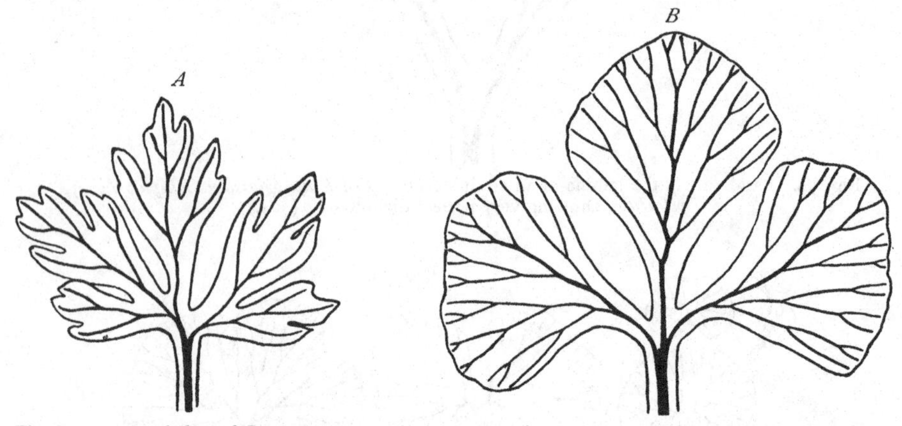

Fig. 81. *A* = cotyledon of *Botrychium virginianum*. (× 4.) *B* = a juvenile leaf of *Helminthostachys*, probably an actual cotyledon. From collection of Dr Lang. (× 4.) Here the ontogeny starts from a stage which appears relatively late in the series of *Osmunda* seen in Fig. 80, or *Anemia*, Fig. 77.

British species of Ferns (Fig. 2, p. 3). Frequently, however, the promoted shank may not be alternately right and left, but continuously repeated on the same side. This gives the type of *helicoid branching*. There may be two different types of this: in one the promoted branch is on the *anadromic* side:—that is the side directed towards the leaf-apex: in the other it may be on the *catadromic* side:—that is the side directed towards the leaf-base. The former is seen in the pinnae of *Pteris semipinnata*, each of which is developed as an ascending helicoid dichopodium (Fig. 82, *E*). It is also illustrated by *Dipteris conjugata*, and *Dictyophyllum exile* (Fig. 82, *B*, *C*). The latter type is seen in the whole leaf of *Matonia pectinata*, in which the right and left halves are each a descending helicoid dichopodium (Fig. 82, *A*). The examples quoted have been selected as bringing the character of the branching prominently forward. But similar features often appear in the details of the

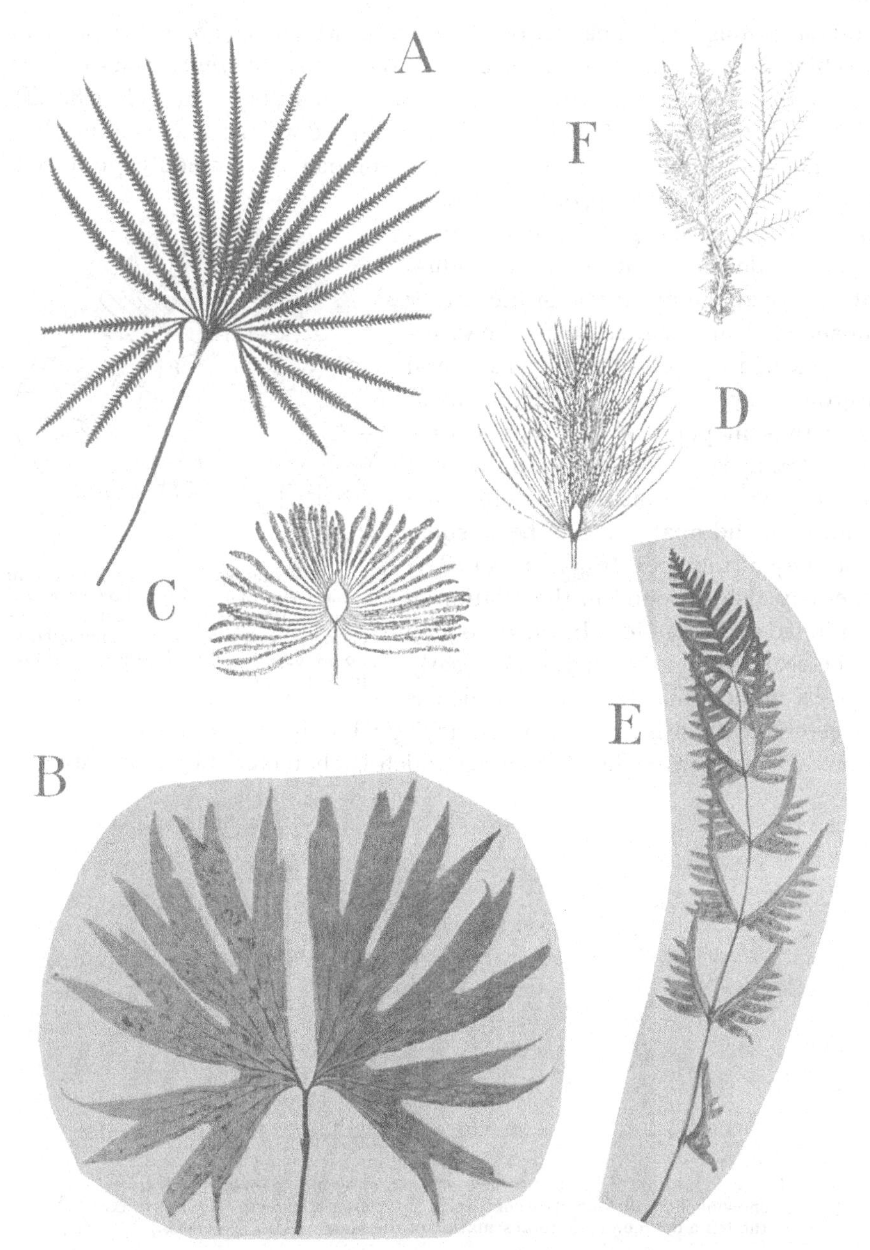

Fig. 82. *A* = leaf of *Matonia pectinata*, R. Br., showing catadromic helicoid structure, much reduced. *B* = leaf of *Dipteris conjugata* (Kaulf.), Reinw.; showing anadromic helicoid structure, much reduced. *C* = leaf of *Dictyophyllum exile*, much reduced, showing anadromic helicoid structure (after Nathorst). *D* = leaf of *Camptopteris spiralis*, much reduced, showing anadromic helicoid structure, spirally twisted (after Nathorst). *E* = leaf of *Pteris semipinnata*, reduced, showing anadromic helicoid structure of the pinnae only. *F* = leaf of *Odontopteris minor*, Brongn., reduced, showing anadromic helicoid structure (after Zeiller).

distal branching and venation of Fern-leaves which do not show helicoid branching throughout. For instance, the leaf of *Pteris semipinnata* as a whole is pinnate, but the pinnae themselves show scorpioid branching (Fig. 82, *E*).

Where the scorpioid structure is strongly developed with a marked phyllopodium, a succession of alternating pinnae is commonly produced, as in the Male Shield Fern, or any other of our ordinary native types. In such cases, if the actual development of the individual leaf be traced, the primordia of the earlier pinnae are found to arise below the growing apex of the leaf, appearing as lateral outgrowths upon it (Fig. 83). The branching is thus in point of fact monopodial. But as the apex of the leaf is approached there is a gradual transition to sympodial dichotomy, and finally it may be to equal dichotomy of the apex (Figs. 84, 85). The same may also be found in the relation of the pinnae and pinnules. In the ontogeny of the leaf there is in fact in the early stages a monopodial branching, which is characteristically a later type of branching: but in the later stages a transition may be seen to that branching which is characteristic of the juvenile

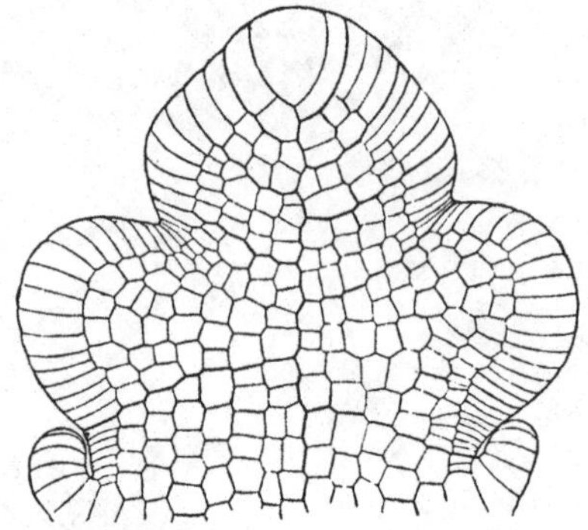

Fig. 83. Young leaf of *Ceratopteris* seen in surface view, after Kny; showing apical segmentation. It is clear that the alternating pinnae, which arise monopodially, do not correspond to the segments cut off from the apical cell.

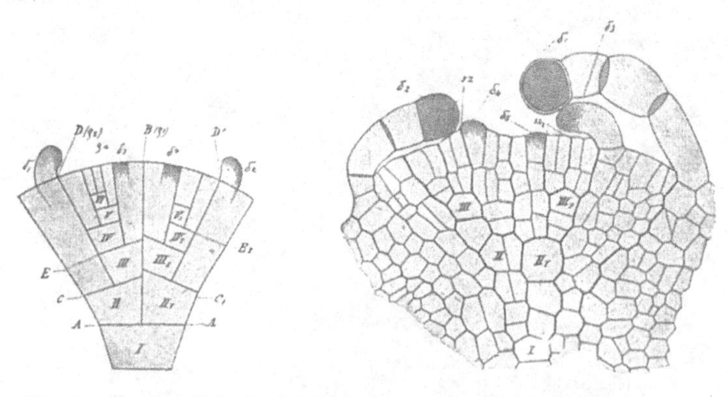

Fig. 84. Portion of the leaf-surface of a sporeling of *Asplenium serpentini*, showing how dichotomy accompanies the marginal growth. (× 190.) To the left a diagrammatic representation of the same. (After Sadebeck.)

leaves. Consequently, if the phyletic history be reflected in these changes in the method of branching, the series would have to be taken in reverse of that actually seen in the development of the individual leaf. The evolutionary succession of steps would then read thus: (i) equal dichotomy;

(ii) sympodial dichotomy; (iii) monopodial branching. This succession probably reflects truly the steps in the phyletic elaboration of the foliar structure in the Filicales at large. A comparison of this monopodial branching of the leaf with the monopodial branching of the axis described in Chapter IV shows that both cases are alike in being referable in origin to dichotomy, and follow on its sympodial development. Accordingly, the general conclusion for the branching of the leaf is, that *monopodial branching, which is characteristic of Angiospermic leaves, and is seen in the lower and first formed pinnae of many Ferns, arrived last in the course of evolution, arising as a modification of dichotomy, which still survives distally in most leaves of Ferns.*

Fig. 85. *Allosorus crispus.* Outline of a leaflet. The branching is clearly dichotomous. The apex has divided into lobes, 1 and 2, of which 1 is the stronger and continues the growth; 2 forms a lateral lobe. Below we have lobes 3 and 4 which have been similarly formed. The leaf-spindle (phyllopodium), *S*, is only a narrower portion of the lamina which is mechanically strengthened later. Magnified. (After von Goebel.)

In some leaves of Ferns, however, the final transition to dichotomy at the apex does not take place at all. The apical growth appears to be arrested before the transition to the more primitive type of branching, but after the monopodial origin of the lower pinnae has been carried out. The phyllopodium then appears to end in a blunt cone, as may be seen in the leaves of *Angiopteris* (Fig. 86), (71). This is precisely what happens in the developing leaves of Cycads, and in many pinnate Angiosperms, in which the phyllopodium does not explain its sympodial origin by any transition to apical dichotomy, as it does in so many Ferns. This arrest of the apical growth of the leaf in certain Ferns, and its habitual adoption in the Cycads and in the Angiosperms, throw into all the greater prominence the continuance of that growth which is habitual in the Ferns themselves: and especially those cases where it is unlimited, as in *Gleichenia* and *Lygodium.* No other series of Vascular Plants, excepting the Pteridosperms, has shown continued apical activity of its leaves in like degree. This marks them off sharply from the Lycopodiales, Equisetales, and Sphenophyllales: in fact it supplies the most distinctive feature of their vegetative system among Vascular Plants.

A scorpioid dichopodium naturally results in a midrib with apparently lateral veins, a condition which is seen in most Ferns. Where there are separate pinnae or pinnules borne upon a rachis, their margins are distinguished respectively as *anadromic*, which is that directed towards the leaf-apex, and *catadromic*, that directed towards the leaf-base. Mettenius (66, iv, p. 2) found that in the venation of Ferns there is some degree of constancy

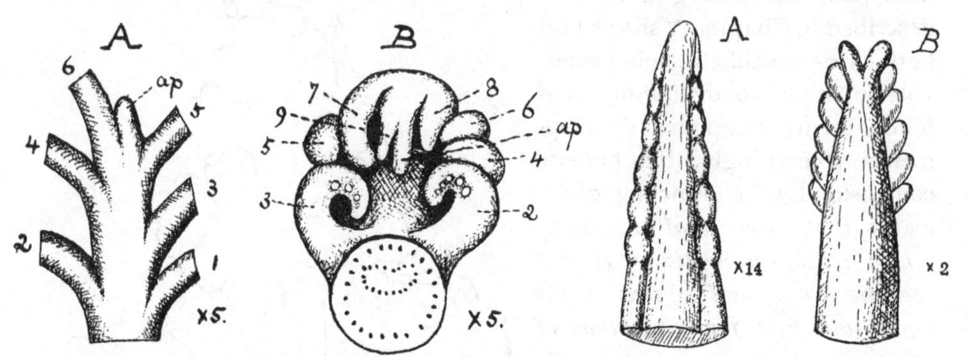

Fig. 86. Young leaves of *Angiopteris*. *A*=apex of leaf bearing six alternate pinnae. The phyllopodium ends abruptly in a cone, *ap*. *B*=apex of a younger leaf with alternating pinnae 2—9, formed by monopodial branching. *ap*=apex. The pinnae 2, 3 have begun to form pinnules, also monopodially.

Fig. 87. Young leaves from seedling of *Cycas*. *A* shows a younger state, where the pinnae are appearing monopodially, on the flanges of the phyllopodium (×14). *B*=a later stage, where the two distal pinnae suggest a reversion to dichotomy (× 2).

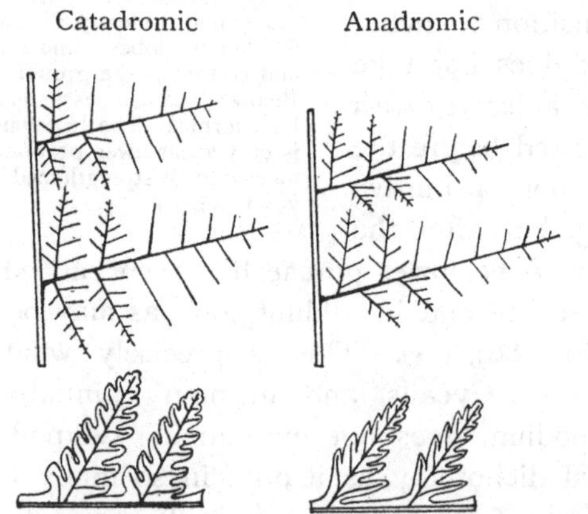

Fig. 88. Anadromic and Catadromic sequence of branchings of veins in leaves of Ferns (after Christensen). Note that the distinction applies to the pinnules. But it is only available for leaves which have branched at least twice.

in the relative position of the first or lowest vein. He designated the venation as anadromic when the lowest lateral vein of a pinna or pinnule lies on the acroscopic side, and catadromic when it is on the basiscopic side (Fig. 88). Potonié indicated that this character had a phyletic bearing (72, p. 110). He drew attention to the preponderance of the catadromic venation among Palaeozoic Ferns, while in the majority of modern types it is anadromic. That this criterion can only be applied in general terms and not in detail follows from the fact that the character is not constant even in the same individual in *Trichomanes* (Mettenius (67), pp. 415, 416). A like inconstancy is seen in the juvenile leaves of *Gleichenia* (Bower (77), p. 684, Fig. 18). It may be illustrated in many other examples, notably in the genus *Aspidium* (*Nephrodium*, *Dryopteris*). At best this character can only be of restricted use. C. Christensen states (83, p. 5, 1920) that "a stable arrangement of the ribs can only be found in leaves that are divided to a certain degree, at least twice. The sequence of the primary pinnae is unavailable." Whatever its value may be in giving minute systematic distinctions, it cannot be used generally.

VENATION

Characters of comparative importance are found in the vein-endings, and the relations of the veins throughout their course. If the origin of the expanded blade be as already suggested, and if all broad leaf-areas of Ferns have resulted ultimately from webbing of narrow dichotomously branched lobes, then the primitive venation should have free endings to all the veins. This is sometimes called an *open* venation, as distinct from a *closed* venation, where the veins may be fused into loops. Any such vein-fusions should be held as derivative, and secondary. Comparison of the leaves of those Ferns which on other grounds are held to be primitive shows that in them the veins end free. For instance, *Botrychium*, *Helminthostachys*, *Marattia*, *Angiopteris*, *Schizaea*, *Gleichenia*, and almost all of the Cyatheaceae and Hymenophyllaceae, have an open venation, with free vein-endings, and no fusions. A first step towards a closed venation is the formation of distal loops. This may result in a continuous intramarginal commissure, such as that seen in *Marsilia* (Fig. 89). Vein-fusions may, however, be initiated in other ways,

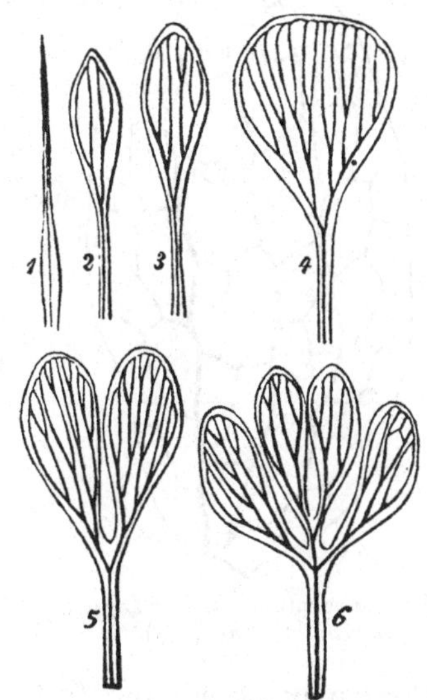

Fig. 89. Successive types of juvenile leaves of *Marsilia*. (After Brann.)

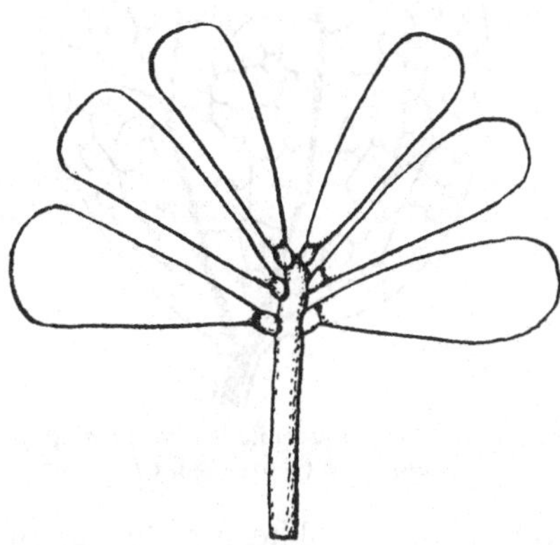
Fig. 90. Abnormal leaf of *Marsilia quadrifolia*, with six pinnae. (After Velenovsky.)

leading to more elaborate reticulation, and finally to small-meshed networks. There is some evidence of ontogenetic progression from an open to a

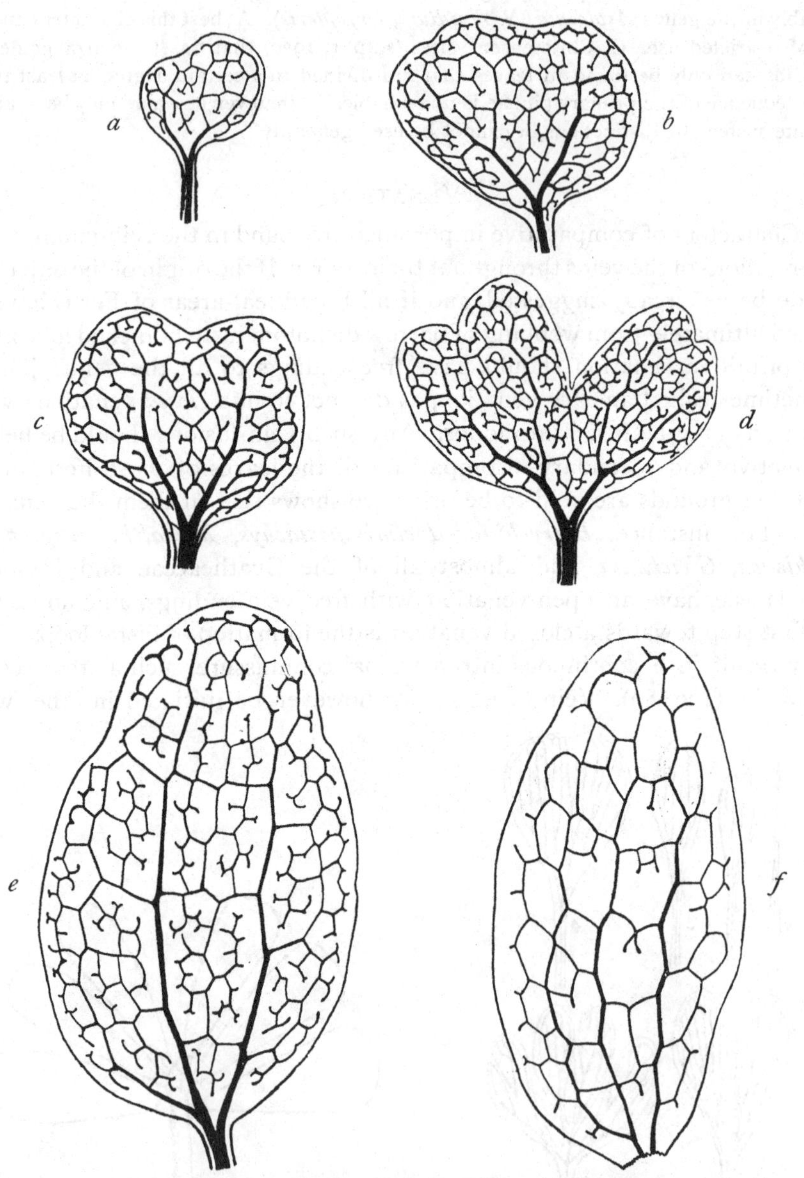

Fig. 91. *a*—*f* = juvenile leaves showing closed, or reticulate venation. *a*—*d* = *Dipteris conjugata* (× 6); *e* = *Cheiropleuria* (× 3); *f* = *Platycerium Veitchii* (× 3).

closed venation. The first leaves of *Ceratopteris* may be without fusions, though these appear in the later leaves. The earliest juvenile leaves of

Dipteris Lobbiana have the meshes imperfectly coupled. But usually where reticulation exists in the adult it is established in the earliest leaves of the individual, as in *Dipteris conjugata*, *Cheiropleuria*, and *Platycerium* (Fig. 91). So far as it goes the ontogenetic progression indicates that an open venation is primitive, and the closed a derivative state.

More cogent evidence comes from the stratigraphical sequence of fossils. Reticulation is unknown in Devonian plants. Potonié notes that Stur does not record a single case of reticulate venation from the Flora of the Culm. Meshwork is first seen in the Middle Coal Period, though many of the

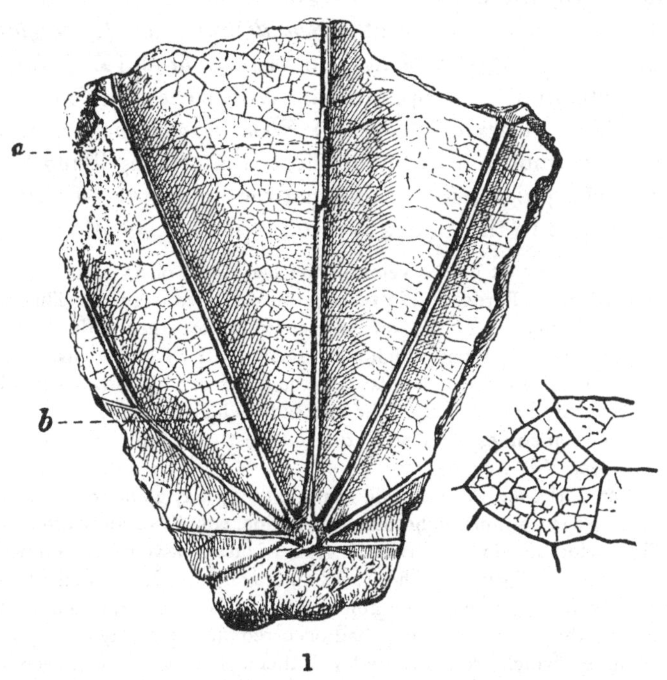

Fig. 92. *Clathropteris egyptiaca*. (Nat. size): *a—b*, pieces of main ribs in grooves. To the right a small piece of lamina showing the venation. (After Seward.)

examples of it that have been quoted are now ranked as Pteridosperms: for instance *Lonchopteris* and *Linopteris*. But here the vein-fusions are few, and the meshes relatively large. It is not till the Mesozoic Period that reticulation of a higher order, that is with a smaller network within the larger meshes, became prevalent. This state is well illustrated by *Clathropteris egyptiaca* Seward, from the Nubian Sandstone, which has a venation that may be matched by many modern Ferns (Fig. 92). From these facts it seems necessarily to follow that *open venation was primitive, and that reticulation with progressively smaller meshes was derivative.* The physiological advantage gained by intimate vein-fusions in promoting equal water-distribution over large leaf-areas gives added probability to this conclusion.

Among modern Ferns it is quite a common thing to find single genera or species that have reticulate venation in families or genera characterised for the most part by open venation. Sometimes the vein-fusions are only occasional, but frequently the reticulum may be of an advanced type. This occurs even in the most primitive groups. For instance, while *Helminthostachys* and *Botrychium* have open venation as a rule, in *Ophioglossum* it is reticulate. All the genera of the Marattiaceae have open venation except *Christensenia*, in which the few broad segments are reticulate. The Schizaeaceae are mostly open-veined, but *Anemidictyon* and *Hydroglossum* are reticulate sections respectively of the genera *Anemia* and *Lygodium*. *Hypoderris* is reticulate among the open-veined Woodsieae, and *Onoclea sensibilis* is reticulate while *Matteuccia* (*Onoclea*) *Struthiopteris* has open venation. Instances might be multiplied which indicate that among living Ferns there has been frequent progression within narrow circles of affinity from the more primitive open to the more advanced reticulate venation.

The detailed arrangement of the veins in the leaf-blade or its parts has been widely used in the comparison of Ferns, and in their systematic arrangement. This has led to a classification and terminology of the various venations, together with elaborate verbal descriptions of those several types which were recognised by Mettenius. It will not be necessary here to enter into these details. It should suffice to illustrate them by a series of figures, as selected by Luerssen (81). The names by which the several types of venation are designated refer sometimes to living Ferns, but mostly to fossils, in which the venation is often very well seen (Figs. 93, 94).

An examination of the several types will show that they may all be traced back along the lines explained above to a simple dichotomous venation, equally or more often sympodially developed. This is combined with various degrees of fusion of veins, sometimes marginal, sometimes nearer to the midrib: in other cases quite generally distributed. In leaves with a broad surface the appearance is sometimes given as though a repetition or duplication of the meshes from the midrib outwards had occurred, and this may often go along with a repetition of the sori which are associated with them. It is particularly seen in *Nervatio Goniophlebii* and *Cyrtophlebii* (Figs. 93, *i*; 94, *j*). Such factors as those mentioned will account for the genesis of even the most complicated reticulate systems, as a careful analysis and comparison will show.

Beyond the fact that these complex venations are characteristic for the most part of relatively late and advanced types of Ferns, their detailed study cannot be used as a consistent or trustworthy basis for the phyletic seriation of the Filicales at large. This conclusion, which robs the subject of much of its interest, naturally follows from the fact now demonstrated for so many isolated genera that a closed venation has originated repeatedly in distinct stocks. It is in fact a widely homoplastic character, determined in great measure by physiological necessity or convenience.

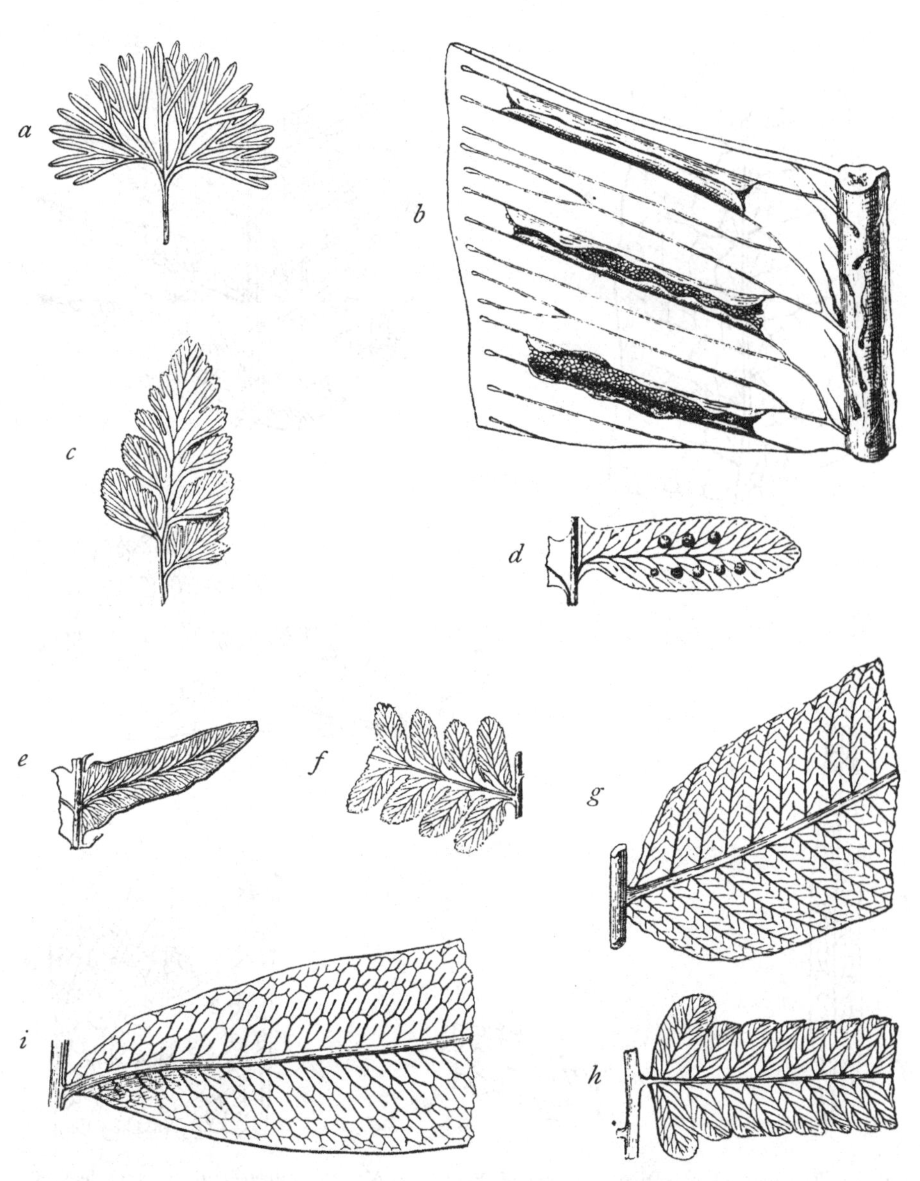

Fig. 93. Examples of nervation in leaves of Ferns. *a*=*Nervatio Caenopteridis* (sterile leaf of *Elaphoglossum peltatum*). Nat. size; *b*=*Nervatio Taeniopteridis* (part of fertile leaf of *Scolopendrium vulgare*). × 2; *c*=*Nervatio Sphenopteridis* (primary segment of *Asplenium adiantum nigrum*). Nat. size; *d*=*Nervatio Eupteridis* (fertile pinna of *Polypodium vulgare*). Nat. size; *e*=*Nervatio Neuropteridis* (fertile segment of *Pteridium aquilinum*). Nat. size; *f*=*Nervatio Pecopteridis* (part of segment of *Dryopteris Filix-mas*). Nat. size; *g*=*Nervatio Goniopteridis* (part of segment of *Meniscium reticulatum*). Nat. size; *h*=*Nervatio Goniopteridis* (part of segment of *Asplenium esculentum*). Nat. size; *i*=*Nervatio Goniophlebii* (part of segment of *Polypodium neriifolium*). Nat. size. (After Luerssen.)

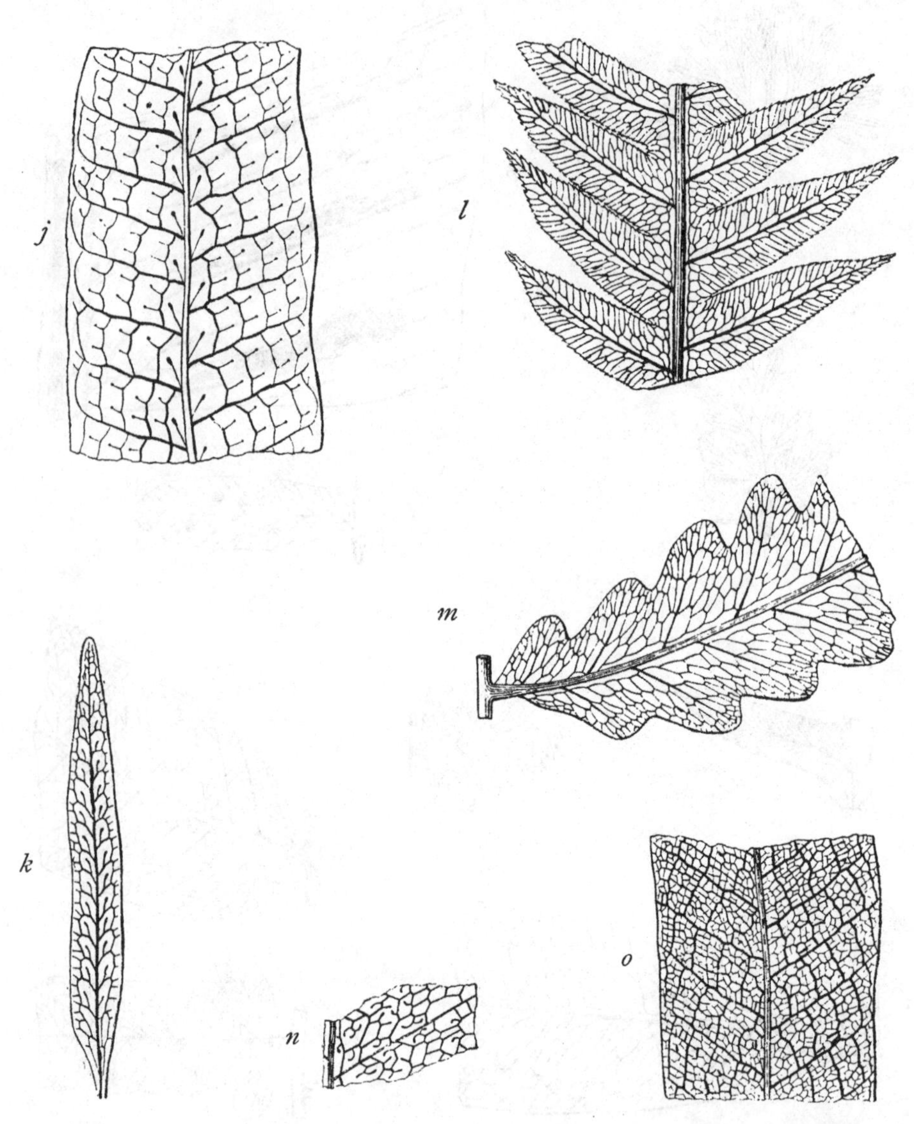

Fig. 94. Examples of nervation in leaves of Ferns. *j*=*Nervatio Cyrtophlebii* (part of leaf of *Polypodium caespitosum*). Nat. size; *k*=*Nervatio Marginariae* (leaf of *Polypodium serpens*). Nat. size; *l*=*Nervatio Doodyae* (part of segment of *Woodwardia radicans*). Nat. size; *m*=*Nervatio Sageniae* (part of segment of *Onoclea sensibilis*). Nat. size; *n*=*Nervatio Anaxeti* (small part of leaf of *Polypodium crassifolium*). Nat. size; *o*=*Nervatio Drynariae* (part of segment of *Polypodium quercifolium*). Nat. size. (After Luerssen.)

Aphlebiae

Under the name of *aphlebiae* certain lateral appendages of the leaves of fossil Ferns and Pteridosperms have been assembled, which though corresponding in position to pinnae or pinnules differ from them in outline, and apparently also in texture. They are irregularly cut or lobed, and they often appear as though they were without venation (Fig. 95). Comparison indicates that they are of the nature of pinnae or pinnules, arrested and specialised for the protection of other parts while young. Their position on

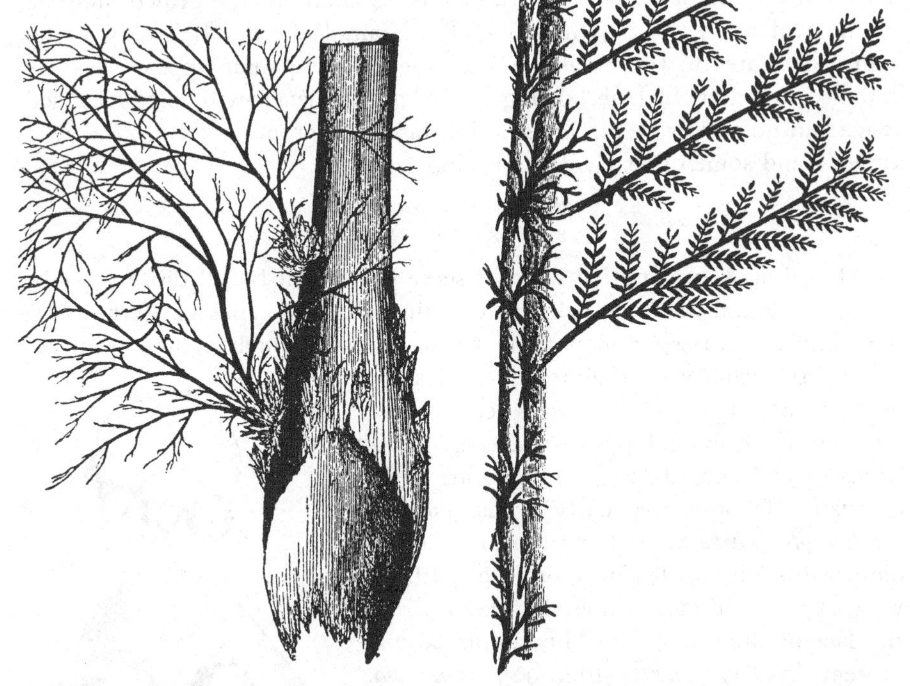

Fig. 95. Left, aphlebiae at the base of a petiole of the living *Hemitelia capensis*. Right, aphlebiae on a leaf of *Sphenopteris crenata*, Lindl. from the Carboniferous Period. (After Schimper, from Velenovsky.)

the surface of the stem of *Ankyropteris Grayi*, as well as on the leaf-base, but still with vascular connection with the stem, suggested to P. Bertrand their description as "sorties hatives," their insertion being at a point lower than that at which the normal pinnation begins. A parallel is to be found in some living species of *Gleichenia* (*G. gigantea*, Wall., is quoted by Potonié), while in *Hemitelia capensis* and in a minor degree in other species of the Cyatheoid Ferns, somewhat similar modifications of basal pinnae are found, which have been described as aphlebiae (Fig. 95, left). There appears to be no sufficient reason for regarding any of these organs as forming a category apart from other segments of the leaf-blade, but as merely specialised examples of them. The basal position often taken by such pinnae is due to the way in which the intercalary growth is distributed in the phyllopodium. Many of the characteristic features of the different types of mature leaves can be thus explained. But the most frequent, as it is also the most marked, result of intercalary growth is the long stipe or petiole, which carries up far above the leaf-base that distal branch-system that constitutes collectively the blade or lamina. Sometimes, however, intercalary growth may be localised between the pinnae, separating them in successive apparent pairs, as in the Bracken, or the Oak Fern, or the large Cyatheoids. Sometimes the growth may be intermittent as in the Gleichenias or the Bramble Ferns. It is only a step from such states to those cases where some of the pinnae appear basal, as in the so-called "aphlebiae," owing to the intercalary growth being localised above and not below them. This is the state seen prominently in *Hemitelia capensis*, and some other Cyatheoids (Fig. 95).

STIPULAR GROWTHS

This simple explanation will not serve to account for all those growths which are found at the leaf-base. Certain primitive types of Ferns show sheathing structures, often described as "stipular," which, arising from the base of the leaf, protect either its own apex or the whole of the next younger leaf. These are seen in the living Ophioglossaceae, Marattiaceae, and Osmundaceae, and they are characteristic of some very early types, such as *Archaeopteris hibernica*. They appear as lateral outgrowths on the leaf-primordium, with or without a commissure connecting them across the face of the leaf-stalk. The commissure is present in *Angiopteris* (Fig. 96), *Marattia*, *Christensenia*, and *Todea*, but it is absent in *Osmunda*, where two distinct lateral flaps are found. There is no sufficient reason to regard these lobes as of pinna-nature. They were probably in the first instance basal growths, and distinct in phyletic origin from those distal branchings that went to form the leaf-blade as we now see it. They are absent from all the more advanced Filicales, but similar structures are found among the Pteridosperms and the Cycads, *Cycas* being without, and *Stangeria* with a commissure.

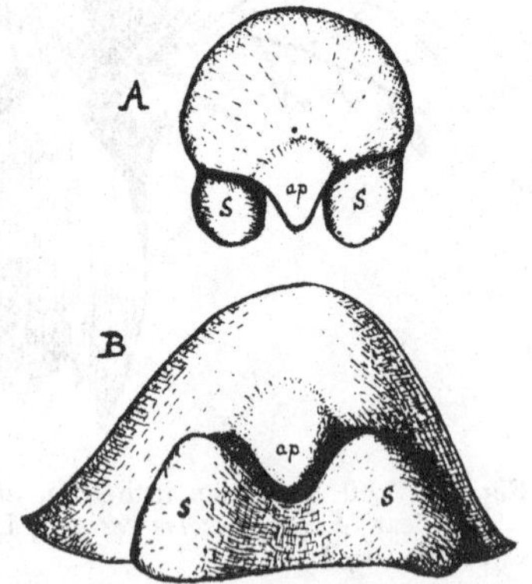

Fig. 96. Young leaves of *Angiopteris*, before the formation of the pinnae. *A*, seen from above: *B*, presenting the adaxial front. *ap* = apex of the phyllopodium: *s* = stipules. (× 10.)

The essential features in the leaf-architecture of the Ferns have been described in general terms in the preceding pages. But many points of importance for comparison have been left over to the detailed description of genera, which will follow in the Second Volume. The leaf-development of Ferns and Pteridosperms stands alone in its complexity and variety among the types seen in Vascular Plants. It is the prototype upon which, by various modifications, the foliar development of Seed-Plants is based. At the same time so many features that are primitive still remain in some of the living Ferns, or are shown in the related fossils, that it is possible to trace with

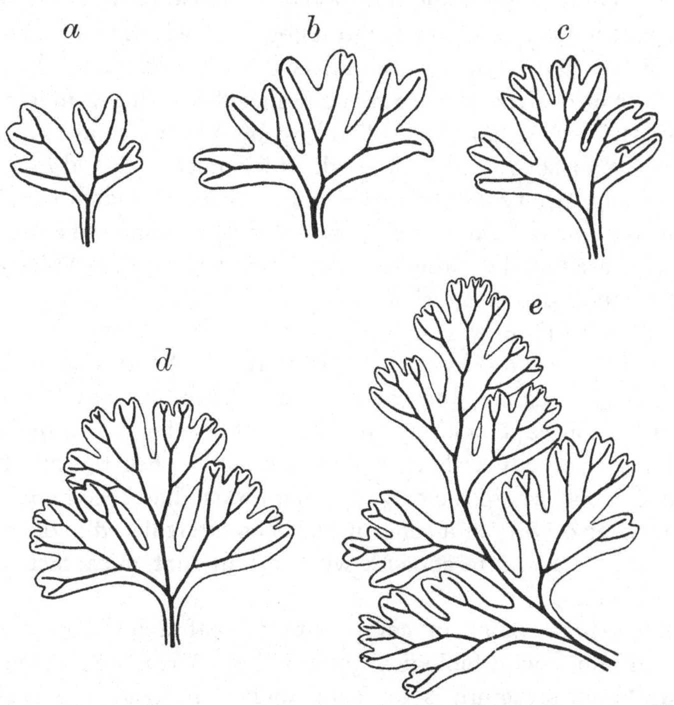

Fig. 97. Series of juvenile leaves of *Cyathea insignis*, illustrating progressive dichopodial development in a relatively primitive type. (× 4.)

some degree of certainty the way in which even the most complex leaf-structures have been built up. On the basis of the comparative analysis given above a primitive type of Fern-leaf may be sketched out as a stalked structure with or without basal protective growths ("stipules"), the distal end dichotomising as a rule in a single plane tangential to the axis which bears it. Narrow lobes are thus formed, each distinct from its neighbours, and each traversed by a single vascular strand. Living examples are seen in the adult leaves of many of the Hymenophyllaceae, or in *Todea superba*, or in the juvenile leaves of *Cyathea* (Fig. 97). Later types would show progressive degrees of webbing of the lobes to form broad expanses with dichotomous venation. But where the blade becomes elongated sympodial development, with or without webbing, makes its appearance. In large leaves the dichopodium becomes massive below, appearing as a direct continuation of the robust leaf-stalk, but fading out distally with transition of the branchings to the primitive equal dichotomy. In such cases the lower pinnae arise laterally from the phyllopodium by monopodial branching, as in the leaves of Flowering Plants. Varying distribution of the intercalary growth throughout the length of the phyllopodium, and differences in the time of its development as well as in the extent and duration of its apical growth and branching, give that great variety of character to its outlines which is seen in the leaf of Ferns. Where the segments are narrow the veins end free. This is sometimes the case also where the segments are webbed. But in more advanced forms than these the veins are apt to be joined by vascular loops, or commissures, so as to constitute a network which is coarser and less complete in relatively primitive types, but more finely reticulate in later and derivative types. The external characters of Fern-leaves when analysed along the lines thus indicated are found to provide material for comparisons which are useful as leading to their phyletic seriation.

One general result of the analysis which has gone before will have special interest in the later discussions. It is that the basal region where the "stipular" outgrowths are, and the distal end where the phyllopodium runs out into the primitive dichotomy, may be recognised as those regions of the leaf which are relatively archaic in their character. But the middle region, and especially that where the origin of the pinnae has become monopodial, is a part of later origin: in a sense it has been intercalated in descent. This conclusion will be found to coincide with certain anatomical facts which will be disclosed later.

The steps in advance in complexity of leaf-architecture thus briefly sketched harmonise with biological probability. A ready way of elaboration of a simple foliar structure is by distal dichotomy, which is illustrated in *Sphenophyllum* and in *Asterocalamites* (Fig. 98), as well as in the Filicales: and it appears that it was widespread among the leaves of early Vascular Plants. A lateral webbing together of the lobes thus produced is progressively illustrated both in the Sphenophylls and in the Calamarians, and it leads to a more effective expanse of photo-synthetic tissue (Fig. 99). But the modern Filicales show it in higher degree than they, in proportion to the larger size of their leaves. A mere fan-like expansion of dichotomy is less effective in producing a large photo-synthetic organ than would be gained by a longer form. This further step is secured by inequality of the forking, which leads to various types of dichopodium with or without lateral webbing. In the latter case a separate course of the veins has served well enough in relatively primitive types. But in those where a widening leaf-expanse has been formed occasional lateral fusions of the veins led to a connected network of coarser meshes. This provides for a more equal distribution of water over the enlarged area. Later types show in their smaller-meshed reticulum a still more efficient conducting system within the expanded lamina. Such reticulation usually appears in those specially widened leaf-expanses characteristic of life under forest shade: as in *Hypoderris*, and *Christensenia*. This type of venation first became prevalent in the Mesozoic Period, thus coinciding with the earliest records of broad-leaved trees. It is significant that these afford a more complete forest-canopy than the narrow-leaved vegetation of earlier

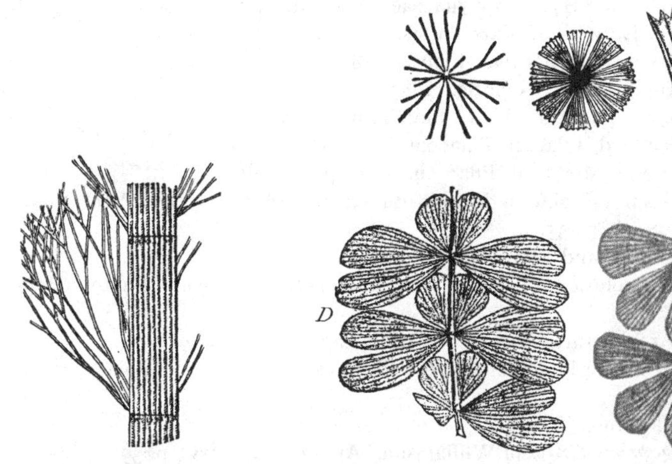

Fig. 98. *Asterocalamites scrobiculatus*, Schlotheim, from the Culm. Fragment of a leafy shoot, reduced to half its natural size. (After Stur, from Zeiller.)

Fig. 99. Types of leaf in the Sphenophylleae. *A*=a leaf-whorl of *Sphenophyllum cuneifolium*, and one leaf of it somewhat enlarged. *B*=a leaf-whorl of *Sphenophyllum tenerrimum*. *C*=*Sphenophyllum verticillatum* (from Potonié's Pflanzen-Palaeontologie). *D*=*Trizygia speciosa*, Royle, from the *Glossopteris*-facies of India. (After O. Feistmantel.)

epochs. Notwithstanding these advances, Modern Ferns show a peculiar conservatism of structure, retaining in many cases the features held as primitive in the sequence thus recognised. But on the other hand they show in their most advanced types such form and structure as can only be matched among the Higher Flowering Plants. Nevertheless that advanced structure can be referred, through the steps of reasonable comparison as above indicated, and with biological probability, to an origin from a simple, single-veined leaf. But this type was amplified by distal dichotomy, by webbing of the resulting lobes laterally, and by fusion of the veins to form a reticulum, associated as these features are with continued apical growth. It is this prevalent conservatism, combined with the various factors of advance, which gives its special interest to the comparative study of the leaves of the Ferns, and of the allied Pteridosperms. In fact these most prominent members of their vegetative system provide external characters of greater value for phyletic seriation than are as a rule yielded by the leaves of Vascular Plants higher in the scale. Furthermore, their leaf-architecture serves to illuminate that of Flowering Plants, which we may well believe to have sprung from some Fern-like, or Pteridosperm ancestry.

BIBLIOGRAPHY FOR CHAPTER V

65. HOFMEISTER. Higher Cryptogamia. Engl. Edn. 1862, p. 208.
66. METTENIUS. Farngattungen, IV. *Phegopteris* u. *Asplenium*. Frankfurt. 1858.
67. METTENIUS. Ueber die Hymenophyllaceae. Abhandl. K. Sächs. Ges. d. Wiss. 1864.
68. SADEBECK. U. d. Entwick. d. Farnblattes. Berlin. 1875.
69. PRANTL. Die Hymenophyllaceen. Leipzig. 1875.
70. PRANTL. Die Schizaeaceen. Leipzig. 1881.
71. BOWER. Comp. Morph. of the Leaf. Phil. Trans. 1885.
72. POTONIÉ. Lehrbuch d. Pflanzen-Palaeontologie. Berlin. 1899.
73. VELENOVSKY. Vergl. Morph. d. Pflanzen. 1905, p. 184, etc.
74. LIGNIER. Essai sur l'Évolution Morph. du Règne Végétale. Bull. Soc. Linn. de Normandie. 1808—9, p. 35.
75. BOWER. Origin of a Land Flora. 1908.
76. TANSLEY. The Evolution of the Filicinean Vasc. Syst. New Phytol. Reprint. Lecture I. 1808.
77. BOWER. Leaf-Architecture. Trans. Roy. Soc. Edinb. Vol. li, 1916.
78. BOWER. Studies in the Phylogeny of the Filicales. Ann. of Bot. Vols. xxiv—xxxii, Nos. i—vii.
79. SEWARD. Fossil Plants. Vol. iii.
80. SCOTT. On *Zygopteris Grayi* of Williamson. Ann. of Bot. xxvi, p. 39.
81. LUERSSEN. Farnkräuter. Rab. Krypt. Flora, iii, p. 11.
82. SADEBECK. Natürl. Pflanzenfam. i, 4, p. 55, etc.
83. CHRISTENSEN. Monograph of the Genus *Dryopteris*. Copenhagen. 1913—1920. Part II. Introduction.
84. GORDON. *Metaclepsydropsis duplex*. Trans. Roy. Soc. Edin. Vol. xlviii, Part I.
85. P. BERTRAND. Progressus, iv, 2. 1912.

CHAPTER VI

CELLULAR CONSTRUCTION

SINCE the growing points of stem, leaf and root are capable of continued growth and multiplication of cells, they are the ultimate source from which the tissues composing those parts arise. Following the analogy of animal embryology it might reasonably be expected that the characters of the meristems would be specially important for elucidating the origin of parts and the morphology of tissues. But it was long ago shown by De Bary (*Comp. Anat.* Engl. Edn. p. 23) that the analogy is misleading, and more recent observations have confirmed the view that apical segmentation is not a secure guide in the morphology of mature tissues. Nevertheless, though the facts may not bear that direct interpretation which was at one time expected from them, and though the apical cell or cells and their segments cannot be held as dominating the genesis of tissues, the comparison of meristems in the Filicales has another, and a distinct importance. It is the object of this Chapter to show how the facts of segmentation may lead to a seriation of Ferns, which can be used systematically. It already forms, in point of fact, the foundation for their phyletic grouping: for it is on segmentation that the generally accepted distinction of Leptosporangiate from Eusporangiate Ferns is based.

The Plant-body of the Pteridophytes is of cellular construction so developed that all the tissues originate by cell-lineage from superficial cells. There is no dermatogen present at the apical points, giving rise to a superficial skin covering the tissues of distinct origin within. At the growing tips superficial cells give rise by anticlinal and periclinal segmentations on the one hand to superficial cells, on the other to the deeper-lying tissues. Thus in not being stratified the apices of Pteridophytes differ from those of the Higher Flowering Plants, in which the apical regions are more or less stratified from the first. In this the former show a character that may be held as primitive.

Those superficial cells of the growing tips which are thus the ultimate parent cells of all the tissues are called the *initial cells*. They vary in number and in form. In the Filicales it is found that their number is small, frequently only a single cell. The form may be prismatic, or it may be conical with two, three, or four sides. Differences in number and form such as these may themselves be used for purposes of comparison and phyletic seriation, as will presently be shown. But before this point is taken up a just opinion as to the real nature and behaviour of these cells must be formed. In the first place, though in cell-lineage they are the ultimate parents of all the tissues, they do not actively dominate the apex. The growth in them is actually slower than at any other point in the near neighbourhood, a fact that is closely related to their very existence. Moreover in the actual scheme of construction initial cells possess negative rather than positive characters.

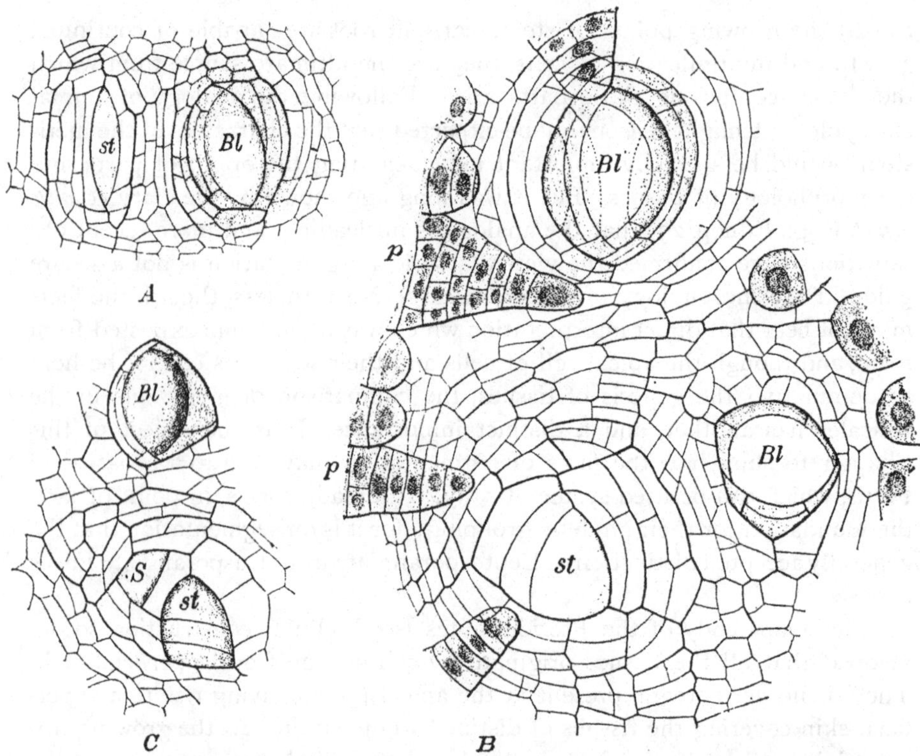

Fig. 100. Apical views of growing stem-apices of dorsiventral Ferns. *st*=initial cell of stem: *Bl*=young leaves: *S*=initial of a lateral shoot: *p*=ramenta. *A*=*Pteridium aquilinum* (L.), two-sided biconvex initial cell of stem, with one young leaf. *B*, *C*=*Polypodium vulgare* L. *B*=three-sided conical initial cell of stem with two young leaves, and the ends of several young ramenta. *C*=three-sided conical initial cell of stem with a lateral shoot (*S*), and a young leaf. (×280.) (After Klein, from Engler and Prantl.)

Sachs has shown how solid meristems conform to a scheme of construction built up of two systems of confocal parabolas the one periclinal the other anticlinal, together with walls also in radial planes. These all cut one another at right angles. He compared the very complete system of partitioning walls in Flowering Plants with those less complete systems in plants like the Pteridophyta, where apical cells are present. From this comparison he concluded that an apical cell "represents merely a break in the constructive system of the growing point: i.e. the apical cell is a spot in the embryonic tissue in which neither anticlines nor periclines nor radial longitudinal walls have yet been formed." So far then from the apical cell dominating the apex its characters are negative. What we shall see is that the degree of negativity of this cellular construction of the apex as expressed in the absence of certain walls provides a basis for comparative treatment, related as this is found to be to the bulk of construction of the organisms compared.

The most delicate and least bulky in construction of the Filicales are those which have been distinguished as the Leptosporangiate Ferns. In these it has been recognised since the time of Naegeli and Leitgeb(86) that a single initial cell is present at the apex of stem, leaf, and root. This is, however, by no means uniform for all the Filicales, as the Eusporangiate Ferns show. It is in point of fact an extreme condition. Nevertheless for clearness of exposition it will be taken first, as being the most definite type of structure of them all, as it is also the most generally known.

The axis of most Leptosporangiate Ferns is terminated by a slightly conical tip, at the distal centre of which is a single apical or initial cell. In most cases this cell has the shape of a three-sided pyramid, of which the base forms part of the outer surface of the plant, while its apex is directed inwards. From the three slightly convex sides segments are cut off in regular succession by walls parallel to them. Consequently the fourth segment will be opposite the first, the fifth opposite the second, and so on (Fig. 100). It is stated that in such a rhizome as that of *Polypodium vulgare* only a few such segments are formed in each year(75). The stem of the Bracken is exceptional in the fact that the initial cell may be only two-sided. It is shaped like half of a biconvex lens, and is set on edge in the creeping rhizome, with its segments coming off alternately right and left (Fig. 100, *A*). A vertical section in either case discloses the obconical form of the initial cell, and the relation of it to its segments is shown in Fig. 101, for *Trichomanes*, in which the cell has the usual three-sided form.

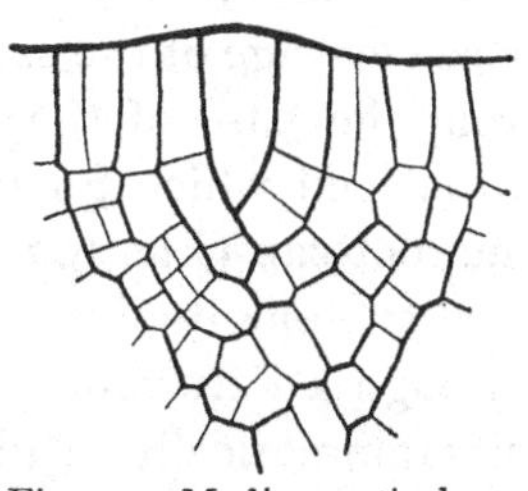

Fig. 101. Median vertical section of the apex of the stem of *Trichomanes radicans*. (× 100.)

The same shape of initial cell, but with the additional complication of segments cut off from its base to form the root-cap, is found in the roots of Leptosporangiate Ferns. A transverse section traversing the apical cell and its latest segments which go to form the body of the root, shows the scheme in ground plan (Fig. 102, *B*). The segments are arranged in three rows corresponding to the three sides of the apical cell, and are separated by zig-zag lines, which are called the *principal walls* (*p*, *p*). In transverse sections below the apical cell these principal walls can still be traced: but a notable additional feature is seen in the presence of *sextant walls* (*s*, *s*), one of which runs radially inwards through each segment, and finally curving before it reaches the centre inserts itself laterally on one of the

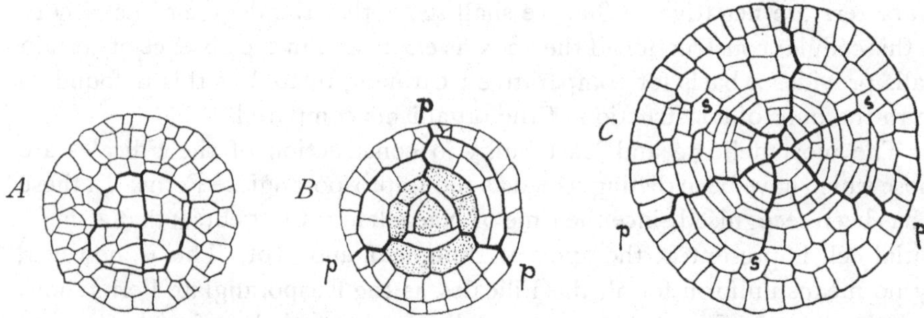

Fig. 102. *A—C*, basipetal series of transverse sections of the root-tip of *Dennstaedtia punctilobula*, after Conard (× 200). *A* traverses the root-cap showing centrally a recent segment divided into quarters and further subdivided; *B* shows the initial cell and three rows of segments separated by the principal walls *p*, *p*; *C* traverses the body of the root at a lower level and shows the principal walls *p*, *p* and sextant walls *s*, *s*. The initial cell and the second series of segments are dotted in *B* to bring them into prominence.

principal walls. Many other minor cleavages occur, but these principal and sextant walls are important for the argument that is to be developed later (Fig. 102, *C*). Lastly, a section transversely through the region of the root-cap, just above the apical cell, shows the method of cross-wise cleavage of the segments from the base of the conical initial cell which go to form the root-cap (Fig. 102, *A*). A median longitudinal section through a Fern-root is shown diagrammatically in Fig. 103, for comparison with these transverse sections, while it also illustrates the details of tissue-generation which are described in most of the larger text-books.

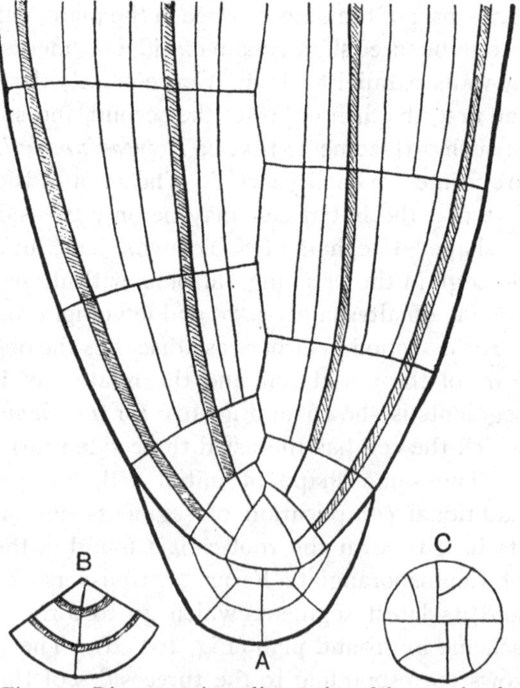

Fig. 103. Diagrammatic median section of the root-tip of a Fern, showing the origin of the various tissues in relation to the initial segmentation. *B* = subdivision of a single segment of the body of the root. *C* = subdivision of a single segment of the root-cap. (After Conard.)

In the Leptosporangiate Ferns each leaf arises from a single superficial cell of the axis, and in some cases a definite relation is suggested between certain segments of the initial cell and this leaf-primordium. Thus in *Polypodium vulgare* each dorsal segment is said to produce a leaf(75). But this is not of general application: moreover even here leaves are not formed from the ventral segments. At first a few rather irregular cleavages occur in the primordial leaf-cell. These lead to the establishment of a two-sided initial, with one of its edges directed to the centre of the apical cone, while the segments form two rows right and left (Fig. 100, *Bl*). The principal walls will here lie as two zig-zag lines approximately in the median plane of the organ. As the leaf thus produced elongates, it takes a flattened form, and usually branches. The initial cell maintains the same orientation, while the middle portion of each segment forms part of a series of marginal cells. These undergoing repeated segmentations by walls parallel to one another form a marked feature of the developing leaves of Ferns (Fig. 107). In branched leaves the segmentation at the leaf-apex does not determine the formation of the individual pinnae, for the segments and pinnae do not correspond. There is reason to believe that such degree of correspondence between segmentation and the formation of parts as is seen in Ferns is not obligatory. The two phenomena may coincide, but they appear to be causally independent of one another.

The rare opportunity of observing apical segmentation in a fossil has been afforded by Dr Kidston in a median longitudinal section of a leaf-apex of *Zygopteris corrugata* (Fig. 104). The drawing shows that a regular segmentation existed, possibly with a single initial cell. But it is impossible to be sure whether or not a plurality of initials may have existed in this case.

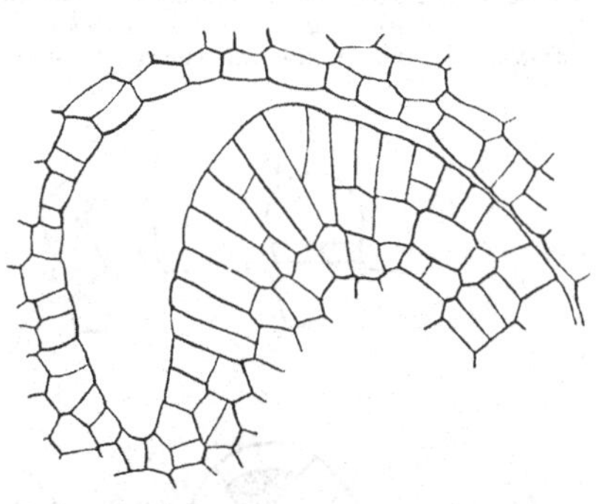

Fig. 104. *Zygopteris corrugata*, Kidston collection (1985). Halifax Hard Coal. Collected by Hemingway. Median section of apical meristem of a leaf. (× 333.)

The leaf of most Ferns is a winged structure, with a central rachis bearing lateral flaps or wings. The development of these is characteristically carried out in the Leptosporangiate Ferns by the very regular and repeated segmentation of the marginal cells above noted. Transverse sections of the growing lamina of *Scolopendrium* give a striking demonstration how the whole lateral wing is referable in origin to a series of such segments cut off alternately from the sides of each of the row of wedge-shaped marginal cells (Fig. 105, *D*).

The leading feature which characterises the Leptosporangiate as distinct from the Eusporangiate Ferns is that in them each sporangium springs from a single cell: this cell growing out from the surface-tissue undergoes regular cleavages, which in themselves present a certain analogy with those of an initial cell (Fig. 106). These will be described in detail later (Chapter XIII):

meanwhile the points of interest are the delicacy of the unicellular origin, and the regularity of the cleavages. It thus appears that in the Leptosporangiate Ferns all the points where tissues originate show a definite

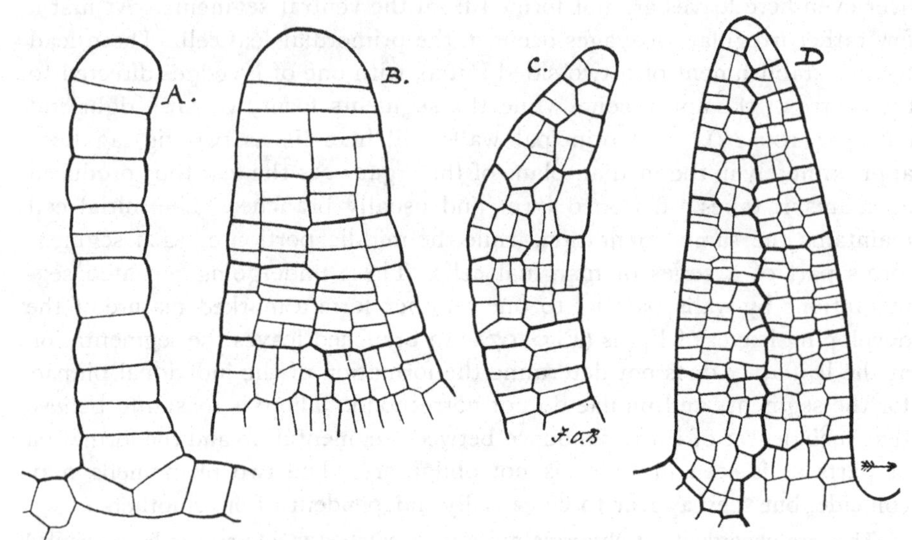

Fig. 105. Drawings of the actual segmentation at the margin of leaves of Leptosporangiate Ferns. *A* = *Trichomanes radicans*; *B* = *Trichomanes reniforme*; *C* = *Asplenium resectum*; *D* = *Scolopendrium vulgare*. The last is the most usual type. *A* and *B* are characteristic of Filmy Ferns.

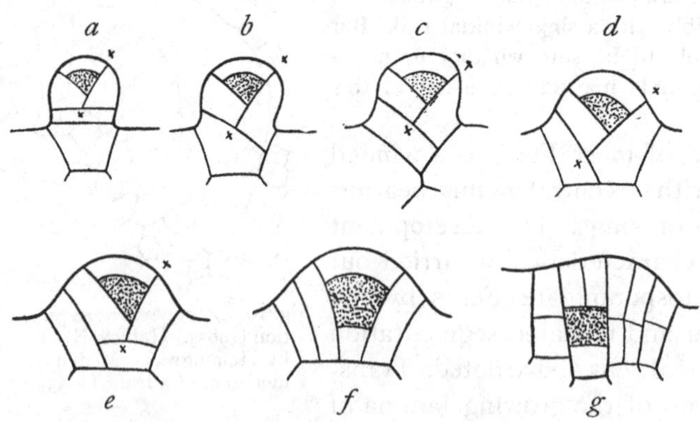

Fig. 106. Diagrams illustrating the segmentation of sporangia in various Ferns. *a* = Polypodiaceae (compare Kny, Wandtafeln XCIV). *b* = *Ceratopteris* (compare Kny, *Parkeriaceen*, Taf. XXV, Fig. 3). *c* = *Alsophila* (compare *Land Flora*, Fig. 334). *d* = *Schizaea* (compare Prantl, Taf. V, Fig. 69), or *Thyrsopteris* (*Land Flora*, Fig. 329), or *Trichomanes* (compare Prantl, Taf. V, Fig. 92). *e*, *f* = *Todea* (compare *Land Flora*, Fig. 295). *g* = *Angiopteris* (compare *Land Flora*, Fig. 284).

method of segmentation. In the apices of the stems, leaves, and roots a single initial cell, with a definite succession of segments cut off from it as it slowly grows, is a constant feature. Such segmentation originates in the first cleavages of the embryo, and it may be held as a regular character of these more delicate types of Ferns.

But there are other Ferns of a more robust type. They are designated Eusporangiate because their sporangia spring not from a single cell with simple cleavages, but from a group of cells with more complex cleavages (Fig. 106, *g*). Moreover this exemplifies their vegetative construction also, which is in all their parts more massive than that of the Leptosporangiates. Since the sporangium may be held as an index of the general construction, the question at once arises whether they differ also from the Leptosporangiate Ferns in the structure of their growing points. The Ferns in question are the Marattiaceae and Ophioglossaceae, while associated with them as intermediate in character are the Osmundaceae. The latter may be taken first. At once it is to be noted that the large sporangium cannot always be referred

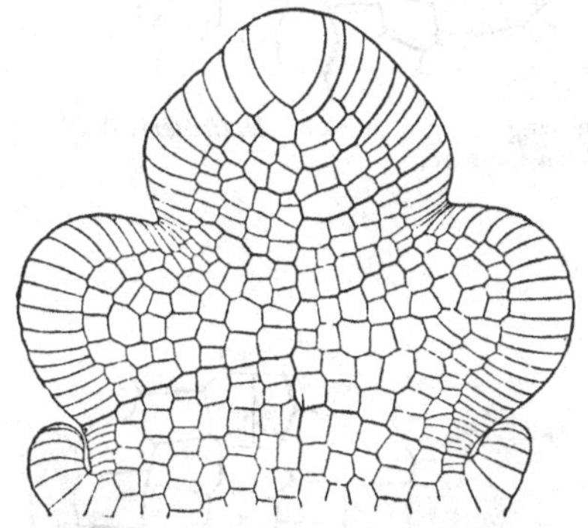

Fig. 107. Young leaf of *Ceratopteris*, in surface view, after Kny; showing two-sided apical cell, and the marginal series continuous round the young pinnae. The latter do not coincide individually with the segments from the apical cell.

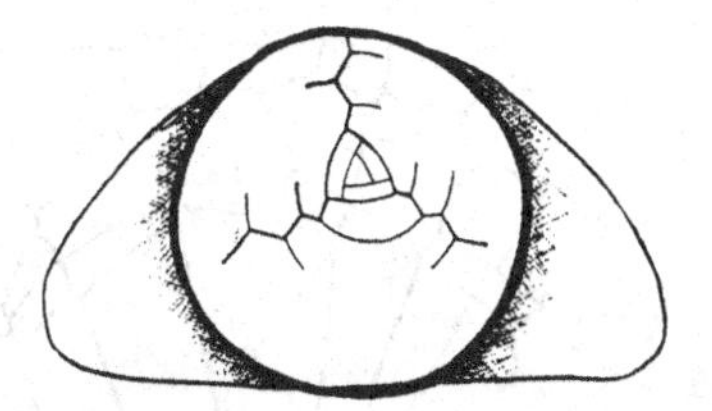

Fig. 108. Apex of leaf of *Osmunda*, showing the three-sided apical cell in its relation to the two latest pinnae. (×26.)

wholly to a single parent cell. It shows cleavages sometimes more nearly of the Leptosporangiate, sometimes of the Eusporangiate type (Fig. 106, *e*, *f*). As to the apical points, the massive axis shows a three-sided conical cell comparable to that of other Ferns (Fig. 10, p. 9). But the leaf both in *Osmunda* and *Todea* has a three-sided instead of a two-sided initial, so placed that one row of segments forms the adaxial side, the other two the abaxial flanks of the leaf (Fig. 108). This more complex segmentation of the Osmundaceous leaf accords with its more robust construction. It also is winged, though the wings are not produced from the middle region of each of two rows of marginal cells. Here two of the three rows of segments contribute to each of the wings. Transverse sections of the developing wing reveal no marginal series as in Leptosporangiate Ferns, but a small-celled meristem

(Fig. 109, *B*). The median line seen in *Todea* may very probably coincide with the limit of the two rows of segments which form the wing.

In the roots of the Osmundaceae, which are commonly more massive than those of the Leptosporangiate Ferns, a single conical initial cell is frequently found. But often there may be a plurality of them, two, three, or four: and their form is sometimes obconical, sometimes with transversely

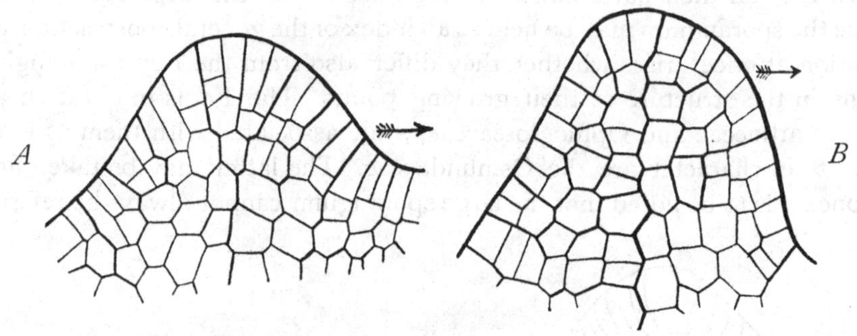

Fig. 109. Vertical sections through the massive wings of the leaves of *Angiopteris* (left) and of *Todea barbara* (right).

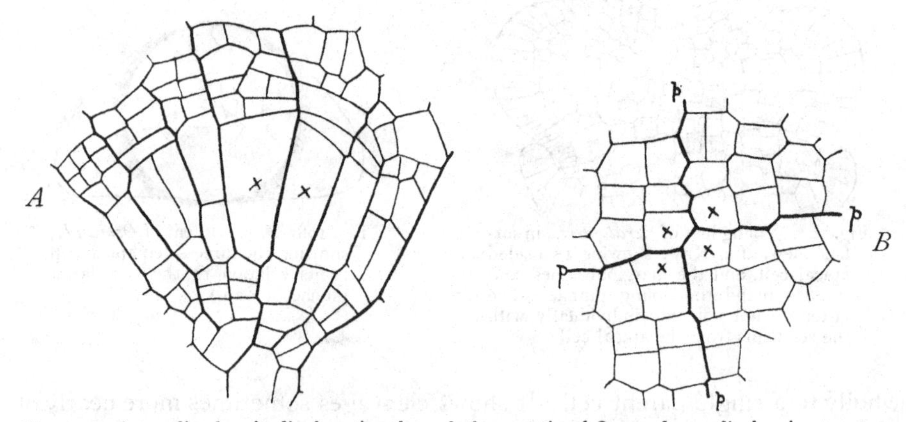

Fig. 110. *A* = median longitudinal section through the root-tip of *Osmunda regalis* showing two truncated initials (*X*), segments being cut off from both ends, as well as from the sides. (× 200.) (Compare Schwendener, *Sitz. k. Preuss. akad.* 1882, Pt. VII, Fig. 1 of *Angiopteris*.) *B* = transverse section through the root-tip of *Todea barbara*, showing four initials (*X*), and the principal walls (*p*, *p*). (× 200.)

truncated apices (Fig. 110, *A* and *B*). Such conditions, which usually show irregularities of segmentation, have been observed both in *Osmunda* and *Todea*. Perhaps the most enlightening of them all is the state seen in certain of their roots where there are three initials (Fig. 111, *A* and *B*). These are separated by walls readily recognised as corresponding to the principal walls of Fig. 102, while walls corresponding to the sextant walls (*s*) also traverse the apical group itself. It thus appears that the gap in the system of construction, which the single apical cell itself represents according to Sachs, is here less complete. The principal and sextant walls do not stop short in the segments already cut off from the initial, but are actually continued upwards so as to take part in defining the three initial cells themselves and their segments. The system of construction of the Osmundaceous root is seen to be more complete than in any of the ordinary Leptosporangiate Ferns. Thus the Osmundaceae, while conforming in some degree to the Leptosporangiate segmentation, show in several details a more complex condition than they.

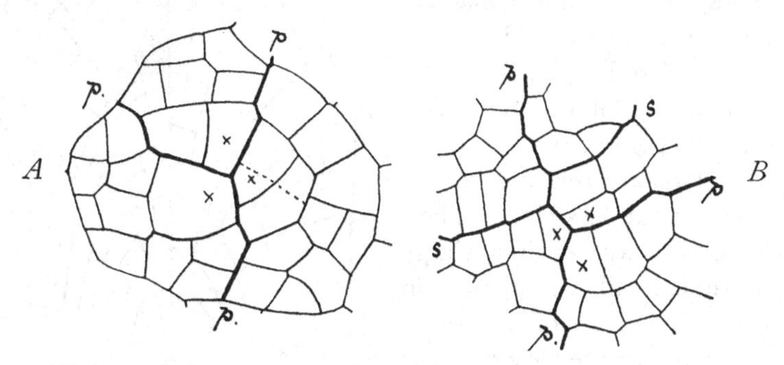

Fig. 111. *A*, *B* = transverse sections of the root-tip of *Osmunda regalis*, showing anomalous structure with three initials (x). Compare the position of these, the principal walls (p, p), and the sextant walls (s, s), with those seen in Fig. 102 of *D. punctilobula*.

Still more is this to be observed in the Marattiaceae. In well-grown stems of *Angiopteris* and *Marattia* a single initial cell has not been found. In its place there are arrangements which indicate three or four initials (Fig. 112). But stems of young seedlings of *Marattia alata* have been found to have a single irregular initial with early transition to more complex structure (Charles (105), Pl. XI). Longitudinal sections show the conical or prismatic initials to be very narrow and deep, suggesting a deeply sunk centre of construction of the system of curves. Similarly in their leaves, though a single initial may appear in those of the sporeling, in older leaves, as in the stem, a plurality of initials is found. In the development of the massive wings of these leathery leaves there is no definite marginal series, but as in the Osmundaceae a small-celled meristem (Fig. 109, *A*). The apices of

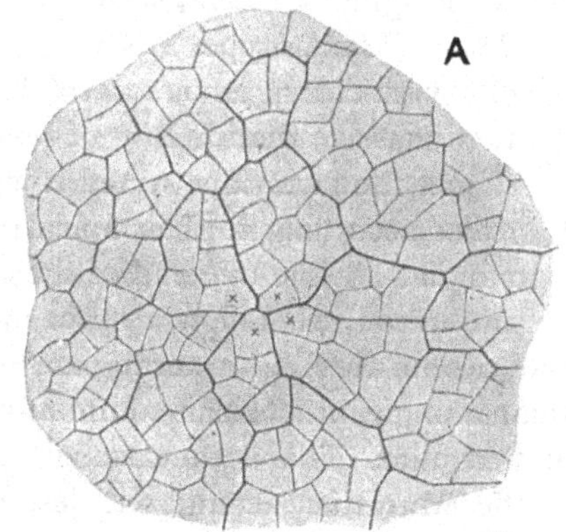

Fig. 112. *A* = apex of stem of *Angiopteris evecta*, seen from above. Apparently there are four initials (X, X). (× 83.)

the roots have been examined by Schwendener. The roots are thicker than those of any other living Ferns, and they contain as a rule four truncated prismatic initials, corresponding in number and form to those occasionally seen in roots of *Osmunda* and *Todea* (Fig. 110, *B*). Here again the centre of construction is deeply sunk. It is the sporangia, however, that give the most marked character to the Marattiaceae. These are much more massive than in the Osmundaceae. They arise by outgrowth of a large number of cells, the cleavages of which form a coaxial system (Fig. 113).

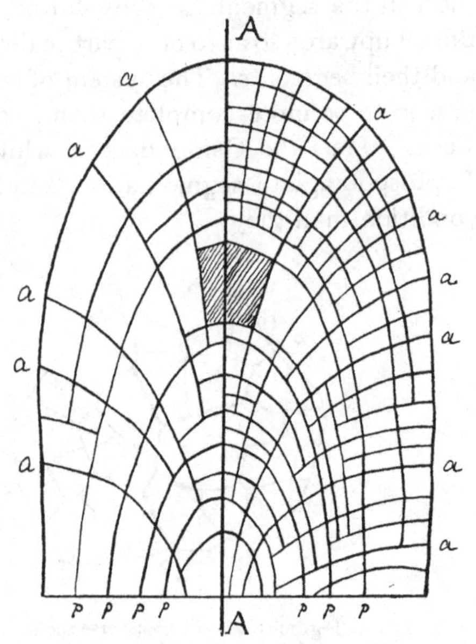

Fig. 113. Scheme of construction of the coaxial or Marattiaceous type of root. AA = common axis; a, a = anticlinal curves; p, p = periclinal curves. Compare Fig. 110, *A*.

The comparisons relating to the Ophioglossaceae are less clear. Their stems and roots have each a single initial cell, though all are massive in construction. The leaves of the young plant of *Botrychium* are described as having a single initial: but neither here nor in *Ophioglossum* or *Helminthostachys* is there any regular apical or marginal segmentation in the adult leaf comparable with that in Leptosporangiate Ferns. The development of the sporangium of *Botrychium* and *Helminthostachys* conforms generally to the Marattiaceous type. It is eu-sporangiate, springing from a numerous body of cells, with cleavages on a coaxial system. In *Ophioglossum* the sporangium is more massive than in any of the other Filicales. (See Chapter XIII.)

Upon the facts thus stated it is possible to seriate the main groups of the Filicales according to their apical meristems, and the segmentation of their sporangia. At the one end of the series will be those in which the cleavages of the initial cell follow a strict and relatively simple rule, as is seen to be the case in the Leptosporangiate Ferns. At the other end those where more than a single initial cell is commonly present, and the cleavages are less simple and regular. This is the case with the Eusporangiate Ferns, most typically in the Marattiaceae, and with less regularity in the Ophioglossaceae, while the Osmundaceae occupy an intermediate position. This series follows and indeed indicates the general constitution of the plants in question. For the Eusporangiate Ferns are relatively massive plants, and typically possess leathery leaves, fleshy axes, and thick roots: all their vegetative parts thus according with the robust type of their sessile sporangia. The Leptosporangiate Ferns with their more regular and simple segmentation are typically more delicate in the constitution of their vegetative organs, their leaves are thinner, and their roots more slender. This accords with their longer-stalked and smaller sporangia. The Osmundaceae take a middle position both in the robustness of their vegetative organs, and in the size of their sporangia.

It may seem to lie near to the hand to suggest that the character of the apical region is determined by the bulk of the organ, and has no other significance. That the relation between bulk and apical segmentation is not simple and direct will be obvious enough to those who have any wide knowledge of segmentation in Ferns. A comparison of sections of the lateral wings of Fern-leaves drawn to the same scale sufficiently proves that the manner of segmentation is not directly dependent on size. *Trichomanes reniforme* (*B*) is the most massive of the four Ferns represented in Fig. 105, and it shows a simple transverse segmentation: while *Asplenium resectum* (*C*), which has the more complicated alternate segmentation, is actually thinner at the margin than (*A*) or (*B*). Again, the apical meristem of the large Tree Fern, *Amphicosmia Walkerae*, with a stem several inches in diameter, shares its regular three-sided segmentation with that of the stems of the most insignificant of Leptosporangiate Ferns (*Ann. of Bot.* iii, Pl. XXI, Fig. 22). The question cannot be summarily disposed of in this way. That there is some relation between bulk and complexity of segmentation was pointed out thirty years ago, and the facts brought forward in this Chapter sufficiently demonstrate it for the main divisions of the Filicales. But it is not simple or direct in the individual. It makes its appearance in the race.

The differences of apical and sporangial segmentation in the Filicales show such a degree of consistency that they cannot be interpreted in any other way than as indicating an hereditary difference of constitution between those more massive and those less massive. The distinction between Eusporangiate and Leptosporangiate types is not merely based on the difference of their sporangia, but on differences of all their parts, of which their sporangia give a trustworthy indication. Such differences are apt to be retained by them, notwithstanding that Leptosporangiates may become massive in certain features, like the stems of Tree Ferns; or that Eusporangiates may become more or less "filmy," as in the Todeas and filmy Danaeas. The more or less massive type of constitution is thus an inherent and hereditary feature.

The next question will be whether the one or the other end of the series, thus indicated by constitution of all the parts, is to be held as the more primitive. The earlier writers taking their stand on comparison rather than on Palaeontological evidence, and seeing some superficial resemblance between the Hymenophyllaceae and the delicate structure of the Bryineae, suggested an affinity between them. This was the view of Linnaeus, Sprengel, Presl, and Bernhardi. Van den Bosch even went so far as to erect the

Hymenophyllaceae into an order, which he styled the Bryo-pterideae, and placed them between the Mosses and Ferns. Later writers at one time shared this view of a Bryophyte affinity with the Hymenophyllaceae, basing it upon the filamentous gametophyte. It was held that we are able even now to follow, at least in part, the phylogenetic development of the sexual generation from the Bryophyta to the Pteridophyta. But it is a palpable departure from the usual methods of Natural Classification to give decisive weight to vegetative characters while the propagative organs, and especially the spore-producing parts, are so divergent as between the Mosses and Liverworts and any Hymenophyllaceous Fern. It is now recognised that similarities in respect of the filmy character of the leaf, the filamentous prothallus, projecting sexual organs, and definite single initials, may all of them find a ready explanation as results of parallel hygrophytic adaptation in races phyletically quite distinct.

Those who advanced these comparisons did not take the Palaeontological evidence sufficiently into account. Already Campbell had argued comparatively in favour of the Eusporangiate Ferns being relatively primitive (96). But over and above the difficulty of comparison there loomed large the impossibility of harmonising a belief in the Leptosporangiate Ferns as primitive with the growing knowledge of the fossils. The dearth of evidence even of the existence of true Leptosporangiates comparable to those of the present day in Palaeozoic times was pointed out. At the same time the existence of numerous fossils then believed to be referable to a Marattiaceous affinity was held to indicate a priority of the Eusporangiate type (102). The comparative study of the development of the vegetative organs and of the sporangium had meanwhile been actively pursued. On the basis of such facts it came to be held as probable that the more delicate structure seen in the Leptosporangiate Ferns was not primitive, but that it resulted from progressive specialisation. The ground was thus open for recognising the Eusporangiate type, whether of Ferns or of other Pteridophytes, as of prior existence. Certain Pecopterids with Marattiaceous sori, the Botryopterideae, and certain forms allied to the most primitive of living Leptosporangiate Ferns, were recognised as the representative homosporous Filicales of the Palaeozoic Period. The early existence of homosporous Ferns, which evolutionary theory would suggest or even demand, is shown to be beyond reasonable doubt. But the ferns which existed in the Primary Rocks appear to have been mostly if not exclusively Eusporangiate.

If this be so then we shall contemplate for the Filicales a progression from a more complex to a simpler but more exact construction, such as is illustrated by comparison of living Eusporangiate with Leptosporangiate Ferns. That such a progression does not stand alone for Ferns is shown by the Lycopodiales, though it is less completely carried out in them. In the genus *Selaginella* the upright types, which on grounds of comparison are held to be relatively primitive, and live in exposed situations, have a relatively robust construction. *S. spinulosa* has from the first a stem-apex with a plurality of initials (*Land Flora*, Fig. 190). The dorsiventral forms are specialised for growth under forest shade. Of these *S. Wallichii* holds an intermediate position with two initials, but *S. Martensii* has only a single initial at the apex of the stem. A similar progressive simplification in relation to shade probably holds for the main series of the Filicales. The relatively primitive Eusporangiatae with their robust habit are characteristic of those early times when life for them was necessarily exposed. They shared their massive sporangia with the Eusporangiate Lycopods, Sphenophylls, Calamarians, Psilotales, and Psilophytales, all of which lived under conditions substantially alike. As vegetation progressed forest shade became more efficient. Shade-loving plants, such as the Selaginellas, the Ferns, and Mosses, became fined down in texture in relation to it. They developed a thinner, sometimes even a filmy structure: their leaf-area was enlarged, and in Ferns it became reticulate: the sporangia grew less robust: and with these changes there advanced that progressive simplification of structure which finds its reflection in the more delicate and exact segmentation of their apical regions, and of their sporangia. If this be the true story, then exactitude of segmentation and the presence of a single initial cell are not necessarily signs of a primitive state, as has been commonly assumed; but a derivative condition, secondarily acquired as a consequence of life under new circumstances where a delicate construction proves practically efficient.

There is reason to believe that when the sporophyte took possession of the land, it soon assumed a relatively robust structure fit to meet ordinary subaerial conditions. This was provided by a complex primary segmentation, perhaps ultimately derived from a transversely segmented filament, by crosswise division, thus giving four initials. In many cases this has persisted, and it is seen to-day in the sporogonia of many Liverworts, such as *Jungermannia*, and *Anthoceros*: in the embryos of Lycopods, of some Eusporangiate Ferns, and of Angiosperms. In certain of these the same construction is perpetuated in the growing tips, as in Eusporangiate Ferns and some Lycopods. But in others a simplification led to more exact segmentation at the apex with a single initial. That is seen to-day in the Leptosporangiate Ferns, the advanced Selaginellas, the stems and roots of the Equiseta, and the sporogonia of the Bryineae. Signs of it are also seen in the apices of some aquatic Angiosperms, for instance in the roots of *Eleocharis palustris*, as demonstrated by Schwendener.

This is not the place to discuss fully the problem of what it is that determines the singularly regular methods of segmentation so beautifully illustrated in the apices of the Filicales. For us the point is that it presents features that may be used in phyletic comparison. But it is worthy of note in this connection that there is no obligatory correspondence between segmentation and the genesis of parts, or of tissues. It is true that there may be such correspondence in special cases. For instance Kny has shown how in *Ceratopteris* one leaf is produced from each segment of the apical cell (90). Klein states that one leaf springs from each of the two dorsal segments in the rhizome of *Polypodium vulgare*, though none arises from the ventral segments (94). The subdivision of the segments of the root in many Ferns are found to correspond exactly to the limits of the stele. These facts appear to suggest habitual correspondence. But the suggestion is negatived by the fact that the lobes of the leaf of *Ceratopteris* do not correspond to the segments of the apical cell of the leaf. Further, in Leptosporangiate Ferns the wings of the leaf are formed from the middle region of each of the two rows of segments: in the Osmundaceae they each arise from the margins of only two of the three rows of segments. Lastly, in the sporangia of ordinary Leptosporangiate Ferns three lateral segments are cut off from the sporangial mother-cell: but in *Dipteris* and *Metaxya*, and in *Cheiropleuria*, as will be shown later, there are only two segments instead of the usual three. Yet the structure and orientation of the annulus remains the same as it is in the allied *Platycerium*, and indeed in most other Leptosporangiate Ferns, where there are three. These examples show that correspondence between segmentation and mature structure may exist, but that the two are not necessarily related. Segmentation would thus appear to be a phenomenon independent causally of the genesis of parts and tissues, though frequently the two may be related.

Returning in conclusion to the main subject of this Chapter, the final decision on the point of the priority of the Eusporangiate or of the Leptosporangiate Ferns in Descent will have to be based on the whole sum of knowledge respecting them. All their parts will have to be treated comparatively in testing its truth, and to this a large part of the present volume will be devoted. Pending the completion of that study the view is here adopted as a provisional hypothesis, based both on comparison and upon the palaeontological record, and harmonising with the facts of cellular construction of the plant-body as described in this Chapter, viz. that the Eusporangiate Ferns are relatively primitive, and appeared first: and that the Leptosporangiate Ferns are derivative and specialised forms, and essentially shade-loving: and that they appeared later in the course of Descent. While many of the former have survived to the present day, the Leptosporangiatae are the typical Ferns of modern times. The provisional hypothesis in fact implies that there has probably been a progression from a more robust type of cellular construction as seen in the Eusporangiate Ferns to a less robust type as seen in the Leptosporangiatae.

BIBLIOGRAPHY FOR CHAPTER VI

86. NAEGELI. Zeitschrift f. Wiss. Bot. II. 1845.
87. SACHS. U. d. Anordnung d. Zellen. Würzburg. 1877.
88. SACHS. Ueber Zell-anordnung. Arb. Bot. Inst. Würzburg. Vol. ii.
89. SACHS. Lectures on the Physiology of Plants. No. xxvii.
90. KNY. Die Entw. d. Parkeriaceen. Dresden. 1875.
91. PRANTL. Unters. z. Morph. d. Gefässkryptogamen. Hymenophyllaceen. Leipzig. 1875.
92. SCHWENDENER. Sitz. d. K. Preuss. Akad. d. Wiss. 1882.
93. DE BARY. Comparative Anatomy. Engl. Ed. 1884. Introduction.
94. KLEIN. Bot. Zeit. 1884, p. 577.
95. GOEBEL. Ann. Jard. Bot. d. Buitenzorg. Vol. vii.
96. CAMPBELL. Bot. Gaz. Jan. 1890, p. 4.
97. CAMPBELL. Mosses and Ferns. 2nd Ed. 1905. Where the literature is fully cited.
98. CAMPBELL. Eusporangiate Ferns. 1911.
99. BOWER. On the apex of the root in *Osmunda* and *Todea*. Quart. Journ. Micr. Sci. Vol. xxv, p. 75.
100. BOWER. Comparative Morphology of the Leaf. Phil. Trans. Part II. 1884.
101. BOWER. Comparative examination of the Meristems of Ferns as a Phylogenetic Study. Ann. of Bot. iii, p. 305.
102. BOWER. Is the Eusporangiate or the Leptosporangiate the more primitive type for Ferns? Ann. of Bot. v, p. 109. Here reference is made to the literature up to its date.
103. BOWER. The origin of a Land Flora. 1908.
104. CONARD. Structure and Life-History of the Hay-Scented Fern. Carnegie Inst. Washington. 1908.
105. CHARLES. Bot. Gaz. Vol. li (1911), p. 94.
106. VON GOEBEL. Vergleichende Entwickelungsgeschichte. Schenks Handbuch. iii, 1884.
107. J. E. CRIBBS. Early stem-anatomy of *Todea barbara*. Bot. Gaz. Oct. 1920.

CHAPTER VII

THE VASCULAR SYSTEM OF THE AXIS

(A) THE PROTOSTELE AND ITS IMMEDIATE DERIVATIVES

THE Vascular System provides the most constant structural characters available for comparison in Land-living Plants. As vascular tissue is frequently well preserved in the fossils it gives a basis for comparison between ancient and modern Pteridophytes. But in dealing with anatomical facts it must be remembered always that in any progressive evolution vascular structure follows, it does not dictate, external form. All the evidence which it yields is necessarily *ex post facto* evidence. On the other hand, the structural effect of a certain development may remain after the formal characters with which it was primarily bound up have been altered, or even wholly removed. Anatomical characters are apt tardily to follow evolutionary progress, and thereafter to persist as vestigia. In them we see illustrated what may be described as phyletic inertia. This gives them a special value for comparison.

The Filicales show a very wide latitude of vascular structure. In no Class of living plants is there so great a complication of the primary vascular tracts. With one or two insignificant exceptions, the vascular system of all Ferns is primary. Secondary thickening is a rare occurrence in them. The physiological problem of their progress to large size has then been to provide, through a primary development, a conducting system sufficient for an enlarging plant-body. Little wonder then that with means so restricted the Ferns have never achieved the largest dimensions.

The scientific basis for a comparative treatment of the vascular system of such a Class should be founded on (i) the facts of actual structure and distribution of the vascular tissues in the adult of the various types ; (ii) the facts derived from the related fossils, presumably also in the adult state ; (iii) the facts derived from the ontogeny of living and possibly also of fossil types. Historically these three avenues of enquiry have been taken up in the order above stated, and frequent misunderstandings have arisen as a consequence of the long delay in developing the last of them. The best method will be to carry along all of these lines of observation and comparison simultaneously. In particular, the study of the ontogeny should always take quite as prominent a place as either of the other two. It is only when these three sources of knowledge have been combined, aided it may be by actual experiment by means of starvation-cultures which may reverse the steps of ontogenetic advance, that a satisfactory theoretical position can be attained. But even then other features must be taken into account also. The conclusions derived from the study of the stele must be interpreted in harmony with the results of still wider study. The biological relations of the vascular system to the tissues which surround it must also be considered, and worked into a conception of the general physiology of the organism: for anatomy cannot stand by itself as a branch of study apart from others. Before any conclusion from mere structural comparison is finally adopted it should be checked and tested according to the whole body of related knowledge. This principle has not always been sufficiently realised or followed by those who have placed evolutionary interpretations on the facts derived from anatomical study.

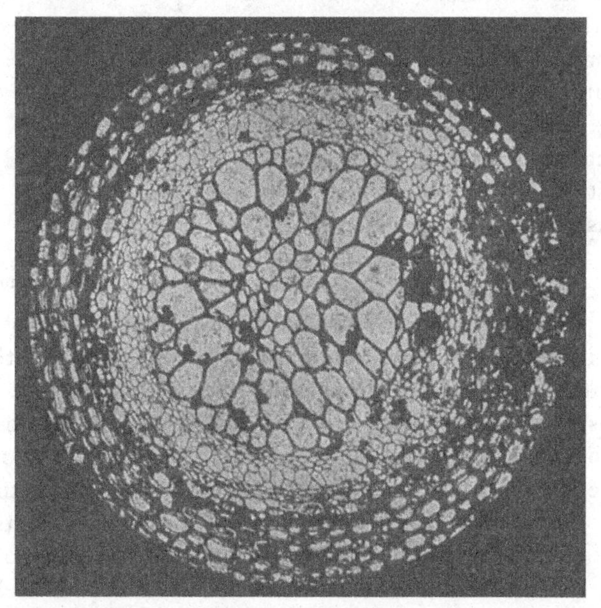

Fig. 114. Transverse section of a fossil Fern, *Botryopteris cylindrica*, showing a protostele with a solid core of xylem, and peripheral phloem.

THE PROTOSTELE

By a general consensus of opinion the non-medullated protostele has been recognised as the primitive stelar type. It is actually represented in all the phyla of primitive Vascular Plants, whether megaphyllous or microphyllous. In the Filicales it is seen generally in the sporeling, and it is permanently maintained in the adult stems of certain Ferns of relatively primitive character. The non-medullated protostele consists of a central core of xylem, often composed only of tracheides, as in *Botryopteris cylindrica* (Fig. 114); but thin-walled parenchyma-cells may be associated with them, especially where the protostele is large, as in *Lygodium*, *Gleichenia*, or *Cheiropleuria* (Fig. 115). The xylem is surrounded by a band of phloem, followed by a single or a multiple layer of cells of the pericycle, and finally the stele is delimited externally by the continuous sheath of the endodermis. The protostele is without intercellular spaces, and the endodermis, being a continuous sheath of cells, serves as an efficient barrier limiting the ventilating system of the cortex internally, and controlling by its living protoplasts all transit of materials into or out of the stele. The parenchymatous cells within this sheath are sometimes spoken of collectively as conjunctive parenchyma. They permeate the phloem, and often also the xylem-core itself. Thus constructed the protostele receives the leaf-trace from each successive leaf which, passing through the cortex, inserts itself upon and fuses with the stele of the stem. The several tissues of the one come into intimate continuity with those of the other. It is important to note that the leaf-trace in typically protostelic Ferns is itself surrounded by endodermis, and its entry is effected without any break of continuity of the enveloping endodermis, and without any marked disturbance of the core of xylem. The root-traces also connect directly with the protostele; but they arise in less definite order and cause even less disturbance of the stelar structure than the leaf-traces do. Thus constructed the simple vascular system serves the whole plant, and for small plants such as young sporelings the non-medullated protostele appears to be suitable and efficient.

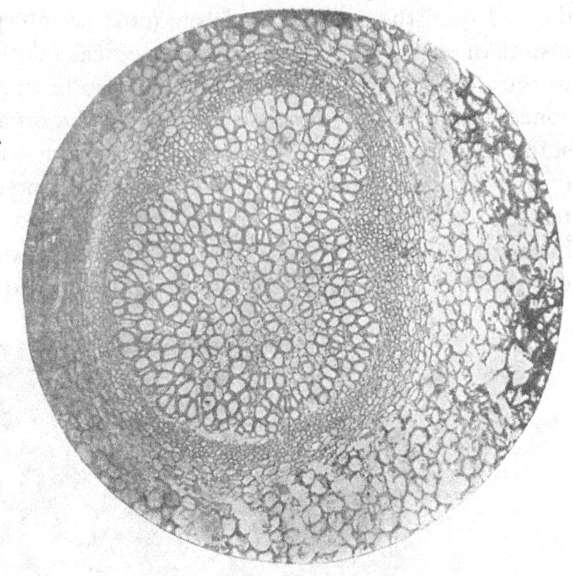

Fig. 115. Transverse section of the protostele of *Cheiropleuria*, showing a leaf-trace being given off from it. (× 45.)

In certain Ferns the protostelic state is maintained throughout the life of the adult. This is so in *Botryopteris cylindrica* (Fig. 114) and other ancient fossils. It is also found in the living genera *Gleichenia*, *Lygodium*, and *Cheiropleuria*, as well as in the whole family of the Hymenophyllaceae (Fig. 116). This is, however, uncommon for the adult stems of Ferns. As a rule in them the stele expands into various complicated forms. Some explanation of the permanent retention of the protostele in such cases as those named may be found in the fact that, for instance, in *Gleichenia* and

Lygodium there is a creeping rhizome, with straggling and climbing leaves of great length. Their leaf-stalks are thin, with a single contracted vascular strand. Thus the stele is not liable to be influenced towards expansion by a broad insertion of the leaf-trace. The case of *Cheiropleuria* is less easily understood, for in certain features it is a more advanced type than these relatively primitive Ferns (Fig. 115). Structurally it is very like *Gleichenia*; and its creeping rhizome and attenuated leaf-stalks may also be held as favouring the retention of the primitive state. These genera may all be regarded as perpetuating in the adult shoot the protostelic structure of the young plant. But they all share the peculiarity of having plentiful parenchyma scattered among the tracheides of the bulky cylinder of wood. This is probably a derivative state, for in *Botryopteris* and other very primitive Vascular Plants the xylem consists only of tracheides.

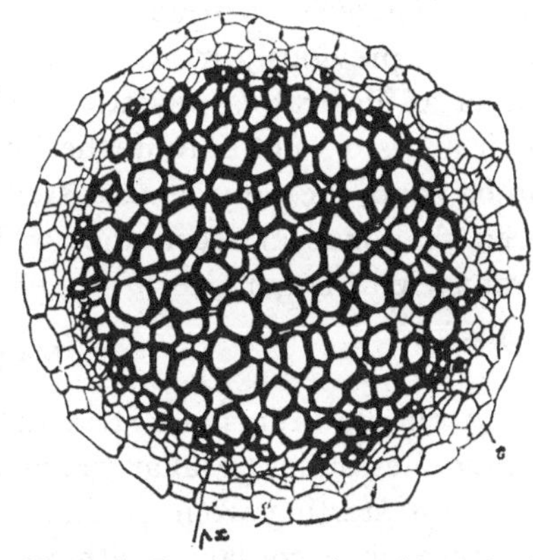

Fig. 116. Transverse section of the stele of *Trichomanes scandens*. *px* = protoxylem; *s* = endodermis. (After Boodle.)

The protostelic state with differences of detail holds for the Hymenophyllaceae. In these Ferns with their filmy, hygrophilous habit, small size, and delicate leaf-stalks, the protostele of the axis is as a rule minute (Fig. 116). But in the larger species it often shows an internal differentiation very like that seen in the Palaeozoic fossil, *Ankyropteris* (Fig. 117), and other Zygopterideae. Here there is a central region of the xylem, associated with parenchymatous cells, which is surrounded by an outer xylem of tracheides only. The similarity between the ancient fossil and the modern Filmy Ferns, in this as well as in

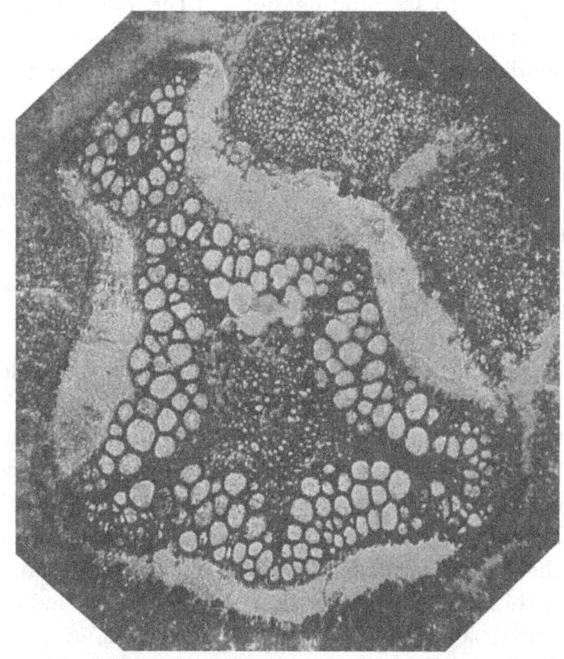
Fig. 117. *Ankyropteris Grayi*; stele, from a section in Dr Kidston's collection. (× 18.) (After Seward.)

other respects, makes it appear probable that the latter have retained the primitive protostelic structure in relation to their delicate hygrophilous habit. There is no need in such cases to assume any reduction from an ancestry previously more elaborate in anatomical structure. They appear rather as illustrating phyletic inertia—retention of the primitive structure of their ancestors not only in the young, but also in the adult state. On the other hand, the very minute stems of *Azolla* and *Salvinia* may well have been simplified in accordance with their aquatic habit from some original protostele of larger size.

MEDULLATION

A simple protostelic structure may suffice for small organisms, or for larger ones under peculiar circumstances of growth; but the great majority of Ferns starting from protostely show in their stelar structure phases of advance, in accordance with the increasing demands made by the more numerous and larger leaves of the growing plant. The same general problem associated with increasing size will have to be met in the ontogeny of any enlarging Fern-Plant. But there is no reason for assuming that the solution need always have been worked out in the same way. The observer ought to be prepared to see modifications of the stele in any of the several phyla, which while tending to the solution of the physiological problem, may differ in the details of execution. The most direct solution is clearly an increase in size of the protostele, and more particularly of its xylem-core. This habitually brings with it an internal differentiation of the stele, and frequently the first step is the appearance of parenchymatous cells, either scattered through the xylem, or constituting a definite *medulla or pith* in its centre. Evidence of the origin of a pith may be traced in the individual ontogeny. It is further illustrated by comparison of adult living forms, and by study of the stratigraphical sequence of fossils.

In many small stems, like those of young plants of *Anemia* or *Schizaea*, there is at the very base a small stele with a solid core of tracheides without parenchyma (Fig. 118, i, ii). Sooner or later as the upward series of transverse sections is passed in review (and the point may vary in individuals of the same species) a mass of parenchymatous cells is found to occupy the centre of the xylem. Often it is completely enclosed by a ring of tracheides between the points where the leaf-traces are given off: the condition may then be described as *soleno-xylic* (Fig. 118, iii, vi). Where the xylem-ring is thin a sector of this ring becomes detached at the departure of each leaf-trace; as the sector passes outwards, a parenchymatous connection is established between the pericycle and the central mass of parenchyma (Fig. 118, iv, v). This central parenchyma may in the ordinary sense of the term be called a "pith," or "medulla." In such cases as *Anemia* and *Schizaea*, as

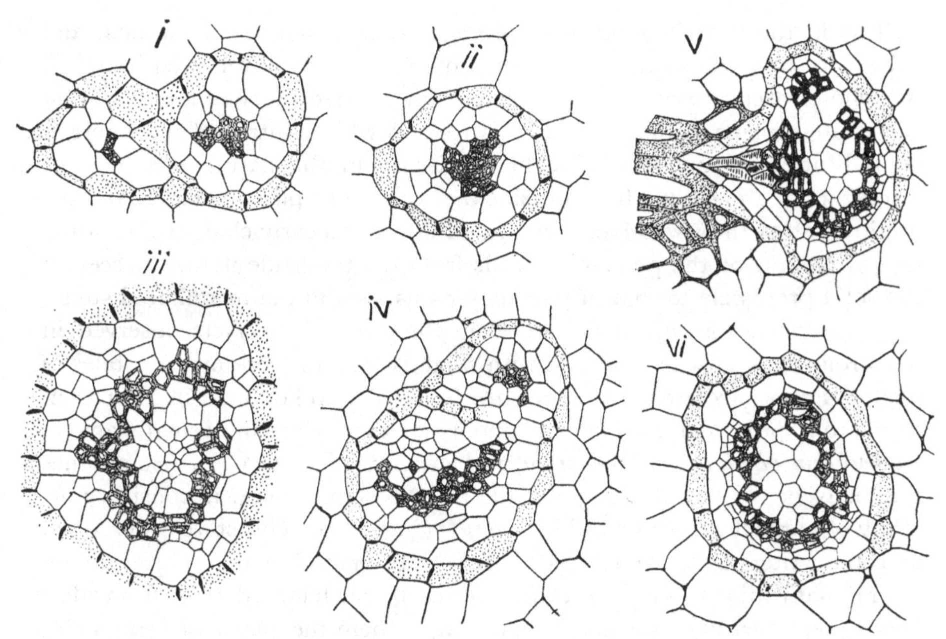

Fig. 118. i–iv. Transverse sections, in succession from below, of the axis of a young plant of *Anemia phyllitidis*. i, protostele gives off leaf-trace without any interruption of the endodermis; ii, protostele higher up; iii, still higher, with central pith; iv, same stele giving off a leaf-trace without interruption of the endodermis; note isolated tracheide in the pith.

v, vi. Transverse sections of adult stele of *Schizaea rupestris*. vi, shows continuous xylem round central pith; v, shows departure of a leaf-trace: to the left a root-trace. Endodermis is dotted. (× 150.)

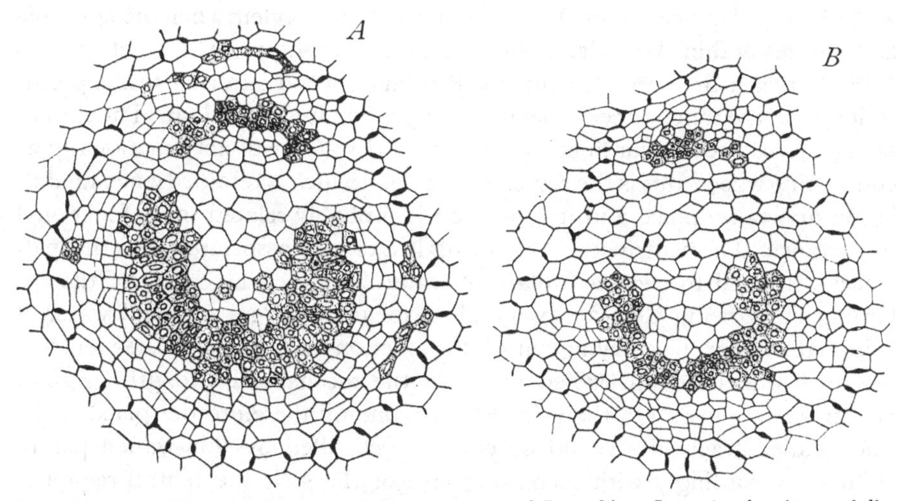

Fig. 119. *A*, *B*. Transverse sections of the young stem of *Botrychium Lunaria*, showing medullation, the departure of the leaf-trace, and the first steps of cambial activity. Note isolated tracheides in the pith of *A*, and the complete endodermal investment of *B*. (× 125.)

well as in the sporelings of *Botrychium*, *Helminthostachys*, *Osmunda*, and *Gleichenia pectinata*, in all of which the ontogeny has been followed in detail, the whole vascular system is completely surrounded by endodermis during the establishment of the pith. The process is wholly intra-stelar. There is no indication of any readjustment of tissues in the nature of a flow or intrusion of parenchyma from outside the stele. The pith appears to owe its origin in all of these relatively primitive Ferns to an early change of destination of certain of the procambial cells from the tracheide-nature as seen in the solid protostele to that of parenchyma as seen in the medullated stele.

This conclusion, which follows from the ontogenetic facts observed in numerous examples, is in accord with what is seen in more mature steles of related Ferns. Isolated tracheides are frequently to be found in the parenchymatous pith near the inner boundary of the xylem in normal stems of *Botrychium virginianum*, *ternatum*, and *Lunaria* (Fig. 119, *A*). Sometimes they may be scattered throughout the pith, in stems where it consists normally of parenchyma only, as in *Osmunda* (Fig. 120). This state is described by some writers as a "mixed pith." It has been found to follow occasionally on traumatic injury, as in *Botrychium ternatum*. Such facts derived from adult stems point clearly to the conclusion that in them the pith is of intra-stelar origin, and that it is a consequence of a change of procambial destination of the central tract of the xylem from development as tracheides to development as parenchyma. The isolated tracheides are then held to be residual cells in which the change has not been perfectly carried out, or has been for some reason reversed. The physiological probability of this will appear from the following considerations. In the larger types of medullated protostele, such as the Ophioglossaceae and Osmundaceae, the protoxylem when recognisable as such lies within the xylem, but near to its periphery. A distinction thus arises between the outer or centrifugal primary wood external to the protoxylem, and the inner or centripetal primary wood lying within it. The former being more directly continuous with the leaf-traces will act functionally as conducting wood: this is the region of the wood which is retained when medullation first takes place, and its retention is probably related to its functional importance. The centripetal wood being less closely associated with the leaf-traces will tend to become a place of water-storage. The centre of the enlarging stele, being thus a storage-place for more or less stagnant water, it is immaterial for the carrying out of this function whether the tissue-elements are thick-walled or thin-walled, provided the mechanical strength be sufficiently maintained. On the principle of economy therefore the place of the thick-walled tracheides would be equally well filled by thin-walled parenchyma. Accordingly with increasing size of the stele the central region of the wood is liable to be converted into parenchymatous pith. Whether or not this is the true physiological reason, it accords with the structural facts leading up to the soleno-xylic state, a condition frequent among primitive Ferns (Ophioglossaceae, Osmundaceae, Schizaeaceae).

Such ontogenetic and comparative evidence of the intra-stelar origin of the pith derived from living Ferns receives full support from comparison with early fossils. The Botryopterideae and Zygopterideae, families of the Carboniferous Period, illustrate the

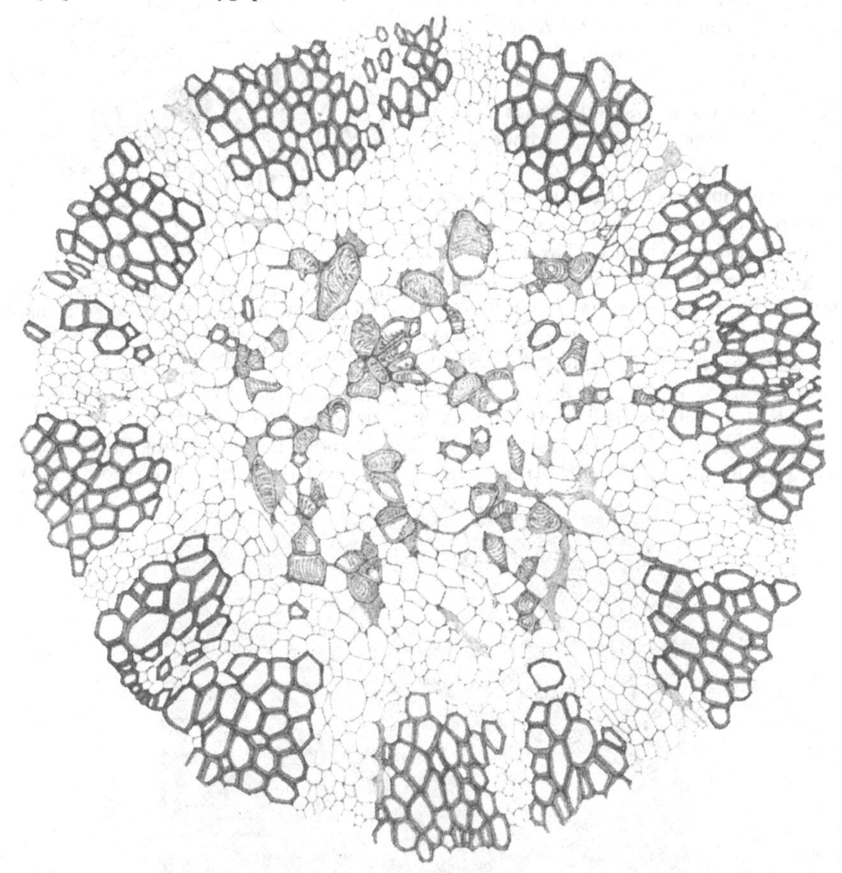

Fig. 120. Transverse section of a stem of *Osmunda regalis*, showing "mixed pith," and the greater part of the xylem-ring. (× about 75.) (G.-V. Coll. slide 1912). (After Gwynne-Vaughan.) The presence of tracheides scattered through the pith appears to have followed on injury. It was held by Gwynne-Vaughan to support the theory that the pith of the Osmundaceae is phylogenetically stelar, not cortical. Compare similar traumatic change in the pith of *Botrychium ternatum* (Bower, *Ann. of Bot.* xxv, p. 544).

initial steps from the solid xylem-core of *Botryopteris forensis*, or of *Tubicaulis* (Fig. 121), leading to the condition seen in *Diplolabis Römeri*, where the xylem is composed exclusively of tracheides, but is differentiated into an inner zone of short reticulated tracheides, and an outer zone of long pitted tracheides. In *Metaclepsydropsis duplex* the inner zone includes also parenchyma, which, with the inner tracheides, forms a "mixed pith" (Fig. 122). This internal differentiation is entirely intra-stelar, and may be held as providing two successive steps towards the establishment of a medulla. It is significant that sporangia having characters resembling those of the Osmundaceae are frequent in the Carboniferous Rocks. Whether the relation of these early Ferns to the Osmundaceae be a close one or not, the stelar structure of *Diplolabis* corresponds to that to be shortly described in *Thamnopteris*, a member of the Osmundaceous Series, while *Metaclepsydropsis* corresponds to what is seen in the more recent *Osmundites Kolbei*. Thus initial steps in medullation in Ferns occurred in the Carboniferous Period.

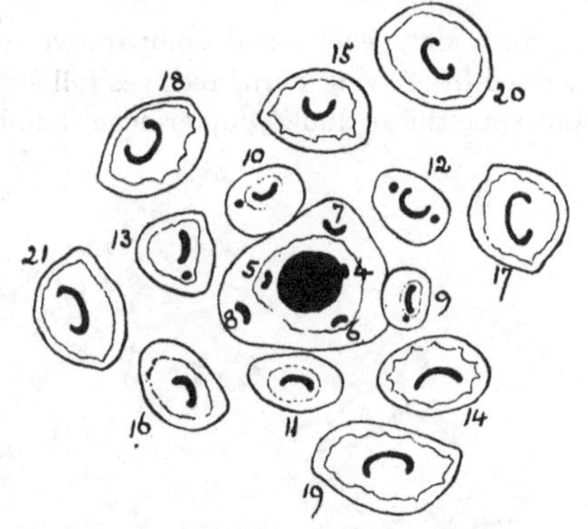

Fig. 121. Transverse section of shoot of *Tubicaulis solenites* Cotta, after Stenzel, from the Permian, showing stem with protostele, and the last four leaf-traces traversing the cortex. The leaves are numbered in their succession. The drawing is simplified by omission of roots, etc.

It is from the Osmundaceous Series of Ferns that the most cogent evidence comes of the origin of medullation. The recognition of the Ferns in question, including the fossils, as a phyletic unity is based primarily upon the structure of the leaf-trace, which is characteristic for them but not for other types of Ferns. The correctness of this recognition has been doubted: but those who object should produce evidence of the existence of Ferns having the Osmundaceous leaf-trace which are not of Osmundaceous affinity. Till this is done the objection cannot be held as valid. On this basis we are justified in recognising the Osmundaceous type by means of its leaf-trace structure as far back as the Palaeozoic Period.

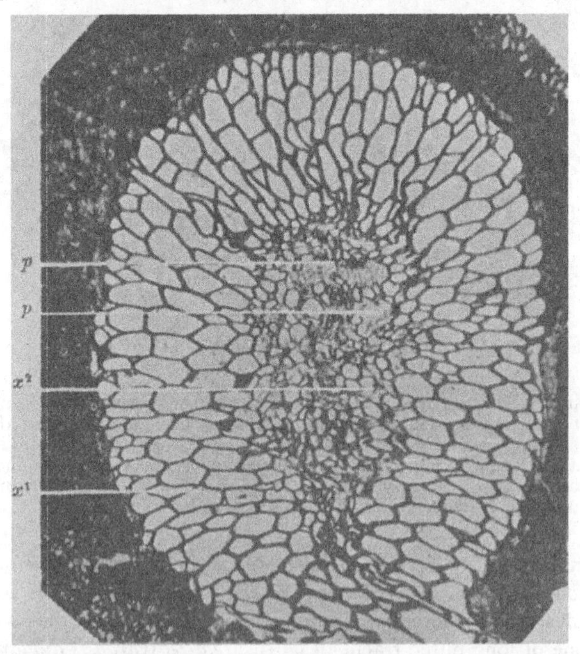

Fig. 122. Transverse section of stele of *Metaclepsydropsis duplex*, to show inner tracheides and parenchyma. x^1 = outer xylem; x^2 = inner xylem; p = conjunctive parenchyma: slide 1109. (× 36.) Lower Carboniferous. (After Gordon.)

Unequivocal evidence of their existence in this Era is supplied by the Russian Upper Permian genera *Zalesskia*, and *Thamnopteris* (Seward, *Fossil Plants*, ii, p. 325). Comparison of their stem-structure with the ontogeny of the modern Osmundaceae offers corroborative evidence of the intra-stelar origin of pith. Five steps in the elaboration of the protostele have been recognised in these Ferns, and they appeared roughly in stratigraphical sequence. They are: (i) a solid xylem-core; (ii) a heterogeneous xylem, but without pith; (iii) a central pith surrounded by a xylem-ring; (iv) the same, but with the ring interrupted; and (v) an interrupted xylem-ring lined within by accessory phloem and endodermis. Of these the first three are involved in medullation: the others will be considered later in this Chapter.

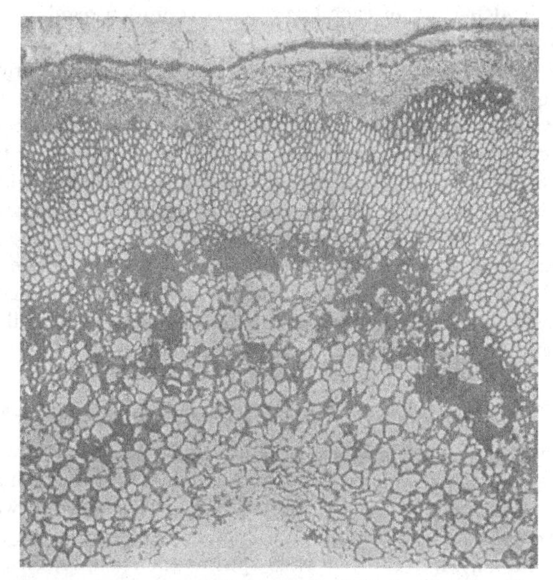

Fig. 123. *Thamnopteris Schlechtendalii*, Eich. Part of a stele. *a* = outer xylem; *b* = inner xylem. Permian. (After Kidston and Gwynne-Vaughan. From Seward.) (× 13.)

(i) Kidston and Gwynne-Vaughan at the conclusion of their comparative Studies wrote as follows (*Fossil Osmundaceae*, iv, p. 466): "We regard the Osmundaceae, as a whole, as an ascending series of forms whose vascular system is to be derived from a primitive protostele with homogeneous xylem." Though this is the condition of every sporeling, no adult example of it was known in any Fern with an Osmundaceous leaf-trace till Dr Marie Stopes described *Osmundites Kidstoni* (*Ann. of Bot.* xxxv, 1921, p. 55). Here the primary xylem appears as a solid xylem-core, but flanked externally by secondary tracheides. It seems probable that the latter, which are present in no other representatives of the family, have solved the physiological difficulty of increasing size without enlarging the primary xylem. Consequently the stele of this Fern retained the state of the sporeling. The fossil is of relatively late occurrence, probably Cretaceous, and its secondary thickening may be held as explaining the survival of so primitive a type to so late a time.

(ii) Two species of *Zalesskia* from the Permian have been found to possess a solid xylem in their relatively large steles, without any pith. But their structure is heterogeneous, the outer xylem being composed of long narrow tracheides for conduction, the inner of short wide tracheides effective for storage. The same holds for *Thamnopteris Schlechtendalii* also from the Permian, and here the distinction between the two regions is more marked (Fig. 123).

(iii) In the more recent *Osmundites Dunlopi* from the Jurassic of New Zealand, there is still a continuous ring of outer xylem: but the inner xylem is replaced by pith. The mode of transition from the one state to the other is indicated by the structure seen in *Osmundites Kolbei* from the Neocomian (or Wealden) of Cape Colony, in which there is a "mixed pith" with wide reticulate tracheides scattered through thin-walled parenchyma. The former resembles what is seen in adult stems of living Osmundaceae: the latter corresponds to what has been shown by Gwynne-Vaughan in an abnormal stem of *O. regalis* (Fig. 120). Another feature of advance seen in *O. Kolbei* is the interruption of the xylem-ring at the departure of each leaf-trace, a condition regularly found in modern Osmundaceae: this will be discussed later.

The fossil Botryopterideae, Zygopterideae, and Osmundaceae thus bridge over in rough stratigraphical sequences the transition from the solid to the medullated protostele. The changes are entirely intra-stelar, and though on a larger scale of size, they correspond to the transition already traced among correlative living Ferns. A similar transition may be followed through a "mixed pith" to a parenchymatous medulla in a roughly stratigraphical sequence in the Lepidodendrons, Sigillarias, and other fossil Pteridophytes. In these also the change is intra-stelar, and may lead to interruption of the xylem-ring. The conclusion follows that not only in Ferns but also in other Vascular Plants there has been a progressive medullation of the xylem. It is effected by conversion of the central region into parenchymatous pith. This change goes along with increase in size of the stele, and may be completed within the barrier of an uninterrupted endodermis.

STRUCTURAL ADVANCES FOLLOWING ON MEDULLATION

Among primitive Fern-steles further structural changes may follow. Sometimes the pith becomes sclerosed. This is characteristic of xerophytic Ferns, such as *Schizaea*, and *Platyzoma*. Its importance may perhaps lie in relation to storage of water in a form in which it cannot readily be exhausted through drought. In others the pith-cells may become storage-places for starch, while the cell-walls split at their angles, forming intercellular spaces, which are absent from the smaller vascular tracts. This condition is seen in the Ophioglossaceae and Osmundaceae, in both of which a ventilated storage-pith is found. In this event, so long as the outer endodermis is an unbroken sheath even at the exit of the leaf-traces, it forms not only a barrier which places the transit of plastic materials under the protoplasmic control of the endodermal cells, but it also cuts off the intra-stelar ventilating system from the cortical, and encloses it as effectively as the ventilating system of submerged plants is enclosed by their epidermis. A careful structural examination of the adult stem of *Osmunda regalis* has revealed no connection between the cortical and the stelar systems of ventilation. Physiologically this isolation of the two systems is a weak point: but it is set right by other means in most Ferns, as will be seen later.

More important from a theoretical point of view than these changes is the appearance of patches of internal endodermis in the pith, which has now been recorded in adult stems of *Botrychium* and *Helminthostachys*, in *Schizaea* and *Platyzoma*, and in *Todea* and *Osmunda*. Since the attempt has been definitely made to show that these sporadic tracts of endodermis are relics of degeneration of some more complex structure, it is well to realise how relatively primitive are the organisms in which they are found. Also that the group of cells showing the endodermal characters frequently occur quite isolated from any other tissue of like nature. An internal endodermis is even to be found sometimes surrounding the pith in the root of *Helminthostachys*, a position in which endodermis is normally unknown. *Here at least it seems clear that the inner endodermis is a new development, and the same is probably true also in all the stems named* (Lang, *Ann. of Bot.* xxix, p. 6, Text-Fig. 1 *B*). Since the plants in which sporadic inner endodermis occurs are all relatively primitive, the natural way to view them is as steps in advance rather than as relics of degeneration: and to see whether there may not be some physiological reason for its presence.

The most instructive examples of the sporadic internal endodermis are in *Botrychium* and *Helminthostachys*, which have been carefully worked over by Lang (*Ann. of Bot.* xxvii, p. 203, xxix, p. 1). In the young plant of *Botrychium Lunaria* it is in the region intermediate between the juvenile and the adult parts that the internal endodermis is found; it is associated with extension of the internodes, and interruptions of the outer endodermis at the exit of the leaf-trace. Here, as Lang has shown, the pith is liable to have open connection with the cortex. The internal endodermis would then be a useful newly organised protection for the conducting tract, its function being to place the contents of the stele under protoplasmic control, though still allowing of gaseous interchange through the foliar gap between the ventilated pith and the cortex. In the adult region, with its greater bulk and shorter internodes, this is less important, and there the internal endodermis is absent. Figs. 124 *A*, *B* show its distribution in two fully analysed examples, and how in a long internode the inner endodermis may form a deep pocket-like tube. But Lang states that in *Botrychium* such pockets are never closed at the base. In *Helminthostachys*, in which the internodes are short, Lang specially notes that it is only in the adult rhizomes with larger leaf-gaps that continuity between the cortical and medullary parenchyma is marked, and it is in these regions that an internal endodermis is often developed. It thus appears that the inner endodermis has a use which would justify its appearance as a new formation.

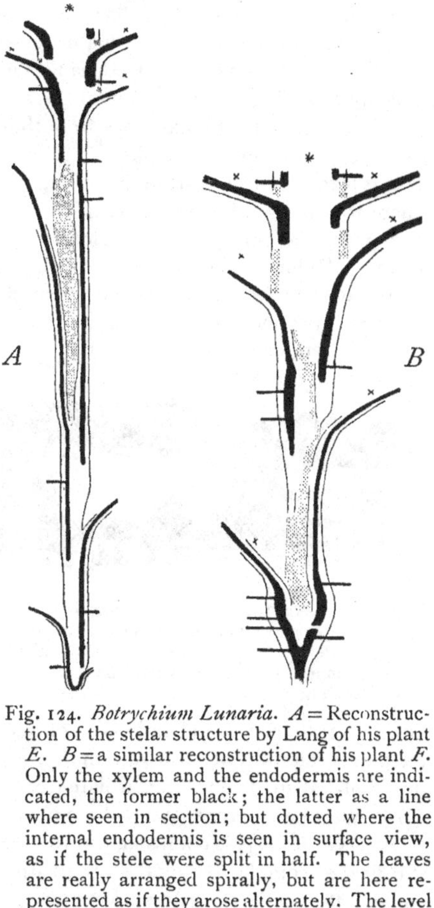

Fig. 124. *Botrychium Lunaria*. *A* = Reconstruction of the stelar structure by Lang of his plant *E*. *B* = a similar reconstruction of his plant *F*. Only the xylem and the endodermis are indicated, the former black; the latter as a line where seen in section; but dotted where the internal endodermis is seen in surface view, as if the stele were split in half. The leaves are really arranged spirally, but are here represented as if they arose alternately. The level of origin of the root-traces is indicated. The proportions have been altered from those in nature so that the stele is represented as broader in proportion to its length than is actually the case. (×) represents axillary buds; (*) the apex.

The structure of *Platyzoma* has been worked out in detail by Thompson (*Trans. Roy. Soc. Edin.* lii, 1919, p. 571). In fully developed specimens of this xerophytic Fern there is a medullated stele with a continuous ring of xylem, from the outer margin of which leaf-traces arise without any interruption of the ring. The medulla is often heavily sclerosed, and between this and the xylem there is a ring of inner endodermis, which is at no point connected with the outer endodermis. In a small plant it was found that the inner endodermis was not constantly present (Fig. 125). As the stele was followed forward acropetally from the broken base, it was found that the pith and inner endodermis decreased until the latter was narrowed to a vanishing point, and the simple medullated protostele was the result. Later the pith expanded and a tubular inner endodermis again appeared. It subsequently narrowed again to a vanishing point, the pith at the same time being reduced. Thus for a second time a simple medullated protostele resulted. On subsequent re-expansion of the pith an inner endodermis again appeared and was continued onwards to the apex. A probable explanation of these changes is that the formation of an inner endodermis is causally related to the expansion of the pith which is often sclerosed in xerophytic Ferns. Its reduction, or even its absence from those zones where the xylem is protostelic, accompanies the reduction of the pith. It is suggested that here the function of the inner endodermis is to establish a further intra-stelar physiological control over the water-storage-pith where it is bulky. But this is unnecessary where the pith is small. The whole may thus be regarded as an upgrade development, involving a formation of an inner endodermis *de novo*. Unfortunately the ontogenetic development is not known in this plant.

It has been seen in the juvenile stem of *Anemia*, and in the relatively simple *Schizaea rupestris*, how medullation is initiated in the protostele. The latter appears to have permanently retained this state of simple medullation. But other species of *Schizaea* show

Fig. 125. Reconstruction from sections of the stele of a small plant of *Platyzoma*. The base is to the left. Outer and inner endodermis are represented by black lines: phloem is hatched, and xylem is black. Pericycle, inner parenchyma, and pith are white. The diagram represents the general arrangement of tissues as would be seen in a median longitudinal section of the stele. (After Dr McLean Thompson.) (× 10.)

further intra-stelar complications. For instance in the adult state of the large *S. dichotoma* the pith is sclerotic, but always with thin-walled parenchyma lining the xylem-ring. Within this thin-walled parenchyma isolated endodermal spindles may arise, quite independent of the outer endodermis, and with no obvious relation to the leaf-insertions (Boodle, *Ann. of Bot.* xv, p. 703; Thompson, *Trans. Roy. Soc. Edin.* lii, 1920, p. 718). The matter is still further complicated by the observation of groups of tracheides either lodged within the spindles, or in the parenchymatous pith outside them. Such observations suggest a high variability of determination of the cells of the procambium.

Schizaea malaccana appears to be a smaller example of structure substantially of the same type (Tansley and Chick, *Ann. of Bot.* xvii, p. 493).

The origin of the pith in the Osmundaceae has already been accounted for (p. 128). But the large upright stocks of this family present further complications which are intra-stelar in their nature. It has been noted how in *Osmundites Kolbei* the outer xylem-ring, which is continuous in *Thamnopteris* (Fig. 123), is interrupted at the departure of each leaf-trace. This is the rule in the adult stems of all living Osmundaceae, though not in the juvenile plants. The ontogeny of *Osmunda*, which was followed by Gwynne-Vaughan and by Faull, reflects the probable phylesis. The young plant has a solid xylem-core, from which the first leaf-traces spring in the manner usual in protostelic Ferns. But as the later traces depart parenchyma appears among the stelar tracheides. Higher up a permanent column of pith is established at the centre of the wood. From such ontogenetic facts it may be concluded that the protostele was primitive for the living Osmundaceae, and the medullated stele derivative. In this medullated region the xylem-ring is interrupted above each leaf-trace by a xylic gap. Through it the pith is connected by a parenchymatous ray with the parenchymatous sheath outside the xylem; and since the leaves are numerous, and closely placed, these xylic gaps overlap, and produce the characteristic dictyo-xylic state of the adult stem of *Osmunda* (Fig. 120, p. 127). All these arrangements are still intra-stelar, for outside the xylem-sheath lie the continuous cylinder of the phloem, the pericycle, and finally the endodermis. In the adult the latter remains unbroken at the departure of each leaf-trace.

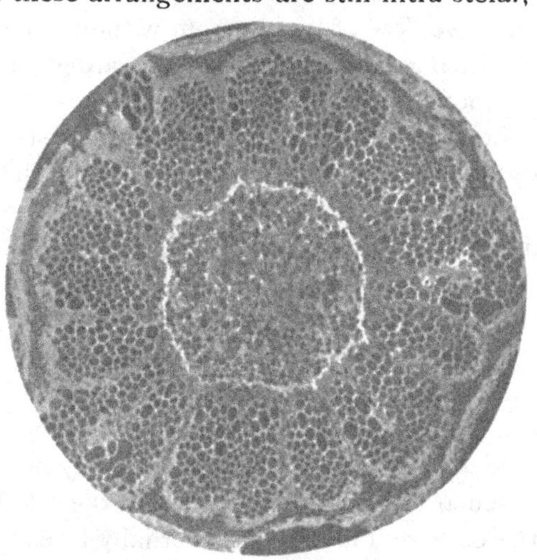

Fig. 126. Stele of a full-grown stem of *Osmunda cinnamomea*, from a photograph by Gwynne-Vaughan. For details see text. (× 25.)

All this appears intelligible as an upgrade development from the protostele, following on the ontogenetic enlargement of the shoot. The result is that in the adult shoot there is a bulky central column of storage-pith, sometimes partially sclerotic, surrounded by a dictyo-xylic ring, with peripheral phloem, pericycle, and endodermis. Outside this again is a densely starchy cortex. The conducting tissues thus lie between two storage-tracts, both of which are ventilated by intercellular spaces, though the conducting tract which completely separates them is not. It is delimited by an external endodermis, but there is no internal endodermis in ordinary stems of the living Osmundaceae.

An internal endodermis does, however, appear occasionally in the living Osmundaceae. It has been noted locally in some few specimens of *Todea hymenophylloides* as a discontinuous sheath, with its suberisation not confined to one layer of cells. There is no internal phloem associated with it. (Seward and Ford, *Trans. Linn. Soc.* vi, 1903.) But in adult stems of *Osmunda cinnamomea* both endodermis and phloem are found in some cases within the ring of xylem: in other specimens of the species there is an internal endodermis without phloem (Fig. 126). The occurrence of the internal phloem is local in the region of a bifurcation, where a "ramular gap" is more or less completely formed. The detail of the gap varies within the genus. In *O. claytoniana* there are no such gaps at all: in *O. regalis* and *T. barbara* the gaps extend to the xylem only: but in *O. cinnamomea* the gaps vary, being in some cases complete, the outer endodermis connecting at the gap with the inner, so that there is free communication between the cortex and the pith: the phloem also extends from the outside of the xylem through the xylic gap, and is continuous as a sheet of internal phloem for some distance down and up the stem. Here again the presence of the inner endodermis, and of the inner phloem as well, may find its elucidation in terms of local change of destination of the procambial cells, there being no evidence of any intrusive flow of tissues from without to explain it.

Such local change may be regarded as a still more advanced stage similar to, though more far-reaching than, that seen in *Platyzoma* and *Schizaea*. Where the complete ramular gap exists the conducting tract is delimited by endodermis on both sides, while the ventilating systems of pith and cortex are placed in communication, a state not arrived at in any other living Osmundaceae than in *Osmunda cinnamomea*.

In the Cretaceous fossil *Osmundites skiegatensis* ventilation throughout the stem was still more fully achieved than in the above cases. The stock of this plant must have been much larger than that of any living member of the family. It had a stele 24 mm. in diameter, with a very wide pith surrounded by a ring of about 50 separate vascular strands. Each leaf-trace at its departure interrupts the continuity not only of the xylem but of the whole vascular ring. Through each gap the pith becomes perfectly continuous with the cortex. Phloem lies internally to each xylem-mass, and it is continuous through each leaf-gap with the external phloem. No layer resembling endodermis can be distinguished in this fossil, so that it is practically impossible to set a definite limit to the stele. Nor is there to be seen any endodermis round the leaf-trace. This large-sized Osmundaceous Fern thus exceeded any living Osmundaceae not only in size, but also in the complexity of its stelar structure. It had actually taken the step to a sort of dictyostely.

This is more clearly shown in the still larger Paraguayan fossil stem, of late but uncertain horizon, *Osmundites Carnieri*, Schuster (*Ber. d. D. Bot. Ges.* xxix, p. 534; Kidston and Gwynne-Vaughan, *Trans. Roy. Soc. Edin.* Vol. l, Part II, p. 476). Here the stele is greatly dilated, being 33 mm. in diameter, that is, about one-third the diameter of the stem. It shows 33 distinct xylem-strands of unusual depth. The stele is surrounded by a line of delimitation, believed to be an endodermis, which is interrupted by the departure of each leaf-trace. This leaves a wide leaf-gap (Fig. 127). Thus the vascular ring is divided up into separate meristeles, each containing a varying number of xylem-strands. There is external phloem, and probably an internal phloem also existed. If it did, and the line of delimitation be really endodermis, then *O. Carnieri* is truly dictyostelic. Physiologically the effect is that the ventilated cortex is directly related to the greatly dilated pith, while the vascular tissue is completely shut in by endodermis. The final result in the adult will then be a disintegration of the stele comparable to that seen in the Leptosporangiate Ferns; but it was achieved by an evolutionary progression quite distinct from theirs.

The disintegration of the stele in the adult stem and the omission of an effective endodermis, which are foreshadowed in the larger Osmundaceae, are both carried out still more completely in the living genus *Ophioglossum*,

Fig. 127. *Osmundites Carnieri*, Schuster. Arrangement of meristeles. The endodermis is shown by dotted lines. (After Kidston and Gwynne-Vaughan.)

and in the Marattiaceae. In the young plant of *Ophioglossum* an outer endodermis has been found delimiting the stele, and in some instances even an inner endodermis has been described, as in *Botrychium* (Poirault(122), Bower, *Ann. of Bot.* 1911, p. 540). But in the adult no endodermis is present in stem or leaf (Campbell, *Eusp. Ferns*, p. 92). The stele of the axis expands into a dictyostele with wide leaf-gaps, which is embedded in sappy parenchyma (Fig. 128). An extreme condition is seen in the tuberous stocks of *O. palmatum*, which may be as much as 2 cm. in diameter, with the sheathless meristeles and petiolar strands scattered through the sappy parenchyma.

A similar disintegration carried even to a higher degree is seen in the Marattiaceae. Here also an endodermis is present in the young sporeling surrounding the integral stele (Brebner; also Farmer and Hill, *Ann. of Bot.*

xvi, pp. 525 and 386). This, however, becomes distended, and breaks up in the adult into a number of meristeles scattered through the massive parenchyma. In *Marattia* and *Angiopteris* their number and polycyclic arrangement form a very marked feature. No effective endodermis is found delimiting these meristeles of the adult stem from the surrounding tissue. It thus appears that while both *Ophioglossum* and the Marattiaceae have a definite and integral stele in the young stem, the adult shows a disintegration and a loss of delimitation of the vascular tissues, which are carried to a higher degree than in any of the other primitive Ferns. Moreover, the sappy stock is traversed throughout by a continuous system of intercellular spaces. The appearance is as though the stelar construction had entirely broken down; but the ontogeny shows that a normal stelar structure underlies the ruins. (See West, *Ann. of Bot.* 1917, p. 361.)

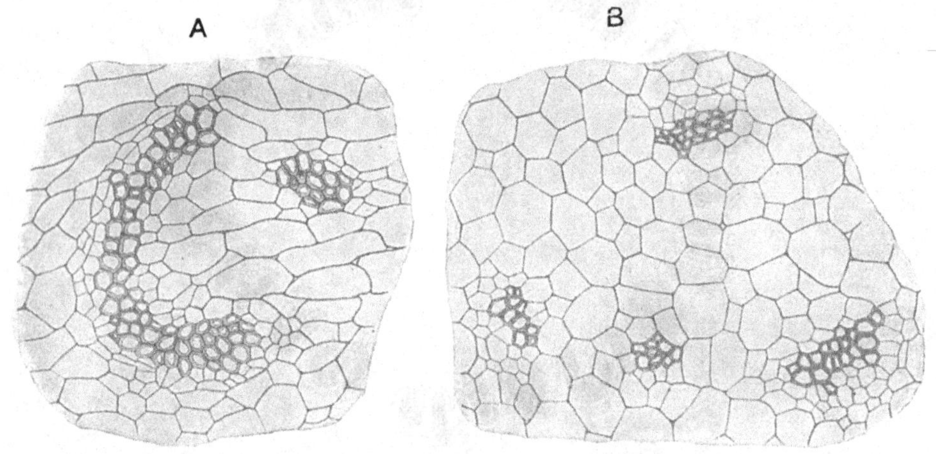

Fig. 128. *Ophioglossum Bergianum*, Schlecht. *A* = transverse section of the stock, showing a large semilunar stele, with wide foliar gap from which a small leaf-trace-strand is passing out. *B* = another section showing the overlapping of foliar gaps. (× 200.) No endodermis is to be seen.

The physiological interpretation of this seemingly anomalous state, and of the structural facts which mark the advance from the primitive protostelic condition to a high degree of stelar disintegration, must be held over to Chapter VIII. Here it must suffice to have recognised the successive steps which appear to have led to that state, whether traced by comparison of adult structure or by following the ontogenetic history.

Secondary Thickening

Lastly, the innovation of cambial activity remains to be mentioned. Though this has played so important a part in the Vegetable Kingdom as a whole, in the Filicales it never established a real hold. It appears early in the young plant of various species of *Botrychium*, for instance in *B. Lunaria*, where it is of relatively feeble development (Fig. 119). It is more extensive in the larger species, and in *B. virginianum* the zone of secondary wood is extensive (Fig. 129). That zone is traversed by parenchymatous rays comparable in their essentials with those of Gymnosperms and Dicotyledons. Externally a zone of cork may also spring from the periphery of the cortex. A similar condition found in a Palaeozoic fossil led Scott to its designation under the name of *Botrychioxylon*. At the centre of its stele is a "mixed pith," and outside this there is a zone of secondary wood. Periderm bounds

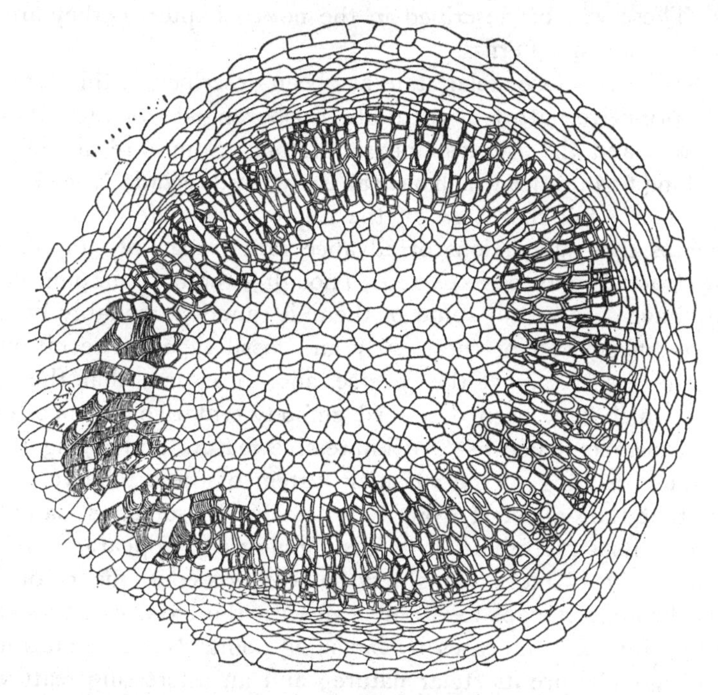

Fig. 129. Transverse section of the stele of *Botrychium virginianum*, showing massive pith: radially disposed tracheides and medullary rays are evidence of the secondary activity of the cambium which is seen surrounding the xylem. (After Atkinson.) (×6.)

the cortex externally. Scott refers the plant to the Botryopterideae, with the nearest relation to *Metaclepsydropsis*, in which secondary wood is also occasionally found. Recently a further example of secondary increase has been described by Dr Marie Stopes in *Osmundites Kidstoni*, probably from the Cretaceous Period (*Ann. of Bot.* xxxv, 1921, p. 55). It shows a combination of a solid somewhat stellate protostele with secondary wood arranged in seven bays between the rays of the star, and it has a typical Osmundaceous leaf-trace. These features supply further structural links connecting the characteristics of the Botryopterideae with those of the primitive Osmundaceae.

The interest of these examples lies not so much in the fact of secondary thickening having been established, as that none of the later Filicales ever appear to have developed so important an innovation to the best advantage. It is seen only in Ferns of primitive type. For some reason which is not obvious Ferns worked out the evolution of all their more modern forms upon the basis of the primary vascular tissues alone. In the great majority of them these were composed of strictly circumscribed tracts, without intercellular spaces, enclosed within continuous and well-defined endodermal sheaths. These will be described in the next Chapter, as they are seen in the Leptosporangiate Ferns.

All the Ferns whose stelar state has been described in this Chapter are relatively primitive. Most of them are Eusporangiate, or are intermediate between this and the Leptosporangiate state. Some are actually fossils, others belong to affinities which have an early fossil record. None but *Cheiropleuria* show the full characters of the Leptosporangiate Ferns. It is from such sources that suggestions of advance rather than of simplification of vascular structure may be expected. Though structural simplification may naturally have occurred in some of them, it cannot reasonably be assumed that it has affected all of these early and fossil types. The rational view would rather be to expect the reverse, and to assume that the apparent anomalies which have been described embody tentative advances upon the simple protostelic structure. An impressive feature underlying the whole series of facts relating to the vascular system of these relatively primitive Ferns is that they all start from a protostele. In some, as Campbell has pointed out, its structure is ill-defined, giving the appearance of a mere sympodium of leaf-traces. This may be held as a consequence of a megaphyllous disproportion of axis and leaf, with the axial constituent of the stele reduced in extreme cases to vanishing point. But that does not alter though it may obscure its stelar nature; and an interesting feature of the facts disclosed is in what different ways the ontogenetic amplification may be carried out. It will be shown in Chapter X that they are natural consequences of increase in size of the conducting system, such as is seen in the individual life in the course of its progress from the sporeling to the adult. On the other hand, *from the phyletic point of view, the nearer the adult structure remains to that of the original protostele the more primitive the plant which shows it is to be held in respect of that important structural feature.*

STELAR THEORY

If an adult stem of a Dicotyledon, a Gymnosperm, a Lycopod, a Horsetail, or a simple type of Fern be cut transversely, a more or less compact column of conducting tissues is found lying centrally and connected outwards with the leaves by strands of the leaf-trace. Often it is seen to be delimited by an endodermis or "phloeoterma" which gives precision to it, and appears to justify the recognition of the whole column as an entity.

The term *stele* was applied to it by Van Tieghem (*Ann. Sci. Nat. Bot.* Sér. 7, 1886, p. 276). There can be no doubt of the existence of such a tract in ordinary adult stems. The question remains how is it to be interpreted? Is the stele a real entity, forming an essential constituent of all axes: or is it a composite structure built up merely of decurrent tracts from the leaves? Or does the truth lie between these extremes? This question has been raised afresh by Campbell (*Amer. Journ. of Bot.* 1921, p. 303).

A provisional answer is suggested by examination of the apical structure of the adult shoot in some Fern with elongated protostelic rhizome and isolated leaves, such as *Cheiropleuria* (Fig. 115). Here the disproportion of size between stele and leaf-trace is great. The former may be followed up to the apical cone as a column of procambium maintaining its identity, while giving off only small fractions of its tissue as a leaf-trace to each isolated leaf. The residual vascular tissue appears to be truly cauline and not referable in origin to leaf-trace tissue. In elaborated vascular systems the compensation-strands and medullary strands, such as are seen in many Leptosporangiate Ferns (Chapter VIII), are clearly cauline. Thus it appears that in adult shoots a part of the vascular system of the axis may be, and often is, not common but cauline. Here then the stele is not merely a sympodium of leaf-traces. But Campbell has pointed out in relation to the sporeling in certain Eusporangiate Ferns that there is "an absence of a cauline stele in the young sporophyte," and he contemplates with Delpino the sporeling plant as being composed of leaves as primary organs, and the so-called stem as being formed by the coalescence of leaf-bases. This may appear as though it were the fact in extreme cases such as certain Ophioglossaceae and Marattiaceae selected by Campbell. But that does not demonstrate that it is the correct view for general application, nor even that it is the true fact for the phyletic history of the selected examples.

The vascular system follows the lines of evolutionary progress, it does not dictate them. If the progression be towards a preponderance of leaf over stem as is the rule in Ferns, the vascular development of the latter will recede, and the former become dominant. Conversely, whereas in Lycopods the stem preponderates, it is quite obvious that the greater part of the axial stele is cauline. It may be held that the shoot in the Filicales fluctuates between a preponderance of leaf over axis, and a reasonable balance between the two. Where, as in the juvenile state of the Ophioglossaceae, the former is seen in its extreme state, there may appear to be no cauline stele: nevertheless in a type such as the adult *Cheiropleuria* the cauline factor may preponderate with converse result. Other Ferns may take intermediate positions. Accordingly it would appear that extreme monophyllous examples, such as *Ophioglossum* (the actually primitive character of which genus may be held in doubt), are not those best suited to serve as a foundation for a general stelar theory. A more illuminating line would appear to be to start from a really ancient fossil type, such as *Botryopteris cylindrica.* It may, however, be submitted that in the present state of uncertainty as to the early evolutionary relations of the leaf and axis it is impossible to speak with confidence, and in general terms, as to the origin of the stele. This much however seems clear: that at an early stage in the ontogeny of Ferns a coherent body of tissue, "partly made up of elements which have a truly cauline origin, serves to connect up adjacent leaf-traces" (West, *Ann. of Bot.* 1917, p. 369). It may be added that the greater the preponderance of the leaf over the axis in the young shoot, the later will be the stage when this composite nature of the stele can be recognised.

NOTE. To avoid needless repetition the bibliography of stelar tissues will be given collectively for Chapters VII, VIII, and X at the close of Chapter X.

CHAPTER VIII

THE VASCULAR SYSTEM OF THE AXIS (*continued*)

(B) THE STELAR STRUCTURE OF THE LEPTOSPORANGIATE FERNS

THE Ferns quoted in the preceding Chapter, excepting only *Cheiropleuria*, are either Eusporangiate, or near derivatives of that type. It will be shown in Volume II that they are all relatively primitive Ferns, and that they illustrate a number of distinct phyletic lines, most of which do not appear themselves to have given rise to any Leptosporangiate derivatives. This is probably the case for the Ophioglossaceae, Marattiaceae, Osmundaceae, and Hymenophyllaceae. But there is reason to believe that the Schizaeaceae do represent with some degree of accuracy a source from which a considerable section of the Leptosporangiate Ferns sprang. In these the sori are typically marginal (Marginales), while others in which the sori are typically superficial (Superficiales) may probably be traceable to some Gleicheniaceous origin. The argument on these points will be developed later: but meanwhile the statement of this general position in advance of the argument will have cleared the ground for their anatomical study. Both the Gleicheniaceae and the Schizaeaceae include permanently protostelic types, while in all Leptosporangiate Ferns the axis of the sporeling is in the first instance protostelic. Starting from this primitive stelar state it will appear from comparison that transitions have occurred to variously elaborated, and finally to disintegrated stelar structure. The types of vascular construction thus achieved in the Leptosporangiate Ferns are known as *Solenostely*, *Dictyostely*, *Polycycly*, and *Perforation*. Each of these will be described and illustrated. But underlying them all are certain features that are constant in the Leptosporangiate Ferns. However complicated the form of their vascular masses may be, they are always *strictly delimited from the surrounding tissues by a continuous sheet of endodermis*. This forms an unbroken barrier exercising a physiological control over transit into or out of the conducting tracts. Another constant feature is *the absence of any ventilating system of intercellular spaces within the conducting tracts*. These structural points indicate a highly specialised condition of the conducting system in the Leptosporangiate Ferns.

SOLENOSTELY

The term *solenostele*, or *siphonostele* as it is sometimes called, is applied to that state in which the stele takes the form of a cylindrical tube, so that in transverse section it appears as a ring of vascular tissue surrounding a parenchymatous pith, and itself enveloped in cortex (Fig. 130). The middle zone of the stelar ring itself is occupied by a band of tracheides, as a rule continuous, with groups of protoxylem embedded at intervals among the metaxylem. This is surrounded on either side—or sometimes only on the outer side—by phloem. The former is described as the *amphiphloic*, the latter as the *ectophloic* condition. Investing these tissues is the parenchymatous pericycle, of one or more layers, and finally there is a complete covering of endodermis within the tube and outside it (Fig. 133). The evolutionary relation of this solenostelic state to the primitive protostele is best studied in the ontogeny of Ferns that are themselves solenostelic, but are closely related to such as remain protostelic even in the adult. The best examples are found in the genus *Gleichenia*.

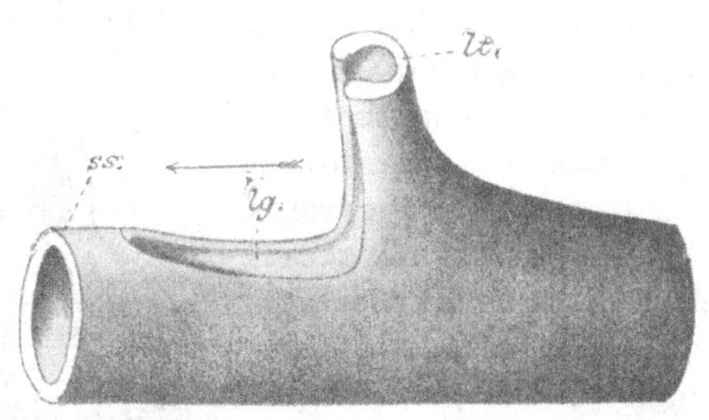

Fig. 130. *Loxsoma Cunninghamii.* Diagram showing the form of the vascular system at a node of the rhizome. *ss* = solenostele; *lt* = leaf-trace departing; *lg* = leaf-gap. The arrow points toward the apex of the rhizome. (After Gwynne-Vaughan.)

The section *Mertensia* of the genus *Gleichenia* includes species that are protostelic throughout the adult plant (*G. flabellata, linearis*, etc.), and others in which the adult stem is solenostelic (*G. pectinata*). All of them have creeping stems with long internodes, a fact which makes them particularly suitable for the study of the origin of solenostely. It will be shown later that the soral condition indicates *G. pectinata* as a relatively advanced species. Thus it is not a mere assumption that the solenostely is an advanced state, but it is backed by evidence drawn from other characteristics as well. Cut transversely through an internode of the long creeping rhizome of *G. flabellata*, the stele is seen to be cylindrical, with a solid xylem-core composed of tracheides and parenchyma, the numerous protoxylems being immersed within its margin (Fig. 131). Cut

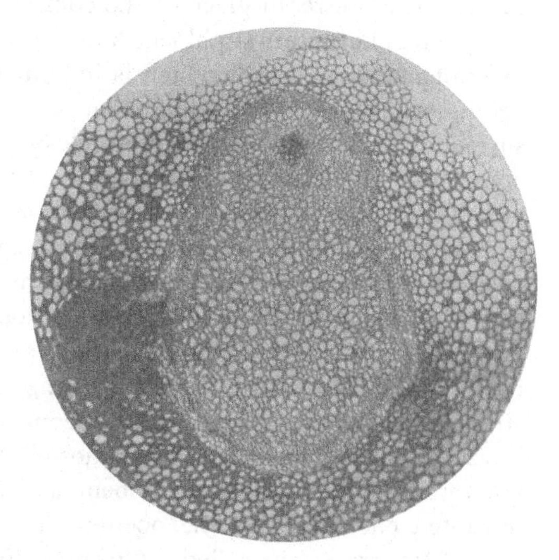
Fig. 131. Transverse section of protostele of the rhizome of *Gleichenia flabellata*, R. Br., from a section by Gwynne-Vaughan, showing a leaf-trace being given off. (× 30.)

through a node the leaf-trace is seen to separate on the upper side as a C-shaped tract of tissue, at first surrounding an involution which is continued downwards as a shallow pocket. This is delimited by internal phloem, and again lined by internal endodermis, while centrally there lies a mass of sclerenchyma (Fig. 132). Passing upwards from the base of this pocket the leaf-trace separates without break of endodermal continuity, as a single C-shaped tract, carrying out with it three protoxylem groups (*A*). The pocket does not here extend into the internode, and consequently the stelar xylem is not indented by the departure of the trace. The condition is typically protostelic.

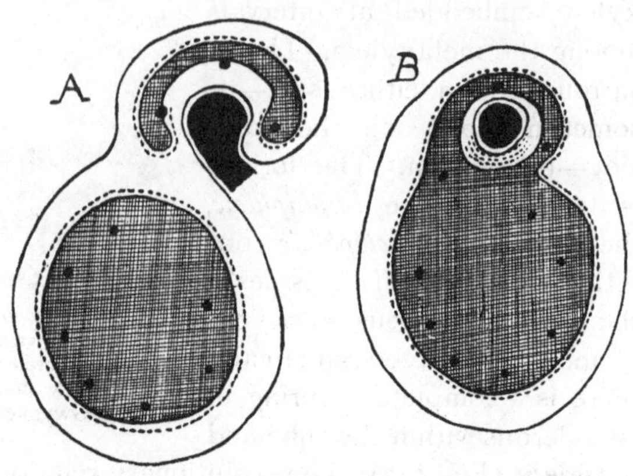

Fig. 132. Transverse sections of the rhizome of *Gleichenia flabellata* showing the detachment of the leaf-trace from the protostele, with its included protoxylems. (After Tansley.)

In *Gleichenia pectinata* a section through an adult internode shows typical solenostely. The outer tissues down to the xylem are as in *G. flabellata*, but towards the centre, in place of the continuous xylem, there appear successively a band of internal phloem, a ring of internal endodermis, and a central mass of medulla or pith, which is indurated. This core of sclerotic pith is continuous through the adult stem, and the hollow cylinder of the stele surrounds it throughout the internodes (Fig. 133). But transverse sections cut at intervals in the neighbourhood of the node show that the leaf-trace passes off in essentially the same way as in *G. flabellata*. Here, however, in place of the shallow pocket there is seen directly above its exit a deep channel, filled by sclerotic tissue, and limited by the still continuous endodermis, around which is a sheet of phloem connected with that outside. A direct continuity is thus established at the departure of each leaf-trace between the outer cortex and the pith, which happens in this case to be strongly sclerosed. The vascular structure of the adult is accordingly that described as the amphiphloic solenostele, since the stele is tubular, and is lined within and without by phloem, and is delimited completely by uninterrupted endodermis. The openings of the ring at the departure of the several leaf-traces are called *foliar gaps* (Fig. 133).

It would then appear probable that an examination of the young sporeling of *G. pectinata* should reveal facts of the ontogeny which will shed light upon the origin of solenostely from that protostelic state so prevalent in other Gleicheniaceae. This is found to be the case (Fig. 134). Starting from the base of the young plant there is first a long protostelic tract, which bears a number of leaves. The structure is in fact that which is seen in permanently protostelic Gleichenias. This first stage is followed by a long medullated stage, the central region of the xylem being replaced by parenchyma. At first there are no xylic gaps at the departure of the leaf-traces, but later the xylem-ring is interrupted, and opposite the xylic gaps thus formed there may be a slight involution of the endodermis. At first no internal phloem is seen, but parenchyma only. Just before the solenostelic stage is reached inner phloem appears as an indefinite and incomplete ring of sieve-tubes lining the xylem internally. The solenostelic condition finally results from

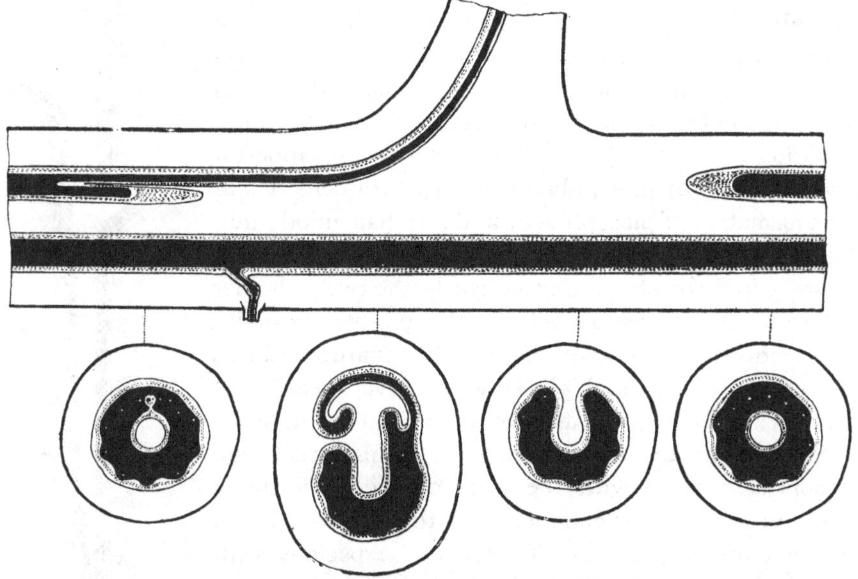

Fig. 133. Diagram illustrating the solenostelic structure, and attachment of the leaf-trace in *Gleichenia pectinata*. The transverse sections show the structure corresponding to the several points indicated. After Boodle.

the following further step, which is described as from below upwards. First an inner endodermis appears, as a tube closed below and widening upwards. It is continuous on the one hand with the endodermis of the next departing leaf-trace: and on the other it is continued as the inner endodermis lining the next higher internode: it is continuous also with the endodermis surrounding the next succeeding leaf-trace. These steps are all illustrated in the reconstruction by Dr Thompson (Fig. 134). The *continuity of the endodermis right across the pith-column* follows on the interruption of the stelar xylem and phloem immediately above the departure of a certain leaf-trace, which is the first showing a true foliar gap. Its effect will be to stop any possible leakage outwards from the otherwise completely delimited vascular system. The result of this will be physiological delimitation of the stelar pith below it from the pith above, while histological continuity is established between the cortex and the pith above that barrier. There is no sign of intrusive flow of the outer tissue inwards, nor of displacement of cells as an initial step from the medullated protostele to the solenostelic state.

This description is written as a simple objective statement of what is seen in successive transverse sections of a young plant of *Gleichenia pectinata*, without any preconceived theoretical interpretation. The species is one with prolonged internodes in the adult: it is thus a peculiarly favourable example for testing the validity of such theoretical interpretations as have been suggested in explaining the origin of solenostely. The structural changes are correspondingly extended, a fact which makes their succession clearer than is usual in Ferns. They support a view of change of procambial destination of cells *in situ* in the formation of pith, phloem, and internal endodermis. They do not give ground for a belief in a flow or intrusion of cortex into the stele. Consequently the pith, whether in the medullated region below the cross-barrier of internal endodermis or above it, is to be regarded still as the product of procambial tissue, however nearly the character which it may assume when mature may resemble that of the external cortex. This conclusion harmonises readily with the facts of the opportunist appearance of internal endodermis in the primitive Ferns described in the previous Chapter, and especially with that within the stele of the root of *Helminthostachys*. Such facts lead to the general position that endodermis is a tissue that can be formed when and where it is required, and its presence in imperfect sheets or groups of cells does not demand explanation in terms of reduction, i.e. of origin from some pre-existent and complete endodermal layers imperfectly developed in the individual under examination.

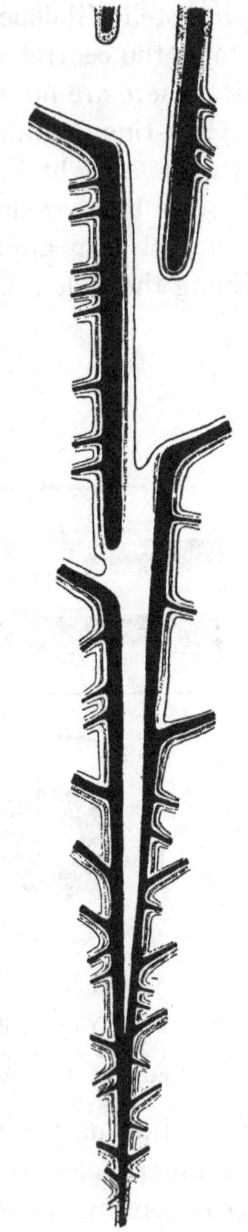

Fig. 134. Plan of stelar construction of a juvenile plant of *Gleichenia pectinata* as seen in median section. For detailed description see text. (After Dr McLean Thompson.)

The typically constructed solenostele of *Gleichenia pectinata* finds its correlative in many others of the Superficiales. It must suffice here to mention *Metaxya*, *Lophosoria*, *Matonia*, and *Dipteris*, in all of which the internodes are long, a feature which conduces to the typical presentment of the solenostele.

The validity of the conclusions based on these ontogenetic facts relating to one of the Superficiales, and the general nature of the facts themselves, may be tested by examination of relatively primitive types of the Marginales, which are believed to represent a parallel evolutionary sequence, though phyletically distinct from the Superficiales. The Schizaeaceae illustrate the steps of stelar elaboration better than any other family of Ferns. *Lygodium* is a permanently protostelic type: *Schizaea* itself has been described above, as illustrating advances in medullation, and the formation of xylic gaps and pockets (p. 133). In *Anemia*, § *Anemiorrhiza*, solenostely is seen, while § *Eu-Anemia* and § *Mohria* are examples of dictyostely. But they all have relatively short axes. For the demonstration of the steps leading to a typical solenostely a Fern with long internodes is preferable. It is found in *Loxsoma*, a genus of Schizaeoid-Dicksonioid affinity, already described anatomically by Gwynne-Vaughan as an example of typical amphiphloic solenostely, showing a foliar gap immediately above the point of departure of each gutter-shaped leaf-trace (Fig. 130). The ontogeny may here again be expected to throw light upon the probable origin of solenostely.

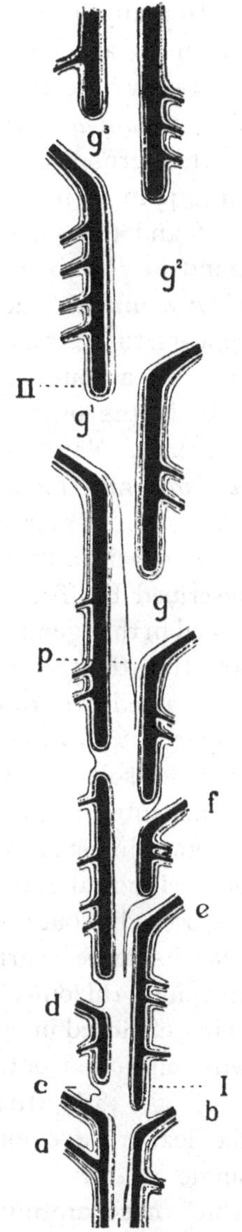

Fig. 135. Plan of the stelar construction in the upper part of a juvenile plant of *Loxsoma Cunninghamii*, as seen in median section. It starts from the medullated stele at the base, and attains a full solenostelic structure in its upper part. a—f = leaf-bases; g—g^3 = leaf-gaps; p = "pocket." For details see text. (After Dr McLean Thompson.)

It starts with a basal region of protostely, which soon passes into the condition where a medulla is present surrounded by a ring of xylem (Fig. 135). Internal phloem makes its appearance with rather irregular arrangement, and extends for a considerable distance, giving the so-called *Lindsaya*-condition. As the leaf-traces pass off in this region (b—f) pockets of variable depth are formed, with involution of the endodermis. These extend into the bulky stelar pith in somewhat the same way as in *Schizaea dichotoma*. Passing upwards a point is reached where a pocket, widening upwards, opens not only into a single foliar gap, but extends onwards into a succession of them (g). Here, as in *Gleichenia pectinata*, a continuous internal endodermis is established, surrounding a central column of pith, which becomes sclerotic upwards, and is connected directly through the successive foliar gaps with the cortex. There is again no indication of sliding growth, or intrusion of tissue from without. All the vascular tracts are completely shut in by an unbroken endodermal sheath. Here also the structural facts as actually presented in a series of transverse sections point to changes of procambial destination of the cells *in situ*, so as to form parenchyma, sieve-tubes, or endodermal cells in place of tracheides present in earlier stages of the individual. A statical change of quality of the cells produced will account for the transition actually observed, from protostely to solenostely, better than any suggestion of dynamical intrusion from without. The steps of the final transition to solenostely in *Loxsoma* are shown in Dr Thompson's reconstruction (Fig. 135), and they resemble those seen in *Gleichenia pectinata* so closely that together they may be accepted as illustrating for Ferns generally the structural passage from the protostelic to the solenostelic state.

In many Ferns certain steps of the ontogenetic progression seen in these examples are apt to be abbreviated, or even omitted: this is especially the case where the internodes are short, as they are in upright stems. In *Acrostichum aureum*, a Pterid-derivative which has an upright stock with short internodes, the stages of the ontogeny are condensed, though on the same plan as just described. The protostelic and the *Lindsaya*-stages are brief, and the solenostely is established earlier. Such condensation is also found in *Paesia podophylla* (*Studies in Phylogeny*, VII, p. 37), and in *Pteridium aquilinum*, according to Le Clerc du Sablon, and Jeffrey, and this appears to be general for the Pterid-affinity. It is this abbreviation which probably accounts for the frequent failure to observe the simply medullated state in the ontogeny of Ferns. It has been telescoped virtually out of existence. Nevertheless it remains in such favourable cases as *Gleichenia pectinata* and *Loxsoma*, and is there comparable with what is seen in the adult state of most Ophioglossaceae and Osmundaceae.

A special interest attaches to that intermediate state which has been described by Tansley as the *Lindsaya*-condition, because it has been retained in that genus as the adult structure, though in most Ferns it is passed over quickly in the course of transition from protostely to solenostely. Sections at an internode of *Lindsaya* show an almost protostelic stele: but an internal island of phloem enclosed by outer xylem lies near its upper surface in the creeping rhizome, the whole being enclosed in outer phloem with unbroken outer endodermis (Fig. 136). In the adult plant the leaf-trace comes off as a single sharply curved strand, which on separating breaks the dorsal arch of xylem covering the internal phloem: thus the latter is continuous with the phloem of the trace, which as it separates is always limited externally by unbroken endodermis (Fig. 137). This condition has been explained by supposing a very shallow pocket to be formed with slight involution of the outer endodermis, but much deeper involution of the outer phloem, so that this extends continuously through the axis. But a better description would be in terms of the ontogeny, as revealed in

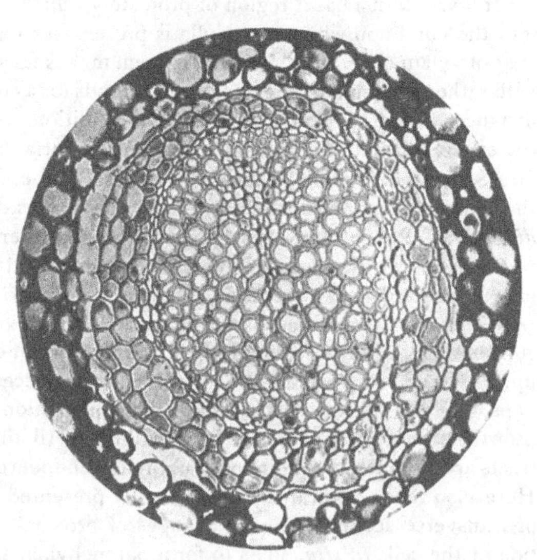

Fig. 136. Transverse section of stele of *Lindsaya linearis*, Swartz. G.-V. collection, slide 992. (× 125.)

reconstructions such as that made by McLean Thompson (*Trans. R. S. Edin.* lii, p. 727, Fig. 5). The *Lindsaya*-condition recurs regularly at an early stage in the ontogeny, and the state seen in *Lindsaya* may be understood as that where no structural advance has been made to complete solenostely: probably this is owing to the restricted size of the species of this genus, which made distension of the stele unnecessary. The last step to typical solenostely is, however, taken in the larger but allied genus *Odontoloma*, which is sometimes merged in *Lindsaya*.

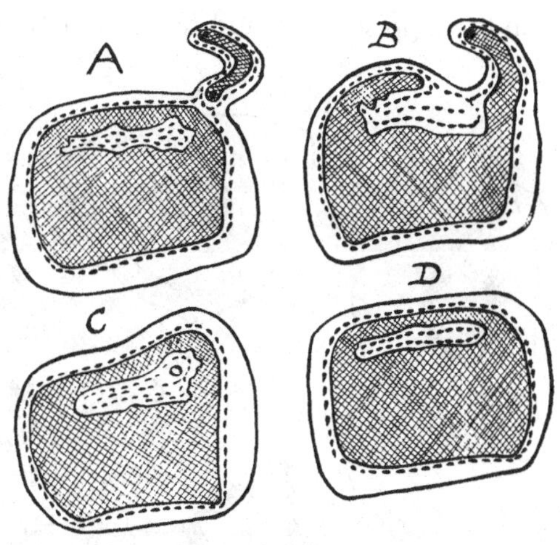

Fig. 137. *Lindsaya scanaens*. A series of sections through a node, showing the internal phloem and the mode of detachment of the leaf-trace. (After Tansley.) Xylem is hatched, phloem dotted, and endodermis a simple line.

It has thus been shown in the ontogeny of Ferns, both Superficiales and Marginales, that a regular succession of stages leads from the protostelic to the solenostelic state. They are, the protostele, the medullated protostele, the *Lindsaya*-stage, and finally the solenostele. But the length of each stage may vary, being greatest where the internodes are long and the adult stem is creeping: and shorter where the axis is upright, and the leaf-arrangement closer.

Dictyostely

A natural consequence of that shortness of the axis which is seen in ascending, or vertical shoots, and of the crowding of the leaves which they bear, is the apparent further complication of the vascular system called Dictyostely. It is specially prominent when the stem is seen in transverse section. Instead of the stelar ring being interrupted only at a single point at a node, as in the solenostelic rhizome, several foliar gaps appear in each transverse section (Fig. 138, *C*, *D*). This involves no other innovation than the shortening of the internodes, and the consequent overlapping of the leaf-gaps: and it is already suggested in certain creeping rhizomes with their leaves closely set (Fig. 139). But the best indication of it comes in those runners which appear as branches on many Ferns. They have usually long internodes at first, but later the leaves are more crowded on their upturned tips, where the axis is proportionately widened to receive them. This is well shown in *Plagiogyria pycnophylla*, in which such branches, called stolons, frequently spring near to the leaf-bases (Fig. 138, *A*—*D*). At first the axis is simply solenostelic (*A*), giving off at intervals a leaf-trace (*B*). Since its foliar gap closes before the next leaf-insertion, only one is seen in any transverse section. But where the internodes are shorter they overlap, so that more than one appears (*C*), and in the widely expanded adult stem several leaf-

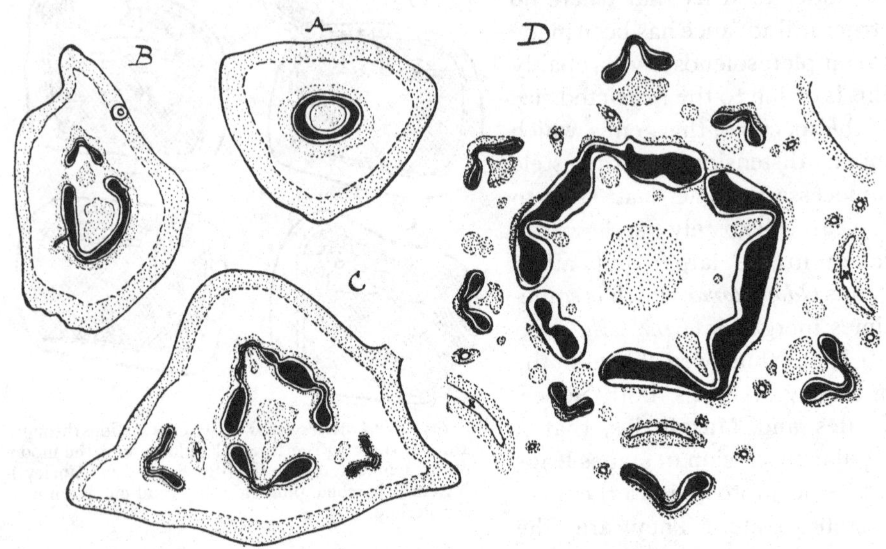

Fig. 138. Transverse sections of rhizomes of successive ages of *Plagiogyria pycnophylla*, showing transition from solenostely to dictyostely. *A*, a young solenostelic rhizome; *B*, same giving off a leaf-trace; *C*, a larger rhizome with two leaf-gaps in section; *D*, vascular system of adult dictyostelic rhizome, with the peripheral sclerenchyma omitted. Sclerenchyma dotted; xylem black. Note the air-spaces (*x*) in *D*, and see text. (*A*, *B*, *C* × 4; *D* × 2.)

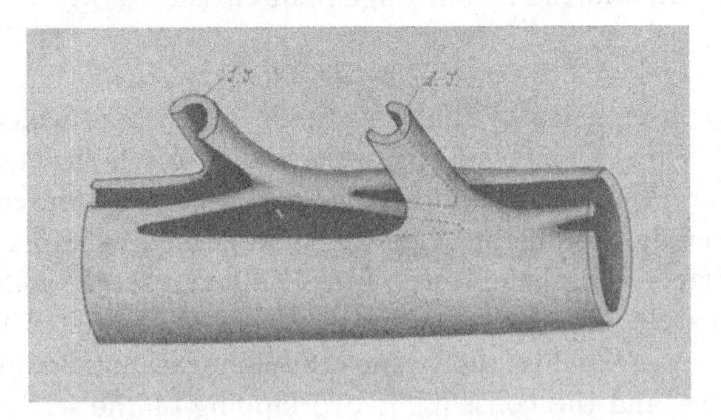

Fig. 139. Vascular system of *Pellaea rotundifolia*, showing the departure of two leaf-traces (*L.T.*), and their relation to the solenostele of the axis. (After Gwynne-Vaughan.)

gaps are seen (*D*). *Plagiogyria* being a relatively primitive type, not far removed from solenostely as Gwynne-Vaughan remarks, it demonstrates the relation of solenostely to dictyostely more clearly than it is seen in more advanced Ferns, such as *Dryopteris Filix-mas*. Here the whole vascular system may be dissected out as a cylindrical network, with large overlapping foliar gaps, from the margins of which the numerous strands of the divided leaf-trace arise (Fig. 140). Clearly several leaf-gaps will appear in any transverse section. The vascular system is represented by a ring of "meristeles," each delimited by unbroken endodermis. From the point of view of ventilation the numerous, wide openings between these meristeles will allow of free communication between the cortex and the central pith, which in such stems is often very bulky.

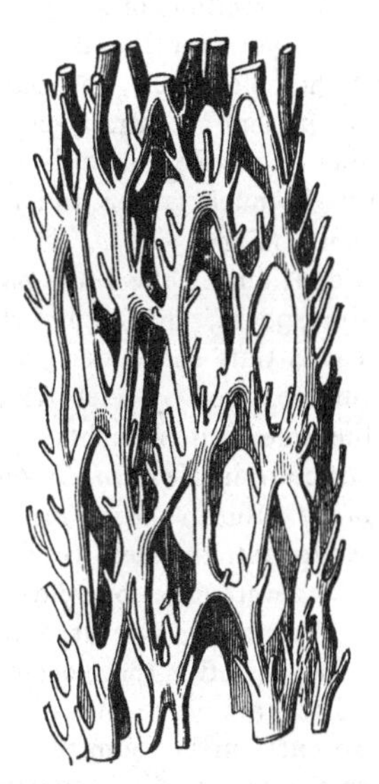

Fig. 140. Vascular skeleton prepared by maceration and dissection from the stock of *Dryopteris*. (After Reinke.) (× 2.)

PERFORATION

In addition to the foliar gaps there are still other means of communication between cortex and pith in many Leptosporangiate Ferns of advanced type. "Perforations" of circular, oval, or elongated outline are found, and they are specially a feature of rhizomatous Ferns where the leaf-gaps would only occur at intervals. Sometimes they are imperfect, appearing as xylic-perforations, with or without involution of the endodermis: sometimes an isolated perforation may occur, as in *Metaxya*: but commonly they are so numerous that in a transverse section the meristeles appear as a ring of small isolated tracts, each surrounded by its complete endodermal sheath. Perforations were demonstrated by Mettenius in *Stenochlaena tenuifolia* (Desv.), Moore (Fig. 141), and in many other rhizomatous Ferns. The typical solenostelic structure is usually present in relatively primitive creeping stems. But in them the continuous stelar tube would form an obstacle to gaseous interchange between cortex and pith. The presence of "perforations" in creeping Ferns in other respects of relatively advanced type may be held as an amendment on that structure, whereby gaseous interchange is ensured. At the same time, as in the lattice leaves of the flowering plant *Ouvirandra*, an increased proportion of surface to bulk of the vascular tracts is secured, a point which will be taken up again in Chapter X.

In the advanced Leptosporangiate Ferns the ascending or upright stem has a vascular system in the form of a tubular lattice-work, embedded in a matrix of "ground-tissue." It has been seen from evidence, both ontogenetic and comparative, that this is derived from the expansion and disintegration of a protostele. The origin of the internal ground-tissue or pith has here been attributed to change of procambial destination. In Gwynne-Vaughan's words, a "theory of transformation" has been adopted. Others have suggested an intrusion of cortical tissue from without, which may be described as a "theory of substitution."

Certain peculiar types of stem-structure which appear relatively late in the ontogeny of a few isolated genera have a special significance in this relation.

The sections of *Plagiogyria* (Fig. 138) have shown that in the younger shoots the leaf-trace separates as a sector of the solenostele, and passes out without complications, except that a centrally-placed strand of sclerenchyma follows its course outwards into the cortex. In older shoots this is much larger, and its tissue is discontinuous at the centre (Fig. 138, *D*, *x*). A semilunar space appears which widens out as the trace passes upwards, and opens to the atmosphere in the angle between leaf and axis. Each adult leaf is thus subtended by a deep involution of the outer surface, lined by sclerotic tissue; but it stops short before the stele itself is reached. Gwynne-Vaughan found a similar structure in *Anemia*, *Cystopteris*, and *Matteuccia Struthiopteris*. But in these Ferns, when adult, the involutions extended into the stele itself, forming deep pockets in the pith, which were lined by epidermis (*New Phyt.* Vol. iv, p. 211).

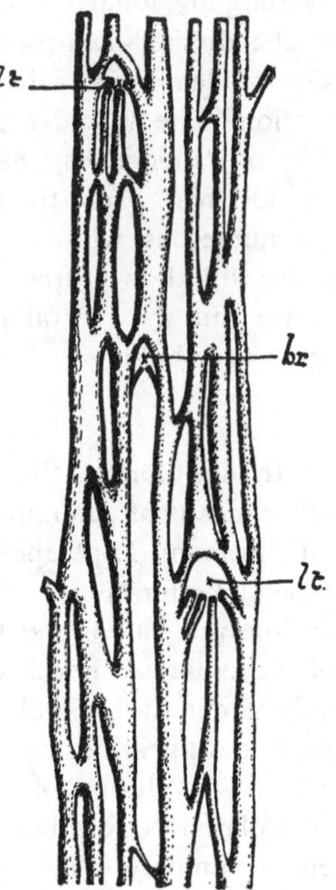

Fig. 141. *Stenochlaena tenuifolia* (Desv.), Moore, after Mettenius. Stelar system flattened into a single plane, showing perforations. *l.t.* = leaf-trace; *br.* = vascular supply to a branch.

In interpreting these involutions it is important to note how isolated is their occurrence. The genera named have no near affinity to one another. Moreover the involutions are absent from the young plant. As Gwynne-Vaughan points out they appear as "in each particular case the latest expression of a long series of advances from the primitive solid stem with its single solid central protostele." The physiological significance is almost certainly in relation to gaseous interchange with the atmosphere, which has thus direct access to the inner tissues, together with enlarged surface of exposure. It seems possible to interpret them in terms either of transformation or of substitution of the inner tissues. A static change of destination like that seen in forming any involution of endodermis would afford as good an explanation of the structure described as some dynamic inflow of tissue from without. The involution may be held to be a last consequence of the transformation of the primitive cylindrical stele into a dictyostele, and not as an indication of the way in which dictyostely arose.

In some creeping Ferns of large size and relatively primitive character the simple solenostelic structure is maintained, the greatly enlarged pith being surrounded by a distended and very thin solenostele. Examples are seen in *Cibotium* (*Dicksonia*) *Barometz*, and in *Pteris laciniata*, where the soft parenchymatous storage-pith may be over an inch in diameter (Fig. 142). It is likewise seen, though on a smaller scale, in *Dipteris conjugata*. Such arrangements when carried to large dimensions are obviously unpractical, and the physiological difficulties raised by the continuous endodermal barrier are met in various ways. One of these is illustrated in *Histiopteris incisa* (Thunbg.), J. Sm. Here at each leaf-insertion a deep lateral infolding of the cylinder encroaches on either side of the pith. This is merely a simple solenostele complicated by corrugation, by self-fusions, and by perforation. In the internodes the corrugation is often continued, with many deep folds (Fig. 143). The effect is greatly to enlarge the surfaces of the thin expanded solenostele, which will naturally aid gaseous and other interchange. But since the endodermis is a closed sheath the effect is less than that of perforation. It may be compared, as regards method, with what is seen in the stellate protostele of *Asterochlaena*: in both cases the corrugation appears in organisms of considerable size (see Chapter X).

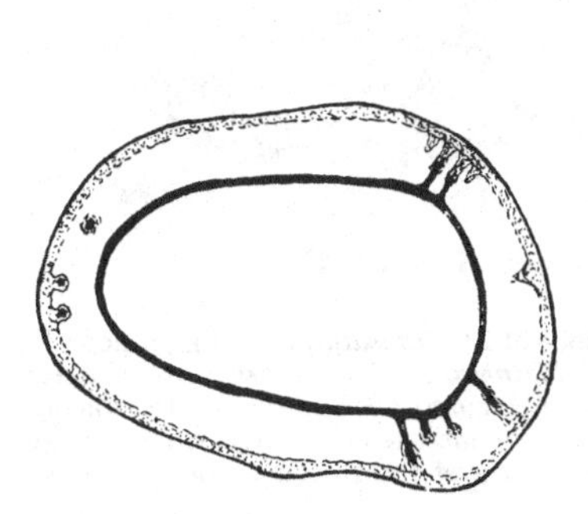

Fig. 142. Transverse section of solenostelic rhizome of *Cibotium* (*Dicksonia*) *Barometz*. Natural size.

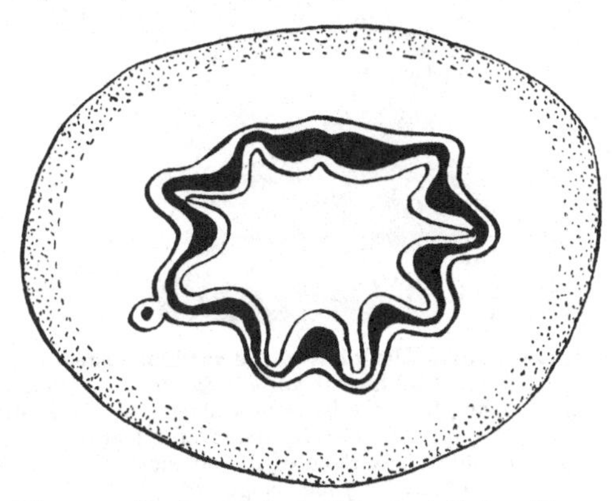

Fig. 143. *Histiopteris incisa* (Thunbg.), J. Sm. Transverse section of internode of rhizome (× 10) showing corrugation of the solenostele. (Gwynne-Vaughan collection, slide 1163, by Tansley.)

POLYCYCLY

In other groups of plants, such as the Cycads and Angiosperms, large parenchymatous masses are commonly traversed by accessory vascular tracts. This is found also in Ferns, giving rise to those states described under the general term of "polycycly," since in conformity with the vascular evolution of Ferns these inner conducting tracts are arranged in more or less regular cycles, one within another. Examples are seen in the Matoniaceae, Cyatheaceae, and Marattiaceae among the Superficiales, and in the Dicksonieae and Pterideae among the Marginales. The occurrence is so sporadic that it can only have been initiated along a plurality of evolutionary lines. As Tansley says, "it is no doubt a response to a need for an increased vascular supply." The need for ready conduction along the axis, as well as to and from the enlarged pith, appears a sufficient explanation of the occurrence of polycycly.

Gwynne-Vaughan first showed the gradual steps which led to polycycly in the Pterideae. In *Dennstaedtia apiifolia* there is a typical solenostele· but there is a thickening of the xylem at the margin of the leaf-gap. In *D. adiantoides* this thickening projects into the pith, and continues as a ridge along the internal face of the vascular cylinder from one node to the next: while in the internode the ridge contains a distinct strand of xylem surrounded by a phloem-sheath of its own (Fig. 144, *A*). In *Dennstaedtia rubiginosa* there are one to three separate vascular strands traversing the internodal pith, but connected with the main vascular cylinder at the nodes (Fig. 144, *B*). In the very large rhizome of *Dennstaedtia dissecta* (Sw.), Moore,

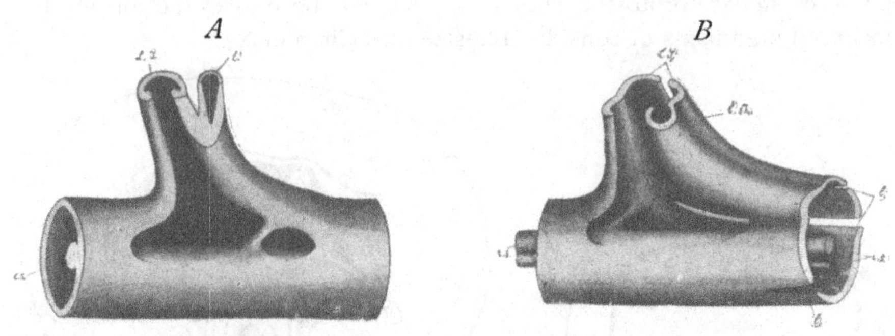

Fig. 144. *A*, *B*. Diagrams of the vascular system of rhizomes of *Dennstaedtia*, including a node in each. *A* = *D*. (*Dicksonia*) *adiantoides*. *B* = *D*. (*Dicksonia*) *rubiginosa*. *L.T* = leaf-trace; *l.sh* = lateral shoot arising from the basiscopic margin of the leaf-trace; *l* = lacunae or perforations of the solenostele not related to the departure of a leaf-trace; *i.s* = internal meristele, initiated from a ridge upon the internal surface of the solenostele. The upper surface of the rhizome would face the observer. (After Gwynne-Vaughan.)

there is an inner solenostele, which shows occasional openings. In *Pteris elata*, var. *Karsteniana*, the internal system is still more elaborate. Within the typical outer solenostele the accessory system may consist of a second cylinder which opens by occasional perforations. This regularly gives off a large flat strand which, passing forward at the node, inserts itself on the anterior margin of the leaf-gap of the outer cylinder. It is sometimes called the "compensation strand." Within the inner cylinder a rod-like strand, or even in large plants a third cylinder, may be found (Fig. 145). This structure is essentially the same as that of the large upright Fern, *Saccoloma elegans*, Klf., long ago described by Mettenius. A series of transverse sections from a large plant shows the detail of separation of the leaf-trace from the outermost cylinder (Fig. 146). The compensation strand is seen to pass off from the inner cylinder (secs. *c* or *g*): it moves outwards (*d*, *e* or *h*—*k*), and finally joins with the outer cylinder so as to close the leaf-gap (*f* or *l*). The acme of these complications hitherto observed is seen in *Pteris* (*Litobrochia*) *podophylla*, in which within three

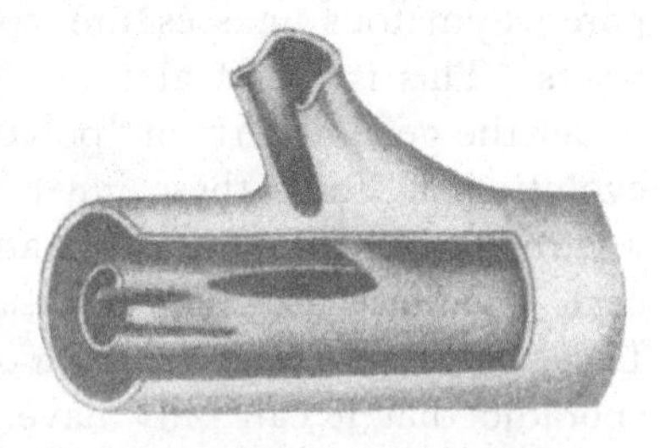

Fig. 145. *Pteris elata*, var. *Karsteniana*. Diagram showing the vascular tissue at the insertion of a leaf. A piece is supposed to be cut out of the side of the solenostele, so as to show the internal vascular system. (After Gwynne-Vaughan.)

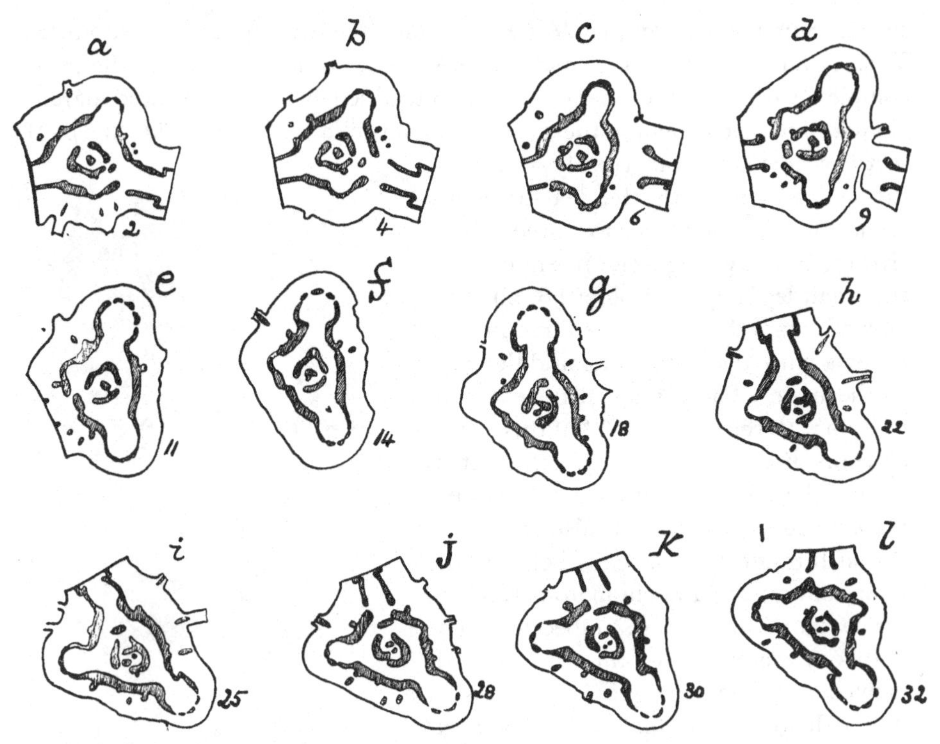

Fig. 146. Series of transverse sections arranged in ascending sequence, showing the relations of the polycyclic axial system to the leaf-traces in *Saccoloma elegans*, Klf. The numbers relate to the actual sections cut. Natural size.

complete cylinders a fourth is initiated (Fig. 147). Polycycly has also been observed in a lateral shoot of *Thyrsopteris elegans* (Fig. 149, 5), in which Fern the sappy stems are said to be sometimes as thick as the thigh. It would be interesting to know whether still more numerous cycles of vascular tissue are present in the largest of them.

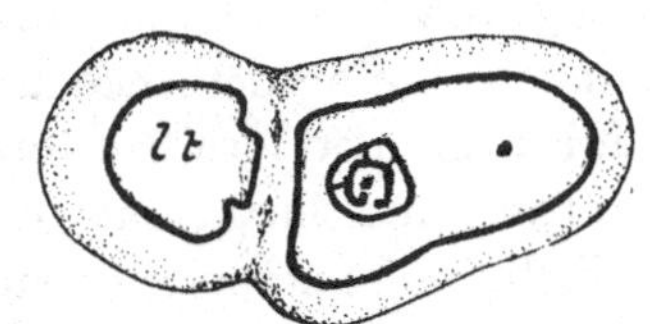

Fig 147. Stem and leaf-base of *Pteris* (*Litobrochia*) *podophylla*, Sw., represented the natural size. It shows polycycly in the axis, to the fourth degree. Note the completed ring of the leaf-trace (*l.t.*); on the right a further leaf-trace is preparing for departure, with an island of unconnected vascular tissue at its centre. This is surrounded by endodermis.

The familiar structure of the rhizome of the Bracken (*Pteridium*) is an example of polycycly in the Pterideae (Fig. 3), complicated by profuse perforation both of the outer and inner cylinders, and by the fact that the inner as well as the outer contributes to the supply of the leaf-trace (see Tansley and Lulham, *New Phyt.* Vol. iii, 1904, and compare their Fig. 67).

Such observations relate to Ferns belonging to the Marginales, and the comparisons are based upon adult material. A similar structure is seen in certain Superficiales, and in *Matonia* the ontogeny has been worked out by Tansley and Lulham. The mature rhizome of *Matonia* shows in the most complex cases three concentric rings embedded in well-ventilated parenchyma, and each having the typical solenostelic structure (Fig. 148). If sections be taken transversely at a node, the connections with the leaf-trace are seen. Fig. 148 shows this for three successive ages. The youngest (*A*) demonstrates the small leaf-gap; the older (*B*, *C*) show how the curved leaf-trace is directly continuous with the outer and middle rings at the node. There may also be a connection with the inner ring; but this occurs at some little distance from the actual node. The result is that the whole system is connected, while there is also communication through the leaf-gaps between the inner and outer parenchymatous tracts in which the cylinders are embedded. The ontogeny shows how this structure is arrived at. The young axis contains at first a slender protostele, which soon expands, and phloem appears centrally in it; but there is as yet no true leaf-gap. The stele soon widens into a solenostele with internal endodermis surrounding a central pith. Meanwhile at the nodes a ridge of vascular tissue projects internally, and, becoming detached as a separate rod, it is continued forwards into the internode further and further at successive nodes, till that of one node eventually connects with that of the next node (Fig. 148, *A*). A continuous central strand is thus produced, which is connected at the nodes with the outer cylinder. The process thus described may then be repeated in this central strand: it also becomes cylindrical, forming a second vascular ring which is still connected at the nodes with the foliar system (Fig. 148, *B*). A fresh strand may then originate from it: this in turn becomes cylindrical in the largest rhizomes, but it still maintains its connection with the middle and outer rings in the neighbourhood of the nodes (Fig. 148, *C*). The whole

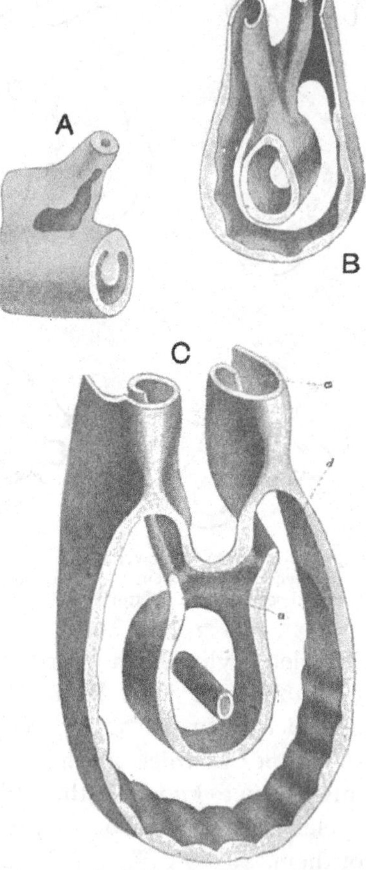

Fig. 148. *Matonia pectinata*, drawings from wax models of the stelar system. *A* = from a young stem showing a node. *B* = from an older stem, showing a node as seen from behind. *C* = still older node seen from the front. ($A \times 25$; $B \times 12$; $C \times 10$.) (After Tansley and Lulham.)

development corresponds essentially with that described by Gwynne-Vaughan for the Pterideae: but as the Matonineae are systematically far apart from these, it appears that polycycly is homoplastic in two at least, and probably in several other phyla.

The very much disintegrated system of vascular strands found in *Platycerium aethiopicum* may be referred to a state of polycycly combined with profuse perforation of the solenosteles (Fig. 149, 8). In the smaller *P. alcicorne* the numerous meristeles are arranged in a simple circle, and a group of strands passes off from them as a leaf-trace to each leaf (Fig. 149, 7). This seems to be

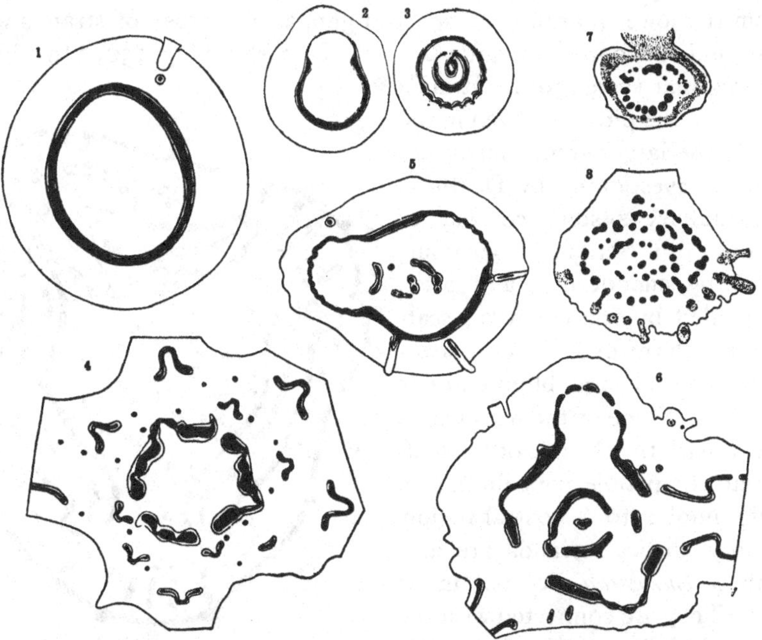

Fig. 149. Series of solenostelic and dictyostelic stems of Ferns, all drawn to same scale. (× 2.) 1, *Metaxya*; 2, *Dipteris conjugata*; 3, *Matonia pectinata*; 4, *Plagiogyria pycnophylla*; 5, *Thyrsopteris elegans*; 6, *Saccoloma elegans*; 7, *Platycerium alcicorne*; 8, *Platycerium aethiopicum*. These drawings show that the disintegration of the stele does not depend on absolute size alone.

derived from a solenostelic state with many perforations, as commonly seen in advanced Leptosporangiate Ferns. In the larger *P. aethiopicum* the very numerous meristeles are disposed in irregular concentric circles, an arrangement referable to a polycyclic state with profuse perforations (Fig. 149, 8). The extremely complicated vascular systems of the Marattiaceae and Psaronieae may also be properly ranked as further examples of a high state of polycycly, greatly disintegrated by perforation. Here they can only be mentioned and reference given to Tansley's account of them (*Lectures on the Filicinean Vascular System*, pp. 82–95), and to the reconstructions of

West (*Ann. of Bot.* 1917, p. 361). But in referring the structure of the Marattiaceae to this source it is to be remembered that after the first ontogenetic stages are past there is no regular endodermis surrounding the vascular tissue; and the same appears to have been the case with the Psaronieae. In this respect these Eusporangiate Ferns differ from all the Leptosporangiates.

Accessory Strands

The structure seen in the dendroid Cyatheaceae (excl. Dicksonieae) is fundamentally dictyostelic, with very broad meristeles, each enclosed within equally broad plates of strengthening sclerenchyma. The leaf-traces come off from the lower margins of each leaf-gap, as a number of strands which take an oblique course outwards through the cortex (Fig. 150). In addition to this vascular system other strands are found in the pith, and even in the cortex in the larger stems, such as *C. Imrayana*, investigated by De Bary. The medullary system consists of numerous thin strands, each surrounded by endodermis, and often accompanied by sclerenchyma, scattered through the pith. They anastomose freely, and show blind endings downwards. Some of them may pass outwards with the strands of the leaf-trace into the petiole, contributing in varying number to its central region. Accessory cortical strands are also seen in *C. Imrayana*, but not in all species. They are connected with the leaf-trace-bundles as strands continuous downwards, sometimes anastomosing, sometimes ending blindly, sometimes joining up with the traces of lower leaves. It would seem probable from the irregularity of their course and occurrence that all these cortical and medullary strands are accessory, in the sense that they have originated in the enlarged parenchymatous tracts independently of the general vascular system, not as a result of branching from it or of disintegration of any former polycyclic state. That such new formations may occur is shown by the occasional appearance of a short and isolated vascular strand, surrounded by endodermis, which ends blindly in both directions, in the centre of the large leaf-trace of *Pteris podophylla* (Fig. 147).

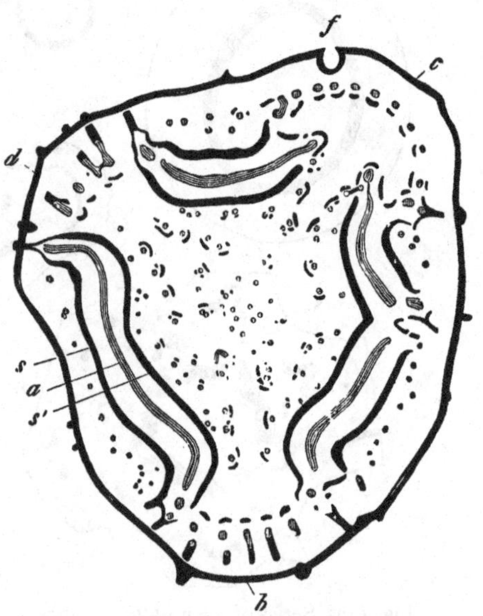

Fig. 150. *Cyathea Imrayana*, Hook. Transverse section of stem. Natural size. At *b*, *c*, *d*, foliar gaps; all the black bands and parts are stereom, all the paler bands are vascular strands in section. *a* = vasc. strands of the main cylinder; *s*, *s'* = outer and inner plates of the sclerotic sheath. (After De Bary.)

Similar blind strands are seen also in the pith of *Schizaea dichotoma*, and of *Platyzoma*. Such examples show how futile it is to attempt to derive all vascular tissue, or all tracts of endodermis, from some pre-existent source: for *P. podophylla* is as complex in vascular structure as any known Pterid.

Accessory strands are also recorded in the pith of large plants of *Ceratopteris thalictroides*, but neither their origin nor their connections are clear.

It is thus seen that in the Leptosporangiate Ferns there is a wide range of stelar structure in the stem. The leading character of it is the disintegra-

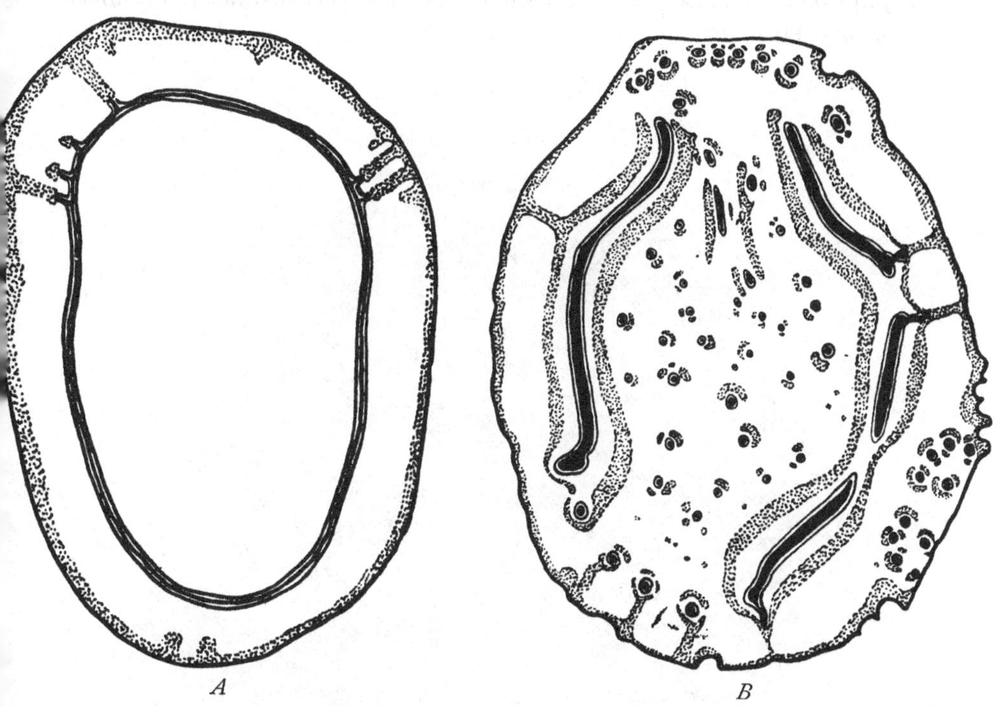

Fig. 151. Transverse sections of stems, drawn to the same scale, showing that stelar complication does not depend directly upon size alone. (× 2.) *A* = *Cibotium Barometz*; *B* = *Hemitelia setosa*.

tion of the stele, so that as the plant enlarges it is broken up into numerous smaller tracts, instead of the single stele itself growing proportionately as a solid cylinder, in the manner seen in so many other plants. Naturally the more complex structure appears in the larger stems. But there is no exact proportion between size and complexity. This is shown by comparing the wide and simple solenostele of *Cibotium Barometz* with the complex solenostele and accessory medullary strands found in the rhizome of *Hemitelia setosa* (Fig. 151, *A*, *B*). But the section of the latter here shown is of no greater size than that of the simpler *Cibotium*. The whole series of the Leptosporangiate Ferns is further characterised by the completeness of the endodermal sheaths, which shut off the vascular tissue of the stem from the surrounding parenchyma; that is, the non-ventilated conducting tracts are strictly delimited from the ventilated tissues that embed them. This feature is not only seen in the juvenile stem, as is the case in the Higher Plants, but it is maintained throughout life, and equally in the largest axes. In this the Leptosporangiate Ferns differ from the larger primitive Ferns, and indeed they stand alone among the larger groups of Vascular Plants, as they do also in the complexity of their stelar subdivisions. It will be seen in Chapter X that physiological reasons can be assigned for these very distinctive peculiarities of their structure.

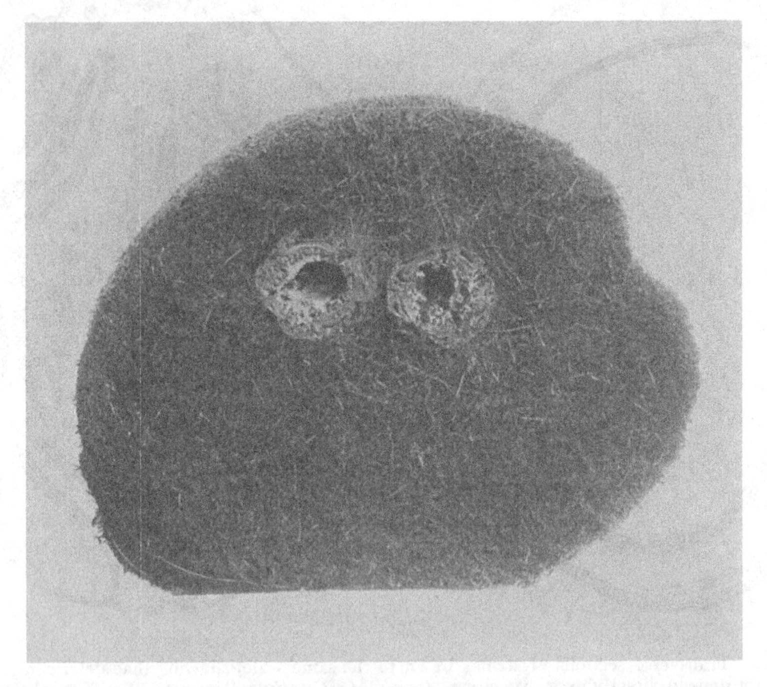

Fig. 152. Transverse section of a large bifurcated trunk of *Cyathea medullaris*, showing the relatively small size of the twin stems, and the large bulk of the adventitious roots that embed them. The two pith-cavities can be followed down through the massive section, and are found to be continuous with that of a single trunk below; this is in itself evidence of dichotomy. Much reduced.

It will be obvious that any stem constructed on the lines described in this Chapter will be mechanically unstable. Since it increases in size from the sporeling to the adult, and there is no secondary thickening, the form of the stem is obconical, and in upright plants the cone is balanced on its apex at the level of the soil. The mechanical, and also the physiological problem will be the same as that of the Palm-type among the Monocotyledons; such problems are solved in both cases by adventitious roots. In most Palms the cone widens rapidly at the base, and the strut-roots arise near the base only. Sometimes, as in *Verschaffeltia*, they spring far up the elongated conical stem. This latter state is seen in Tree Ferns. Their stem gradually increases from its narrow base upwards, till the full size of the adult tuft of leaves is reached. It then continues approximately as a cylinder. The lower conical region is, however, buttressed by the development of innumerable adventitious roots, which, growing down, interlacing, and developing with sclerotic cortex, constitute not only an efficient mechanical support, thus making up for the weakness of the slender obconical stem, but also yield the increased physiological supply needed by the enlarged area of leaves (Fig. 152). With such aid from the adventitious roots a primary development of the shoot, which would otherwise have been an unworkable proposition, becomes mechanically and physiologically sufficient for upright stems, without resort to secondary increase. Most Ferns, however, have a creeping habit, so that in them the mechanical difficulty does not arise.

From the phyletic point of view the argument to be based on the stelar structure seen in Leptosporangiate Ferns is the same as that for the more primitive Ferns described in Chapter VII. *Those which in the adult state are structurally nearest to the protostele are held to be relatively primitive in respect of that feature; those which have departed furthest from it are regarded as the most advanced.* But since similar advances are found in Ferns divergent in other respects, it appears that they are liable to homoplastic development. Hence *similarities in advanced structure cannot themselves alone be held as trustworthy indications of affinity.*

NOTE. To avoid needless repetition the bibliography of stelar tissues will be given collectively for Chapters VII, VIII, and X at the close of Chapter X.

CHAPTER IX

THE VASCULAR SYSTEM OF THE LEAF

FROM the study of certain sporeling-leaves of the Filicales given in Chapter IV, some idea will have been gained of that increasing complexity of the leaf which is shown in the development of the individual plant. It may be held that an advance similarly carried out in the race, combined with the "webbing" of the finer segments, has led to that confirmed megaphyllous state which is characteristic of Ferns. Photo-synthesis and Transpiration are the chief functions of the expanded blade. These functions impose a demand for conduction roughly proportional to the area of exposed surface, so that an elaboration of the vascular supply of the leaf-stalk may be expected to follow progressive expansion of the blade. Moreover this effect in the enlarging leaf will need to be harmonised with the stelar structure of the stem, so as to balance the dimensions and the physiological activity of the leaves which it supplies. Comparative study of the stele of the axis shows clear evidence of its adjustment in size and construction to the physiological demands. This also is seen in the ontogeny from the sporeling onwards. Though the phyletic sequences are naturally less clear than the ontogenetic, a like adjustment can be traced in them also. This being so, it will be to the base of the leaf that we shall look for the characters that are archaic, while the upper leaf with its complex branchings, from which the initiative in its evolutionary advance as a photo-synthetic organ has clearly come, may be expected to give evidence of structural innovations. That this expectation is justified by the results of comparative examination of the vascular system of the leaf in Ferns is the very general opinion of anatomists.

For the purpose of anatomical description the leaf may be divided into four successive regions. They are: First, the region of insertion of the vascular supply upon the stele of the stem, and its passage through the cortex into the leaf-base. This is designated the region of the *leaf-trace.* Secondly, the region of passage outwards through the stipe or petiole, and its continuation upwards in monopodial types through the midrib or rachis; this is the region of the *phyllopodium.* Thirdly, the region of insertion of the supply to the several pinnae upon the upper part of the phyllopodium, each of the branches being designated a *pinna-trace.* Fourthly, the ultimate branchings of the strands in the flattened expansion, which are designated collectively the *venation.* It will be found that the facts relating to each of these have their comparative value, but the most important region for phyletic comparison is the first, that is, the conservative basal region.

The Leaf-trace

The *leaf-trace* in types recognised on other grounds as relatively primitive is *undivided*, consisting of a single vascular tract enclosed habitually by endodermis. The form of it as seen in transverse section varies from the circular to some flattened or variously curved outline. An almost cylindrical leaf-trace with one enclosed tract of protoxylem is seen in *Botryopteris cylindrica* (Fig. 153, *A*). In this case it closely resembles the vascular supply to a

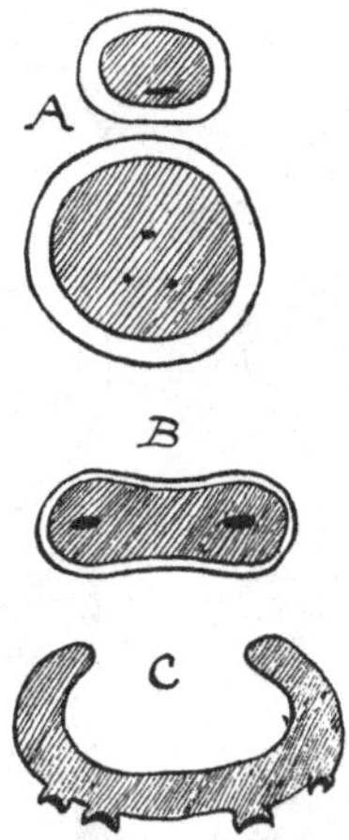

Fig. 153. *A* = diagram of stem-stele and petiole-trace of *Botryopteris cylindrica*, showing undivided protoxylem in he oval leaf-trace, which is the smaller and upper tract in Fig. *A*, the lower and larger being the axis. (After Miss Bancroft.) *B* = petiolar meristele of *Clepsydropsis*; *C* = foliar trace of *Anachoropteris*. (After P. Bertrand.)

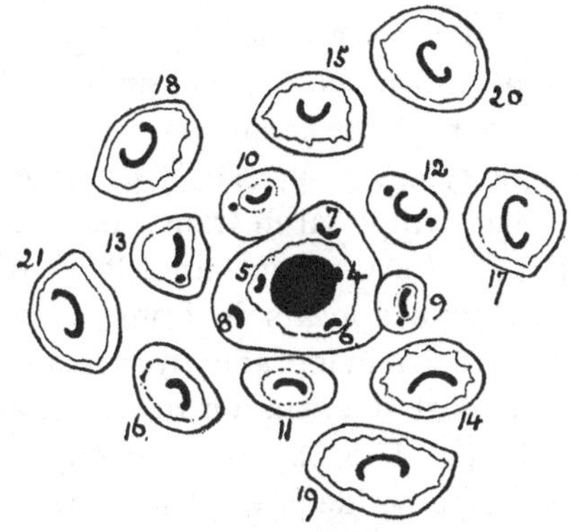

Fig. 154. Transverse section of shoot of *Tubicaulis solenites*, Cotta, after Stenzel, showing stem, with protostele, and the last four leaf-traces traversing the cortex. The leaves are numbered in their succession. The drawing is simplified by omission of roots, etc.

branch of the axis itself, and a similar condition is seen in the leaf-trace of many living Hymenophyllaceae. Oval or slightly flattened leaf-traces are found in many Botryopterideae and Zygopterideae (Fig. 153, *B*), and in the Ophioglossaceae: and the same type is seen at the extreme base in the Osmundaceae (Fig. 155, 2). The outline of the trace in Ferns that are rather more advanced than these varies according to the form of the upper leaf, and as this is usually flattened, the leaf-trace takes the shape of a still more flattened strap. As the trace passes outwards through the cortex it may become curved, as seen in section, and it is noteworthy that the concavity may be abaxial or adaxial. The former appears in certain Botryopterids and Zygopterids, a fact which suggested to Professor P. Bertrand the descriptive name

"Inversicatenales" for the whole series of ancient forms named by Seward the Coenopterideae (Figs. 153, *C*, 154). Further up in the leaf of many of these ancient forms the vascular tract expands at the curved ends, giving a very characteristic appearance in the case of the Zygopterideae. They thus afford supply to the four rows of pinnae which their leaves bore (Fig. 73, p. 82). But in the vast majority of primitive Ferns which had flattened leaves, and only two lateral rows of pinnae, the undivided leaf-trace shows an adaxially concave section, and often it may be sharply folded into a complicated horse-shoe form. The steps leading to this state are illustrated, as seen at successive levels from below upwards, in the leaf-trace of *Thamnopteris Schlechtendalii*, as it departs from the stele of the axis (Fig. 155). Here the xylem of the leaf-trace first appears as a protuberance on the surface of the xylem of the stem, and contains a single xylem-group (1). It separates as a mass elliptical in transverse section (2). Further out an island of parenchyma appears in front of the protoxylem, and it gradually increases until at last it opens on the adaxial side (3, 4). The bay of parenchyma thus formed widens until the xylem resembles a crescent with curved ends, while the protoxylem elements spread over the margin of the bay (5), and there divide into two, and later into several groups (6, 7). At first the outline of the whole trace remains elliptical: but further out in the cortex of the stem a depression appears on the adaxial side, which widens till the trace becomes curved into a broad arch, while its ends are incurved. The xylem has meanwhile thinned out, and numerous protoxylem-groups are now seen along its concave margin (8, 9). It is believed that these stages may be taken as indicating the changes undergone in the phylogeny of the adaxially-curved leaf-trace so generally representative of the Filicales. But in the more advanced types the cylindrical or oval region at the base is frequently omitted, the trace being ribbon-shaped or curved from the first. The whole progression may probably be related to the flattened form of the leaf-blade, and to the progressive webbing of its segments laterally.

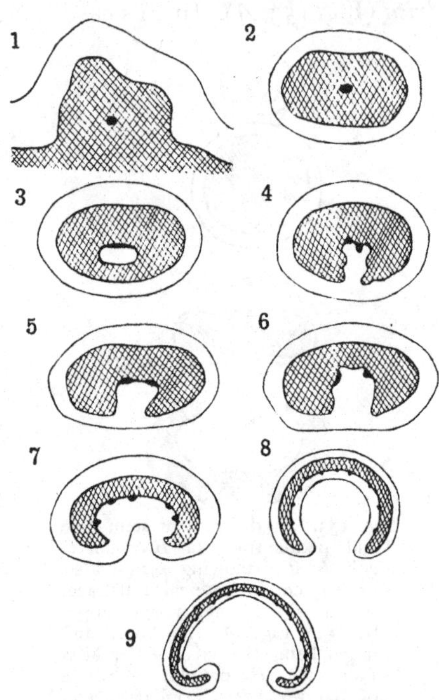

Fig. 155. Diagrams illustrating the departure of the leaf-trace in *Thamnopteris Schlechtendalii*, Eichd. (After Gwynne-Vaughan.) Compare text.

In the most primitive types such as the Botryopterideae and early Osmundaceae the leaf-trace, however widened it may be above, is contracted downwards as in *Thamnopteris* to something like the cylindrical form, and inserted without causing any depression or gap upon the protostele of the axis. But in more advanced types the full width of the leaf-trace may be continued downwards to the actual point of insertion, while the stele of the axis, dilated sometimes with medullation, or more commonly having solenostelic or dictyostelic structure, is by such means raised to a size suitable for receiving it. The dilatation of the stele is closely related to the width of the leaf-trace, and both have followed from the demands of an enlarging lamina; the effect of this has thus spread downwards in the course of Descent. Where solenostely has been arrived at, as described in Chapter VIII, the leaf-trace comes off bodily as a sector of the vascular ring, leaving a wide leaf-gap as

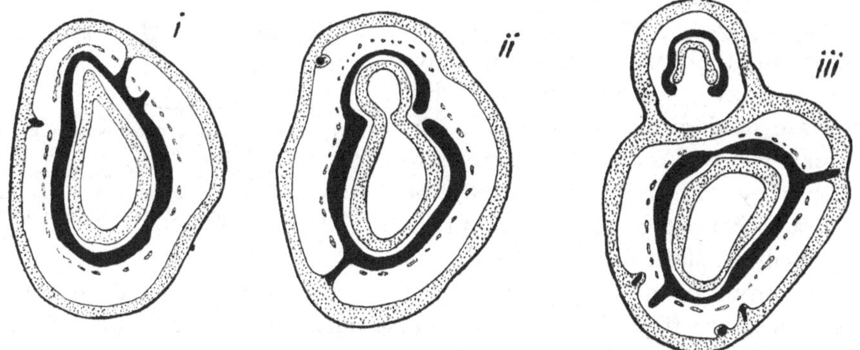

Fig. 156. i, ii, iii, acropetal succession of sections of a "runner" of *Lophosoria quadripinnata* (Gmel.), showing the origin of a leaf-trace: i, the solenostele is still complete, but the leaf-trace is indicated by a projecting curve; in ii, the leaf-trace has separated at one margin, thus opening the leaf-gap; in iii, the leaf-trace has passed off, and the gap is again closed. Slightly enlarged, sclerenchyma dotted, vascular tissue black.

evidence of the disturbance which it causes, and incidentally allowing communication between cortex and pith (Fig. 156). The trace itself may widen out upwards into a broad, more or less gutter-shaped sheet, which in many relatively primitive Ferns is quite continuous. It is completely enclosed in all Leptosporangiate Ferns by endodermis, and in proportion as its margins are near together it approaches structurally to the state seen in a solenostelic stem. Occasionally even the open channel of the gutter-shaped trace is closed, as may be seen in the large leaves of *Pteris* (*Litobrochia*) *podophylla* (Fig. 147, Chapter VIII), so that a vascular ring is completed also in the petiole.

On the other hand, in Ferns of relatively advanced type the leaf-trace may be *divided* into two or more separate vascular tracts; but these are still so arranged as to represent more or less clearly the underlying horse-shoe curve. In simple cases the trace is divided in a median plane into two equal portions, as in *Athyrium Filix-foemina*, *Trismeria*, or *Asplenium adiantum-*

nigrum (Fig. 157, *a*), and many others. Often, however, the subdivision is carried further, and a number of small oval or circular strands appear arranged in a curve, while two adaxial strands, which are usually larger than the rest, correspond to the free margins of the horseshoe. *Dryopteris Filix-mas* is a common example of this last state, which holds for most of the advanced Polypodiaceous Ferns (compare Fig. 161, 4, 6). A specially high degree of subdivision of the leaf-trace, accompanied by further complications arising from the highly divided polycyclic structure of their stems, is found in all of the living Marattiaceae. It may appear strange that Ferns so antique in other characters should have this advanced type of leaf-trace. But they are modern survivors, and in the ancient *Psaronius* and *Megaphyton*, to which they are probably related, the leaf-trace consisted of a continuous, highly-curved sheet of vascular tissue.

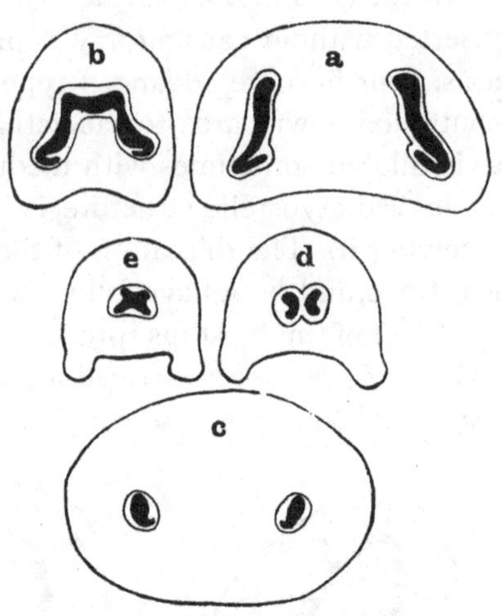

Fig. 157. Modifications of meristele in Fern-leaves. *a*, *b*, *Athyrium Filix-foemina*, *a* = at base. *b* = higher up. *c*, *d*, *e*, *Asplenium adiantum-nigrum*, *c* = at base, *d*, *e*, at points higher up. (After Luerssen from Rabenhorst.)

The protoxylem-groups in the adult leaf-trace are mesarch at the extreme base, but as the trace flattens in its course outwards they pass towards the adaxial face of the xylem (Fig. 155). Their number varies, and it often increases as the meristele passes upwards into the petiole. The attempt has been made to establish definite types of leaf-trace in living Ferns, according to the number of the protoxylem-groups; the primitively monarch with only one, the primitively diarch with two, and the primitively triarch with three (Sinnott, *Ann. of Bot.* 1911, p. 167). To the first the Osmundaceae and Ophioglossaceae are assigned; to the second the Marattiaceae; to the third all remaining Ferns. It is true that speaking generally, and from examination of the adult leaves, there is a larger number of protoxylem-groups at the leaf-base in the advanced than in the more primitive Ferns, and that the single protoxylem is the most primitive of all. But all Ferns show this character in their first sporeling-leaves. In fact Ferns generally start their ontogeny from the archaic state with a single protoxylem in the leaf. Some primitive Ferns retain this throughout life, but most of them pass on quickly to larger numbers (Fig. 155, 7, 8, 9). The Marattiaceae will serve as an example. Sinnott assigns two protoxylems to these Ferns. But Campbell (*Eusporangiatae*, 1911) has shown that if sections are made of the cotyledons there is only one vascular strand in each, with a very minute tract of xylem. This is seen in *Angiopteris* (*l.c.* Fig. 130), *Danaea* (Fig. 131), and in *Kaulfussia* (Fig. 137, *D*). Still there is an underlying truth in Sinnott's generalisation, which has its interesting bearing on ontogenetic comparison.

The distance through which the *protoxylem* may be traced downwards into the axis varies according to the proportions. Where the axis is short with crowded leaves it may be impossible to follow the protoxylem onwards into the stele. This is a natural consequence of the fact that "protoxylem" is the structural expression of development during extension of tissues. If little or no extension occurs, an absence of xylem structurally defined as "protoxylem" may be expected.

A comprehensive view of the leaf-traces of Ferns leads then to the generalisation that the primitive leaf-trace was structurally very like the protostele of the axis, with a single protoxylem-group. That losing its cylindrical form it became flattened, with various modifications of outline. That in the majority of Ferns, and especially in the Leptosporangiate Ferns, it adopted the form of an adaxially concave strap, curved like a horse-shoe as seen in transverse section, and with a plurality of protoxylem-groups on the adaxial face of the xylem. But that in the most advanced types of them this strap became disintegrated as a "divided leaf-trace," though its parts still show by their arrangement the underlying horse-shoe curve. Thus the characters of the leaf-trace provide important features for phyletic comparison.

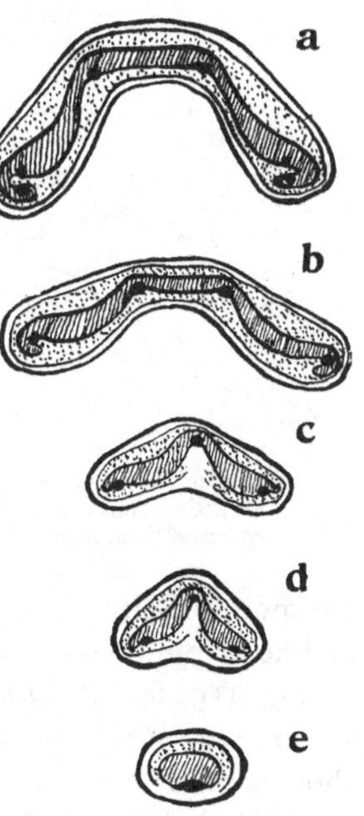

Fig. 158. Acropetal series of sections of the meristele of *Loxsoma*, showing its modifications towards the apex of the leaf. (After Gwynne-Vaughan.)

The Phyllopodium

In Chapter V the origin of the phyllopodium has been traced. In the great majority of Fern-leaves it extends from the leaf-base to the distal tip, appearing below as the petiole, and above as the rachis or midrib. It is traversed throughout by the vascular supply. As the leaf-trace passes outward to the petiole the meristele very frequently dilates, especially in relatively primitive forms (compare Fig. 155). Attaining its greatest size and complexity near to the base of the petiole, it is gradually attenuated upwards to the leaf-tip. Where there is an undivided trace there is often little to record. This is the case in *Loxsoma*. Here the curve of the trace is of rather open form, the xylem showing the usual adaxial hooks. There are six protoxylem-groups, of which two occupy the adaxial angles of the curve (Fig. 158, *a*). Higher up the curve flattens (*b*), and the strands of the pinnae are nipped off from its margins. Higher again the hooks disappear, the protoxylems becoming fewer until only three are left, one at the end of each arm, and one in a median position (*c*). Further up the curve again contracts, the meristele taking a triangular outline (*d*), and finally an oval form, with completely collateral structure and a single protoxylem (*e*). This may be taken as the history of a foliar strand in any relatively primitive example. An essentially similar structure is seen in *Schizaea*, *Anemia*, *Plagiogyria*, some *Davallias*, *Onychium*, *Cheilanthes*, and *Pellea*, all being Ferns of moderate size. A like form is also found in the pinna-supply of larger Ferns. It may be taken as a very general type for

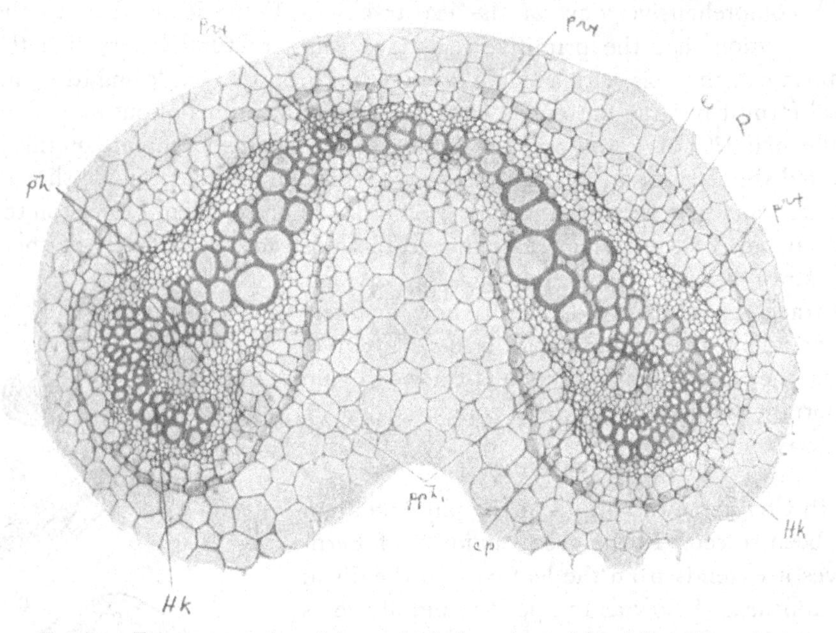

Fig. 159. Transverse section of the meristele from the petiole of *Davallia speluncae*. *e*=endodermis; *p*=pericycle; *ph*=phloem; *pph*=protophloem; *prx*=protoxylem; *cp*=cavity parenchyma; *Hk*=adaxial hooks. (After Gwynne-Vaughan.)

the more primitive Leptosporangiates where the leaves are small, or for the pinnae or pinnules of larger Ferns. The detail of such a meristele is shown in Fig. 159, for *Davallia speluncae*, which corresponds very nearly to that seen in stage (*b*) of the above figure for *Loxsoma*. *Athyrium* and *Asplenium* show another type of change by separation of the trace in a median plane into two halves, at the base of the petiole (Fig. 157). A like condition is seen in *Gymnogramme japonica* (Fig. 160), and in *Matteuccia*, *Struthiopteris*, *Onoclea*, *Dryopteris phegopteris*, *Scolopendrium*, and many others (see *Studies*, IV). This has been styled the *Onoclea-type* by Bertrand and Cornaille. Further up the phyllopodium the two halves unite, sometimes reconstructing the normal curve (Fig. 157, *b*), sometimes forming strands with xylem-masses of various outline (Fig. 157, *e*), but clearly related to the usual curve, of which they may be regarded as modifications. The basal splitting of the curve is

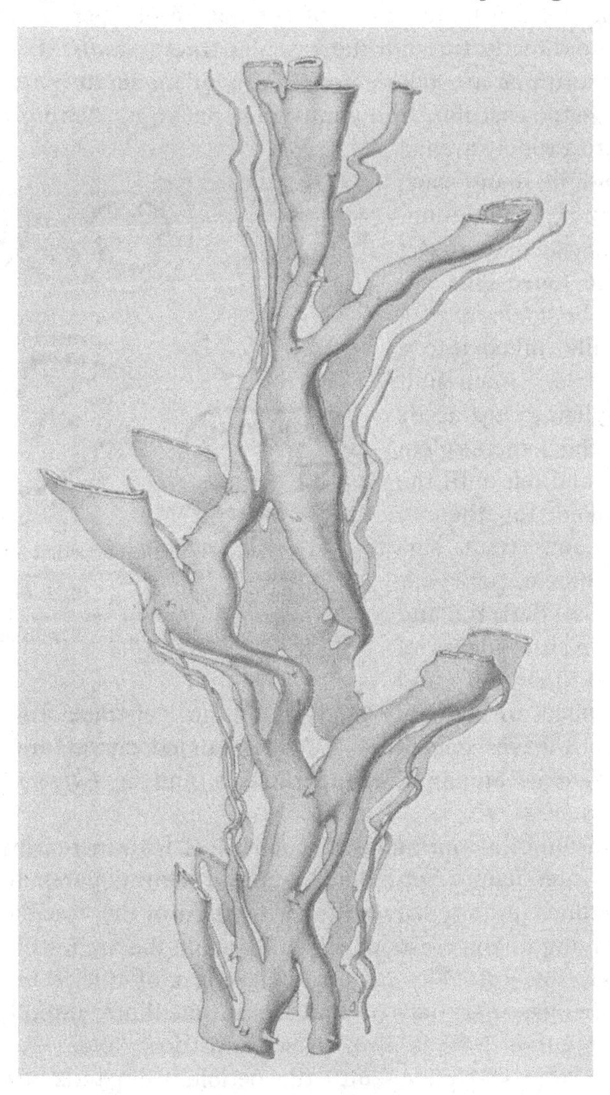

Fig. 160. Reconstruction of the vascular system of the axis of *Gymnogramme japonica*, showing how a stele with alternate leaves, each with bi-partite leaf-trace, may be further disintegrated by irregular perforations. (After McLean Thompson.)

probably in relation to ventilation, and its effectiveness is readily seen from the reconstruction of *Gymnogramme japonica*, where a long slit opens directly into the central pith. Two somewhat similar slits appear about half way down the petiole of *Plagiogyria semicordata*, though in *P. pycnophylla* there are none. The meristele is thus divided into three as seen in transverse section. The slits close downwards before the enlarged leaf-base is reached. Such slits may be considered to be of the nature of "perforations," allowing interchange directly through the vascular tract (*Studies*, I, p. 431).

These illustrations are taken from Ferns of moderate size. But where the leaf is large the vascular supply must also be large. A simple expansion of the trace into a widely arched curve is found in many such cases (Fig. 161, 1, 3). The number of the protoxylems increases and the whole horse-shoe appears thus to be made up of a number of units linked into a continuous chain. Each unit has a protoxylem-group as its centre, from which metaxylem extends right and left until the units are linked together to form a continuous tract, surrounded by phloem, pericycle, and endodermis. Bertrand and Cornaille gave the name of "divergent" to these units, and they have applied an elaborate analysis of the leaf-trace from this point of view to Fern-leaves generally. Such expanded curves are seen in the petioles of *Metaxya* among the Superficiales, and in *Thyrsopteris* among the Marginales (Fig. 161, 3, 5).

Fig. 161. Transverse sections of petioles, all drawn to the same scale. (×2.) 1, *Dipteris conjugata*; 2, *Dipteris Lobbiana*; 3, *Metaxya*; 4, *Phlebodium aureum*; 5, *Thyrsopteris*; 6, *Alsophila australis*. These show that while greater size leads to vascular disintegration, there is no definite proportion.

The expansion of a continuous vascular sheet with increasing size raises difficulties of interchange between cortex and central parenchyma. They are met sometimes in large leaves by corrugation of the vascular tracts, the protoxylems lying at the crests of the folds while the arches of metaxylem curve convexly inwards (Fig. 161, 3). The effect of this is to enlarge the surfaces of the vascular barrier, and so to facilitate interchange. But commonly the curve itself is also thrown into deep folds. As the arched leaf-trace of a large leaf passes into the petiole it expands, and two deep lateral involutions appear, which have obvious relation to the two ventilating tracts at the outer surface (Fig. 161, 5, 6 *pp*). In most Ferns of large size the petiole is strengthened by peripheral sclerenchyma impervious to gas-interchange. But laterally two lines, or in some cases interrupted areas roughly in linear succession, run up the leaf-stalk, and onwards to the lamina, where they are superseded by the flange-like wings of the flattened blade. They are particularly well seen in the Bracken. Stomata are found on their surface, and below it soft, well-ventilated tissue. They are in fact *pneumathode-areas* extending along the leaf-stalk and its branches. It is in relation to these that the deep involutions lie, so that there is ready access from the outside to the deeper-lying tissues for purposes of ventilation.

In addition to this the vascular tract itself is frequently broken up by perforations, which usually originate at points midway between the protoxylem-groups. Here the xylem may first be interrupted, forming xylic gaps. It is only a step further to the interruption of the meristele. The perforations may appear first on the abaxial curve (Fig. 162), but more commonly at the base of the involutions. The latter is seen in *Thyrsopteris* (Fig. 161, 5), and in *Histiopteris incisa*. In both regions the interruption is very completely carried out in the large leaves of the Cyatheaceae, so that the meristele has the appearance of being resolved into its constituent units, according to the analysis of Bertrand and Cornaille (Fig. 161, 6). In all of these units, however, the complete endodermal sheath is maintained, in sharp contrast to the breaking up of the meristele in the Marattiaceae and Ophioglossaceae, where the sheath is absent. The effect is that while the protoplasmic control over the conducting strands is maintained, intercommunication between the external and internal parenchyma of the petiole is fully provided for, even in the largest of the Leptosporangiate Ferns.

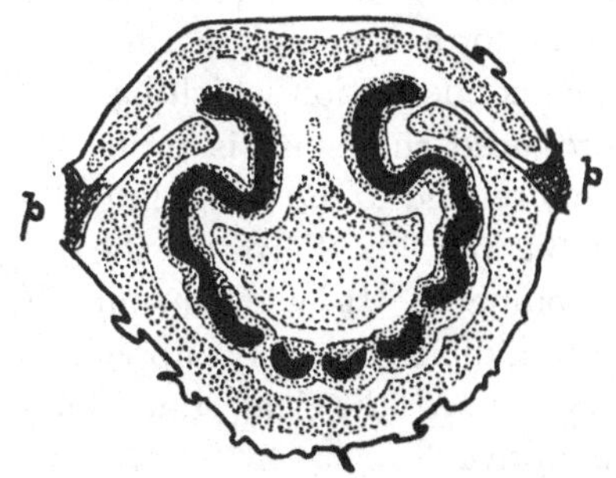

Fig. 162. Transverse section of the petiole of *Saccoloma elegans* (× 4). The vascular strands are black, sclerenchyma dotted: the clear areas are ventilated parenchyma, connecting with the pneumathodes *p*, *p*.

Perforations of the meristele naturally explain the divided leaf-trace. It has been seen that they frequently occur close to the base of the petiole. If they are extended downwards through the cortex of the stem to the very insertion of the foliar meristele, the effect is that in relatively advanced types the leaf-trace is divided from the first; it may be into two straps as in *Asplenium* or *Trismeria*, or it may be into numerous strands as in *Dryopteris*, and most of the Polypodiaceous Ferns. In relatively primitive Ferns, however, such as *Cibotium Barometz* and *Saccoloma*, the leaf-trace is still undivided at its base. But after traversing the cortex and entering the petiole it soon breaks up into a number of separate strands (Fig. 163), thus indicating structurally the intermediate position of these Ferns.

In contrast to these effects of expansion certain Ferns show a structure in the phyllopodium which points to contraction of the meristele. In place of a wide curve the leaf-trace is compact and almost cylindrical, while its details indicate not a primitive cylinder, such as that of *Botryopteris*, but actual condensation from a curved source. A good example is found in sections of the leaf-base of *Gleichenia dicarpa*, one of the section *Eu-Gleichenia* (Fig. 164). The meristele appears almost circular: the xylem shows the usual horse-shoe curve, but the two curved margins are fused together so as to form a complete ring. The condensation which has produced this state is probably related to the straggling habit of this Fern of the savannahs. It widens out into a dilated curve in upper regions of the leaf, thus indicating that the contraction is local and adaptive.

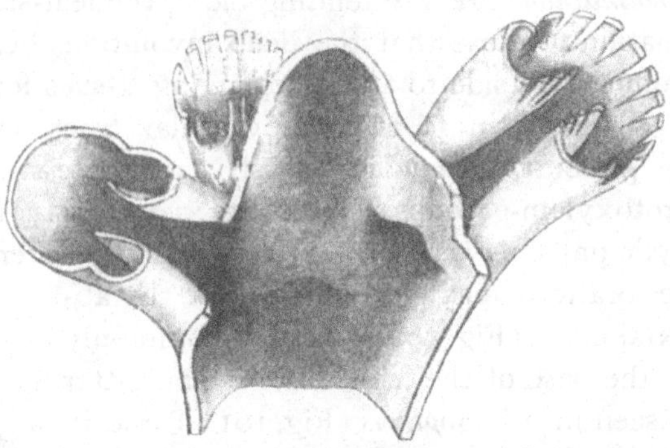

Fig. 163. Portion of the vascular system of the stem of *Cibotium* (*Dicksonia*) *Barometz*, seen from within, and showing the departure of three leaf-traces, which become disintegrated as they pass into the petiole. (After Gwynne-Vaughan.)

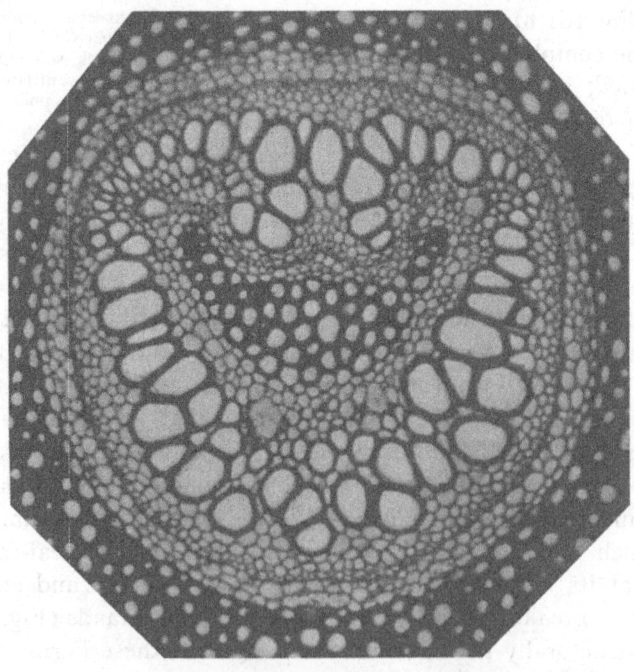

Fig. 164. Transverse section of the base of the petiole of *Gleichenia dicarpa*, showing the pseudo-stelar structure resulting from contraction of the horse-shoe-like xylem, till its margins fuse. (Photograph by Dr Kidston from section by Gwynne-Vaughan.)

Other climbing or straggling Ferns show consolidation of the vascular trace similar to that in *Gleichenia*, though differing in the details. In *Odontosoria* (*Davallia*) *fumarioides* (Sw.), J. Sm., which is also a plant of the savannahs, being grouped with *O. aculeata* (L.), J. Sm., as "bramble ferns," the meristele of the leaf below the lowest pinnae has a form as in Fig. 165, *e*, *f*. The two sides of the curve are massive and close together, while the adaxial hooks of xylem are short and stout in strong leaves, but in weak ones they are much reduced, and higher up they are obsolete. In *Lygodium scandens*, which is a successful climber, the outline of the meristele is oval, and there are no adaxial hooks, while the xylem of the two sides of the curve is completely fused (Fig. 165, *g*). From comparison of different specimens all

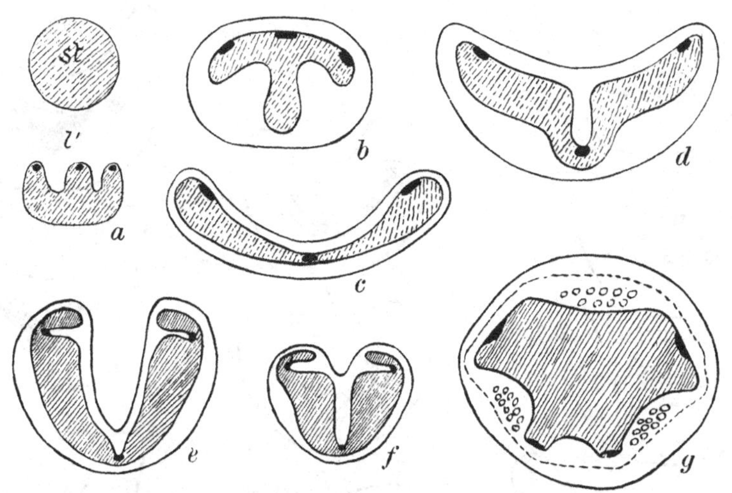

Fig. 165. Transverse sections of meristeles in the petioles of various Ferns, to show their modifications of outline. *a* = *Botryopteris*; *st.* stem, *l'.* leaf: *b* = *Schizaea*: *c* = *Anemia*: *d* = *Pellaea*: *e*, *f* = *Davallia fumarioides*: *g* = *Lygodium*. Compare text. The adaxial face is in all cases directed upwards.

stages between the open C-shaped trace and the condensed state of *Lygodium* may be found. As regards the orders of living Ferns it seems probable that their petiolar traces are all variants of the Osmundaceous type, having the form of an adaxially curved C. The traces of Gleicheniaceae, Cyatheaceae, Dicksoniaceae, and their derivatives conform fairly readily to this type. Bertrand and Cornaille have analysed those of the Hymenophyllaceae and Marattiaceae with a like result. In the Schizaeaceae, *Anemia* has a very typical C-shaped trace, but difficulty had hitherto been found with *Lygodium*. Gwynne-Vaughan's posthumous memoir shows that a comparison with the less specialised climber, *Odontosoria*, gives for it also an interpretation in terms of the Osmundaceous meristele (111). The final conclusion is that the meristele of the leaf of the living orders of Ferns is open to biological modifications, which are often homoplastic; but underlying these there is the leaf-trace of the Osmundaceous type.

The Pinna-trace

If the theory of leaf-architecture given in Chapter V be correct, and if all the more complex leaf-forms are ultimately referable in origin to dichotomy, then the branching of the vascular supply at a forking should give the clue to origin of the supply to the pinna. A good test case will be the regular dichotomous forking of a leaf of *Schizaea*. A series of transverse sections below a fork of the leaf of *S. dichotoma* will show how the single meristele here

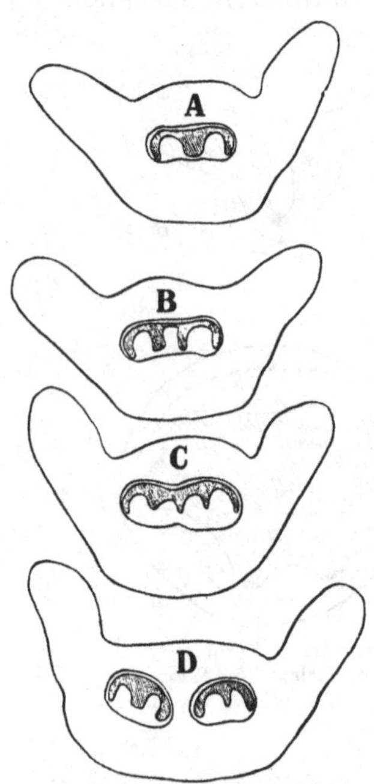

Fig. 166. Successive sections of the leaf-stalk of *Schizaea dichotoma*, showing dichotomy of the meristele. (×12.) The adaxial side is uppermost.

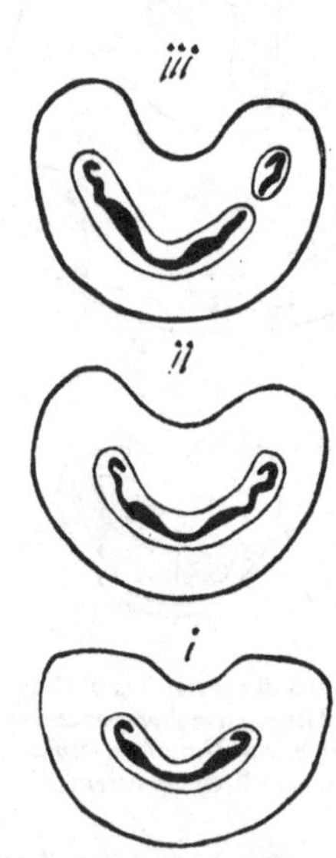

Fig. 167. i, ii, iii, diagrams illustrating the marginal type of pinna-supply, as in *Pteris umbrosa*, R. Br. (After Davie.)

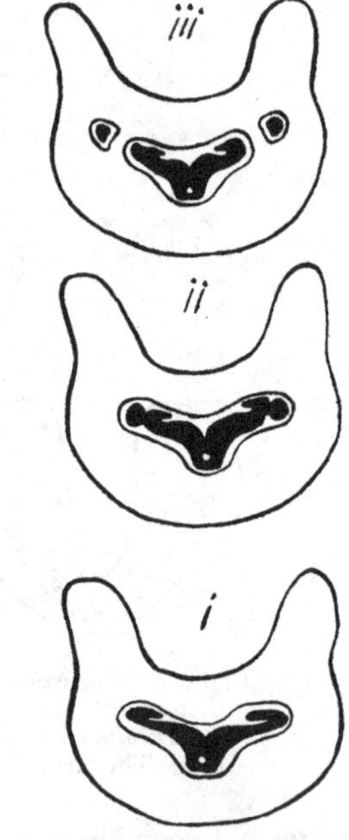

Fig. 168. Diagrams illustrating the extra-marginal type of pinna-supply, as in *Dryopteris vivipara* (Raddi), C. Chr. (After Davie.)

divides into two approximately equal parts by a median constriction (Fig. 166). Various degrees of inequality of these parts may be seen in passing from one specimen to another. A sympodial development naturally results in inequality of the branches, and the pinna-trace will be the smaller as a rule. The constriction would then nip off laterally a smaller from a larger portion of the meristele. In many Ferns, and especially in those where the leaves are relatively small, the pinna-trace is still isolated in this way, by abstriction from the margin of the meristele. However unequal the division may appear, it may still be held as an unequally developed dichotomy. Such "marginal" origin of the pinna-trace is seen in *Anemia* and *Loxsoma*; and a good example is shown by *Pteris umbrosa*, R. Br. (Fig. 167). This may be held to be the primitive method of supply of the lateral pinna, and it is characteristic of leaves of moderate size. But in many leaves of large size the meristele is strongly curved, and in these the origin of the pinna-trace is not by abstriction from the margin, but from a point or points on the abaxial or convex side of the curve. This has been described as "extra-marginal," and a simple example of it is seen in *Dryopteris vivipara* (Raddi), C. Chr. (Fig. 168). Either of these types may be further complicated in those Ferns where the meristele is widened out and laterally grooved in the manner already described for leaves of large size (Fig. 161, 1, 3). In them there may be vascular connections of the pinna-trace with both abaxial and adaxial curves of the meristele. This is illustrated for the marginal type in *Lonchitis pubescens*, Willd. (Fig. 169), and for the extra-marginal type in *Lophosoria pruinata* (Fig. 170).

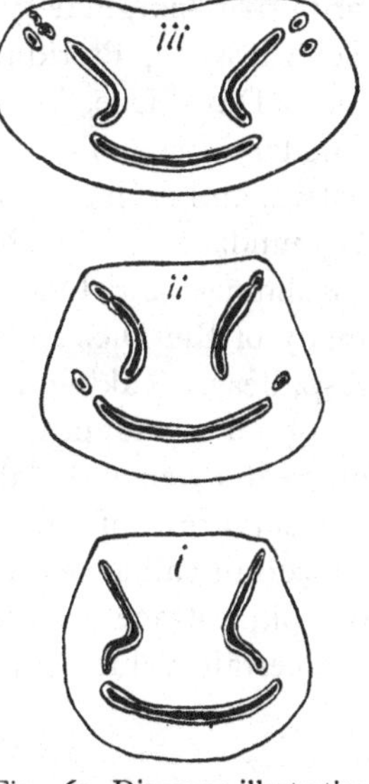

Fig. 169. Diagrams illustrating the marginal type of pinna-supply with a "reinforcement" derived from the abaxial curve of the leaf-trace, as in *Lonchitis pubescens*, Willd. (After Davie.)

There may be a good deal of variety of detail in these attachments, and where the leaf-trace is broken up into separate strands their recognition may be difficult. But they clearly follow lines of convenience, the ends to be gained being first a direct and effective supply for the pinna, and secondly the maintenance structurally of a means of conduction upwards to the higher region of the leaf. The supply to each pinna is most readily secured in small leaves where the leaf-trace is not strongly curved by simple abstriction from the margin of the meristele. But where it is strongly curved, as in larger leaves with numerous pinnae, the attachment is most convenient on its convex abaxial face. A direct supply upwards is at the same time secured by the continuance of the marginal tract of the trace forwards, which thus forms a direct channel of supply to the higher pinnae.

The connections of the pinna-traces might hardly be expected to supply reliable phyletic characters. In point of fact both the types above described may be found in the same leaf in *Trismeria trifoliata*, where commonly the supply to the larger, lower pinnae is extra-marginal, while that to the smaller upper pinnae is marginal. This is exactly what might be anticipated if the sympodial construction of the leaf be really true, and its progressive modification according to convenience be as above stated. As the transition occurs towards the apex to the primitive dichotomy, there should also be a transition to the primitive marginal origin of the leaf-trace; and this is what may actually be observed in the leaf of *Trismeria*.

Comparison of the pinna-traces in various groups of Ferns has shown, nevertheless, that there is greater uniformity than might have been expected. The more primitive marginal type is found constantly in the Schizaeaceae, Pteridinae, and Polypodiinae, and in many Davallieae, Asplenieae, Gymnogramminae, and Cheilanthinae. The extra-marginal is probably derivative, and is characteristic of the larger leaves of the Osmundaceae, Gleicheniaceae, Hymenophyllaceae, Blechninae, and Onocleinae; while it occurs also in many of the Dicksonieae, Cyatheae, Woodsieae, and Aspidieae. Taking these facts together, the extra-marginal type appears to be essentially a consequence of greater size and elaboration of the leaf, with which the curvature of the meristele is so closely related. In certain alliances, however, the type of origin of the pinna-trace appears to be so far fixed as to give it a certain value for phyletic comparison.

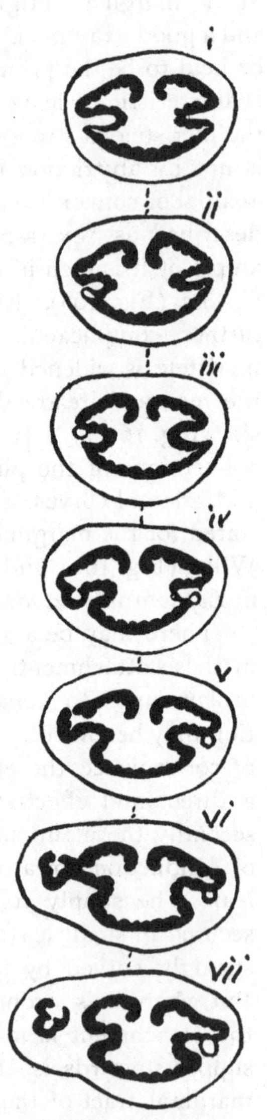

Fig. 170. Series of transverse sections through the rachis of *Lophosoria*, successively from below upwards, showing the origin of a pinna-trace. (×4.)

VENATION

The course of the veins in the flattened blade has already been discussed, and its value for purposes of comparison considered. Little need be said on the anatomical detail of the vascular system as it approaches the distal end or margin of the leaf, and resolves itself into the ultimate veins. These remain surrounded by endodermis, which encloses a larger or smaller mass of xylem directed towards the adaxial surface, and of phloem directed towards the abaxial. Technically the ultimate strands are collateral, and mechanical tissue is often associated with the strand, in girder form. In fact the characters of the ultimate veins in Ferns resemble in essential points of structure those in the lamina of other leaves, or that of the trace of the primordial cotyledon.

It thus appears that the vascular system of the leaf offers characters which are of value for phyletic seriation. They run substantially parallel to those of the axis, and there is reason to believe that the advancing complexity of structure of the blade has had its influence downwards, in promoting an increased complexity in the structure of the leaf-base and of the axis. The most archaic region of the leaf is believed to be the base, and *particular comparative importance attaches to the leaf-trace itself.* In the most ancient types it resembled the axial protostele. Progressive steps, such as are seen in the ontogeny of the leaf-supply in *Thamnopteris*, are believed to indicate correctly the phyletic dilatation of the cylindrical trace to the horse-shoe curve so characteristic of Ferns (Fig. 155). Further steps led to the gradual disintegration of the curved trace, and *the phyletic primitiveness or advance may be judged according to the degree of its disintegration.* The curve may, however, be contracted secondarily into a compact form in such Ferns as *Gleichenia*, *Lygodium*, or *Odontosoria*, in accordance with their climbing habit. The apparently similar simplicity of their leaf-traces is probably homoplastic, and does not give valid ground for phyletic argument.

The origin of the pinna-trace from the vascular supply of the phyllopodium is a less reliable feature than that of the leaf-trace from the stele. *The marginal origin of the pinna-trace is held to be relatively primitive, the extra-marginal being derivative.* But the latter may arise homoplastically, as a convenient adjustment in large leaves, and may even be departed from in the apical region of an individual leaf where it is present lower down. Nevertheless, in general terms the marginal origin has priority in evolution, and is characteristic for certain alliances, especially where the leaves are relatively small. The value of the venation for comparison lies rather in the study of it in surface view than in the anatomical details. *It is thus seen that in the anatomy of the leaf the most valuable features for comparison are found in the basal region, and in the leaf-trace itself.*

THE ROOT

The vascular system of the root in Ferns shows greater uniformity than that of the shoot. The roots are as a rule fibrous, and of small size; and their stele is correspondingly simple in structure. There is a broad cortex often densely sclerotic surrounding the relatively small stele, which is bounded on all sides by endodermis. Internally to this is the pericycle, sometimes consisting of only a single layer of parenchymatous cells, but often more, especially in xerophytic Ferns (Fig. 9, p. 8). The xylem is commonly diarch, as in *Pellaea*; but it may be monarch in the roots of *Ophioglossum*, while in the large roots of the Marattiaceae and some Ophioglossaceae the number may be as high as six (*Helminthostachys*), or more (*Angiopteris*). Beyond such variations in number, which appear in general to be correlated with size, the construction of the root is along lines familiar in other vascular plants. The attachment of the roots is commonly upon the stele or the meristeles of the axis; but it may also be upon the basal part of the leaf-supply, especially where the leaves are crowded, as in the case of *Dryopteris Filix-mas*. These characters give little help in comparative study.

BIBLIOGRAPHY FOR CHAPTER IX

108. DAVIE. The pinna-trace in the Ferns. Trans. Roy. Soc. Edin. Vol. l, No. 11. 1914.
109. DAVIE. On the leaf-trace in some pinnate leaves. Trans. Roy. Soc. Edin. Vol. lii, No. 1. 1917.
110. DAVIE. Comparative list of Fern pinna-traces. Ann. of Bot. 1916, p. 233.
111. GWYNNE-VAUGHAN. Some climbing Davallias. Ann. of Bot. 1916, p. 495.
112. BERTRAND & CORNAILLE. Études sur la Structure des Filicinées actuelles. Lille. 1902.
113. PAUL BERTRAND. L'Étude anatomique des Fougères anciennes. Progressus Rei Bot. Vol. iv, p. 182.
114. SINNOTT. The Filicinean Leaf-Trace. Ann. of Bot. 1911, p. 169.
115. GWYNNE-VAUGHAN & KIDSTON. Origin of the adaxially-curved Leaf-Trace in the Filicales. Proc. Roy. Soc. Edin. Vol. xxviii, 1908, No. 29.
116. BANCROFT. *Rachiopteris cylindrica*. Ann. of Bot. 1915, p. 513.
117. STENZEL. Die Gattung *Tubicaulis*. Bibliotheca Botanica. 1889. Heft 12.
118. TANSLEY. Evolution of the Filicinean Vascular System, p. 114. New Phytologist Reprints. No. 11. Cambridge. 1908.

CHAPTER X

SIZE A FACTOR IN STELAR MORPHOLOGY[1]

THE vascular system of the shoot in the Filicales has always commanded attention because of its high complexity. It will be shown, as the phyletic treatment of the Ferns is developed, how greatly the varying details which have been described in general terms in the foregoing Chapters may be made use of in their classification according to Descent. But this is only one aspect of the interest which their structure arouses. The question remains why such unusual vascular systems should have been developed at all. When the condition of Ferns as growing organisms is considered, and the limitation which their peculiar structure sets upon the performance of their functions, it will appear that increase in *Size*, carried out under certain structural restrictions, has been a decisive factor in leading to their extraordinary vascular development. According to the Principle of Similar Structures, so long as the same form and material are maintained, the surface and the resistant strength vary as the square of the linear dimensions, and the bulk, and consequently the weight, as the cube. It follows from this that what may be possible for a small structure may be quite impossible for a large one. The mechanical application of the principle is already familiar with reference to the animal body. For instance, the columnar legs of the elephant are held to be the inevitable sequel to its large size and consequent weight, while the thin arched legs of insects are only possible where the body itself is small and light. Botanists have, however, been slower in applying the principle to the study of plants. It is true that the question of the practicable limit of size of trees has been discussed from this point of view, and it is recognised that a change either of material or of method of construction would be necessary for effective growth beyond the limits already reached by some of them. But the principle is also applicable to other points of construction, such as the size and constitution of individual cells, and even to the forms of chloroplasts; as well as to various problems of distribution of tissue-masses, and to the physiological functions which they perform. For in many cases these depend upon the proportion which the surface bears to the bulk of the organ.

[1] Chapter X is based upon an Address on "Size, a neglected factor in Stelar Morphology," which was delivered at the Anniversary Meeting of the Royal Society of Edinburgh, Oct. 25, 1920, and published in the Society's *Proceedings*, Vol. xli, p. 1. In that Address the subject was treated generally, for Vascular Plants at large. Here a special application has been made of the Principle of Similar Structures in elucidating the peculiarly complex Vascular System of the Filicales.

The adult stems and roots of most plants are approximately cylindrical. The same is the case as a rule for their conducting tracts also. The cylinder is one of those solid forms in which the proportion of external surface to bulk is exceptionally low. Any deviation from the cylindrical form, either by external projections or by involutions of surface, necessarily leads to increase in the proportion of surface to bulk. The surface varies only as the square of the linear dimensions, but the bulk as the cube. It follows therefore that in carrying out any of those physiological functions of living organisms which depend upon surface, as do all those of the acquisition and interchange of materials, the actual size of the part which exercises that function is a matter of the greatest moment. It may be assumed that, if other things be equal, such as the structure and quality of the tissues that form the surfaces in question, the rate of interchange by diffusion of soluble gases or salts through a tissue-surface will be directly proportional to the area of the diffusing surface. But the demand will probably vary as the bulk enclosed by that surface. If that be so, then for each such function there will be a limit of size beyond which its exercise with sufficient rapidity will become impossible if the form be maintained, or if the quality of the surface-tissue through which the transit occurs remains the same. This suggests that the larger the plant is the more dependent it will be upon its form and detailed structure, not only for its stability, but also for the performance of its functions of absorption and transit of liquids and gases. This will apply not only to the external surface, but also to those internal surfaces which limit one tissue-tract from another. Not only the outer surfaces, but also the limiting surfaces of the internal tissue-tracts should then be carefully examined, both as to area, and as to their detailed structure.

In point of fact stems and roots are only approximately cylindrical. Fluctuations of size either by increase or by decrease are common. But the most general and the most important of them all is that primary increase of dimensions which is found in the stems of most plants as they pass from the juvenile to the adult state. For the moment only the *primary* increase is meant: all secondary or cambial increase may be ruled out of this discussion, however interesting its problems may be. Here the intention is to concentrate upon those problems which any land-living plant, *having no cambial increase*, must face as it passes from the juvenile to the adult state. The facts of ontogenetic development in plants which, like the Ferns, have no secondary growth provide the most cogent evidence of the effect of increase in size upon internal structure. In Ferns the first leaves are small: the later leaves are successively larger. The stem which bears them is relatively small at its base, but as larger leaves are formed the stem which bears them becomes proportionately larger, till the adult size is reached (Fig. 171). The same is the case with the stele that lies within. The form of the stem of the young plant is then not cylindrical, but it is a gradually enlarging cone. Consequently problems depending on the proportion of surface to bulk, whether of the stem as a whole or of the stele within, will be progressively changing in each successive transverse zone from the juvenile to the adult region. It may be anticipated that at some point of size a critical proportion of surface to bulk will be reached, where the interchanges between stele and cortex will demand some alteration of structure if they are to be satisfactorily carried out.

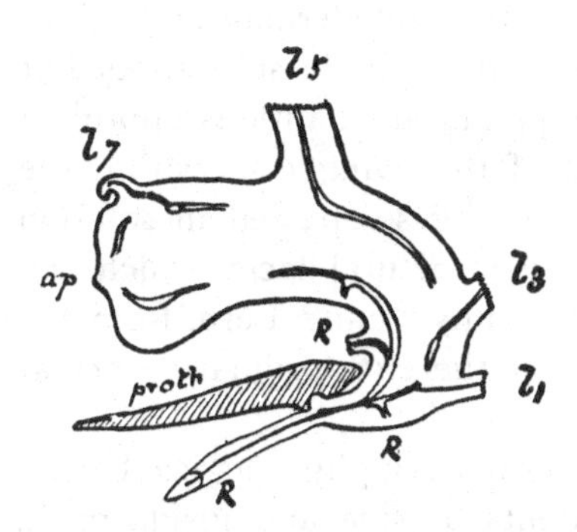

Fig. 171. Median longitudinal section through the prothallus and embryo of *Polypodium vulgare*. (×6.) Leaves l_1, l_3, etc.; R=roots; ap=apex of stem. The drawing shows the widely expanding conical stem—small at the base, where it is protostelic; larger above, where it is dictyostelic.

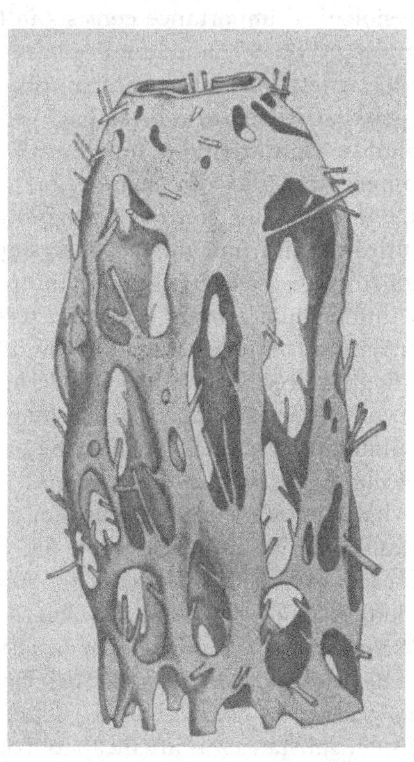

Fig. 172. *Deparia Moorei*, Hook. Reconstruction from sections of the vascular system of a plant which had been "starved" by unsuccessful culture. The normal dictyostelic state is shown below, with wide foliar gaps, and perforations. The stele narrows upwards, approaching the solenostelic state. (After McLean Thompson.) (×12.)

Conversely, however, an axis or root may diminish progressively in bulk from the base upwards. In a Fern that has been starved by unfavourable culture the size of its stem will be less than it is in the region developed under normal conditions, and the internal tissues follow suit, with simplification of their structure (Fig. 172). For them the problem, so far as it depends upon size, would be progressively simplified, and the evidence of this appears in their structure. The same result is seen in the roots of Palms. (See Cormack, 125.)

In the young stems of vascular plants, and in those of the Ferns in particular, the conducting tracts are strictly delimited from the surrounding tissues by endodermis. The same holds also for their roots. This endodermis forms not only a morphological, but also a physiological, boundary that when normally developed is without any gap or imperfection. Its physiological importance consists in the fact that the existence of even the simplest type of endodermis places the contents of the conducting tract under strictly protoplasmic control. All the lateral walls of its cells are so specialised in substance that instead of being permeable like ordinary cellulose walls, they are impermeable to fluids. Thus all possible leakage is stopped, and the only channel for transit of substances into or out of the stele is under the control of the protoplasts of the endodermal cells. This control applies not only to salts, sugars, and other soluble substances, but also to gases. Since in young and primitive plants the mantle is unbroken by intercellular spaces, even the respiration of the living cells within the barrier can only be conducted by interchange of gases passed in solution through the cells of the endodermis. These structural facts, which can be verified by sections of the stem or root of any young Fern, form the foundation of a theory which may account for the extraordinary vascular developments seen in these plants.

Evidence of the effectiveness of the endodermis as a physiological barrier is afforded by comparison of the cell-contents outside and inside of it. Marked cases may be found of difference in size of the starch-grains on either side of the barrier. This is well seen in the storage-rhizomes of *Pteridium*, or of *Helminthostachys* (Fig. 173), and a still more striking case is seen in *Acrostichum aureum*. Such facts indicate that the endodermis controls the passage of soluble sugar. That it is an effective barrier to the passage of such substances as are incapable of penetrating the protoplasm, but whose passage through the walls can be followed by their colour or by staining reactions, has been shown by de Lavison (*Rev. Gen. de Bot.* 1910, p. 225), and by Priestley (*New Phyt.* 1920, p. 192). Having such evidence before us pointing to the endodermis as a selective screen, or even an effective barrier to transit between the outer tissues and the conducting tract, the constant diminution of the proportion of surface to bulk as the stele increases in size becomes a matter of the utmost importance. The conical form of the stele, noted above for the stem of all young Ferns, starts from the minute stele of the sporeling. The stele expands upwards as a support to the successively larger leaves of the established plant. Often the increase is rapid, especially in Ferns with short internodes. In each case a limit must ultimately be reached where the facility for interchange through the endodermis will cease to suffice for the needs of the tissues within. This facility for interchange will then become a "limiting factor." Either some means of increasing the surface-area of the stele, and so of increasing the means of transit, must be supplied, or the conical enlargement of the stele must be checked, and the later-formed regions of the stele will then be cylindrical. The increase cannot be continued indefinitely in the form of a cone. But on the other hand, any deviation from the simple conical form, by involution or

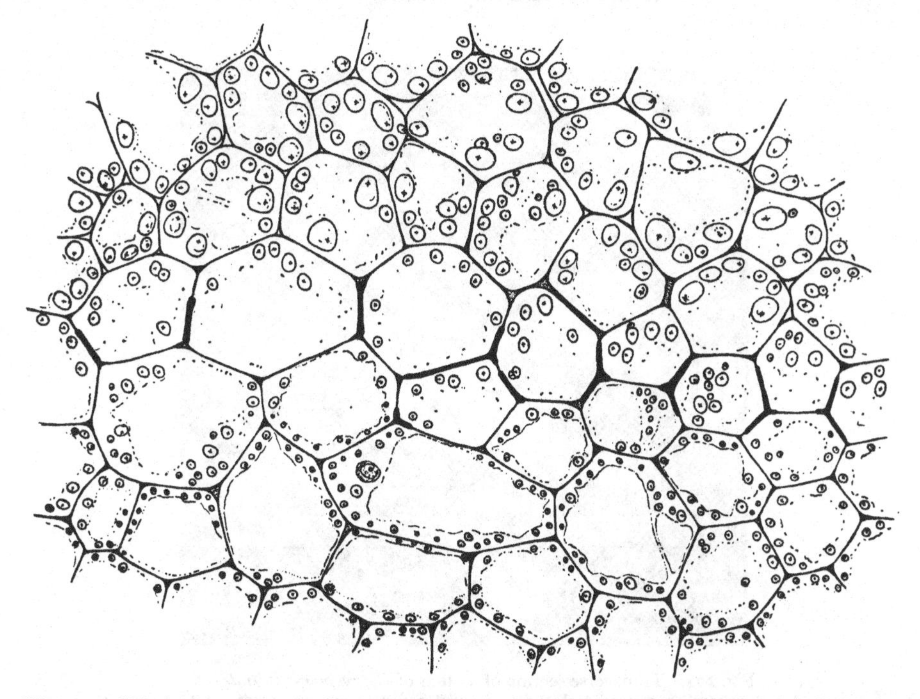

Fig. 173. *Helminthostachys zeylanica*: part of transverse section of root (Gwynne-Vaughan Collection, slide 589; × 66). The endodermis, recognised by the characteristic structure of its radial walls, marks a boundary between the outside cortex, with large starch-grains (here above), and the inner conjunctive parenchyma (here below), with small grains. (Drawn by Dr J. M. Thompson.)

by excrescence, will give an increase of the proportion of surface to bulk, and thus tend to overcome the difficulty. We may proceed now to see how these demands following on increase in size have been met in the stems of Ferns, putting upon the facts already stated in Chapters VII—IX an interpretation in terms of the proportion of surface to bulk.

It is generally admitted that the *protostele* is the most primitive stelar type. It is present in the juvenile stage of all Ferns, and it is permanently retained in the adult stems of some of them (Fig. 174). The stele receives the trace of each successive leaf, and it is important to note that its entry is effected without any break of continuity of the endodermal envelope, which thus forms a gas-tight barrier surrounding the whole vascular system. It is found in the adult stems of *Gleichenia*, *Lygodium*, and *Cheiropleuria*, and is believed to have been present in *Botryopteris*. It is also characteristic of the Hymenophyllaceae. All of these are relatively primitive types with stems of moderate size. In *Botryopteris cylindrica* the stele is about ·5 mm., in *Lygodium* 1 mm. to 2 mm., in *Cheiropleuria* about 1 mm. and in *Trichomanes scandens*, one of the larger Hymenophyllaceae, it is ·5 mm. in diameter. In

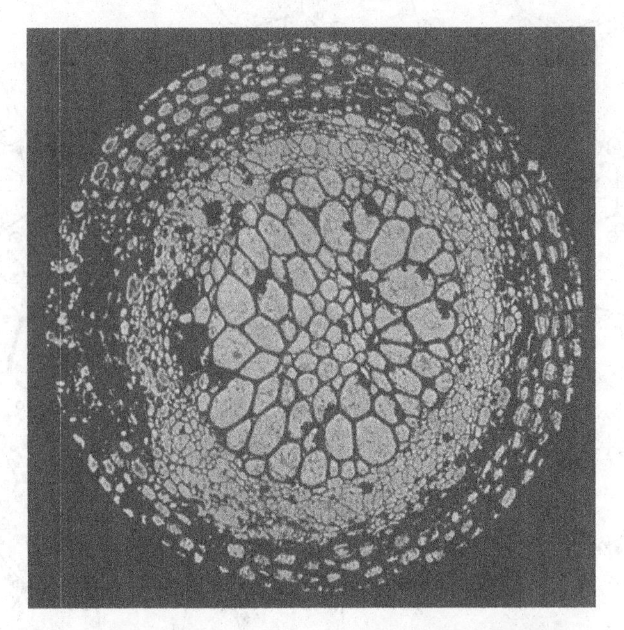

Fig. 174. Transverse section of a stem of *Botryopteris cylindrica*, showing a protostele with a solid central core of xylem, and peripheral phloem. The endodermis is not clearly shown in this fossil Fern.

Gleichenia, and probably in all the others, its form is conical at first, but after reaching a certain size the stele retains that size through life, as a cylinder traversing the cylindrical rhizome. The limiting factors have come into play, one of which is believed to be the proportion of surface of the stele to its bulk. When the stele attains larger size, as it did in certain related fossils while still retaining its protostelic state, it is seen to have undergone a modification of form. For instance in *Ankyropteris Grayi*, which is 2—3 mm. in diameter, it is corrugated, the insertion of the leaf-traces projecting, and the surfaces between them being hollowed (Fig. 175, ii, iii): moreover the curvatures of the hollows are deeper in the larger than in the smaller specimens. A still more extreme case of this is seen in *Asterochlaena laxa*, which may be as much as 15·5 mm. in diameter (Fig. 175, iv). Here the stele is thrown into deep involutions of the surface. It is obvious that this form will give a very greatly increased proportion of surface to bulk. It seems natural to conclude that in such cases the more elaborate form of the stele has made the larger size possible, by overcoming the limiting factor. But notwith-

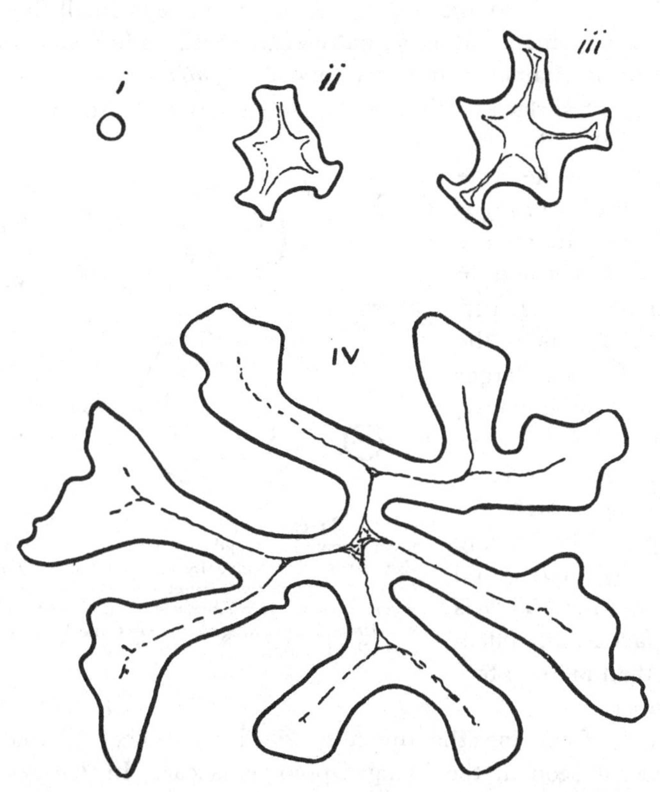

Fig. 175. Outlines of xylem of steles, all drawn to the same scale (× 5), to show approximately relative size. i, *Botryopteris cylindrica*, diameter ·65 mm.; ii, *Ankyropteris Grayi*, diameter 2·0 mm.; iii, *Ankyropteris Grayi*, diameter 2·5 mm.; iv, *Asterochlaena laxa*, diameter 12·0 mm. The elaborateness of outline increases with the size.

standing the complicated outline, and the well-known differentiation of the xylem of these fossils, their steles are still of the nature of protosteles. Their non-medullated structure is maintained.

In other primitive Ferns, as a larger size of the stele is attained in the growing plant, a change of internal structure appears, leading to *medullation*. Since the leaf-traces are inserted peripherally, it is in the outer xylem that the water-transit will be most active. As the stele enlarges the water in the central region will tend to stagnate, and thin-walled cells will serve for its storage as well as thick-walled tracheides would do. This is probably the *rationale* of the condition of "mixed pith," and of the formation of a parenchymatous medulla. Its intrastelar origin has been traced in Chapters VII and VIII. It has there been shown how the relatively small stem of the living Osmundaceae may have a ventilating system in the storage-pith quite distinct from that of the storage-cortex. Where the size is small that may be a workable arrangement: it is actually seen in the adult stem of *Todea barbara*, 3 mm. in diameter, and in *Osmunda regalis*, 5 mm. In such cases the proportion of surface to bulk of the stele contained within is relatively high (Fig. 176). But the case is different for the large steles of *Osmundites skidegatensis* (25 mm. in diameter), or of *O. Carnieri* (35 mm. in diameter) (refer to Fig. 127, Chapter VII). In them the problem raised by their larger size is solved by breaking down the barrier, leading thus to the completion of a common ventilating system for cortex and pith. The limiting factor is met by interruption of the endodermis. This is in point of fact a more effective device than mere extension of its area.

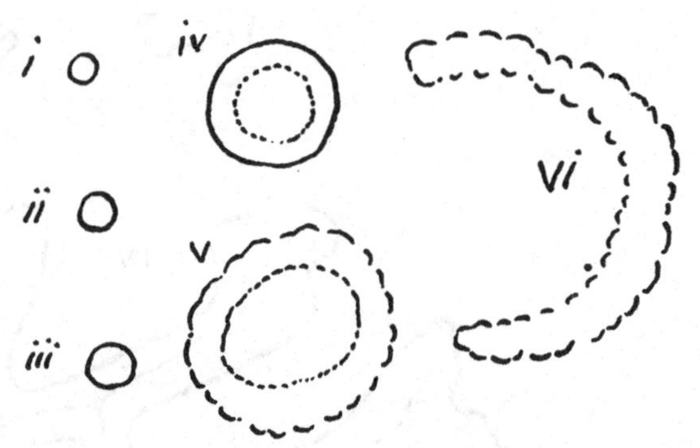

Fig. 176. Traces of the actual size of steles of living and fossil Osmundaceae, all to the same scale, i.e. approx. nat. size. iv, v, vi supplied by Dr Kidston. i = *Todea barbara* (3 mm.); ii = *Osmunda cinnamomea* (4 mm.); iii = *Osmunda regalis* (5 mm.); iv = *Thamnopteris schlechtendalii* (13 mm.); v = *Osmundites skidegatensis* (25 mm.); vi = *Osmundites Carnieri* (35 mm.).

A method of solving the difficulty similar in effect to that of the Osmundaceae is seen in the living Ophioglossaceae. In *Botrychium* and *Helminthostachys* the young plant has a complete endodermal barrier, as in other Ferns, shutting off the vascular system from the cortex. As the plant advances and the stele enlarges a pith is formed which serves for storage. Intercellular spaces appear in it, but the internal ventilating system is at first shut off from the cortical, and it remains so till the plant is well advanced. As the stem enlarges still further free communication is established by foliar gaps (see Fig. 124, Chapter VII), which open outwards to the cortex, but they also open inwards to the pith. This has the disadvantage of laying open the conducting tract, and destroying the completeness of the endodermal control. But it solves the difficulty of communication between the outer and inner tissues, which becomes more acute as the stem enlarges.

The advantage of ventilation here gained is probably greater than the disadvantage that follows from laying open the conducting tract.

This exposure of the conducting tracts is carried much further in *Ophioglossum*. Here the endodermis is discarded early: the pith dilates with large leaf-gaps, so that the ground-parenchyma appears traversed by unprotected vascular strands. The same is the case with the Marattiaceae, though here the number of the strands is much higher, traversing the distended, sappy stock. After the brief juvenile stage is past, there is as a rule no endodermal barrier in these Ferns (compare Fig. 128, Chapter VII). Consequently no such question of the proportion of surface to bulk arises. But on the other hand by discarding the endodermis the conducting tracts have lost that protoplasmic control which the endodermis gives. This state may serve for semi-xerophytic plants, such as the Ophioglossaceae and Marattiaceae, with sappy stocks and leathery leaves, and sluggish fluid-transit. But it would not serve for plants where fluid-transit needs to be rapid, and in particular for those where the leaf-structure is delicate.

The Leptosporangiate Ferns, which are mostly delicate hygrophytes, comprise the vast majority of the living species of Ferns. They have taken a different course of structural development, in which the endodermal barrier is strictly maintained in its complete form, while intercellular spaces are as a rule absent from their vascular tracts. They show in their peculiar vascular structure to what shifts a plant is put as it increases in size by primary and not by cambial activity, maintaining meanwhile its vascular system under complete endodermal control. All of these Ferns start from the protostelic state. It appears from comparison along phyletic lines, parallel but yet distinct, that a disintegrated stelar structure replaced the protostele. The successive steps of this may be seen with varying degrees of clearness in the successive stages of the individual life, those steps appearing as the stele enlarges. According to the reasoning already brought forward, it is on the question of enlargement that the problem of proportion of surface to bulk of the stele turns. The modifications of form of the stele seen in the advanced Leptosporangiate Ferns may be held as the means of its solution.

Since the difference in vascular construction between the more primitive Eusporangiate Ferns and the later Leptosporangiate Ferns turns upon the frequent discontinuity of the endodermis in the one, and its being uninterrupted in the other, it would be well to examine this tissue in detail. In the Filicales two types of endodermis are recognised. That which has been styled *primary* is characterised by the presence of the well-known Caspary-band, which appears as a rule in the middle zone of the radial walls, though frequently it may lie rather nearer to the central than to the peripheral side of each cell. Since the Caspary-band is suberous, and extends continuously all round each cell, it makes the endodermis impervious to the transit of fluids or soluble substances, except through the protoplasts of the cells, which during life are able to control their passage. Moreover, as the cells are all fitted together so as to form a continuous sheet, even the primary endodermis as well as the secondary will form an efficient gas-barrier, cutting off the outer ventilating system of the cortex from the vascular tissues within its circle.

Seen in surface view the Caspary-band appears relatively broad, and it is found to be pitted (Fig. 177, i), but no protoplasmic continuity appears to be established either between the endodermal cells and cells of the cortex, or on the other hand with cells of the vascular strand. The endodermis is thus a structurally isolated tissue, which places the control of transit inwards and outwards upon the efficiency of the protoplasts of its cells. The less highly specialised type of endodermis which is called *primary* is found to be characteristic of the Ophioglossaceae, the Osmundaceae, and the Marattiaceae, and it is found also in the Hymenophyllaceae. These are all families of Ferns which on grounds of general comparison may be held as relatively primitive.

The *secondary* type of endodermis is found in all the other Filicales, and in general in the Phanerogams. It is characterised by the deposit of a layer of suberin on the tangential wall-surface, in direct contact with the cytoplasm. The deposit may be formed in some Ferns

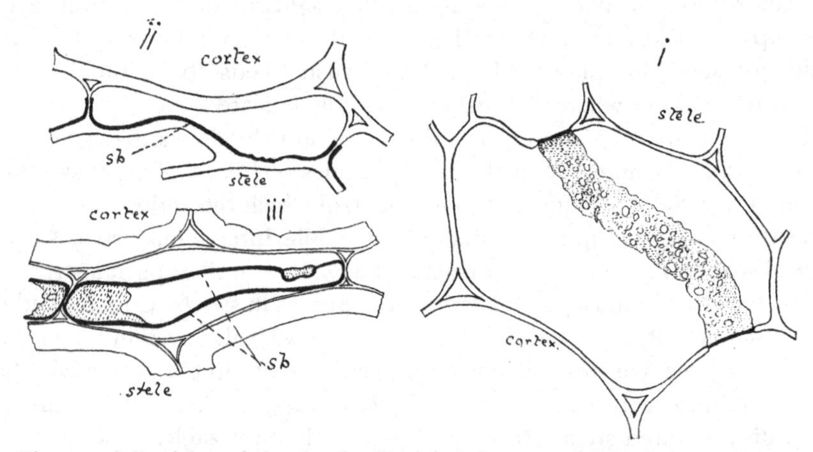

Fig. 177. Cells of the endodermis, after Rumpf. i, from root of *Ophioglossum vulgatum*, showing the Caspary-band in section, and in surface view, with pits. ii, from root of *Asplenium esculentum*, showing suberisation on the inner tangential wall. iii, from root of *Matteuccia Struthiopteris* with suberisation (*sb*) deposited all round the inner surface of the cell, but separated by the method of preparation. (× 1000.)

only on the inner tangential wall (Fig. 177, ii): but in others it may be formed also on the inner face of the wall, a condition which is stated to be general for the Phanerogams (Fig. 177, iii). It has been demonstrated for the Schizaeaceae and Gleicheniaceae, and also for the Cyatheaceae and Polypodiaceae. Thus it exists in certain Ferns which are relatively primitive as well as in those which are more advanced. There is also a *tertiary* condition of the endodermis common in Phanerogams, in which lignified deposits giving mechanical strength are formed lining the endodermal cells internally. But this is not known to occur in any of the Filicales.

It is notable that the Caspary-zone of cell-wall is widest in plants with a primary type of endodermis, as it is seen for instance in *Ophioglossum*, or *Helminthostachys* (Fig. 173). But it is narrower in those which have the secondary type, such as the Leptosporangiate Ferns and the Phanerogams. It has been seen that the endodermis is a tissue structurally isolated from those within and without. But there is some evidence that the cells of the endodermis are in close relation with one another tangentially. This follows partly from the observation of pitting in the radial walls, and notably in the Caspary-band itself (Fig. 177, i): partly from the fact that in many cases when the protoplasts of the endodermal cells are contracted they leave the tangential walls, but remain very firmly attached to the radial walls in the region of the Caspary-band. This appears to indicate a closer relation of the cytoplasm to the cell-wall in that region than over the rest of the wall-surface. It suggests that the endodermis may act not as individual cells, but as a living sheath, more or less independent of the tissues within and without: that is, as a whole, with a common coordination of its constituent cells. The character of the endodermis is assumed very early in the ontogeny. In roots its first appearance coincides very nearly with the appearance of the first tracheides, that is, at about the same level as the formation of the first root-hairs.

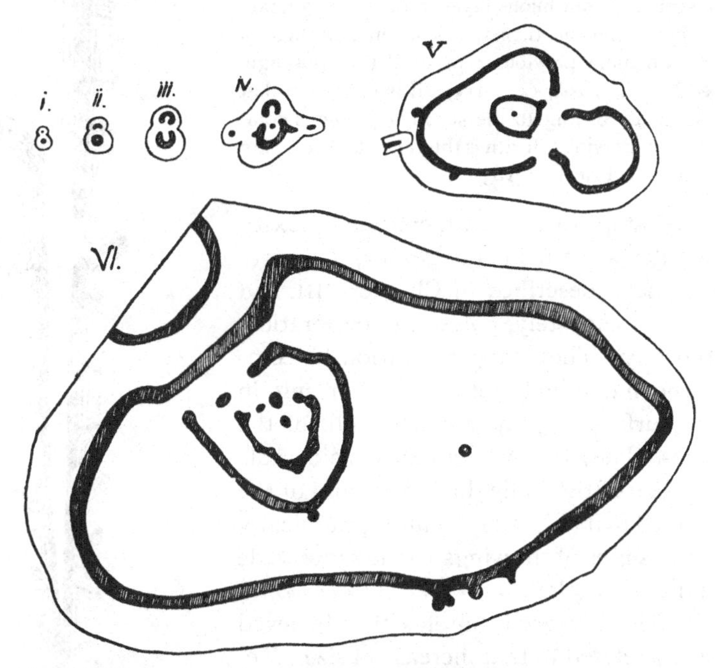

Fig. 178. Series of transverse sections of the stem of *Pteris* (*Litobrochia*) *podophylla*, all drawn to the same scale, showing the great increase of stelar complexity as the conical stem expands upwards. (× 4.)

In position it is first recognisable opposite the strands of phloem. Thence it spreads laterally and extends along the root to its insertion upon the vascular system of the shoot.

In the Eusporangiatae an endodermis is always present in the root, and it is usually present in the young rhizome, but it is incomplete or absent in the adult stock of the Ophioglossaceae and Marattiaceae. Though present in the smaller Osmundaceae as a complete investment, it is incomplete in the largest types. Often in these primitive Ferns it is not restricted to one clearly differentiated layer of cells. It is not continued into the leaf of Eusporangiate Ferns, though it is so in the Osmundaceae and Hymenophyllaceae, which appear to take an intermediate place in this structural detail between the Eusporangiate and the Leptosporangiate Ferns, as they do in so many other features. On the other hand, in the Leptosporangiate Ferns not only does the endodermis form an unbroken sheath to the vascular tissue of root and axis, but it also extends throughout the leaf, completely investing the veins even to their distal ends. Moreover, as their endodermis is of the secondary type, the conducting tracts would appear to be very completely shut in by it. It seems, however, highly probable that means of transit through it for fluids and soluble substances exist which have hitherto escaped detection. Whether or not this is the fact, it is at least indicated structurally that the endodermis of the Leptosporangiate Ferns is a more complete protection to the conducting tracts, and at the same time a more impervious barrier than in the Eusporangiate. But while the endodermis can be seen as a continuous layer in Leptosporangiate Ferns, even below the sori themselves, it remains there of the primary and more pervious type until the sporangia are matured (see Fig. 208, *C*, p. 215): it is only then that its development according to the secondary type is completed, and the impervious sheath is thus closed. (Compare Bäsecke, 160; and Priestley, 183.)

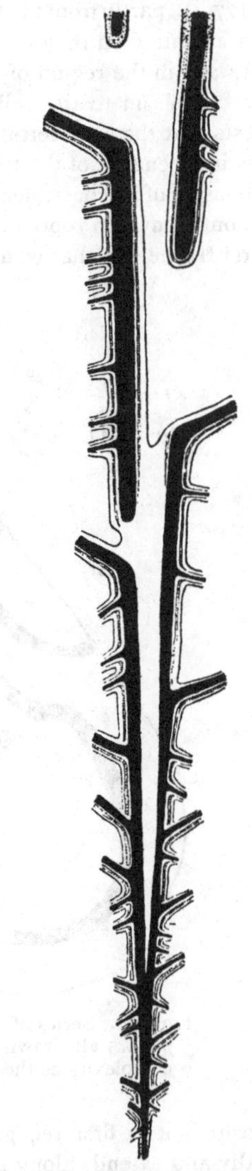

Fig. 179. Plan of stelar construction of a juvenile plant of *Gleichenia pectinata*, after Dr J. M. Thompson, showing in median section the way in which the stele enlarges conically upwards, and widens into a solenostele, with leaf-gaps. Rather above the middle of the figure the endodermis is continuous across the pith, thus separating the lower, "intrastelar," from the upper, "extrastelar," pith.

The chief steps in the advancing complexity of the vascular system in the Leptosporangiate Ferns have been described in Chapter VIII, and are known as solenostely, polycycly, perforation, and dictyostely. They may be variously combined in the individual stem, and all result in increase of surface in proportion to bulk of the stelar tissues. They follow on a very considerable increase in size of the individual stem, and of the stele contained in it (Fig. 178). This is graphically shown in the series of drawings to the same scale of the stem of a plant of *Pteris* (*Litobrochia*) *podophylla*. The increased complexity is believed to be causally related to that increase in size. Medullation may precede it, as it does in *Gleichenia pectinata* (Fig. 179). In others no previous medullation may be apparent. The reason for this is probably to be found in the importance of establishing early and complete endodermal control over the conducting system, combined with internal aeration. *Both are secured by the establishment of an endodermal barrier extending across the medulla* (Fig. 179). The structural difference between the parenchyma within the sheath ("intrastelar pith") from that outside it ("extrastelar pith") is probably a consequence not of tissue-origin, but of development of the cells under the

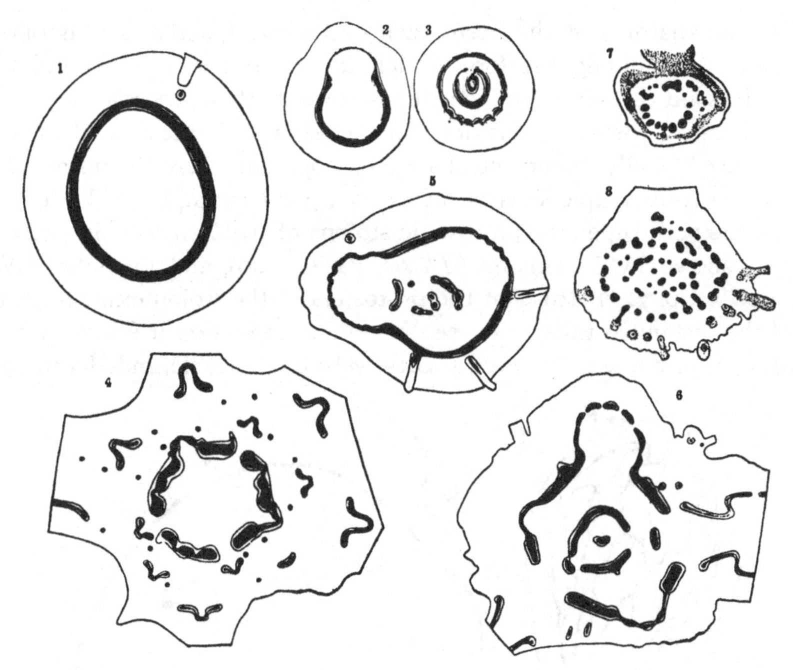

Fig. 180. Series of solenostelic and dictyostelic stems of Ferns, all drawn to same scale. (× 2.) 1, *Metaxya*; 2, *Dipteris conjugata*; 3, *Matonia pectinata*; 4, *Plagiogyria pycnophylla*; 5, *Thyrsopteris elegans*; 6, *Saccoloma elegans*; 7, *Platycerium alcicorne*; 8, *Platycerium aethiopicum*. These drawings show that the disintegration of the stele does not depend on absolute size alone.

conditions on the one hand of endodermal control, and on the other of its absence. The final effect of the stelar changes in the growing stocks of advanced Leptosporangiate Ferns is to break up the vascular tracts, as seen in transverse section, into relatively small, strictly circumscribed, circular or oval masses, each with a relatively large proportion of surface to bulk (Fig. 180). The physiological difficulty following on increase of size is thus fully met in the stems of Leptosporangiate Ferns. As explained in Chapter IX, a similar disintegration of the leaf-trace also follows on increase in size of their petioles (Fig. 181).

Fig. 181. Transverse sections of petioles, all drawn to the same scale. (× 2.) 1, *Dipteris conjugata*; 2, *Dipteris Lobbiana*; 3, *Metaxya*; 4, *Phlebodium aureum*; 5, *Thyrsopteris*; 6, *Alsophila australis*. These show that while greater size leads to vascular disintegration, there is no definite proportion.

If, as the anatomy of the Ferns seems to suggest, actual size is one of the factors determining the form which the stelar tissues take, and that increase beyond certain dimensions tends towards those peculiarities which are seen in them, then tuberous development should lead to such disintegration. More especially should the change be apparent where the normal part shows a relatively simple stelar structure. A good example of this is seen in the tubers borne upon the protostelic stolons of *Nephrolepis* (Fig. 182). It has been shown by Lachmann (*Thesis*, Paris, 1889), and by Sahni (*New Phyt.* Vol. xv, p. 72, 1916), that the protostele of the stolon expands at the base of the distended tuber. As seen in transverse section it first acquires a central mass of phloem, followed successively by pericycle, endodermis, and

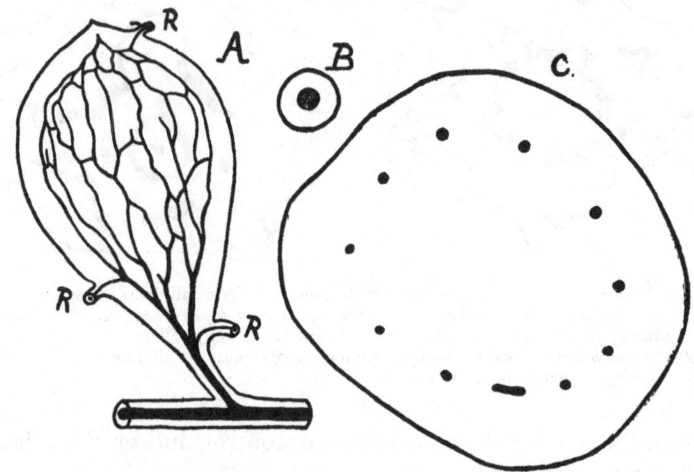

Fig. 182. *Nephrolepis cordifolia.* *A*, stolon bearing a tuber, in which the protostele breaks up into a cylindrical network, contracting again at the apex. *R* = root. (After Sahni.) *B*, transverse section of protostelic stolon. (× 5.) *C*, transverse section of tuber (also × 5) showing ring of meristeles each limited by endodermis. Diameter of stolon, 1·6 mm. Diameter of tuber, 11·0 mm.

ground-parenchyma. In fact, it becomes solenostelic. As the base of the tuber expands further the ring breaks up by irregular perforations, as it does in the leafy shoots of many Leptosporangiate Ferns. But here there are only perforations; since no leaves are borne on the stolons there are naturally no foliar gaps. A network of meristeles is thus formed, each limited by a complete endodermis, and it is arranged as an expanded ring (Fig. 182, *C*). At the distal end where the tuber contracts again the network narrows down through stages of condensation the reverse of the previous disintegration. In a given case the diameter of the stolon was 1·6 mm., and of its protostele ·6 mm. The diameter of the tuber was 1·1 cm., and of its ring of meristeles ·74 cm.; that is nearly fifteen times that of the original protostele. It thus appears that, while complete endodermal control is maintained, when the stolon of

Nephrolepis dilates into a tuber the same features of stelar expansion appear as in the conically enlarging axis of many Leptosporangiate Ferns. This suggests that the increase in size in both cases is at least one factor determining the structural change, while the reversal of that change which follows on the apical contraction of the tuber strongly supports that conclusion.

The triumph of the Leptosporangiate Ferns, which thus show disintegration of the stele in stem and leaf in all their more advanced types, is witnessed by their 6000 living species spread widely over the face of the earth. They are essentially the Ferns of the present day. That triumph has been won structurally by a compromise, effected without cambial increase in the enlarging stem. Their conducting stele has enlarged with the conically enlarging shoot. It has maintained its endodermal barriers complete, and has met the difficulty of physiological interchange consequent on that enlargement by various steps of moulding and disintegration of the stele. This has given the necessary proportion of surface to bulk of those conducting tracts even for stems as large as those of the Tree Ferns.

The structure seen in those Ferns which both comparison and the fossil record stamp as primitive, viz. the Ophioglossaceae, Osmundaceae, and Marattiaceae, is susceptible of elucidation along similar lines. Their success as measured by genera and species, and by geographical spread, is only partial. Their less perfect structural adaptation to the demands of increasing size may well have been one among the factors which have led to this result. Its most marked feature is the opening of the endodermal sheath as greater size is attained. This imperfection of their conducting tracts has probably imposed upon them that semi-xerophytic character which their foliage habitually shows. From the comparative point of view these Ferns may best be regarded as evolutionary indicators. They point the way towards, but never fully attain, that more perfect adjustment of structure to the requirements of increasing size without cambial activity, which is realised with such successful results in the Leptosporangiate Ferns.

Actual size may thus be held to be an important factor in determining the appearance of features which have been habitually used in comparison. The effect of this should be to impose caution in drawing conclusions from characters so pliable. At once the door is opened for frequent homoplasy in stocks phyletically quite distinct from one another, in any one of which increase in size of the individual or the race is possible. The appearance of polycycly in the Matonineae, in *Thyrsopteris*, *Saccoloma*, and in various Pterids, may well be held as homoplastic; that is, as isolated consequences of increase in size rather than as any index of affinity of Ferns otherwise isolated and diverse. More cogent still is the case for dictyostely, or for perforation, the appearance of which seems quite general in stocks distinct one from another. From this point of view the recognition of the Principle of Similar Structures as applied to the vascular system of Ferns may react finally upon their phyletic seriation. But none the less, when once this reservation has been made, the characters of the vascular system remain as the most important structural features for the phyletic treatment of the Class. Subject to certain limitations, the degree of disintegration of the stele, or of the leaf-trace, may still be held as giving a trustworthy indication of the degree of phyletic advance of the Fern that shows it.

POSTSCRIPT TO THE CHAPTERS ON VASCULAR ANATOMY

The attempt thus to place the comparative study of the vascular system of the Filicales upon a physiological-anatomical-phyletic footing is plainly out of harmony with other views which have been advanced upon grounds of purely anatomical comparison. It has already been stated that anatomy cannot stand alone as a branch of enquiry, but must be coordinated and harmonised with other lines of comparative study. Least of all is the anatomy of the adult structure in the mature state sufficient. The adult should always be interpreted in the light of the ontogeny. It must also be stated that to anticipate for the results of anatomical enquiry a logical sequence is wholly to misunderstand its place in Biological Science. Evolution has followed lines of opportunism, not of logic. In its study this should be constantly borne in mind.

No attempt has been made in the foregoing Chapters to argue out the differences of opinion which have arisen, especially those relating to the origin of the pith. All that has been done has been to state the case as it appears to follow from observations of the ontogeny as well as of the adult structure, in a number of examples specially selected as being themselves relatively primitive: this method should be checked by comparison based on a study of other criteria than that of structure alone, as well as by reference to geological sequence. But in order to assist readers desirous of forming an independent opinion on the validity of the views here put forward, and of comparing them with the views of other authors, a list of Literature bearing on the question of the stelar morphology in Ferns is appended, and it has been made as comprehensive as possible.

BIBLIOGRAPHY FOR CHAPTERS VII, VIII, AND X

119. DE BARY. Comparative Anatomy. Engl. Edn. Oxford. 1884.
120. VAN TIEGHEM. Sur la Polystélie. Ann. Sci. Nat. Bot. Sér. 7. 1886.
121. LE CLERC DU SABLON. Ann. Sci. Nat. Bot. Sér. 7. Vol. ii, p. 1. 1890.
122. POIRAULT. Rech. sur les Crypt. Vasc. Ann. Sci. Nat. Bot. Sér. 7. Vol. xviii, p. 113.
123. HARVEY GIBSON. *Selaginella*. 1894. Ann. of Bot. viii, p. 171.
124. ZENETTI. *Osmunda*. Bot. Zeit. 1895, p. 75.
125. CORMACK. Polystelic roots of certain Palms. Trans. Linn. Soc. Vol. v, 1896, p. 275.

126. BOWER. Studies on Spore-producing Members. II. Ophioglossaceae. Dulau. London. 1896.
127. JEFFREY. Report Brit. Assoc. Toronto. 1897.
128. FARMER & FREEMAN. Ann. of Bot. xiii, 1898, p. 421.
129. JEFFREY. *Botrychium virginianum*. Trans. Canadian Inst. 1898.
130. SEWARD. *Matonia pectinata*. Phil. Trans. 1899. Vol. 191, p. 171.
131. JEFFREY. Morphology of the central stele in Angiosperms. Trans. Can. Inst. Vol. v, 1900.
132. BOODLE. Hymenophyllaceae. Ann. of Bot. xiv, 1900, p. 455.
133. GWYNNE-VAUGHAN. Solenostelic Ferns. I. Ann. of Bot. xv, 1901.
134. BOODLE. Schizaeaceae. Ann. of Bot. xv, 1901, p. 359.
135. BOODLE. Gleicheniaceae. Ann. of Bot. xv, 1901, p. 703.
136. LANG. *Helminthostachys*. Ann. of Bot. xvi, 1902, p. 23.
137. BREBNER. *Danaea*. Ann. of Bot. xvi, 1902, p. 517.
138. FARMER & HILL. *Angiopteris*. Ann. of Bot. xvi, 1902, p. 371.
139. TANSLEY & LULHAM. *Lindsaya*. Ann. of Bot. xvi, 1902, p. 157.
140. JEFFREY. The structure and development of the stem in Pteridophyta and Gymnosperms. Phil. Trans. 1902. Vol. 195. Reviewed by SCOTT. New Phytologist. Vol. i, No. 9.
141. SEWARD & DALE. *Dipteris*. Phil. Trans. 1901. Vol. 194, p. 487.
142. FAULL. Anatomy of the Osmundaceae. Bot. Gaz. xxxii, 1901.
143. SEWARD & FORD. Anatomy of *Todea*. Trans. Linn. Soc. Vol. vi, 1903, Part v.
144. SCHOUTE. Die Stelär-Theorie. Jena. 1903.
145. BOODLE. Further observations on *Schizaea*. Ann. of Bot. xvii, 1903, p. 511.
146. TANSLEY & CHICK. Structure of *Schizaea malaccana*. Ann. of Bot. xvii, 1903, p. 493.
147. GWYNNE-VAUGHAN. Solenostelic Ferns. II. Ann. of Bot. xvii, p. 689.
148. RUMPF. Rhizodermis, Hypodermis, und Endodermis d. Farnwurzel. Bibliotheca Botanica. Stuttgart. 1904. Here the earlier literature on endodermis is fully quoted.
149. CHANDLER. On the arrangement of the vascular strands in the seedlings of certain Leptosporangiate Ferns. Ann. of Bot. xix, 1905, p. 406.
150. GWYNNE-VAUGHAN. On the possible existence of a Fern-stem having the form of a lattice-work tube. New Phyt. 1905, p. 211.
151. JEFFREY. Morphology and Phylogeny. Science. N. S. 1906, p. 291.
152. KIDSTON & GWYNNE-VAUGHAN. On the Fossil Osmundaceae.
Part I. Trans. Roy. Soc. Edin. Vol. xlv, 1907.
Part II. „ „ „ „ Vol. xlvi, 1909.
Part III. „ „ „ „ Vol. xlvi, 1909.
Part IV. „ „ „ „ Vol. xlvii, 1910.
Part V. „ „ „ „ Vol. l, 1914.
153. CONARD. Structure and Life-History of the Hay-scented Fern. Carnegie Inst. Washington. 1908.
154. JEFFREY. Review of Kidston and Gwynne-Vaughan. Fossil Osmundaceae. Bot. Gaz. Jan. 1908, p. 395.
155. TANSLEY. Lectures on the Evolution of the Filicinean Vascular System. New Phytol. Reprints. No. 2. 1908.
156. JEFFREY. On foliar gaps in the Lycopsida. Bot. Gaz. Nov. 1908, p. 395.
157. BOWER. Origin of a Land Flora. Macmillan. London. 1908.
158. COULTER. Vascular Anatomy and the Reproductive Structures. American Naturalist. Vol. 43, 1909, p. 219.

159. BOODLE & HILEY. Vascular Structure of some species of *Gleichenia*. Ann. of Bot. xxiii, 1909, p. 419.
160. BÄSECKE. Beitr. z. Kenntniss d. phys. Scheiden. Bot. Zeit. 1908, p. 25. Here a very complete list of literature up to date is given.
161. FAULL. The stele of *Osmunda cinnamomea*. Trans. Canad. Inst. Vol. viii, 1909.
162. DE LAVISON. Rev. Gen. de Bot. 1910, p. 225.
163. JEFFREY. The Pteropsida. Bot. Gaz. 1910, p. 412.
164. BOWER. Studies in the Phylogeny of the Filicales.
    I. *Plagiogyria*. Ann. of Bot. 1910, p. 429.
    II. *Lophosoria*. „ „ „ 1912, p. 269.
    III. *Metaxya*. „ „ „ 1913, p. 443.
    IV. *Blechnum*. „ „ „ 1914, p. 363.
    V. *Cheiropleuria*. „ „ „ 1915, p. 495.
    VI. Ferns showing "Acrostichoid" condition. Ann. of Bot. 1917, p. 1.
    VII. The Pteroideae. Ann. of Bot. 1918, p. 1.
165. SINNOTT. Foliar gaps in the Osmundaceae. Ann. of Bot. xxiv, 1910.
166. CHARLES. On the anatomy of the sporeling of *Marattia alata*. Bot. Gaz. Feb. 1911, p. 81.
166 *bis*. BLOMQUIST. Vasc. Anat. of *Angiopteris*. Bot. Gaz. 1922. Vol. lxxiii, p. 181.
167. CAMPBELL. The Eusporangiatae. Carnegie Inst. Washington. 1911.
168. GWYNNE-VAUGHAN. Some remarks on the anatomy of the Osmundaceae. Ann. of Bot. xxv, July, 1911.
169. BOWER. On the primary xylem, and the origin of medullation in the Ophioglossaceae. Ann. of Bot. xxv, 1911.
170. BOWER. Medullation in the Pteridophyta. Ann. of Bot. xxv, 1911.
171. LANG. On the interpretation of the vascular anatomy of the Ophioglossaceae. Mem. and Proc. Manchester Lit. and Phil. Soc. Vol. 56, Part II, 1911—1912.
172. LANG. Studies in the Morph. and Anat. of the Ophioglossaceae. I. Branching of *Botrychium Lunaria*. Ann. of Bot. xxvii, 1913. III. Anatomy of *Helminthostachys*. Ann. of Bot. xxix, 1915.
173. P. BERTRAND. L'étude anatomique des Fougères. Progress. Rei Bot. iv, p. 182. 1912.
174. GWYNNE-VAUGHAN. On a mixed pith in an anomalous stem of *Osmunda regalis*. Ann. of Bot. xxviii, 1914.
175. JEFFREY. Anatomy of Woody Plants. Chicago. 1917.
176. THOMPSON. *Deparia Moorei*. Trans. Roy. Soc. Edin. 1915. Vol. l, No. 26.
177. THOMPSON. *Stromatopteris*. Trans. Roy. Soc. Edin. 1917. Vol. lii, No. 6.
178. THOMPSON. *Platyzoma*. Trans. Roy. Soc. Edin. 1917. Vol. lii, No. 7.
179. THOMPSON. Rare and primitive Ferns. Trans. Roy. Soc. Edin. 1918. Vol. lii, No. 14.
180. THOMPSON. New Stelar Facts. Trans. Roy. Soc. Edin. 1920. Vol. lii, No. 28.
181. SAHNI. Tubers of *Nephrolepis*. New Phyt. Vol. xv, p. 72. 1916.
182. WEST. Study of the Marattiaceae. Ann. of Bot. 1917, p. 361.
183. PRIESTLEY. The mechanism of root-pressure. New Phyt. 1920, p. 189. Here the more recent memoirs on the structure and physiology of the Endodermis are quoted.
184. CAMPBELL. The Eusporangiate Ferns and the Stelar Theory. American Journ. of Bot. 1921, p. 303.
185. BOWER. Size, a neglected factor in Stelar Morphology. Opening address. Proc. Roy. Soc. Edin. 1920. Vol. xli, p. 1.

CHAPTER XI

DERMAL AND OTHER NON-VASCULAR TISSUES

WHILE in Ferns, as in other Vascular Plants, priority of importance attaches to the vascular system by virtue of its greater "phyletic inertia," other tissues which envelope it make up a great part of the body of the sporophyte. They may be grouped as *Ground-Parenchyma*, *Sclerenchyma*, and *Epidermis*. Associated with the last are the Dermal Appendages, that is, *Hairs* and *Scales*, which arise as outgrowths of the surface. Exclusive of the dermal appendages these tissues prove to be more directly variable according to external conditions than the vascular system, and accordingly they are less trustworthy for purposes of phyletic comparison.

PARENCHYMA AND SCLERENCHYMA

The Ground-Parenchyma consists for the most part of living cells of more or less rounded form, with walls of very variable thickness, and with large intercellular spaces. Usually it is functional in the leaf-blade as photosynthetic tissue, and in the stem as storage-parenchyma. The last is naturally most marked where a seasonal leaf-production occurs, the leaves dying off and being renewed annually, as in *Pteridium*. Starch is the common storage-material, sometimes with mucilage (compare Figs. 131, 173). The intercellular spaces often show in great perfection pectose-rods traversing them, in the narrower angles even extending from wall to wall. These are particularly well seen near the leaf-base in the Marattiaceae. The cell-walls are frequently colourless, but in the leaf-stalk and stem of Ferns they often show varying tints of yellow or brown, even when they are relatively thin. Where the walls are thick the colouring may be deep brown or almost black, while the cytoplasm and storage-contents may still be seen in them. There is, in fact, no sharp distinction between thin-walled and sclerotic cells. Moreover the distribution of the hard tissue-masses is often irregular, giving the impression that in the ontogeny any cell may be determined for development either as a thin-walled or as a sclerotic unit. The form of the sclerotic cells varies from rounded or oblong stone-cells to elongated fibres. For instance the former are seen in the stony rind of the rhizome of *Pteridium* (Fig. 3): they are heavily sclerosed, with pitted walls, and without starchy contents. But it is not so much the nature of the walls as their form which marks them off from the parenchyma of the cortex, into which they merge internally. The sclerotic fibres compose those deeply seated tracts of brown sclerenchyma, whose spindle-shaped cells are marked by oblique slit-like pits in the lateral walls. Since the slope of the pits is in the same direction in all the cells those of adjoining cells appear to cross, giving the characteristic marking of such tissues (Fig. 183).

The distribution of the sclerotic tissues in Ferns appears to be ruled according to two distinct functions, mechanical resistance and water-reserve. It is difficult to dissociate these in individual cases, and they probably overlap. A mechanical use clearly lies with the sclerotic band, often pale-coloured, which is found close to the periphery of leaf-stalks, giving the "stramineous" appearance of systematists (Figs. 156, 161). It is well seen in *Pteridium*,

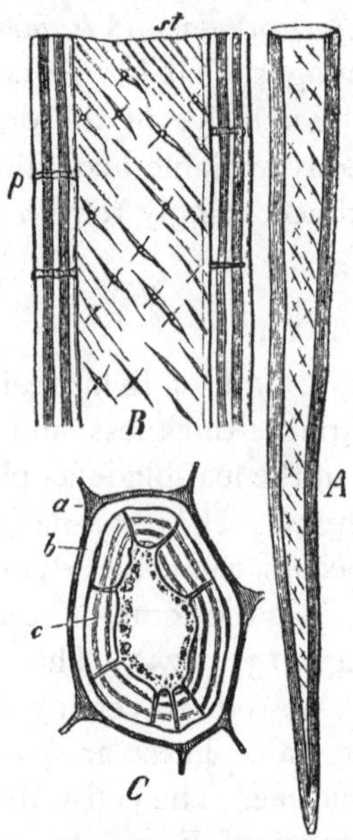

Fig. 183. *Pteridium aquilinum*. *A*, half of a brown-walled sclerenchyma fibre from the stem: *B*, part of one of these more highly magnified (× 550): *p* = one of the slit-like pits seen in section: *C*, transverse section: *a*, limiting lamella: *b*, *c*, inner layers of the wall. (From De Bary. After Sachs.)

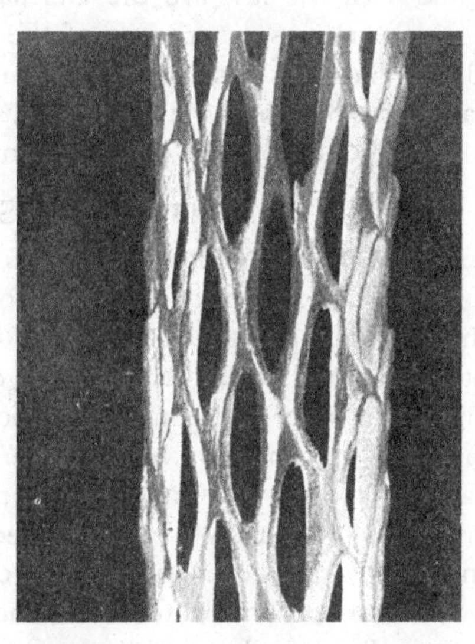

Fig. 184. Outer sclerotic sheath which invests the dictyostele of *Dicksonia*, with the edges of the foliar gaps turned outwards, thus giving added rigidity to the structure. Reduced. (Compare Figs. 150, 151.)

and in *Dryopteris Filix-mas*. The outer stony covering of the rhizome of *Pteridium*, or of *Plagiogyria*, is probably effective in meeting soil-pressures (Figs. 3, 138). The thick and hard outer cylinder of the stems of Tree Ferns certainly gives great strength. Their internally-lying masses of sclerenchyma are also mechanically effective. No one can examine the form of the sclerotic sheath surrounding the dictyostele of a Tree Fern, and note the outward-curved lips round each foliar gap, without realising that it is constructed on the principle of a corrugated metal sheet, thus gaining great mechanical effect with little cost of material (Fig. 184). Lastly, in the leaves the veins are often accompanied by sclerotic bands, giving a girder-construction effective for strengthening the expanded blade. Such examples witness to the mechanical use of sclerenchyma in Ferns.

But on the other hand, less definitely formed sclerotic masses may be found closely following the vascular tracts in stems and roots, but so disposed that mechanical effect is probably not their sole function. Their close relation to the vascular tissues suggests that they act also as reservoirs for water of imbibition, as in the case of many xerophytic Flowering Plants. More especially is this view upheld by the fact that in strongly xerophytic Ferns, such as *Notholaena*, *Cheilanthes*, *Pellaea*, or *Platyzoma*, the proportion of sclerenchyma is very high. It is in their roots that this is most marked, for there a thick sclerotic band surrounds the stele, being densest just at its boundary (Fig. 9, p. 8). It thus seems probable that the sclerenchyma serves a double function in Ferns: but it is so directly adaptible to circumstances and variable in occurrence that its features have little value for phyletic comparison.

The epidermis calls for little remark. When mature it resembles structurally that of non-specialised Flowering Plants. Chlorophyll is often present in its tabular cells, and the stomata appear at the level of the epidermal layer. Many hygrophilous Ferns have stomata of the "aquatic" type, opening directly by a wide aperture into the air-chamber below. These non-specialised characters go along with the late differentiation of the epidermis at the apex of stem and leaf. It is only after the segments derived from the initial cell or cells have undergone numerous tangential divisions that the dermatogen can be recognised, and traced on as giving rise to the epidermis.

Hairs and Scales

All the families of Ferns bear appendages springing from the epidermis. Even the Hydropterideae possess them in their youngest parts, and the Ophioglossaceae also, though they appear to have naked surfaces when mature. In some of the earliest fossils hairs provide characteristic features. Biologically they give protection to the young developing parts. According to their structure the dermal appendages furnish reliable features for comparison. The simplest of them are unbranched *hairs*, which arise by outgrowth of single superficial cells, and cell-division is only by walls transverse to the axis of their growth. The result is a linear series of cells (Fig. 185, i). Not uncommonly the distal cell develops as a gland. Frequently it secretes resinous material, or essential oil soluble in alcohol, as in *Nephrodium molle*, and many others. This is often deposited between the outer and inner layers of the cell-wall. Others again contain mucilage secreted in droplets within the cytoplasm, as in the Osmundaceae, and in *Blechnum* (Fig. 185, v). The mucilage swells with water, and the secretion of many hairs flowing together covers the part with a protective, water-containing sheath. In others again the secretion is produced externally, as in species of *Gymnogramme*, *Notholaena*, and *Cheilanthes*, giving golden or silver effects. Here the enlarged distal cell forms rods of the secretion radiating outwards, which are soluble in alcohol or ether. They have the effect of preventing the leaf-surface from being wetted by water. A very interesting example of glandular hairs was brought to my notice by Mr Boodle, in the case of *Notholaena trichomanoides*, in which hairs with glandular heads and white powdery secretion appear on the margins of the leaf. Hairs of a similar nature and size also fringe the prothallus, and they are thus common to both generations (Fig. 186).

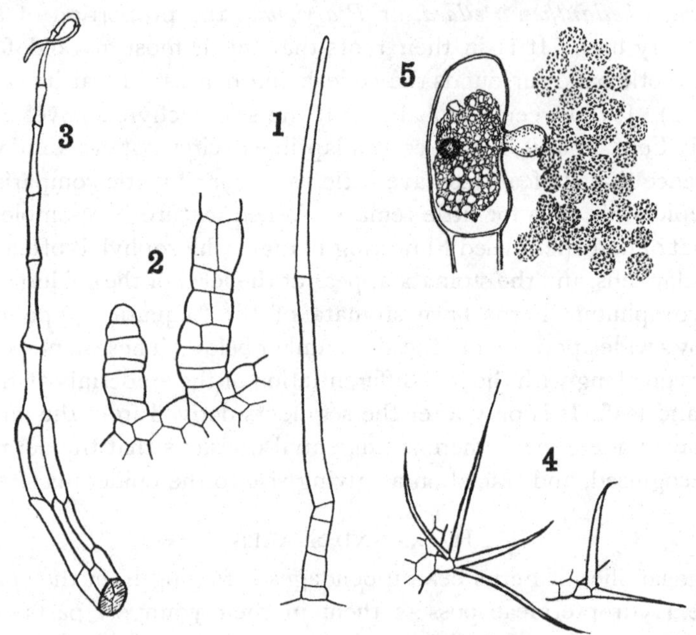

Fig. 185. 1. Simple hair of *Matonia*, after Seward. 2. Hairs of *Dipteris conjugata*, with occasional longitudinal divisions near the base, after Seward. 3. Hair with massive base, of *Dipteris Lobbiana*. 4. Stiff stellate hairs of *Trichomanes alatum*. 5. Mucilaginous hair of *Blechnum occidentale* discharging mucilage from its distal cell. After Gardiner and Ito.

Such *simple linear hairs*, sometimes branched, are found characteristically in many types of living Ferns. For instance in the Ophioglossaceae, Osmundaceae, Hymenophyllaceae, and Dicksonieae, all relatively primitive groups. Notably such hairs are present in certain primitive Ferns recognised as synthetic types: e.g. *Loxsoma*, *Plagiogyria*, *Lophosoria*, and *Metaxya*. Such facts, combined with their simple form, indicate that *the linear hair is a primitive feature*. Sometimes, however, longitudinal divisions may appear in the cells of the young developing hair, especially in its basal region. If they radiate the result is a stiff and massive spindle-shaped bristle, as in *Dipteris* (Fig. 185, 2, 3). This may be held as a derivative from the simple hair as seen in *Matonia* (Fig. 185, 1). Very striking bristles of this type are found in *Zygopteris*: they are attached by a narrow base, and are transversely septate, while each segment may divide by radial cleavages. In other cases these secondary walls may run parallel to one another, and a *flattened scale* is the result. Thus *Zygopteris* shows in this, as in certain other features, interesting signs of advance, though it is itself an early fossil, the affinities of which are certainly primitive.

A very curious structure is presented by the so-called "equisetoid" hairs of *Botryopteris forensis*, figured by M. Renault. They are borne on the petiole. An examination of them in slides belonging to Dr Kidston showed

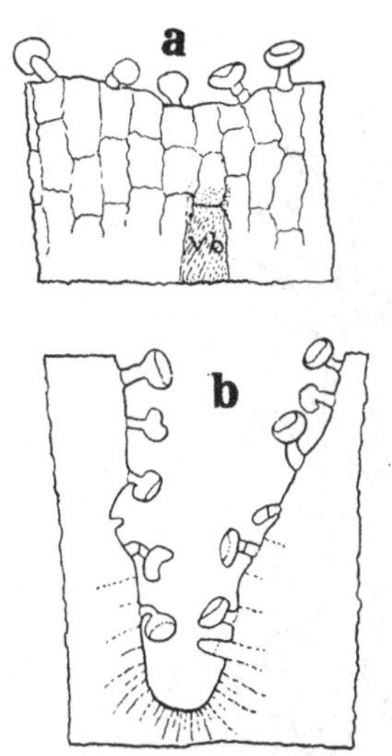

Fig. 186. Glandular hairs, which are alike on sporophyte and gametophyte of *Notholaena trichomanoides*. a = margin of cotyledon: b = apex of prothallus. ($\times$ 85.)

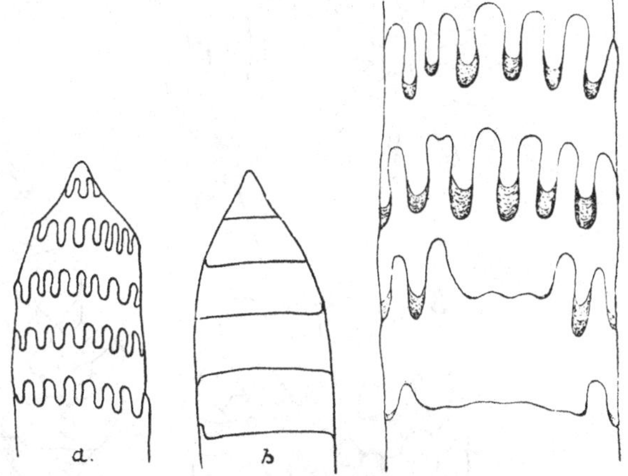

Fig. 187. "Equisetoid" hairs of *Botryopteris forensis*, Autun. *a*, seen from without, *b*, seen in optical section ($\times$ 82). Kidston Collection, slide 1821. To the right a similar hair ($\times$ 164). It is seen in obliquely tangential section, and shows in its upper part the sinuous margins of the septa: below the septa are seen in section, the middle of each being plane, and the margins sinuous. Kidston Collection, slide 1820.

that they are relatively large stiff hairs, transversely septate, and each is seated on a multicellular emergence projecting from the surface of the petiole, in the same way as in some species of *Gleichenia* and in the Cyatheaceae. The "equisetoid" character arises from the fact that the margins of the transverse septa are thrown into deep and rather regular corrugations, with the result that upturned processes of equivalent size, each appearing to have protoplasmic contents, are seen seated at the upper limit of each lower cell just below the septum, but not segmented off from it. These processes interlock with similar downward processes from the upper cell. Each septum of the hair being thus frilled, their characteristic appearance follows (Kidston Collection, slides 1818, 1820, 1821, Fig. 187). Increased mechanical strength, with enlarged area of surface between the cells, results from this unusual development. But the outer surface of the hair appears to be smooth, and its form is cylindrical.

In *Trichomanes* very characteristic hairs appear. Several stiff spines radiate from a basal stalk. A similar condition is presented in *Hymenophyllum* (Fig. 185, 4): in both of these genera the hairs may point in all directions. But in some species of *Polypodium* and in *Platycerium* the branches radiate only in a plane parallel to the surface that bears them (Fig. 188, 1, 2). It is but a step from such *stellate hairs* to a *peltate scale* with

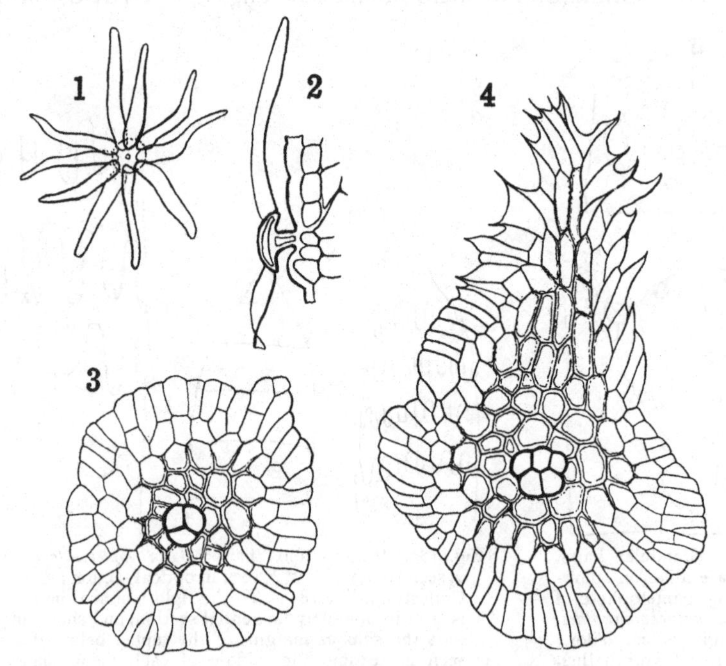

Fig. 188. 1, 2. Stellate hairs of *Polypodium lingua*. 3, 4. Peltate scales of *Polypodium incanum*. (×65.)

central attachment, such as is seen in *Polypodium incanum* (Fig. 188, 3). By stronger growth of the cells on one side of the scale the elongated form characteristic of most of these protective scales is reached. Another way in which protective scales are produced is by the repetition of longitudinal divisions parallel to one another, while the hair in which they occur widens into a flattened surface. The result in either case is that a flattened expansion is formed, known as a *ramentum*, common in the more advanced Leptosporangiate Ferns (Fig. 188, 4). It is a particularly prominent feature in the Cyatheae, distinguishing them sharply from the Dicksonieae, which have only simple hairs. Frequently the distal cell, and sometimes marginal cells, also are glandular (Fig. 189). This, together with the whole structure and development, indicates that the scale is derivative from the simple hair. Such scales overlapping closely form a very perfect covering to the surface that bears them. The protection may be permanent in Ferns exposed to drought, such as the whole plant of *Polypodium incanum*; or they may be restricted to the rhizome, as in *Phlebodium aureum* (Fig. 190). They may fall away in the adult after having protected the parts while young, as in *Dryopteris Filix-mas*.

In some families of Ferns the difference between simple hairs and scales may be used to confirm a distinction between primitive and derivative genera, based upon other features than this. For instance, in the Marattiaceae, as will be shown later, it is probable that *Danaea* and *Christensenia* with their synangial sori are derivative genera as compared with *Angiopteris*: and as Campbell has shown (*Eusp. Ferns*, p. 150, Fig. 126), both of these genera bear small peltate scales, while *Angiopteris* has simple hairs. A particularly good instance of the character of the dermal appendages supporting comparisons based on other features is seen in the Schizaeaceae. In this family simple hairs are prevalent. It is only in *Mohria*, which is in so many features a relatively advanced member of it, that broad scales are present (Prantl, *l.c.*, p. 37).

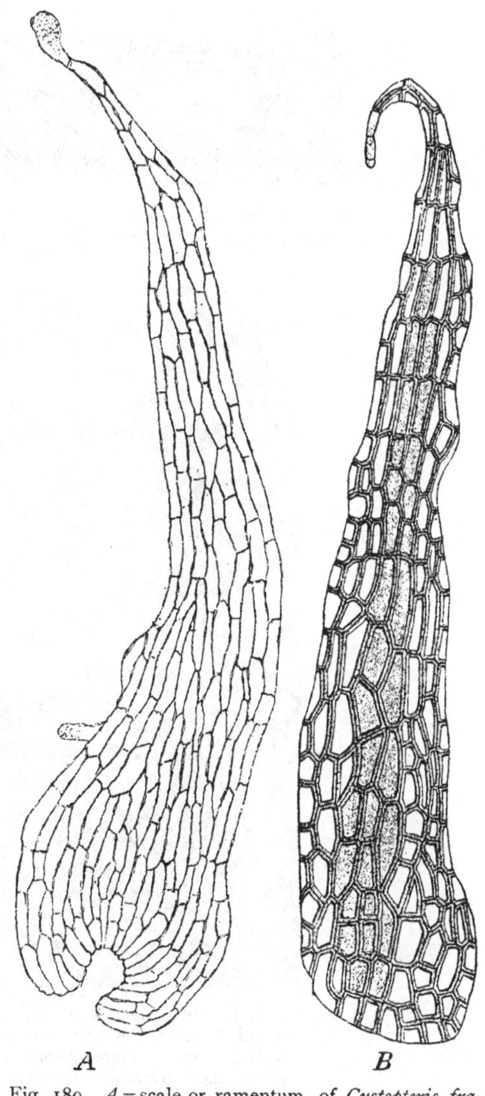

Fig. 189. *A* = scale or ramentum, of *Cystopteris fragilis*: *B* = of *Asplenium viride* (× about 20). (After Sadebeck, from Engler and Prantl.)

Several distinct types of dermal appendages may often be found on the same plant, soft, simple, or branched hairs being scattered among the larger scales in cases where the latter occur. Scales are specially developed upon rhizomes and about the leaf bases in Ferns, while they may be absent from the upper leaves. They attain large size in the Cyatheaceae, where they are borne on massive peg-like outgrowths, which remain as woody spines after the scales themselves have fallen away. This gives a very characteristic

Fig. 190. Vertical section through the scales covering the rhizome of *Phlebodium aureum*, showing their elaborate overlapping. (× 35.)

Fig. 191. Young leaves of *Hemitelia grandifolia*, showing broad dermal scales borne on massive emergences, which persist as woody spines after the scales have fallen away, giving the "armature" frequent in Cyatheaceous Ferns.

"armature" to many species of the family: but it is not constant in them all (Fig. 191). The spines rank as emergences. A very similar condition is found in *Gleichenia pectinata*: but in place of each broad scale a divergent tuft of stiff hairs is seen, suggesting a primitive state from which the Cyatheoid scale may have originated (Fig. 192).

PNEUMATOPHORES

The frequent sclerotic strengthening of the surface-tissues of the stem and leaf-stalk in Ferns has led to the establishment of *ventilating areas*, providing for the necessary gaseous interchange which the dense sclerotic bands would otherwise prevent. These areas are often visible as pale lines or patches. A familiar example of the former is seen in the Bracken, where lateral lines corresponding in position to the margins of the dorsiventral organ are traceable down the petiole, and onwards upon the rhizome. Sections show that here the sclerenchyma is replaced by highly ventilated parenchyma, covered

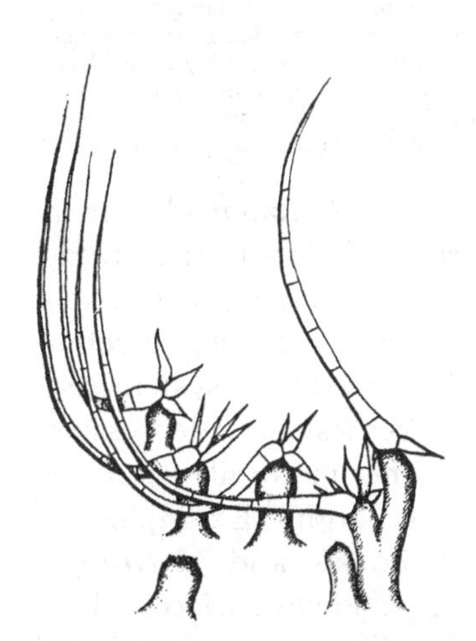

Fig. 192. Hairs from base of the leaf of *Gleichenia pectinata*. Each is seated on a massive emergence, from which it is easily detached, the emergence remaining as in Cyatheaceae. (× 50.)

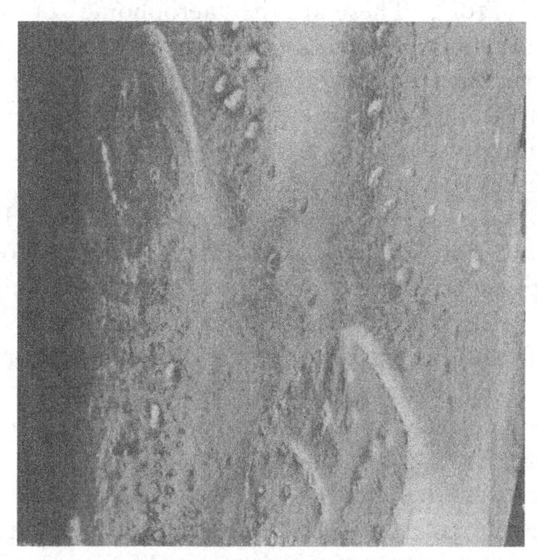

Fig. 193. External surface of stem of *Alsophila crinita*, Hk., cleared of scales, showing two leaf-scars, and the large lenticel-like pneumathodes, or pneumatophores, which traverse the outer sclerenchyma. Reduced.

by epidermis bearing stomata. Similar ventilating areas are seen in the large petioles of *Alsophila*, *Saccoloma*, and *Thyrsopteris* (Figs. 161, 162). They correspond in position to the deep lateral involutions of the vascular tracts. In the large Tree Ferns a somewhat similar perforation of the sclerotic rind appears at the base of the leaves, and on the surface of the stem itself. Round or oval areas irregularly arranged are seen in the adult state filled with pulverulent tissue (Fig. 193). In the young condition the tissue was lacunar parenchyma, covered by epidermis with stomata. These areas serve for ventilation through the sclerotic covering, in much the same way as do the lenticels of Flowering Plants.

In certain Ferns where the young parts are completely covered by dense mucilage secreted from their hairy investment, the ventilation of the young parts would be seriously prevented, were it not that it is secured by outgrowths of the nature of *pneumatophores*. These are well seen in *Plagiogyria pycnophylla* projecting beyond the mucilaginous sheath which covers the young leaves. They remain to the period of maturity of the leaf, and may still be seen functional in the upper region, one at the base of each pinna (Fig. 194). These are the "aerophorae" of Mettenius (*Farngattungen*, ii, 1858). Similar structures are found elsewhere, particularly in the very mucilaginous Ferns of tropical rain-forests, such as *Dryopteris* (*Polypodium*) *Thomsonii*, Jenm., and *decussata* (L.), Urban: also in *Nephrodium stipellatum*, Hk. (see Haberlandt, *Pflanzenanatomie*, Aufl. iv, p. 400). Their occasional occurrence points to their being special adaptations, which cannot be held of high value for phyletic discussion.

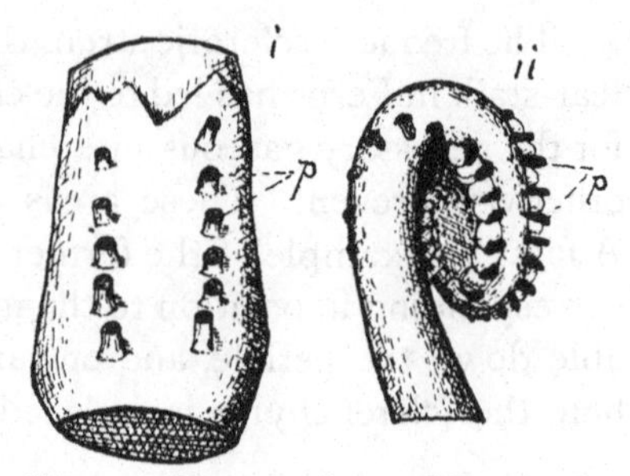

Fig. 194. Pneumatophores of *Plagiogyria*. i, shows the persistent base of the leaf of *P. glauca*, the abaxial face bearing two rows of them (*p*). ii, the circinate apex of the leaf of *P. pycnophylla*, showing the pneumatophores (*p*) alternating with the pinnae, and projecting through the covering of mucilaginous hairs.

The same is the case with those areas of surface specialised for secretion of water, which are in direct communication with the vascular system, and consist of a group of thin-walled epidermal cells, sunk in a shallow depression, immediately above a bundle-ending. These are either uniformly scattered over the surface, or placed in series along the leaf-margin. Examples are seen in the leaves of species of *Polypodium*, *Dryopteris*, and *Nephrolepis* (Fig. 196). Lastly, it must suffice to mention those glandular surfaces which secrete nectar, of which examples are seen in the Bracken at the base of each pinna (Fig. 195). All of these, however biologically important in the individual case, are but sporadic occurrences.

It thus appears that the dermal and other non-vascular tissues of the sporophyte in Ferns are generally of so variable a nature that they give but doubtful material for phyletic comparison. They stand in that respect far behind the vascular system, which shows such a high degree of constancy that it gives a relatively secure ground for comparative argument. Reservation must, however, be made in the case of hairs and scales. The view that the simple hair is primitive and the expanded scale derivative follows partly from the evidence of individual development, but much more securely from the fact that simple hairs are present often exclusively in Ferns whose other features, as well as their fossil relatives, stamp them as primitive. It will be found that this is perhaps the most trustworthy criterion derived from tissues other than the vascular. In fact it may be stated generally and with con-

Fig. 195. Part of an adult leaf of *Pteridium aquilinum* (L.) Kuhn, natural size. *n* = nectary, in form of a small often coloured swelling. (After Potonié.)

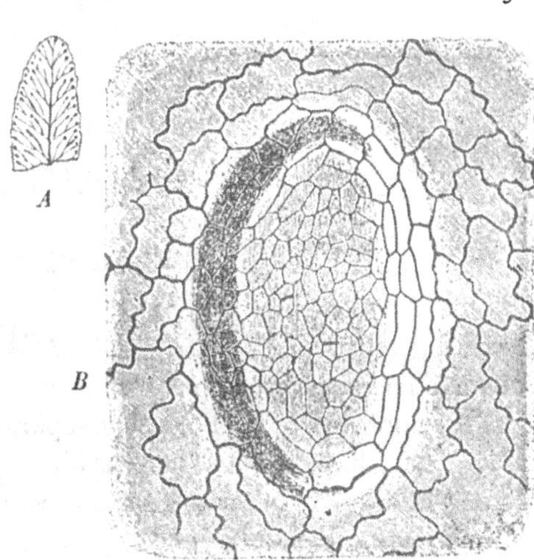

Fig. 196. Pinna of *Polypodium vulgare*, L., seen from below, natural size, showing the water-glands at the ends of the veins. *B* = a water-gland enlarged about 80. (After Potonié.) Both from Engler and Prantl.

fidence ***that simple hairs are a primitive feature, and that branched hairs and in particular flattened scales are an indication of advance from that simple state.*** Evidence of this advance parallel with other indications of advance in other characters may be found in various phyletic sequences, and fully substantiates the correctness of the conclusion thus stated.

BIBLIOGRAPHY FOR CHAPTER XI

186. DE BARY. Comparative Anatomy. Engl. Edn. Oxford. 1884. Hairs, pp. 54, 88. Sclerenchyma, Chap. X, p. 417.
187. PRANTL. Schizaeaceen. Leipzig. 1881, p. 37.
188. KUHN. Die Gruppe der Chaetopterides. Berlin. 1882.
189. RENAULT. *Botryopteris forensis.* Flore Foss. d'Autun et d'Épinac. Part II, p. 33.
190. CHRIST. *Elaphoglossum.* Denkschr. d. Schweiz. Naturforsch. Ges. XXXVI, Zurich, 1899.
191. SEWARD. *Matonia.* Phil. Trans. 1899. *Dipteris.* Phil. Trans. 1901.
192. ENGLER & PRANTL. Natürl. Pflanzenfam. i, 4, p. 59. 1902.
193. BOWER. Studies in the Phylogeny of the Filicales, *passim.* On pneumatophores. Studies. I. *Plagiogyria.* Ann. of Bot. 1910, p. 427.
194. CAMPBELL. Eusporangiatae. 1911, p. 150, Fig. 126.
195. HABERLANDT. Physiological Plant-Anatomy. Engl. Edn. 1914. Chap. IX, p. 423.

CHAPTER XII

THE SPORE-PRODUCING ORGANS

(A) THE SORUS

It has been seen that increase in number of individuals may be carried out in the sporophyte in Ferns by separation of parts of its body (*Sporophytic budding*, Chapter IV). The result is a mere repetition of the sporophyte. Such budding is not a constantly recurring event in the life-cycle. Moreover the way in which it is effected varies in different cases to such an extent that its details do not form a suitable basis for a wide comparative argument. It is otherwise with the alternative mode of *Increase by Spores.* There is reason to believe that the formation of carpospores has recurred during Descent, in one form or another, in each completed life-cycle since sex was established. Each spore detached from the parent plant is a potential new life. Not only the continuance of the species, but also its spread, are secured by spores. The larger their number the more effectively are those ends achieved. A recognition of this principle should underlie all comparative study of the sori of Ferns. Recently it has been suggested that the larger the number of tetrad-divisions in an organism where, as in Ferns, the chromosomes are numerous, the larger will be the number of possible combinations of the characters which they bear, and the greater the variability. Such advantages as these may be held as in some degree accounting for the very large output of spores seen in homosporous Ferns. From this point of view the high productivity of Ferns, their variability, and their cosmopolitan distribution become intelligible. They are at once the most productive and the commonest of all homosporous Pteridophytes of the present day. But a primary limit to the number of their spores is imposed by the capacity of the plant to nourish them. Two other considerations influence successful propagation by spores. They must be properly protected till they are mature; and when ripe they must have the opportunity for free dispersal. *Nutrition*, *protection*, and *free dissemination* are the three conditions dominating successful propagation by spores. The phylesis of Ferns illustrates progressive changes of method by which these ends are effectively secured. Those changes depend upon differences in the form, structure, and position of the spore-producing organs. Such considerations as these give the spore-producing organs a premier place among the distinctive characters of Ferns.

The sorus of *Dryopteris Filix-mas*, described in Chapter I, will serve as an average example for Ferns such as are neither of the most primitive nor of the most advanced type. The whole group of organs which forms the sorus is seated above a vascular strand, whose end dilates as it enters the swollen receptacle. Nutrition is thus secured. Each sorus consists of numerous sporangia protected by an indusium. These together form an apparently coherent entity. But the only constituent of it that is constant for all Ferns is the individual sporangium, which ruptures so as mechanically to eject and scatter the spores. The indusium is absent altogether in many Ferns. Often the sorus is not

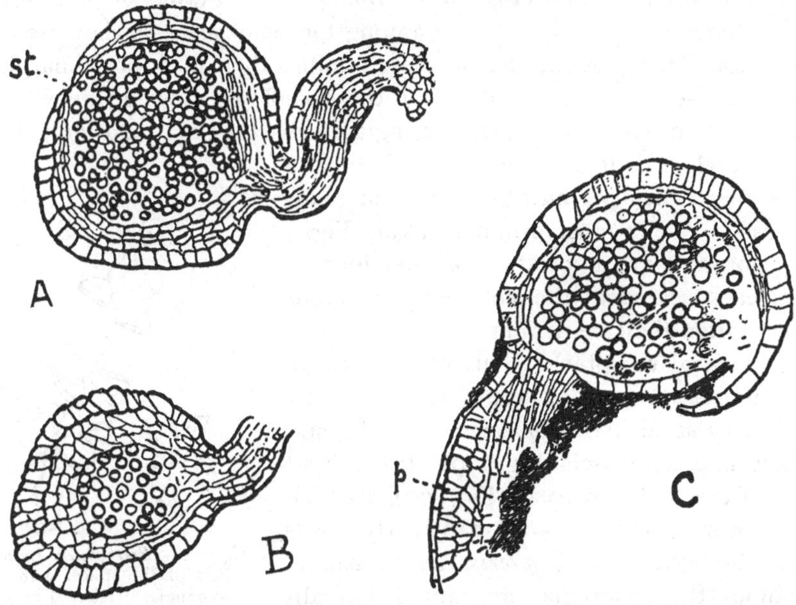

Fig. 197. *Stauropteris oldhamia*, Binney. *A* = sporangium in nearly median section, attached terminally to an ultimate branchlet of the rachis: *st* = stomium. Scott Coll. 2213. *B* = sporangium in tangential section attached to a short piece of the branchlet. Scott Coll. 2207. *C* = sporangium with wall burst, attached as before: *p* = palisade tissue of branchlet. Scott Coll. 2219. All figures × about 50. (From sketches by Mrs D. H. Scott. Specimens are from Shore, Littleborough, Lancs.)

strictly circumscribed as a definite unit. In extreme cases each sporangium may be isolated, and unprotected. But in most Ferns numerous sporangia are grouped together upon a single spot above a vein. Their crowded grouping secures mutual protection while young, and this may be aided by various accessory growths included under the general term "indusium." Such a group of organs is called a "*sorus*." But its individuality is not always maintained. It will be shown that in the course of Descent fusions of sori, or disintegrations of them, have occurred. In several distinct phyletic lines the identity of the sorus may be altogether lost, the sporangia being spread uniformly over the leaf-surface. This is called the "Acrostichoid" condition. It thus appears that the sorus cannot be held as a definite morphological unit for all Ferns. It is only a congeries of spore-producing units. *The essential spore-producing organ is the Sporangium itself.*

CONSTITUTION OF THE SIMPLE SORUS

The simplest way in which the sporangia are borne is singly, and their position is then either terminal or marginal. If the term sorus be here applied at all, each sporangium would be a "monangial sorus." This is seen in primitive living types, and in certain fossils. The solitary distal sporangium of *Stauropteris* is an example (Fig. 197). Among living Ferns *Botrychium*, one of the Ophioglossaceae, and *Mohria*, among the Schizaeaceae, show isolated sporangia, and it is notable that in these families their position is marginal, or we might describe them as distal on the tips of the more or less webbed branchings. In such cases the sporangia themselves are relatively large (Fig. 198), while their position on a vein-ending secures their effective nutrition. But in many early fossils, and in most primitive living Ferns, many sporangia are associated together forming groups, each seated in the same way on a vein-ending. Often they are arranged rosette fashion, round a central receptacle. This constitutes the radiate-uniseriate type of sorus. It is seen in a rudimentary state in *Zygopteris*, where the sporangia form distal tassels (Fig. 199). It is characteristic of the Gleicheniaceae (Fig. 200), *Matonia*, and the Marattiaceae; also of many early fossils, such as *Corynepteris, Scolecopteris, Asterotheca*, etc. Sometimes the sporangia are united laterally so as to form a coherent synangium. This is illustrated in the fossil *Ptychocarpus unitus* (Fig. 201), and in *Marattia* and *Christensenia* among living Ferns. The confluent sporangia of *Ophioglossum* may also be quoted as an example of essentially the same phenomenon, which may be held as derivative. It probably owed its origin, and finds its biological justification, in the mutual protection which the sporangia thus acquire.

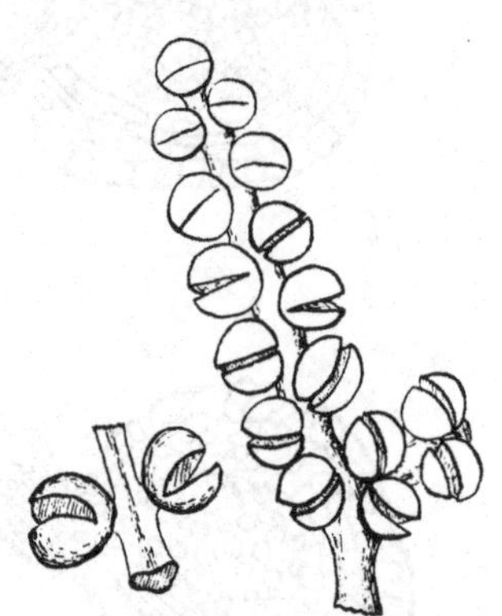

Fig. 198. *Botrychium Lunaria.* Part of the spike, with open sporangia. (After Luerssen.) Enlarged.

Occasionally a reduction in number of the sporangia in the radiate-uniseriate sorus may be traced by comparison of related species. For instance, in *Gleichenia flabellata*, the basal sori of the pinnule may have five or six sporangia; but the distal sori only three (Fig. 200). In § *Eu-gleichenia* the number is usually two in each sorus, sometimes only one. It seems probable that this low number is the result of reduction in these attenuated xerophytic Ferns. On the other hand, by an elongation of the sorus along the course

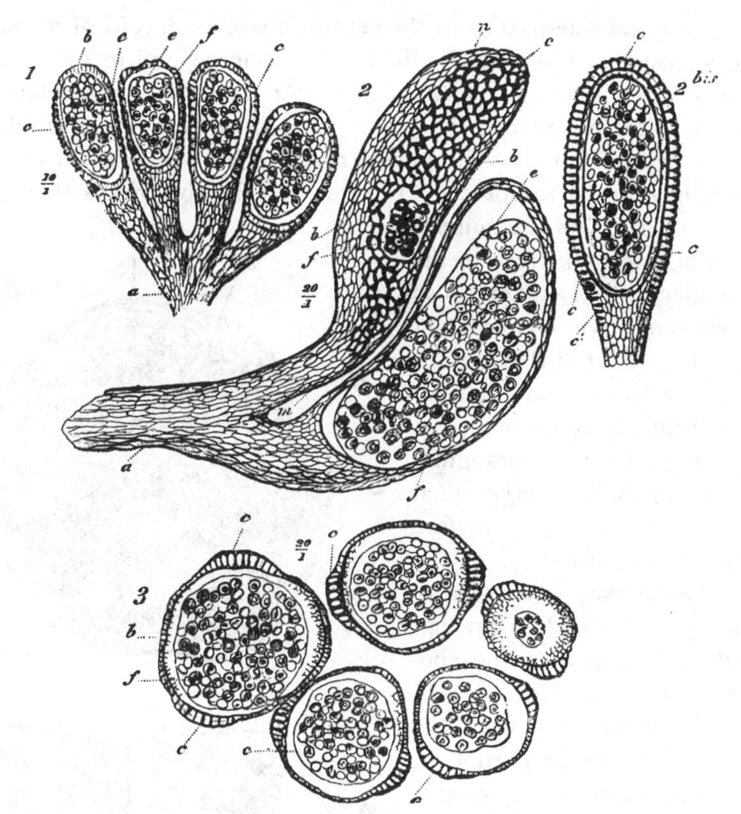

Fig. 199. *Zygopteris* (*Etapteris*). 1, group of four sporangia on a common pedicel (*a*). × 10. 2, two sporangia on pedicel. The upper shows the annulus (*c*) in surface view, with spores exposed at *f*: the lower in section. × 20. 2 *bis*, sporangium cut in plane of annulus. 3, group of sporangia in transverse section. × 20. Lettering common to the figures. *a* = common peduncle: *b* = sporangial wall: *c* = annulus: *e* = tapetum (?): *f* = spores: *m* = pedicel of individual sporangium: *n* = probable place of dehiscence. (All after Renault.) (From Scott's *Studies in Fossil Botany*.)

of the vein, room may be found for an increased number of sporangia. This is illustrated in the Marattiaceae, where in *Angiopteris* and *Marattia* the sorus is oval with relatively few sporangia (Fig. 202); but in *Archangiopteris* and in *Danaea* it is more elongated, and the number of sporangia is great (Fig. 202). Thus, while preserving the same type of simple and uniseriate sorus, the number of sporangia may be either reduced or increased.

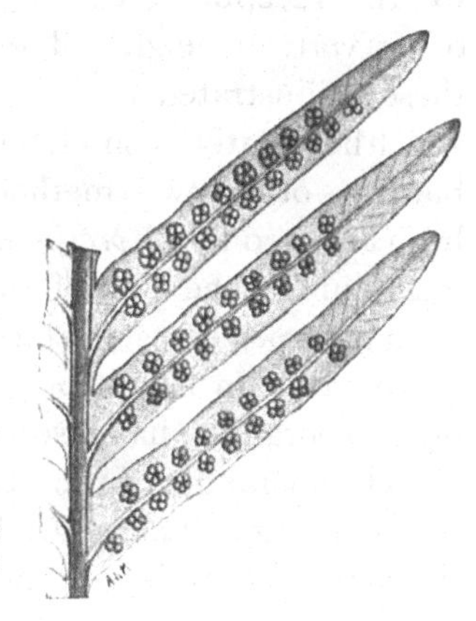

Fig. 200. *Gleichenia flabellata*, Br. Midrib and three pinnules, showing the arrangement and constitution of the sori, with a variable number of sporangia.

In the simplest uniseriate sori the central space is devoid of sporangia, a condition seen constantly in the living Marattiaceae, and in most species of *Gleichenia*. But in *G. dichotoma* and *pectinata* sporangia are found inserted on the apex of the receptacle. They vary in number from one upwards, and may form a second tier above the basal rosette. But all of them originate simultaneously on the receptacle. As in other primitive Ferns there is no succession in time. As the number of sporangia is thus increased, the available space is fully taken up so that the sporangia will be in contact laterally. The effect of this may be seen in *G. pectinata*, where their sides are flattened by mutual pressure (Fig. 203). But the sporangium of *Gleichenia* opens by a longitudinal slit, and elbow-room is therefore required for the discharge of the spores. The increase in number of the sporangia will tend to prevent this, and in *G. pectinata* the sporangia are sometimes found undischarged. Its sorus has reached the practicable limit of proportion of size and number of the sporangia to the area of their insertion on the receptacle. The mechanical deadlock may be met either by diminution of the size of the sporangium, or by increase of the area occupied by the receptacle, or by increase of its vertical height. The first of these is illustrated in *G. dichotoma*, which has relatively small sporangia; but the other two methods have been adopted by *G. pectinata*, which marks in fact the limit of number of sporangia per sorus attained in the family. In other related Ferns, however, the increase in height of the receptacle, together with decrease in the size of the sporangia, has resolved the difficulty, and led to the Gradate Sorus.

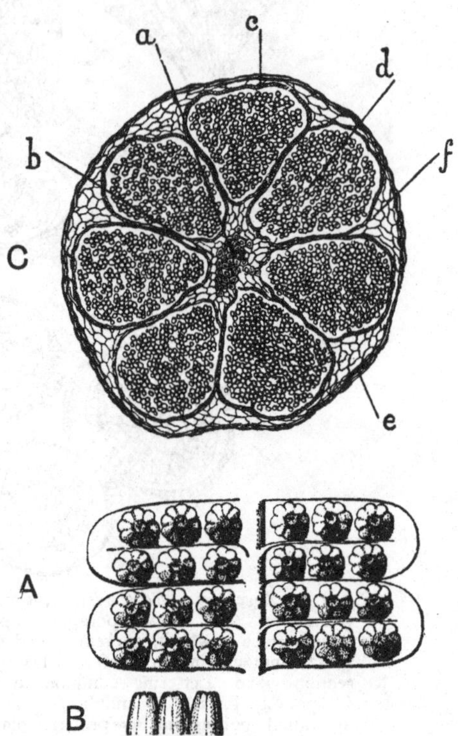

Fig. 201. *Ptychocarpus unitus*. Fructification. *A* = part of a fertile pinnule (lower surface), showing numerous synangia. *B*, synangia in side view. (*A* and *B* × about 6.) (After Grand'Eury.) *C* = a synangium in section parallel to the surface of the leaf, showing seven confluent sporangia. *a* = bundle of receptacle: *b* = its parenchyma: *c* = tapetum: *d* = spores: *e*, *f* = common envelope of synangium. × about 60. (After Renault.) (From Scott's *Studies in Fossil Botany*.)

The sporangia of the Ferns in which simple sori are present are relatively few and large. The physiological point which they all have in common is the simultaneity of origin of all the sporangia in near proximity to one

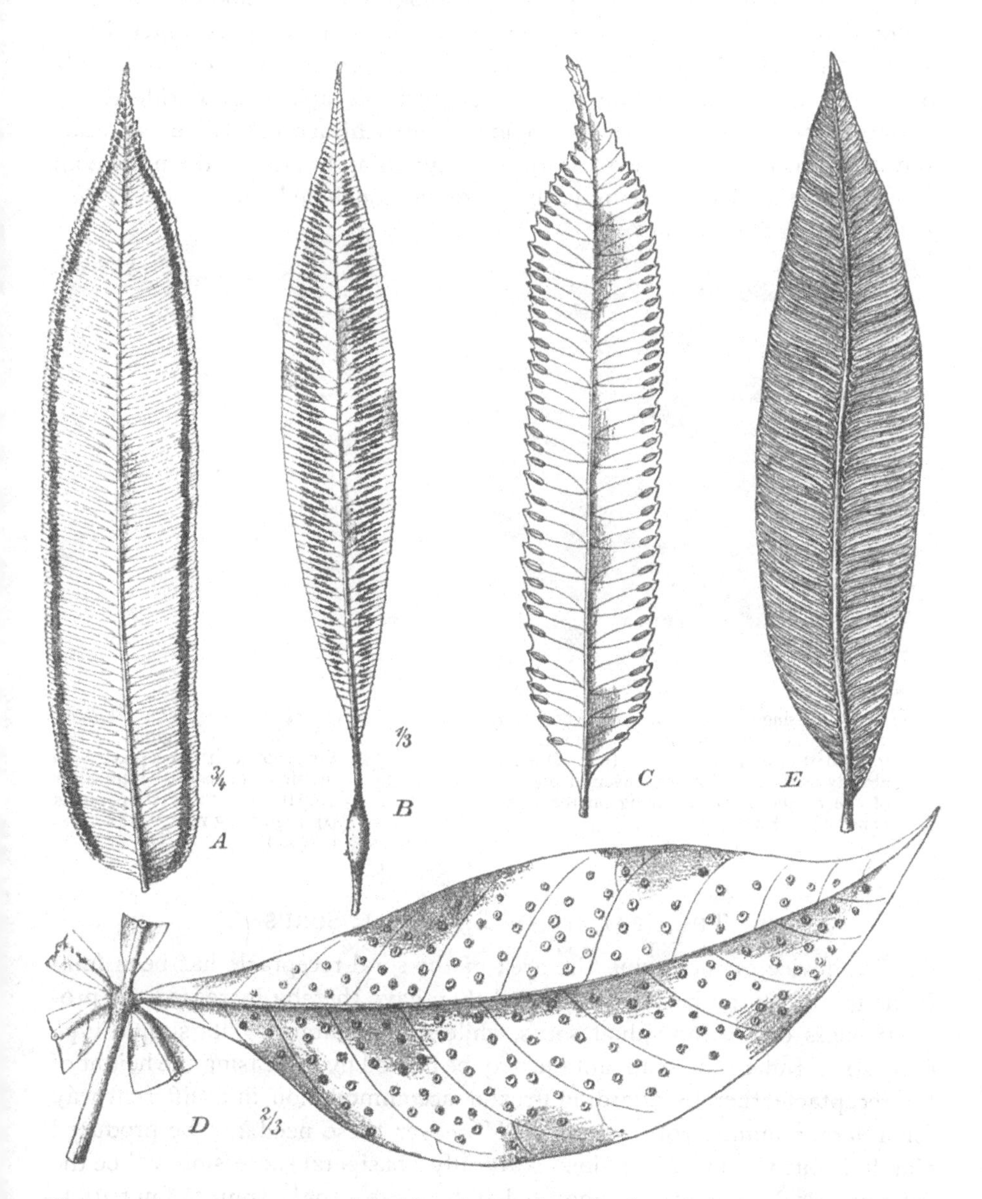

Fig. 202. Pinnae of five genera of the Marattiaceae, all of them lateral. *A* = *Angiopteris crassipes*, Wall.: *B* = *Archangiopteris Henryi*, Christ and Giesen.: *C* = *Marattia fraxinea*, Sm.: *D* = *Christensenia aesculifolia*, Bl.: *E* = *Danaea elliptica*, Sm. *A*, *C*, *D*, *E*, after Bitter: *B*, after Christ and Giesenhagen. (From Engler and Prantl, *Nat. Pflanzenfam.*)

another. The effect of this is that the physiological drain of spore-production is imposed simultaneously by all the sporangia upon the part which bears them. *Such Ferns as show the simultaneous scheme have been styled the* SIMPLICES, and they include the Botryopterideae and Zygopterideae, the Marattiaceae, Osmundaceae, Ophioglossaceae, Schizaeaceae, Gleicheniaceae, and Matonineae. These are all primitive types, and many of them have an early history as fossils. It may, therefore, be concluded that *this condition of the sorus is itself primitive.*

Fig. 203. A single sorus of *Gleichenia pectinata*, showing the sporangia so closely packed as to be flattened against one another. One has already dehisced. Note the inverted position of one of the outer sporangia on the right-hand side. Enlarged.

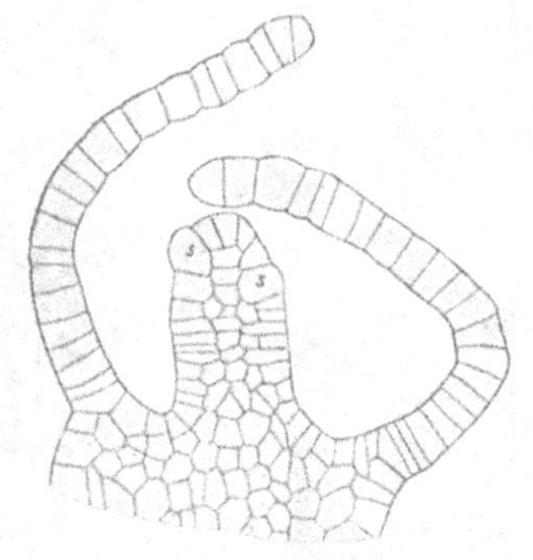

Fig. 204. *Hymenophyllum Wilsoni*, Hk. Sorus in longitudinal section, showing the receptacle with divisions indicating intercalary growth, and the first sporangia (*s*) originating near to the apex. (× 100.)

THE GRADATE, OR BASIPETAL, SORUS

The device of increasing the area of the soral receptacle has been fully made use of by the Marattiaceae, and they have thereby increased the productiveness of the individual sorus, while still maintaining its simple type (Fig. 202). But a similar advantage may be gained by increasing the height of the receptacle, thereby affording greater accommodation in a different way for a larger number of sporangia. Moreover these need not be produced simultaneously but in succession. Naturally a basipetal succession will be the most practical, so that the youngest shall be nearest to the source of nutrition, while the oldest will be most exposed, and shedding of the spores easy. The drain of nutrition will by this means be spread over an extended period. This is the physiological *rationale of the Gradate Sorus.* The receptacle, which may be cylindrical (Hymenophyllaceae), or laterally compressed (*Cibotium*, *Thyrsopteris*), retains the power of intercalary growth (Fig. 204).

Upon this the first sporangia appear at or near to the distal end: they are followed by a basipetal succession of them (Fig. 205). In later stages distal sporangia may be found already mature, while towards the base successively younger sporangia may be seen. This type of sorus is characteristic of the Gradatae; it is seen in the Dicksonieae and Loxsomaceae, and it finds its climax in the Hymenophyllaceae. It is found also in the Cyatheae, and Onocleinae. The sporangia of all of these have an oblique annulus. The spore-members from each sporangium are large in some of the Gradatae (Hymenophyllaceae), but they show great fluctuations. The Ferns with gradate sori are all of a middle position as regards their general characters,

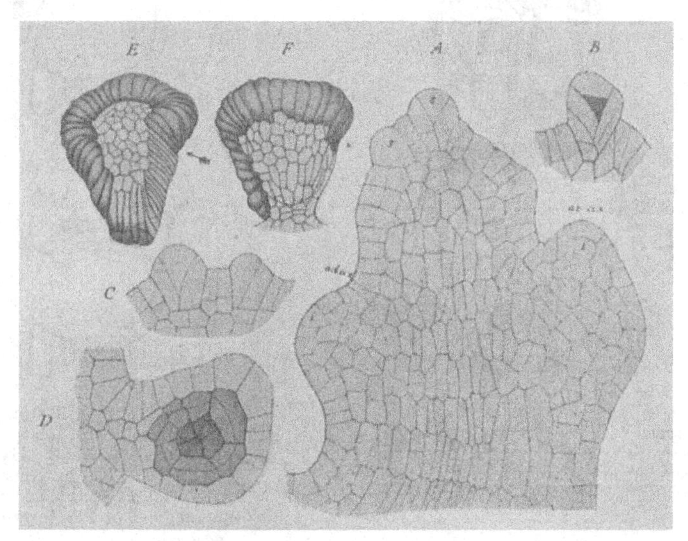

Fig. 205. *Thyrsopteris elegans*, Kze. A = longitudinal section through the young sorus, showing the two-lipped indusium, i, i, and sporangia, s, s, seated on the receptacle, the oldest being at the distal limit of it. C = two young sporangia. B = one rather more advanced. D = a sporangium with tapetum and sporogenous group shaded. E, F = mature sporangia. A—$D \times 200$. E, $F \times 50$.

and their fossil correlatives date back to the Mesozoic Period. They do not, however, represent the main body of recent Ferns, though they include the largest of them. The gradate sorus may accordingly be held as derivative from the simple type, and it brings physiological advantages which lead to a high spore-output in the largest of the Ferns that possess it.

The difficulty of dissemination of the spores is, however, increased by the close aggregation of the sporangia upon the prolonged receptacle. That difficulty is met by the oblique annulus. It has been seen that in *Gleichenia*, where there is a median dehiscence of the sporangium, elbow-room is required for the mechanical ejection of the spores, and that this is only possible where the sporangia are loosely arranged. *Loxsoma* is the only gradate Fern which is known to retain the median dehiscence. There the annulus is incompletely indurated: it merely opens the distal end of the sporangium, and the spores are shaken out (Fig. 206). But in the Hymenophyllaceae, as in other Gradatae, there is oblique lateral dehiscence. The sporangia have a definite orientation, being placed relatively to one another as seen in Fig. 207, so that the distal surface of each annulus has freedom to alter its form independently of the adjoining sporangia.

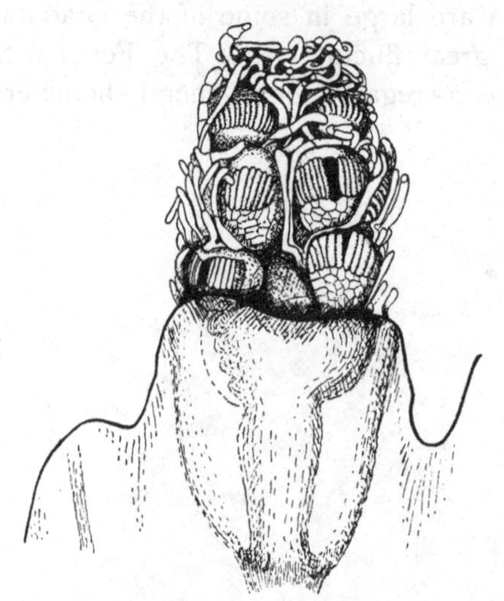

Fig. 206. Mature sorus of *Loxsoma Cunninghamii* with the receptacle elongated so as to raise the sporangia above the cup-shaped indusium, where they open by a median dehiscence. (After von Goebel.) (× 25.)

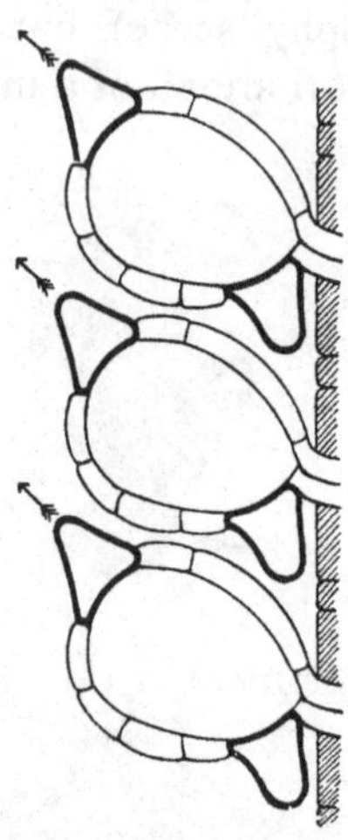

Fig. 207. Diagram illustrating the relative position of the sporangia on the receptacle in the Hymenophyllaceae. It was constructed from Prantl's section of a mature sporangium of *Trichomanes speciosum*.

The Mixed Sorus

It has been found as the result of examination of the remainder of the Filicales, which constitute the great majority of living genera and species, that the sorus is of the type called "mixed." Here the sporangia of different ages are aggregated together without any definite sequence. In development younger sporangia are interpolated promiscuously between those already present. There is no definite orientation of the sporangia, and their stalks are usually long. The annulus is almost always vertical, and it is interrupted at the insertion of the stalk. The number of spores per sporangium very rarely exceeds 64, while lower numbers are frequent. There may be very great differences in the number, position, and individuality of the sori, and in the presence or absence of an indusium. It is upon these characters that the classification of the *Mixtae* has principally been founded.

The origin of this state has probably been along not one, but many lines of Descent. Indications of how it came into existence are given at several quite distinct points in the system. For instance among the Dicksonioid-Davallioid series the former have gradate, the latter mixed sori. *Dennstaedtia apiifolia* and *punctilobula* have a markedly gradate sorus: but an intermediate state is seen in *Dennstaedtia rubiginosa* (Fig. 208, *A*), where there are still some signs of the basipetal sequence; but younger sporangia are

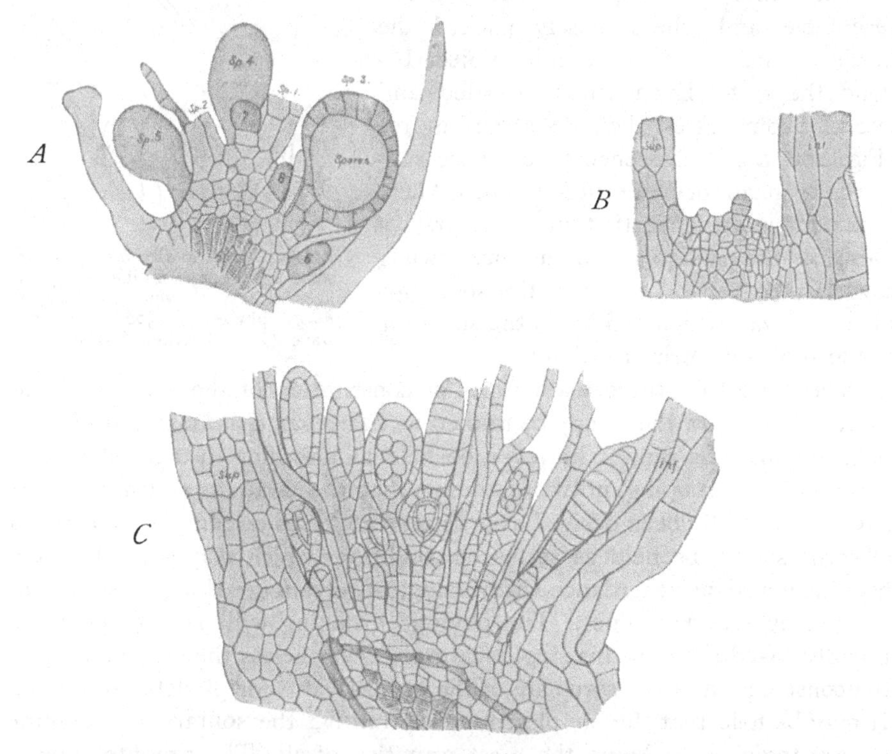

Fig. 208. *A*. Sorus of *Dennstaedtia rubiginosa*. Cut vertically, and showing a mixed condition in a sorus originally basipetal. *B* = *Davallia Griffithiana*, Hk. Young sorus in vertical section showing the first formation of sporangia. *C*. Old sorus of the same, showing sporangia of different ages intermixed. All × 100.

interpolated without order among those pre-existent, while the receptacle, which in gradate sori is convex or elongated, is here relatively flat. These characters approach those of the fully mixed sorus with flattened receptacle seen in *Davallia* (Fig. 208, *C*). Another example from a quite distinct affinity is seen in *Dipteris conjugata*. But here the progression is directly from the simple sorus as seen in *D. Lobbiana*. In the latter all the sporangia appear simultaneously: but in *D. conjugata*, which is an advanced species in other respects, the sporangia are produced in succession, so that those of different ages may be side by side (Fig. 209). The change to the mixed state thus indicated in *Dipteris* has been perpetuated in a number of Ferns derivative from the Dipteroideae (*Studies*, v), such as *Leptochilus tricuspis*, *Neocheiropteris*, and probably *Phlebodium*, and others. There is no reason to think that the gradate condition ever intervened in such cases.

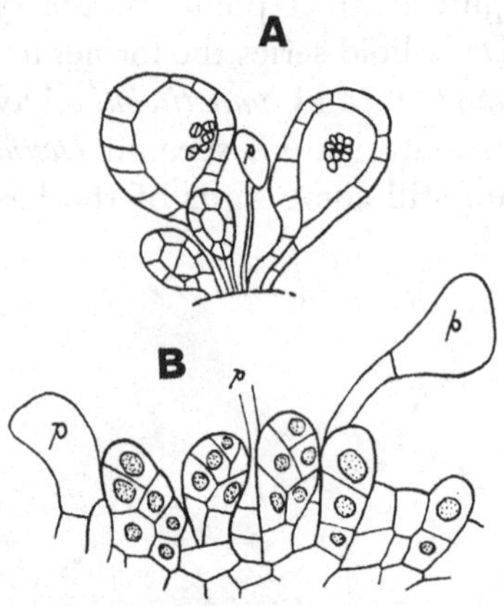

Fig. 209. *Dipteris conjugata*; young sori showing sporangia of different ages in juxtaposition. *B* = younger sporangia, *p* = paraphyses. (× 300.) *A* = older stage. (× 100.) (After Miss Armour.)

The "mixed" sorus has usually a flat receptacle, and where closely packed the heads of the mature sporangia project beyond the rest. Even when crowded and laterally compressed their distal end is free (Fig. 208, *C*). Dehiscence takes place outwards by means of the vertical annulus. There is abundant evidence that in the course of Descent the oblique annulus has swung towards the vertical plane in the sporangia of the *Mixtae*, which facilitates the shedding of the spores in their crowded sori.

There are thus three main types of constitution of the sorus, and the Ferns which show them may be ranked as *Simplices*, *Gradatae*, and *Mixtae*. Such distinctions cannot, however, be held as affording in themselves any true phyletic grouping, notwithstanding that the Gradate and Mixed sori have undoubtedly been derived from those recognised as Simple. The types of sorus should be held as states or conditions which may severally have been achieved in accordance with biological advantage along a number of distinct evolutionary lines. The phyletic grouping will have to be more broadly based than on this single feature. Nevertheless the facts relating to the construction of the sorus are of the utmost value for phyletic argument. It may be held that the simple sorus is primitive, the solitary sporangium, or monangial sorus, being the most primitive of all. The gradate type is derivative from the simple, and therefore more recent. The mixed sorus is the most recent, and may be traced either directly from the simple sorus, or indirectly from it, through the gradate type.

The position of the Sorus

In most Ferns the position of the sorus is upon the abaxial or lower surface of the leaf: each is seated upon a vein, frequently upon a vein-ending. In a very large proportion of them the sori form two more or less distinct rows, one on either side of the midrib of the pinna or pinnule. This is seen in *Polypodium vulgare*, *Dryopteris Filix-mas*, etc. In many other Ferns, however, the position may be either actually on the margin of the leaf, or in near relation to it, as in *Pteridium* or the Hymenophyllaceae. The two positions, the superficial and the marginal, graduate into one another by many intermediate steps. But the position held by the sori of a species, genus, or even of still larger groups, is as a rule definite, showing that it is not a readily fluctuating character, but is so far constant in the individual species or genus that it may be depended on for comparative purposes.

So marked is the constancy of position of the sorus in genera and species that the few exceptions which occur become notable. The most familiar is that of a variety of *Polystichum aculeatum* from the uplands of Ceylon, described as *Aspidium* (*P.*) *anomalum*, Hk. and Arn. (Hooker's *Sp. Fil.* Vol. iv, p. 27). Here normally constructed sori appear on the upper instead of the lower surface of the leaf. Since no intermediate stages appear, either in the individual or in related species or varieties, the natural conclusion is that there has been a transfer bodily of the factors of normal initiation of sori into an abnormal position. Such an instance as this, being anomalous, should not be held to affect the general argument on the position of the sori.

The question will naturally arise whether the marginal or the superficial position of the sorus was the prior state. A general probability that the superficial is relatively late and derivative arises from the fact that it is prevalent in modern Ferns. It is physiologically intelligible that parts so important shall not be unduly exposed: moreover the downward direction would favour the dissemination of the spores. But more direct evidence is afforded by reference to those Ferns which by comparison and by fossil history are held as primitive, viz. the Simplices. This will give a general basis of fact upon which a reasonable opinion may rest. Further comparison may be made of the position of the sorus in genera, species, and varieties related to one another. And in particular of its position in the successive ontogenetic stages of the individual. From such sources it may be shown that a transition from the marginal to the superficial position has occurred repeatedly in the evolution of Ferns, and that it may be traced even in the ontogeny of some of them.

Among the Simplices, the Botryopterideae, Zygopterideae, Ophioglossaceae, Schizaeaceae, and *Osmunda* show a distal or marginal position of the sporangia. The Marattiaceae, Gleicheniaceae, Matonineae, and the genus *Todea* bear their sporangia superficially. *Osmunda* itself bears the sporangia as a rule in tassels at a margin of the attenuated fertile region of the leaf. But von Goebel points to those sporophylls that are half modified into trophophylls: here the insertion of the sporangia is superficial, following the veins, in a manner similar to those of normal specimens of *Todea* (Fig. 210, 207, p. 1139). This supports his general statement that the more the sporophyll is developed as a dorsiventral foliage leaf the more are the sporangia restricted to the lower surface. The facts for the Simplices are in general accord with this; for the sporophylls of the Ferns with marginal sori or sporangia are relatively narrow, while those of the Marattiaceae are broad, and those of the Gleicheniaceae and Matonineae share this feature, though in less degree. The sporophyll of *Todea barbara* also is broad compared with the normally attenuated fertile region of *Osmunda*. Thus on *prima facie* evidence the general view may be stated as a working hypothesis that *the primitive position of the sporangium or sorus was distal or marginal, and that the superficial position is derivative, having been acquired in relation to an increase in surface of the sporophyll.* But before acceptance as a substantive view relating to the evolution of Ferns, it will have to be tested by a comparative study of development, and by preference the test should be made in genera or species nearly related one to another.

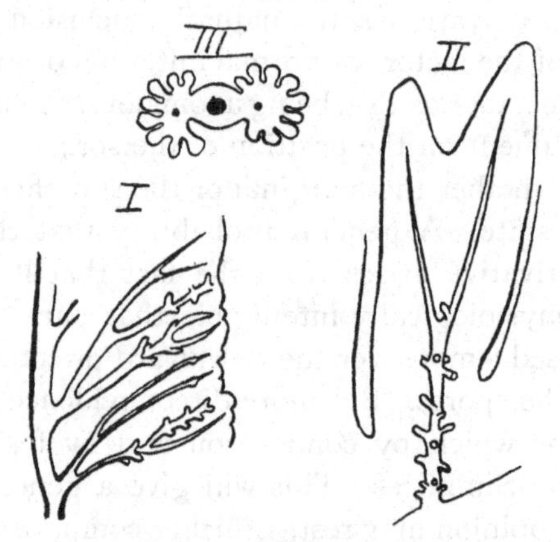

Fig. 210. *I. Osmunda regalis*, part of a metamorphosed sporophyll with aborted sporangia attached superficially. *II. Todea pellucida*, part of a pinnule with sporangial stalks. *III.* Transverse section of a fertile pinnule of *Osmunda* showing normal insertion of sporangia. (All after von Goebel.)

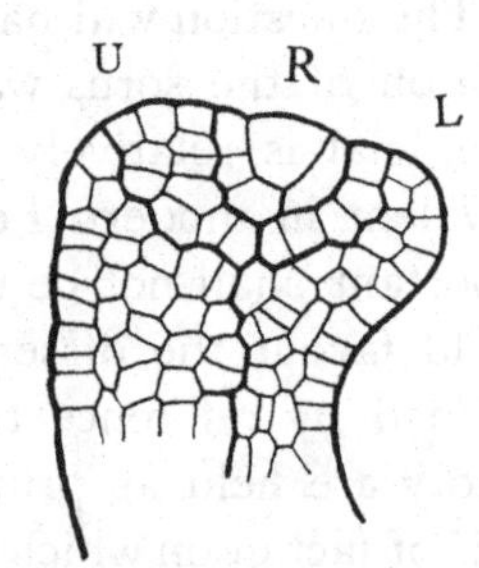

Fig. 211. Vertical section through young sorus of *Dicksonia Scheidei*. (×200.) *R*=receptacle; *U*=upper, and *L*=lower, indusium.

The marginal origin of the receptacle and of the sporangia has been demonstrated developmentally in many examples. Prantl showed it in the Hymenophyllaceae and Schizaeaceae (*Unters. zur Morph. d. Gefässkrypt.* I, II, Leipzig, 1875, 1881). It may be demonstrated fully for the Dicksonieae. A section through the very young sorus of *Dicksonia* (*Cibotium*) *Scheidei* shows how the regular marginal segmentation is continued directly into the receptacle (Fig. 211, *R*). The upper and lower indusial flaps (*U*, *L*) arise by intramarginal surface-growths: neither of them is a continuation of the leaf-surface itself. At first the receptacle, which is here the actual leaf-margin, is flattened, and lies evenly between the two indusial flaps (Fig. 212). Upon its extreme ridge a row of sporangia arises simultaneously (*B*, *C*): they are followed by others successively below the margin. This succession is more readily seen in *Thyrsopteris* (Fig. 205). Here the fertile segments being very narrow the receptacle itself is almost cylindrical, as it is in *Loxsoma*, and in the *Hymenophyllaceae*.

The marginal position of the sorus is sometimes maintained till maturity, as in the Ferns just named. But in most cases where the origin of the sorus is marginal, as the development proceeds it may be more or less directed by unequal growth towards the lower (abaxial) surface. This is seen in the "monangial sori" of the Schizaeaceae. It has been demonstrated fully for

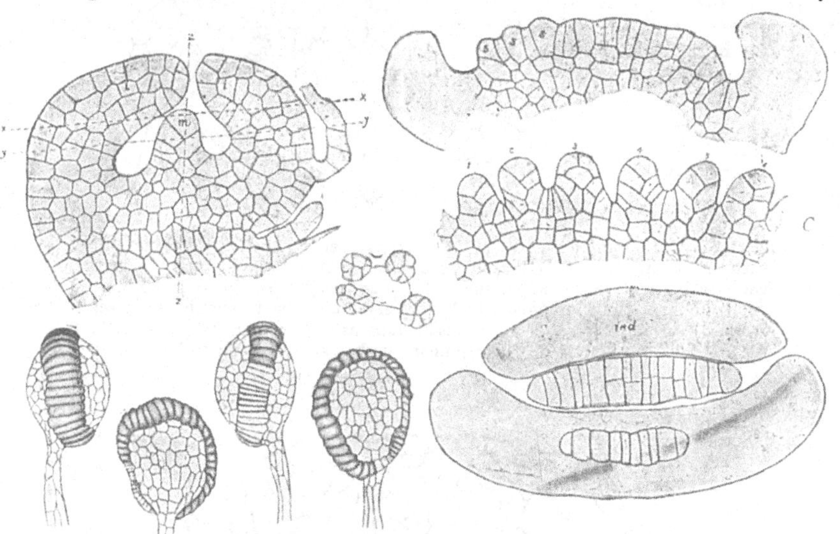

Fig. 212. *Dicksonia Scheidei*, Baker. A = section through a young sorus perpendicular to the leaf-surface; i, i = indusium; m = cell of marginal series. B = section of sorus parallel to the leaf-surface, as along a line, i, i, in Fig. A, showing the receptacle bearing sporangia, s, s. C = a similar section bearing older sporangia. D = transverse section of a young sorus showing the two lips of the indusium (*ind.*), and receptacle between them, as along a plane, y, y, in Fig. A. A section of the receptacle in the plane, x, x, in Fig. A, is superposed on the lower indusial lip. The central figure shows sporangial stalks cut transversely. A—D × 200. E, F, G, H, sporangia of *Dicksonia Menziesii* from four different aspects. × 50.

Schizaea rupestris (Fig. 213), and the results become very evident in *Mohria*, where the sporangium appears as though superficial after the first stages are past, owing to a secondary development of an indusial flap at its base, forming a false margin (Fig. 214). A like diversion to the lower surface as the part matures—but here it is of the whole sorus—is seen in *Dicksonia* (*Cibotium*) (Fig. 212). Such ontogenetic changes of position are common. They may be seen to lead in Ferns of the Dicksonioid-Davallioid affinity to a "phyletic slide" of the sorus to a definite initial position on the lower surface of the leaf, as the following examples taken from closely related Ferns will show[1]. In the first place Conard has found that the origin of

[1] As to this relationship, see *Studies*, VII, *Ann. of Bot.* 1918, p. 50, in which the conclusion of Prantl is adopted, *Arb. K. Bot. Gart. Breslau*, I, i, p. 16.

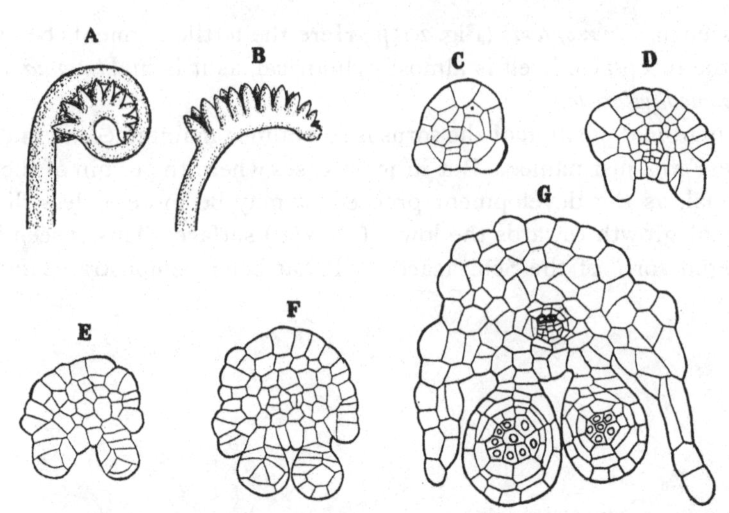

Fig. 213. *A*, *B* = Young leaves of *Schizaea rupestris*, showing circinate vernation, with the pinnae reflexed to the convex (abaxial) side. (× 3.) *C*, *D*, *E* = Transverse sections of very young pinnae of *Schizaëa rupestris*, showing the marginal origin of the sporangia, which are very soon turned towards the lower (abaxial) surface: in *D*, *E* the indusial flaps are appearing, right and left; *F*, *G* = Similar sections of older pinnae, with sporangia and indusia more advanced. (× 60.)

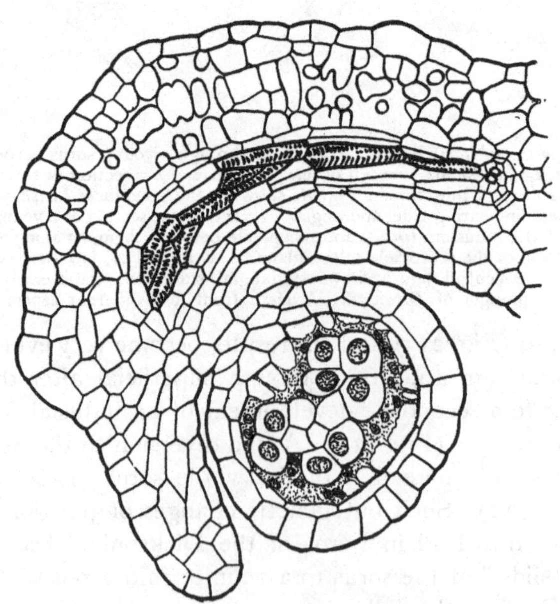

Fig. 214. Longitudinal section through the apex of a pinna of *Mohria caffrorum*, showing the apparently superficial position of the sporangium, which is actually marginal in origin. This appearance is due to the active growth of the indusium from its base, forming a false margin. (× 75.)

the sorus in *Dennstaedtia punctilobula* (Michx.), Moore, is marginal, as it is in *Cibotium*. The same is the case also in *D. dissecta* (Sw.), Moore (Fig. 215).

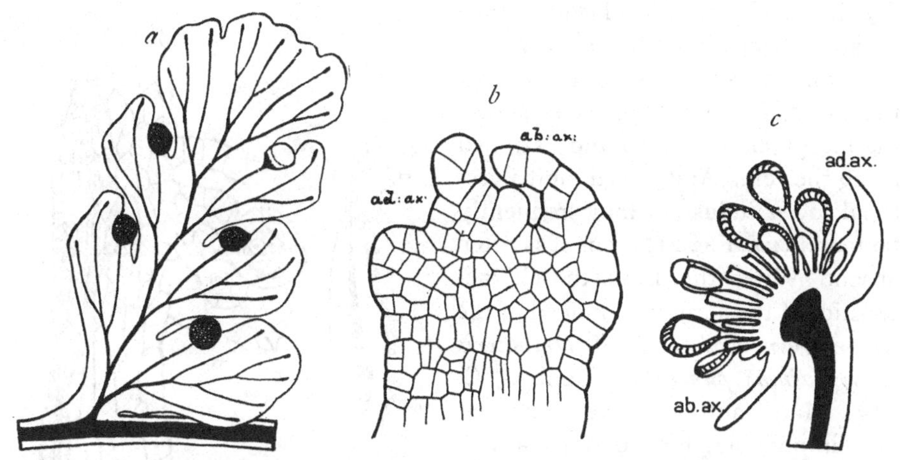

Fig. 215. *a* = Pinnule of *Dennstaedtia dissecta* seen in surface view, showing the marginal sori on the apex of the anadromic branches of the veins. (× 4.) *b* = Sorus very young, cut in vertical section, showing the marginal receptacle and superficial indusial flaps. (× 150.) *c* = Mature sorus in vertical section. (× 35.)

Each sorus is seated on a vein-ending, but when mature it is directed obliquely downwards (*a*). A vertical section through the mature sorus gives the relation of the indusial flaps to the receptacle in which the vein terminates. As the sorus matures it assumes the mixed character (*c*). Both the indusia are well developed; but in the related *Hypolepis nigrescens*, where the downwards curvature of the whole sorus is more marked, the abaxial indusium is absent: obviously it is not necessary for protective purposes when the strong curvature brings the sorus close to the lower leaf-surface, and it is accordingly aborted (*Studies*, VII, Fig. 38) The case is, however, slightly different in *Hypolepis repens*, where the pinnule is broad, and the sorus is seated upon a vein-ending at some distance apparently from the leaf-margin (Figs. 216, 217). In other respects the relation of the parts is as before: but the sorus is more extended along the vein, while the vascular tissue reaches beyond the receptacle into the upper indusium; this is now flattened

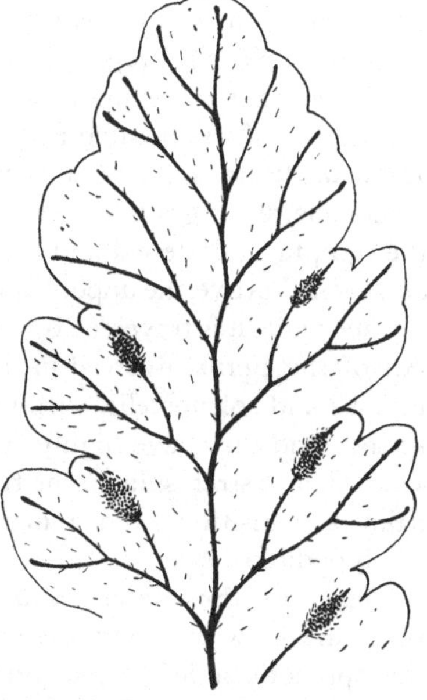

Fig. 216. Pinnule of *Hypolepis repens*, seen in surface view. (× 10.)

in the plane of the pinnule, and appears as though it were a minor lobe of it. In some forms of *H. repens* it still curves over, and protects the sorus. The lower, or abaxial, indusium is not always absent altogether, as in *H. nigrescens*. A vertical section through the sorus, following the vein, shows the receptacle extended and flattened along the vein. At its basal limit a vestigial lower indusium may frequently be found (*v.i.*, Fig. 217), but it is often absent. When this is so, the characteristics of the sorus, and indeed of the whole Fern, are those of *Dryopteris* (*Polypodium*) *punctata* (Thunbg.), C. Chr. Many authors have remarked how impossible it is to draw a line between *H. repens* and *D. punctata*. But if the evolutionary sequence be as described, and the lower indusium is either reduced or absent, there is no obligation to draw any line at all between them, for they represent stages in a natural progression.

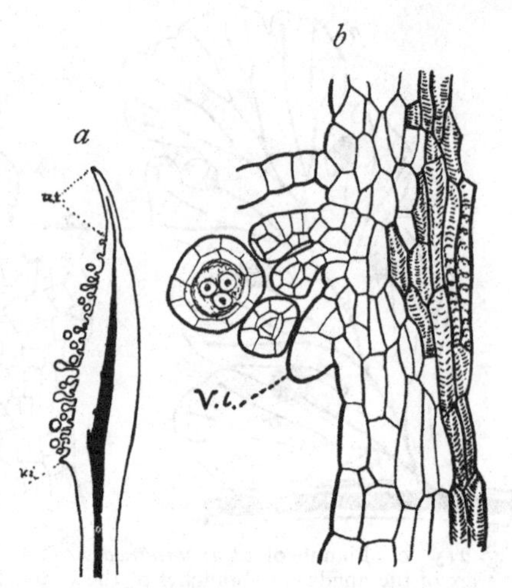

Fig. 217. Young sorus of *H. repens*. *a*=sorus cut vertically, showing the elongated receptacle; *u.i.*=upper indusium, traversed by a vascular strand; *v.i.*=vestigial lower indusium. (× 15.) *b*=a small part of the soral surface including the vestigial indusium (*v.i.*), more highly magnified. (× 160.)

A sequence involving a "phyletic slide" of the sorus from the margin to the surface of the pinnule may thus be traced, starting from a Dicksonioid-Dennstaedtioid source with marginal sorus, a two-lipped indusium, conical receptacle, and a gradate succession of sporangia. The sorus becomes mixed, the receptacle flattened and curved downwards, the lower indusium reduced and even abortive, the upper being flattened out, and assimilated to the laminar surface, while it is traversed by a vascular strand extended from the receptacle. All of the Ferns involved in the series that shows these changes are rhizomatous and solenostelic, with undivided leaf-trace. They all bear hairs, not scales; and they have highly divided leaves with open venation. In fact the series demonstrates, in Ferns that have always been recognised as akin by habit, a transition from a marginal "Dicksonioid" sorus to a superficial "Polypodioid" type.

The case thus described at length for a series of nearly related Ferns runs parallel with others distinct from them by Descent, which also show the "phyletic slide" of the sorus from the margin to the surface of the leaf. It is well seen in the Pterids. Here, in *Pteridium*, the sorus originates as in the Dicksonioid Ferns from the actual margin, the marginal initials forming directly the receptacle (×) upon which the sporangia are borne, while the adaxial and abaxial flaps of the indusium appear as superficial growths back from the margin (Fig. 218, *A*, *B*). But in the mature state the whole sorus appears bent round to the lower surface (Fig. 218, *D*), though both of the indusial flaps are still retained. In *Histiopteris incisa* varying intermediate conditions are found between a marginal and a superficial origin of the receptacle (*R*), while the lower (abaxial) indusium is not represented, and the upper (adaxial) flap takes in some degree the position and character of a leaf-margin (Fig. 219). The next step, that of the actually superficial origin of the receptacle, is seen in *Pteris* itself (*P. serrulata*, L. fil.); here the marginal segmentation of the young leaf leads directly to the formation of the adaxial

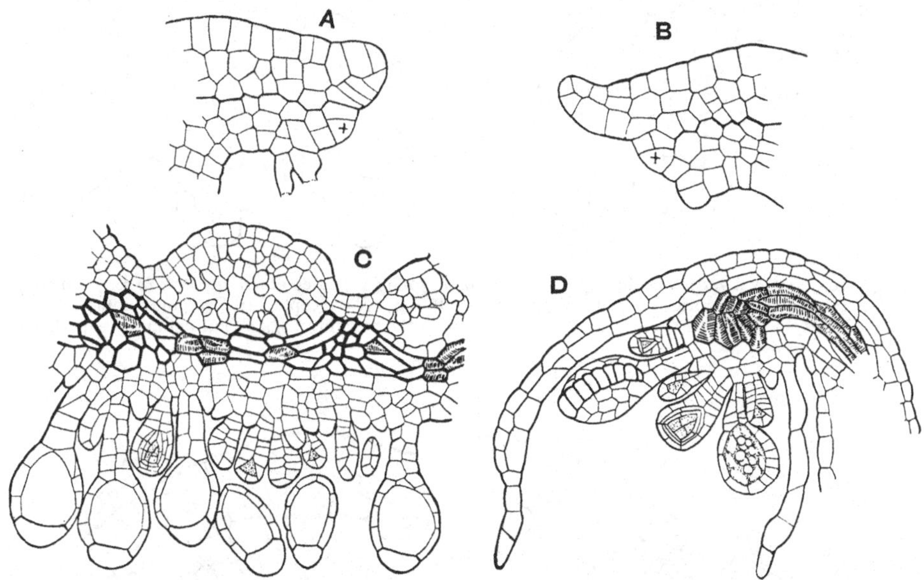

Fig. 218. *A*, *B*, vertical sections through the margin of the pinnule of *Pteridium aquilinum*, showing how the receptacle (×) originates directly from the marginal segmentation, while the indusial flaps are of superficial origin. (× 180.) *C*=vertical section of the fusion-sorus of *Pteridium* following the line of the vascular commissure. (× 90.) *D*=section of a more mature sorus than in *A* or *B*, but in a similar plane showing the mixed character. (× 90.)

flap, which serves as an apparent continuation of the expanded leaf-surface; the sporangia appear to arise from the lower surface of the leaf, without any definitely convex receptacle (Fig. 219). The steps of change in developmental origin of the sorus thus seen in these three Ferns indicate a transference of origin of the receptacle from the margin to the lower surface of the leaf, and the change is demonstrated in Ferns so closely allied that until recently they were all included in the old and comprehensive genus *Pteris*. A like change has also been indicated by Christ, and by von Goebel for the Davallieae (Christ, *Farnkräuter*, pp. 2, 287, etc.; von Goebel, *Organographie*, II. Aufl. p. 1140). Thus it appears to have occurred repeatedly in the Descent of

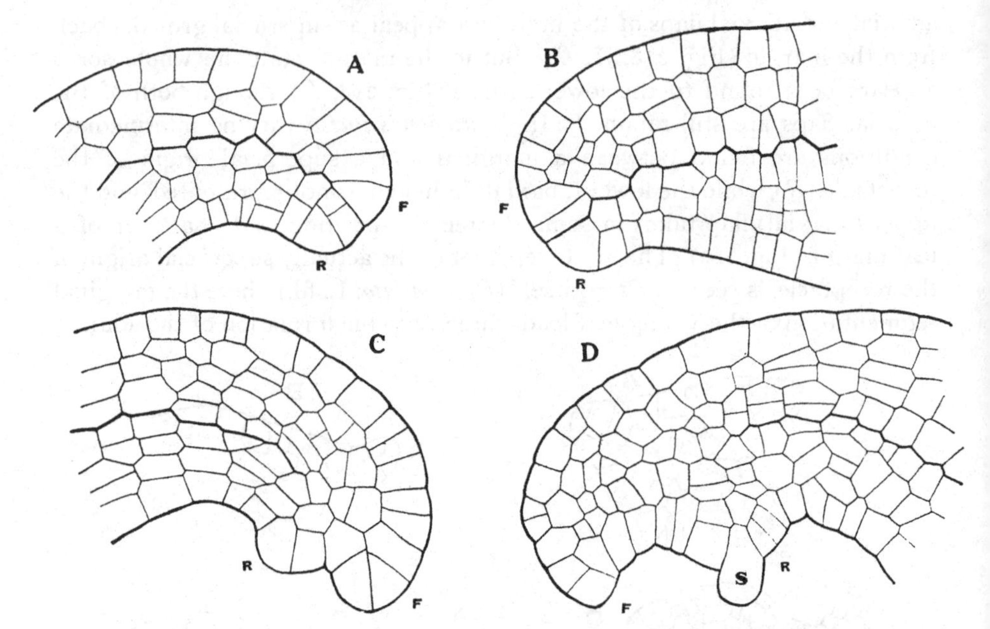

Fig. 219. Sections vertically through the margins of young pinnules of *Histiopteris incisa*, showing the relation of soral origin to the marginal segmentation. *R*=receptacle; *S*=sporangium; *F*=indusial flap. It is seen that the receptacle is close to the margin in *A*, *B*, but definitely superficial in *D*. (× 200.)

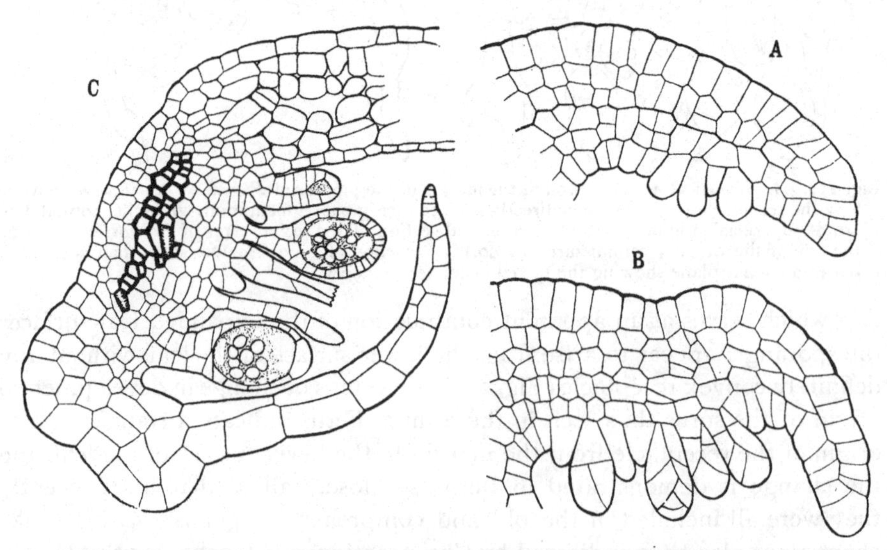

Fig. 220. *A*, *B*=vertical sections through young sori of *Pteris serrulata*. (× 150.) *C*=similar section through a nearly mature sorus of *Pteris cretica*. (× 75.) The indusium is here a direct continuation of the marginal growth, and the flattened receptacle and sporangia arise from the lower surface.

Ferns now living. Its relation to a widening leaf-expanse has already been noted. This suggests that a like progression, executed very early in their evolution, may probably account also for the superficial position of the sori in the Marattiaceae and Gleicheniaceae. The facts observed in the living, but very ancient Osmundaceae, combined with the prevalence of the marginal position in most of the early types of Ferns, support the suggestion. *Accordingly it may be held as probable that a distal or marginal position of the single sporangium was general in the first instance for the Filicales; but that as the leaf-blade expanded the marginal sorus, whether monangial or composed of many sporangia, slid to the lower surface, and that this has happened along many distinct phyletic lines, sometimes early, sometimes late in their Evolution.*

The deeply intramarginal position of many sori of Davallioid Ferns, especially where they are seated on a vein-ending as in *Arthropteris* and others, suggests a possible explanation of the anomalous sori found on the upper surface of *Deparia Moorei* (see J. M. Thompson, *Trans. R. S. Edin.* Vol. l, p. 837, where reference is made to other writers). Normally the sori are marginal, as they are in the allied *D. prolifera.* But occasional sori are found on the adaxial surface of the leaf, without intermediate positions bridging the gap. It is possible that such sori are initiated in a new and independent position, as in *Aspidium* (*P.*) *anomalum* (see p. 217). But the fact that intramarginal sori seated on vein-endings are frequent in the Davallieae, to which family *Deparia* should be assigned, suggests that its anomalous sori may be referred to a like origin with them.

LOSS OF INDIVIDUALITY OF THE SORUS

The Sorus, made up as it usually is of a compact group of constituent parts, viz. the receptacle, sporangia, and indusial flaps, appears in so many Ferns to be strictly circumscribed that the natural tendency is to regard it and to describe it as an entity: and this view is accentuated by the recognition of the changes in its position just described. For isolated examples of Ferns it may be legitimate to hold it as an entity. It is stamped as such by nutrition. Its limited outline appears to be dictated by the limit of convenient radius of nutritive supply from a vein-ending, marginal or superficial, on which it is seated. But if this be the real cause, then it may be expected that a bifurcation of a vein will produce twin sori: or that a fusion of veins, as seen in so many of the later reticulate types of Ferns, might be expected to result in fusion-sori: or these again may break up into fragments in case the vascular arches were again interrupted. It will be seen that all of these states can be illustrated in living Ferns: and other facts will be adduced which show that the sorus is not a morphological entity for Ferns at large.

A primitive state for Ferns was no doubt that in which the leaf or segment was narrow. Here it is commonly found that a single row of sori, marginal or superficial, is disposed on either side of the midrib, and this arrangement of them appears to be fundamental. It is seen in such primitive types as the

Gleicheniaceae, Hymenophyllaceae, and Dicksonieae. It may be retained with modifications even where the leaf-surface is widened, as it is in the Marattiaceae: but then the sori are liable to be extended along the course of the veins. This is seen in slight degree in *Angiopteris*, and *Marattia* (Fig. 202, *A*, *C*); but more plainly in *Archangiopteris*, and *Danaea* (*B*, *E*), where the sori stretch almost the whole distance from the midrib to the margin, having apparently kept pace with the widening of the pinnae. But *Christensenia* is a type with the leaf-blade greatly widened, probably in relation to life under forest shade; and the sori are dotted over the expanded surface, having each a circular form (Fig. 202, *D*). At times the individuality of the sorus appears to have been lost, owing to fission, as suggested by the frequent occurrence of sori in pairs, or of forms pointing to incomplete fission (Fig. 221, *a*—*e* below). This may account phyletically in part or in whole

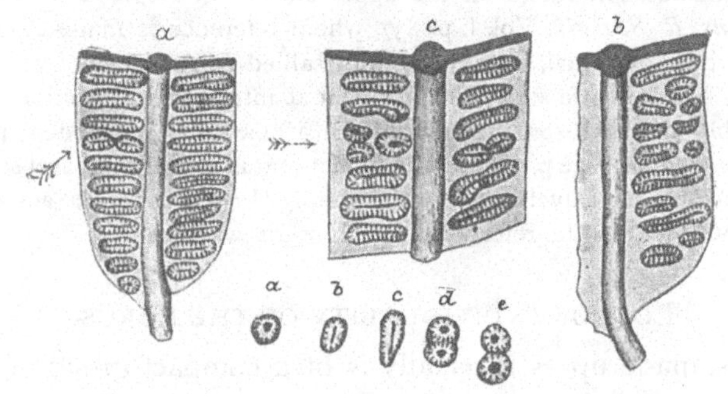

Fig. 221. *a*, *b*, *c* (above), *Danaea alata*, Smith. *a*=fertile pinna with many normal sori: the arrow indicates an abnormal fission; *b*, *c* show more frequent abnormal fissions resulting in irregularly formed sori distributed over a slightly enlarged leaf-surface. (× 2.) *a*—*e* (below), sori of *Christensenia aesculifolia*, Blume, showing states of partial or complete abstriction.

for the numerous scattered sori of the genus. Similar states may be seen also in *Danaea* (Fig. 221, *a*—*c*). The genus *Dipteris* further illustrates the relation of the sori to a widening leaf. In *D. Lobbiana* the segments are narrow, and a single row of sori lies on either side of the midrib. In *D. quinquefurcata* the segments are broader, and the sori are more widely spread, with many signs of fission. In *D. conjugata* the broad lamina is covered with many minute sori, giving again frequent signs of fission (Fig. 222, *A*, *B* and compare *Land Flora*, Figs. 344—346). A slightly different state is seen in *Metaxya*, for there not uncommonly two or more sori may be borne on each vein of the wide pinna. These suggest in each case a fission of the single sorus usually present on each vein (Fig. 223). It is not asserted that such fission is the only mode of increase of the sori borne upon an enlarged leaf-area: but it is at least one factor. Those Ferns which are thus broad-leaved, with many scattered superficial sori, appear to have diverged far from the original linear leaf-form, and the series of superficial or ultimately marginal sori on either side of the midrib; nevertheless we conclude that this was an ancestral feature for them all. It is thus seen that the individuality of the sorus may be lost by *fission*, and especially in Ferns where the leaf-area has been widened, as it so often is in relation to growth under forest shade.

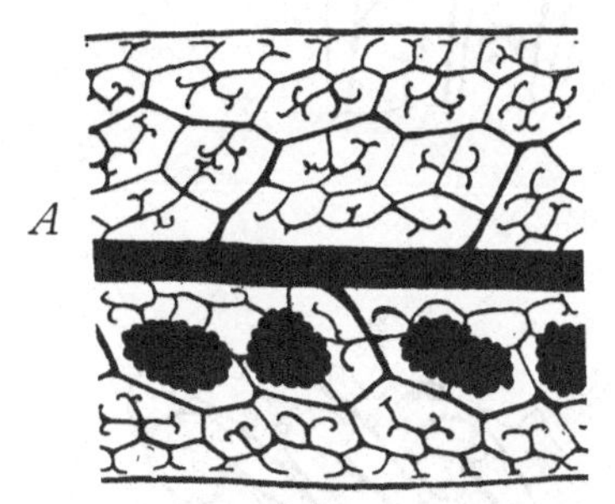

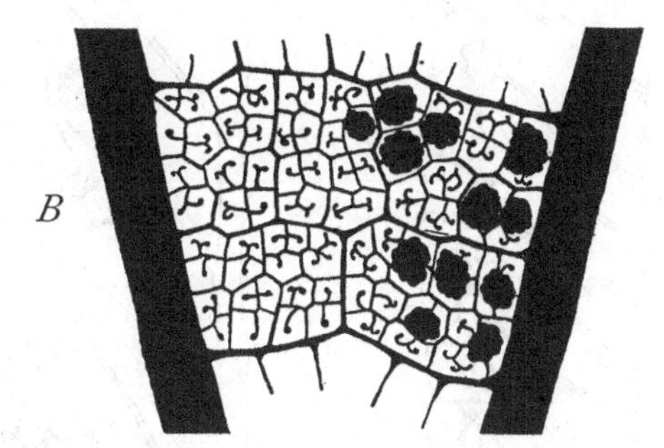

Fig. 222. *A*. Part of a fertile pinna of *Dipteris Lobbiana*, with sori left on the one side of the midrib: they have been removed on the other side, and show that there is here no special vascular supply to the receptacle, only a rather dense plexus of the usual veins. (× 8.) *B*. Fertile area of leaf of *Dipteris conjugata*: the sori have been removed on the left, showing no marked vascular development under the receptacle. (× 8.)

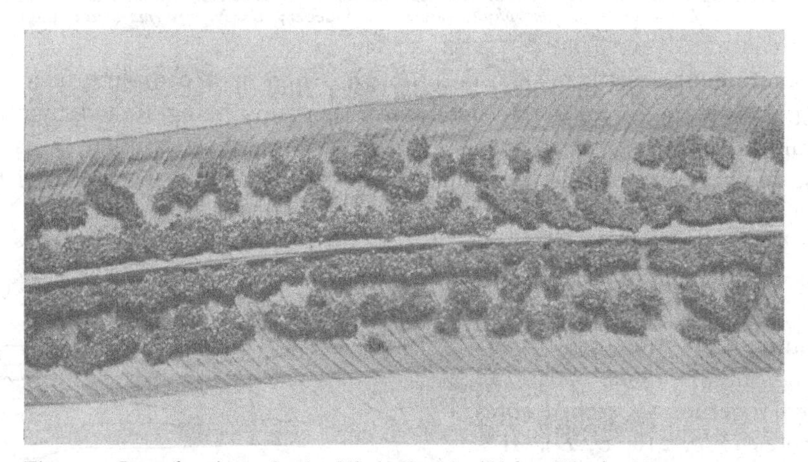

Fig. 223. Part of a pinna of *Alsophila blechnoides* (Rich.), Hk. (= *Metaxya rostrata*, Presl), showing the relation of the sori to the veins: more than one sorus may be borne on a single vein. (× 2.)

On the other hand, where sori are numerous their individuality is apt to be lost also by *fusion*, a change which has appeared in several distinct phyletic lines, and especially in Ferns where the sori are borne in linear sequence. Von Goebel (*Organographie*, II. Aufl. p. 1143) has shown early steps in lateral fusion of the intramarginal sori of *Saccoloma*, where the receptacles and lower indusia remain distinct, but the upper are united to form a continuous flange (Fig. 224). Christ has also indicated a more complete fusion in *Nephrolepis acutifolia*, where "coenosori," or *fusion-sori*, may extend along the whole margin. Similar fusions of marginal sori, more or

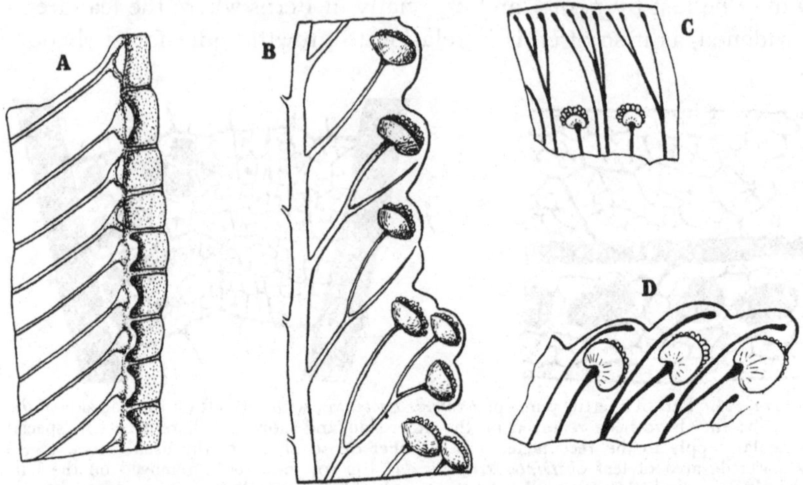

Fig. 224. Group of sori, which have become apparently intra-marginal by the upper indusial lip simulating the rest of the leaf-surface. *A*=*Saccoloma elegans*, after von Goebel; *B*=*Microlepia platyphylla*, after von Goebel; *C*=*Nephrolepis acuta*, after Christ; *D*=*Nephrolepis cordifolia*, after Christ.

less continuous, are seen also in *Lindsaya*; there is a complete fusion of marginal sori in *L. sagittata*, and *lancea*, the veins being linked together distally by commissural arches (Fig. 225). But the most familiar cases are those of the Pterideae and Blechneae. In the former the receptacle of the sorus, which is continuous at or near the margin of the leaf, is traversed throughout its length by a vascular tract parallel to the margin. It is made up of the endings of the veins, which are linked together by arched commissures (Fig. 218, *C*, *D*). If a section follows the vein outwards the appearance is as though there were a simple marginal sorus:

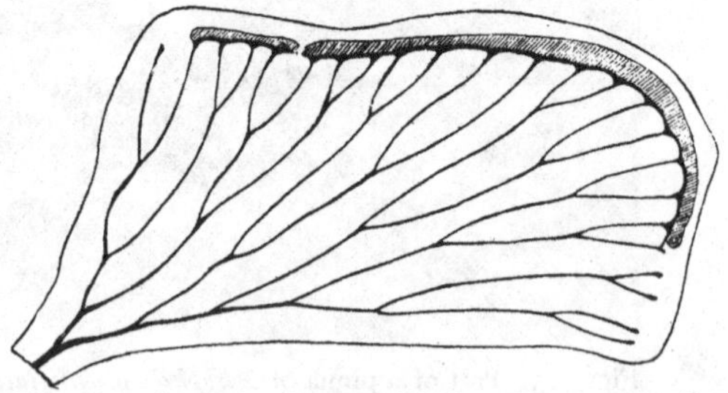

Fig. 225. A single pinna of *Lindsaya lancea*, showing the sori fused laterally to form an almost continuous intra-marginal series. (× 4.)

but if it traverses the space between two veins it cuts the commissure only (Fig. 220, *C*). Such structure is clearly that of a fusion-sorus. The physiological advantage of linking up the distal ends of the veins is obvious, while it gives added space for the accommodation of sporangia.

The condition seen in the Blechneae is more complicated, and it illustrates how plastic is the development of the sporophyll. These Ferns appear to have sprung from some type like that of the genus *Matteuccia*, where the sporophyll is narrow, bearing on either side of the midrib a superficial row of separate, non-indusiate sori, protected by the strongly recurved margins of the narrow pinnae (Fig. 226). In the simpler types of *Blechnum* represented by the old genus *Lomaria* the leaves are dimorphic, and the sporophylls narrow as in *Matteuccia*. The sori are linked together to form an intramarginal chain, with continuous vascular supply below the receptacle,

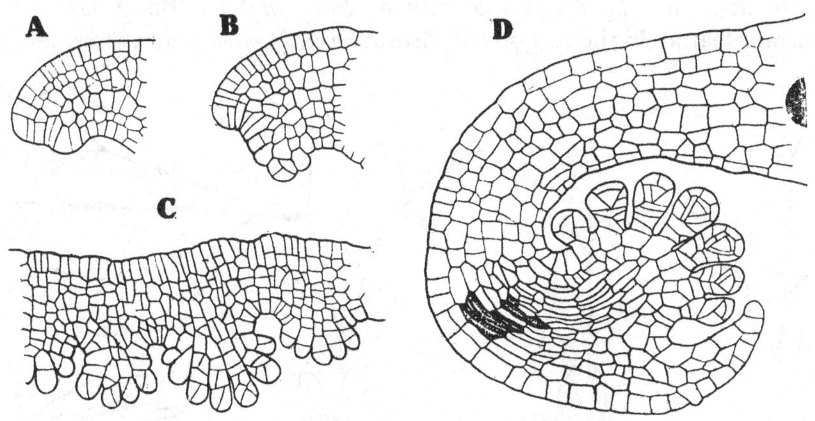

Fig. 226. Development of the sorus of *Matteuccia intermedia*. *A*, *B*, *D*, vertical sections through the pinna-margin, showing successive stages of the superficial, gradate sorus, protected by the curved margin. *C*, a section parallel to the margin traversing a succession of the distinct, non-indusiate sori. (× 125.)

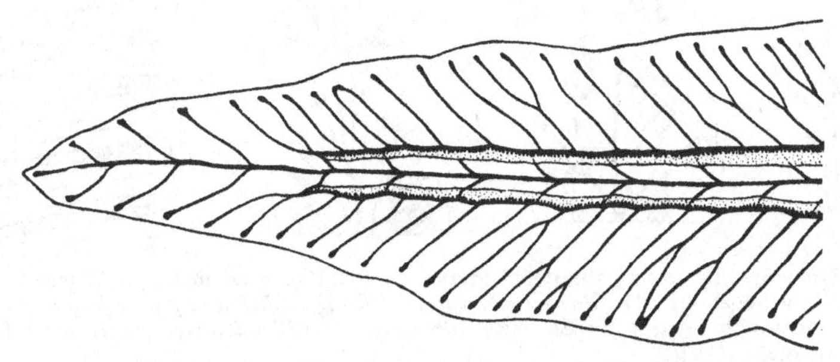

Fig. 227. *Blechnum longifolium* (× 3), showing the relation of the fusion-sorus to the venation, and its protection by a continuous "indusium" which is phyletically the leaf-margin. The expanded leaf-surface is a new formation, increasing the photo-synthetic area.

resulting from commissures which connect the veins laterally together. The whole of each *fusion-sorus* is covered by the incurved margin, which acts as an indusial protection. In the more complex types constituting the old genus *Blechnum*, but now included in *Eu-Blechnum*, the fusion-sorus is constituted as before, but an extra growth has arisen from the convex adaxial surface, in the form of a flange supplied with vascular strands, and forming a wide and efficient, but secondary photo-synthetic area (Fig. 227). The physiological advantage of this is clear, since it provides for the self-nutrition of the sporophyll. Morphologically it has the effect of shunting the *phyletic* margin, now acting as an indusium, from the actual margin to the lower surface of the pinna (for development of this see below Fig. 237).

Among the Blechnoid Ferns, however, a further change is seen—the reverse of fusion, viz. the *disintegration of the fusion-sorus*. The fusion-sori become interrupted, but not necessarily so as to resolve them again into their original constituents. This appears occasionally in *Blechnum* itself. But it has become a constant feature in the sori of *Woodwardia* and *Doodya*, where short lengths

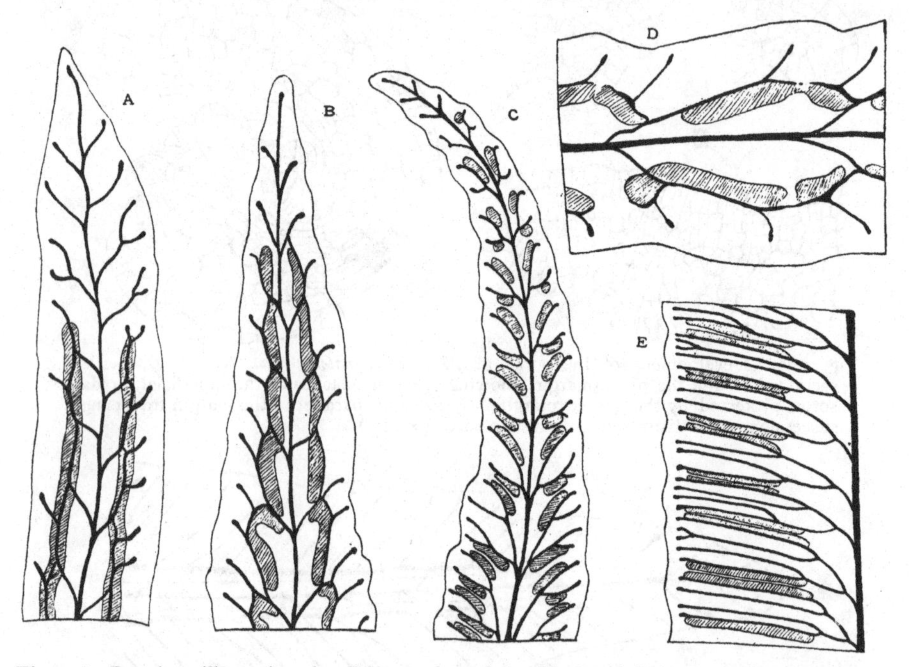

Fig. 228. Drawings illustrating the disintegration of the Blechnoid fusion-sorus (shaded), and its relation to the venation. *A* = *Blechnum boreale*; *B*, *C*, *D* = *Blechnum punctulatum*, var. *Krebsii*; *D*, drawn on a higher scale; *E* = *Scolopendrium vulgare* (= *Phyllitis scolopendrium*). *A*, *B*, *C*, *E*, slightly enlarged. *D* × 6.

of the fusion-sorus appear separate from one another, each covered by its own indusial flap. Steps in this remarkable process of disintegration of the fusion-sorus may be seen in *Blechnum punctulatum*, Sw. var. *Krebsii* Kunze, from S. Africa (Figs. 228, 229). Normal specimens of the species have the usual fusion-sorus. But in var. *Krebsii* the pinnae are relatively broad: the fusion-sori become wavy, and are thrown into strong curves (Fig. 228, *B*); finally they are interrupted in various ways (*C*, *D*). The most interesting are those where the isolated portions face one another, giving a disposition corresponding to that seen in *Scolopendrium* (Fig. 228, *E*). It can hardly be doubted that this genus is a Blechnoid derivative, and that its peculiar sori arose by *disintegration of the fusion-sori of the Blechnum type*, following on a broadening of the sporophyll. Such examples show that the individuality of the sorus is not constantly maintained, fissions and fusions being of frequent occurrence.

Fig. 229. Portion of leaf of *Blechnum punctulatum*, Sw. var. *Krebsii*, Kunze. Photograph slightly enlarged so as to show the soral arrangements.

Another way in which the individuality of the sorus may be obliterated is by the spread of the production of sporangia over the free area of the sporophyll. The sporangia are not always restricted to a placental region above the veins, but they may appear between the veins, even covering the whole leaf-surface. This may go along with a disappearance of the indusium. The condition is described as "*acrostichoid*," and the Ferns showing it used to be grouped under the generic name of *Acrostichum*. It is, however, palpable enough that the Ferns so grouped were diverse in their characters, and that by these they are naturally related to other well-known types. Increasing knowledge of their anatomy confirms this, and developmental study has demonstrated that the non-soral acrostichoid state does not in itself indicate any real affinity of the plants which show it (Frau E. Schumann, *Flora*, 1915, p. 201, *Studies*, IV, VI, VII). It is now clear that the spread of the sorus and loss of its individuality have involved derivatives from at least six well-known types. It has been shown (*Studies*, IV) that *Stenochlaena* and *Brainea* may be traced as derivatives from the Blechneae by such a spread of sporangium-formation over the leaf-surface (Fig. 230). Similarly *Acrostichum praestantissimum* and *aureum* may be referred to the Pterideae:

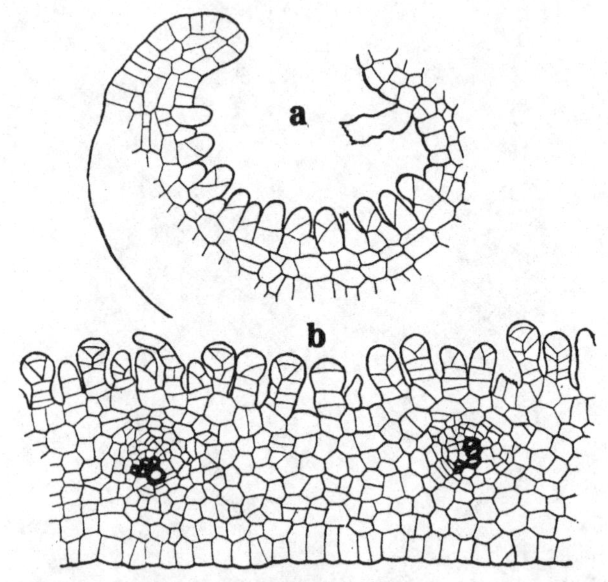

Fig. 230. Vertical sections through fertile pinnae of *Stenochlaena sorbifolia*, with the lower, fertile surface directed upwards. (× 125.)

Syngramme and *Elaphoglossum* to the Metaxyeae: *Trismeria* to the Gymnogramminae: *Polybotrya* and *Stenosemia* to the Aspidieae: and *Cheiropleuria*, *Leptochilus tricuspis* and *Platycerium* to the Dipteridinae. Other Acrostichoid species described under the name of *Leptochilus* are referred by Frau Schumann to an origin from *Meniscium*, and to certain species of *Polypodium* (*Flora*, 1915, p. 222, etc.).

The Acrostichoid derivatives of the Dipteridinae present a peculiar feature in the vascular supply to their fertile region. In most Acrostichoids there is no special vascular provision for the nutrition of the sporangia. But in the Dipterid-derivatives it is different. Their condition appears to be arrived at by an elaboration of the sorus, which meanwhile maintained its individuality and its vascular supply, but its receptacle became extended; it is not a case of the sori undergoing dissolution. The first step is shown by *Cheiropleuria* (Fig. 231, *a*). Here the receptacle with its vascular supply is enlarged and may even be branched. Usually it is restricted to a single areola of the main venation.

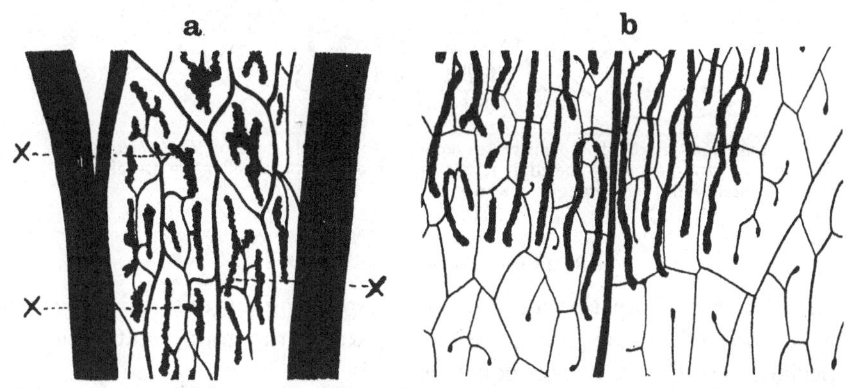

Fig. 231. *a* = part of a sporophyll of *Cheiropleuria*, seen as a transparency, showing the vascular tissue. The steady lines are the venation: the irregular patches are the storage-xylem of the soral receptacles. At the points (×) these have crossed the veins at a lower plane than that in which the veins lie. (×8.)

b = part of a sporophyll of *Platycerium angolense*, showing the origin of the receptacular xylem from the ends of the blind veins. The receptacles are more elongated than in *Cheiropleuria*, and pass to a lower level in the mesophyll, there extending frequently across the course of the original venation, which is here represented by thin steady lines. (×4.)

But occasionally, being extended in a plane close to the lower surface of the leaf, the receptacle oversteps one of the veins, passing in a lower plane to the next areola. This mode of extension, which is not a very marked feature on *Cheiropleuria*, becomes very prominent in *Platycerium* (Fig. 231, *b*), and still more so in *Leptochilus tricuspis* (Fig. 232). Thus there came to be two parallel vascular systems in the fertile region of these Ferns, viz. the normal venation, and the system of the receptacle which ramifies in a lower plane. This condition has been called "*diplodesmic*" (Fig. 233). An apparently parallel condition is seen in some Polybotryas; but here it arises by lapping of the extended sorus backwards on to the upper surface. The physiological result is the same, but the method of its origin has been different (Fig. 234).

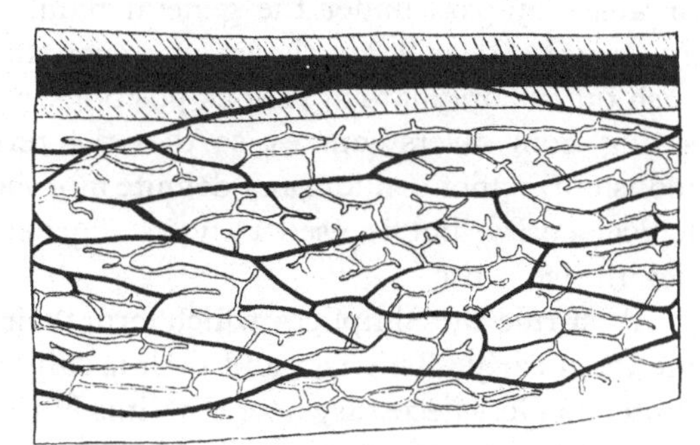

Fig. 232. *Leptochilus tricuspis*. Part of the soral region of the sporophyll seen as a transparency. The heavier lines represent the normal venation, which is nearer the upper surface: the lighter lines are the receptacular system extended in a plane nearer the lower surface. (×6.)

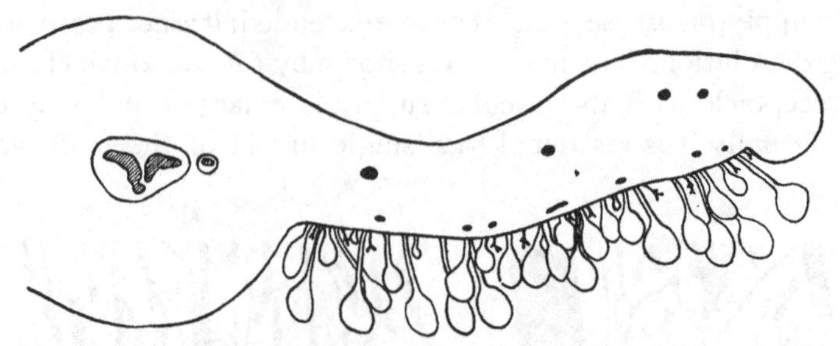

Fig. 233. Part of a transverse section of a fertile leaf of *Leptochilus tricuspis*, showing the diplodesmic structure, and branched hairs associated with the sporangia. (× 16.)

PROTECTION OF THE SORUS

In all Ferns protection of the spore-producing organs is in the first instance afforded by the juxtaposition of parts in the young state, and this is specially effective as a consequence of the circinate vernation, by which the distal end of each leaf is enfolded within the older parts. The lower parts of the leaf are more exposed, and this may explain the frequent absence of sporangia at the base. *Osmunda regalis* is an example of this, and it is a common feature in *Nephrodium*. In addition to such general protection, special developments of the most various kind appear in relation to the sorus. When these take definite form as protective organs they pass under the general name of *Indusium*. It will be shown, however, that such organs originate in various positions, and spring from divers sources, so that the term indusium cannot connote any definite morphological entity, but is used only in a general descriptive sense.

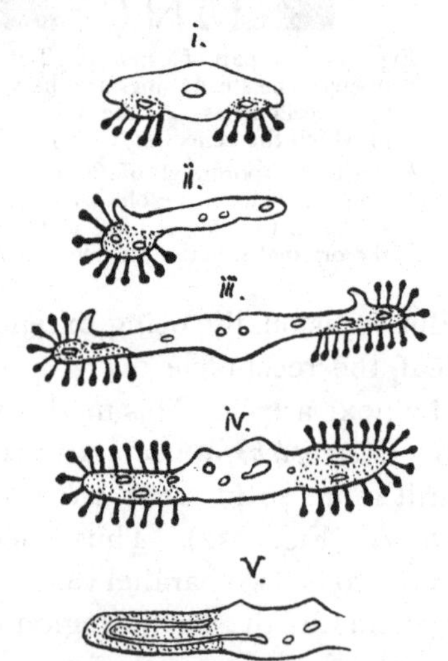

Fig. 234. Various forms of the sori in *Polybotrya*. Diagrammatic: after Frau Schumann.

As a rule the Simplices, which form their sporangia simultaneously, initiate them early. This fact and their usually massive character make indusial protection less necessary here than it is in the Gradatae and Mixtae, in which more delicate sporangia are produced in a succession spread over a long period. Sometimes there is mutual protection by fusion of the sporangia to form massive synangia, as in *Marattia* and *Ophioglossum*. These features help to explain the usual absence of special protection in the Simplices. A few hairs may be found round the base of the sorus in *Angiopteris* and *Marattia*; in *Danaea elliptica* the sori are partially sunk into the tissue of the blade; in *Helminthostachys* irregular lobes are borne on the distal end of the sporangiophore, forming an imperfect covering to the sporangia while young; a somewhat similar, but much more efficient covering, is found in *Matonia*, where the receptacle is continued upwards into a leathery, umbrella-like indusium, completely covering the sporangia while young. On the other hand, *Botryopteris* and *Zygopteris*, the Osmundaceae and Gleicheniaceae, and most of the Marattiaceae and Ophioglossaceae, have no special protection for their sporangia: this may be held as a primitive state.

It is otherwise with the Schizaeaceae. The sporangia in all of them are marginal. In *Schizaea*, *Anemia*, and *Mohria* a strong growth on the upper, or adaxial, side just below the young sporangium forces it towards the lower surface, and the resulting flap appears as a continuation of the leaf-surface. It curls round and protects the sporangium (compare Figs. 213, 214). It is the same in *Lygodium*, except that here the growth appears on both sides of the sporangium, forming a sort of pocket (Fig. 235). As it matures the pocket differentiates into an upper lip which appears as though it were a continuation

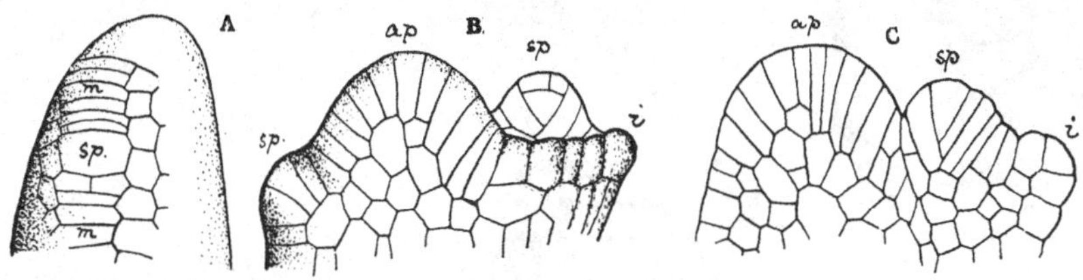

Fig. 235. Origin of marginal, "monangial" sorus of *Lygodium*. A=margin of young pinnule in surface view; m=marginal cells; sp=sporangium. B=a similar pinnule in surface view; ap=apex; i=indusium. A and B after Prantl. C=the same seen in section. After Binford.

of the blade of the leaf-segment, and a membranous lower lip. The two completely enclose the single sporangium. The name indusium has been applied to these growths. That of *Lygodium* may be accepted as the prototype of the form of indusium which has prevailed along the whole of the Dicksonioid-Davallioid-Pterid series, and it is found also in *Loxsoma*, and the Hymenophyllaceae. In all of them the indusium is basal, more or less distinctly two-lipped, and it arises as it does in all the Schizaeaceae by growth of superficial, not of marginal cells. The development of the indusium of this type, together with the abortion of the inner lip, has already been discussed and illustrated above (p. 222). Its persistence in marginal and gradate sori finds its natural explanation in the fact that a basal protection is important for the youngest sporangia, which are the lowest on the receptacle. It is when the marginal position passes into the superficial and the sorus itself assumes the mixed character that the protection being no longer essential, the lower indusium is abortive, while the upper may merge into the leaf-surface.

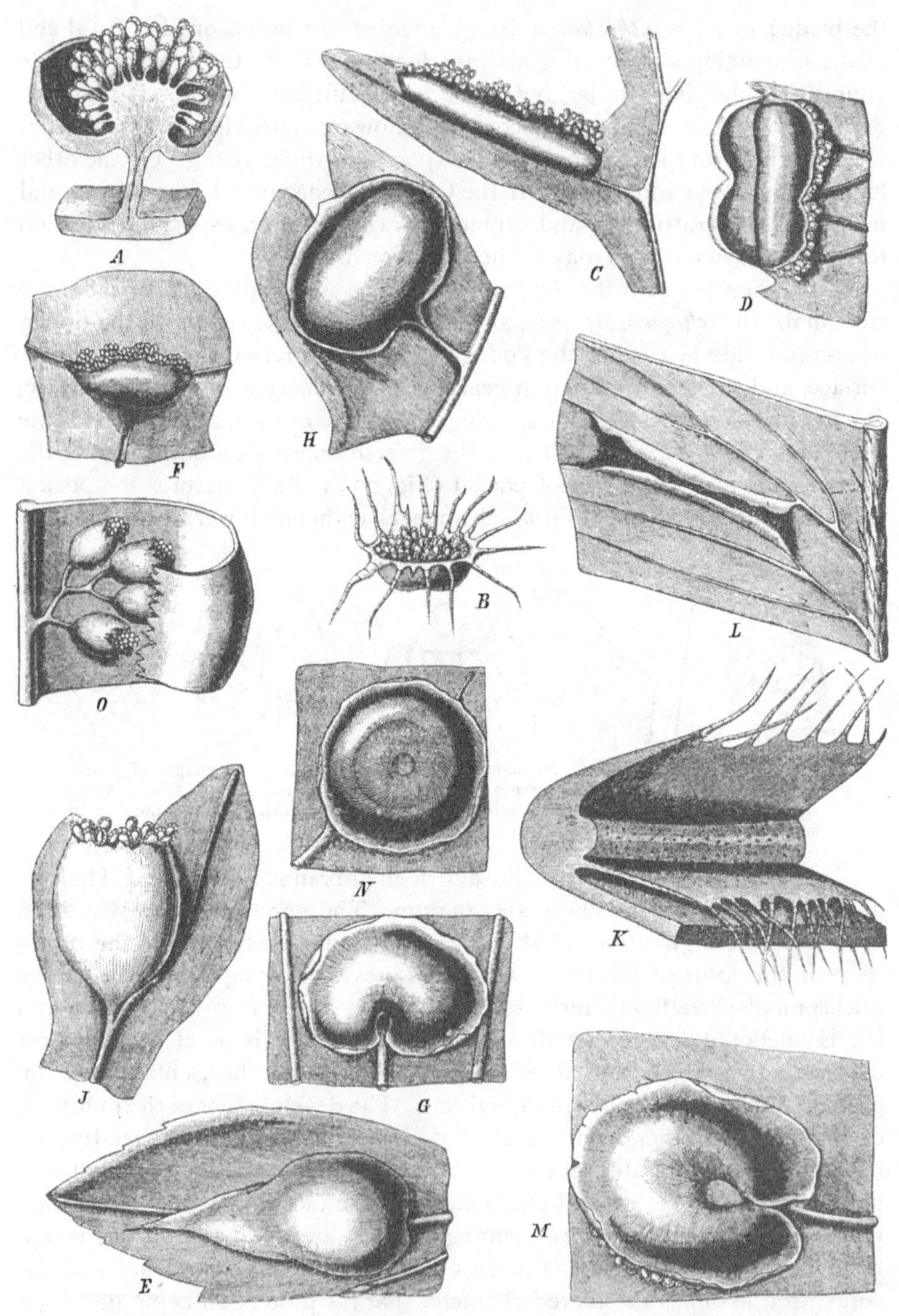

Fig. 236. Types of indusium in the Polypodiaceae. *A* = *Peranema cyatheoides*, Don, with inferior indusium, centrally attached, on the stalked sorus. *B* = *Hypoderris Brownii*, J. Sm., indusium inferior, centrally attached. *C* = *Asplenium bulbiferum*, Forst., indusium lateral, with linear attach-

Of the Simplices with superficial sori the Marattiaceae and Gleicheniaceae are without any definite indusium. But among the Gradatae there are Ferns with sori in the same position, some without and some with a basal indusium. The Cyatheae and Onocleinae provide examples. The physiological use of the basal indusium is the same as before. The question naturally arises, is the indusium here the strict homologue of that in the Dicksonia-Davallia series, or is it of independent origin? Sir Wm. Hooker held that it here represents a specialised ramentum, an opinion combated by Mettenius. Von Goebel (*Organographie*, II. Aufl. p. 1148) holds that it is a derivative of the indusium of the marginal sorus; but without basing the comparison on grounds wider than the soral characters themselves. For reasons based not only on the sorus but upon a general comparison, which will be detailed in the phyletic section of this work (Vol. II), it is held that the basal indusium of the Cyatheae and Onocleinae is of different origin from that of *Dicksonia*, and appeared after the sorus had assumed the superficial position.

Gleichenia, with no indusium and median dehiscence, has as we have seen increased the number of its sporangia in *G. dichotoma* and *pectinata* to the point of deadlock as regards dehiscence. *Lophosoria*, with many points in common, but with slightly oblique dehiscence, is in nearly the same position. But *Alsophila* and *Matteuccia* have met the difficulty by adoption of a gradate sorus and oblique dehiscence, though there is no protective indusium (Fig. 226). There is no reason here to assume abortion to account for its absence, since in *Gleichenia* and certain related fossils it appears to have been primitively absent. In *Hemitelia*, *Onoclea*, and *Cystopteris*, however, a partial, that is a one sided, indusium is present (Fig. 236, *E*, *O*), and in *Cyathea* there is a very complete basal indusium. *Peranema* and *Diacalpe* appear to be transitional links between these Ferns and the Aspideae, all having superficial sori. There seems to be no reason to hold any other view than that the indusium in all of them is a new formation, phyletically distinct from that of the Schizaeoid-Dicksonioid Ferns, but serving the same biological purpose. This is the conclusion that naturally follows from a comparison on a wider basis than the sorus alone affords: in fact these Ferns would appear to have arisen from a non-indusiate Gleichenioid source with superficial sporangia.

---

ment, introrse. *D*=strongly modified protective margin of *Cassebeera triphylla*, Kaulf., for comparison with the introrse indusium. *E*=*Cystopteris fragilis*, Bernh., indusium attached laterally at a point, extrorse. *F*=*Diella falcata*, Brack., indusium attached laterally along a short line, extrorse. *G*=*Nephrodium Filix-mas*, Rich., indusium superior, attached laterally at a point, extrorse. *H*=*Humata heterophylla* (Desv.), Sm., indusium attached laterally with linear insertion, and joined to the leaf-surface, extrorse. *J*=*Davallia canariensis*, Sm., indusium lateral, attached on three sides to the leaf-surface, extrorse. *K*=*Pteridium aquilinum* (L.), Kuhn, two indusia, one extrorse the other introrse, fringed. *L*=*Scolopendrium brasiliense*, Kze, paired indusia over a pair of sori, the one extrorse the other introrse. *M*=*Fadyenia prolifera*, Hook., indusium superior, centrally attached, elongated-kidney-shaped. *N*=*Aspidium trifoliatum* (L.), Sw., indusium superior, centrally attached, circular. *O*=*Struthiopteris germanica*, Willd., indusium reduced, since there is also a broad covering margin present. (After Diels. from Engler and Prantl.)

Given this basal indusium, in Ferns with superficial sori, its modifications so as to form the reniform type of *Dryopteris* (Fig. 236, *C*), or the peltate indusium of *Polystichum* (*N*), are easily intelligible, and may be illustrated by intermediate steps in Ferns of near affinity in other characters (von Goebel, *l.c.* p. 1150). These comparisons may be extended to include the conditions seen in *Didymochlaena*, *Athyrium*, and perhaps even *Asplenium* (Fig. 236, *C*). But this origin of the peltate indusium of *Polystichum* is probably quite distinct from that of *Matonia*, however similar it may appear to be in form and in function.

Protection to the sorus is also afforded by overlapping of the leaf-margin, which itself thins off to membranous texture. A simple case of this is seen in *Matteuccia intermedia*, C. Chr., which may be held as representing a starting point for the sequence *Blechnum*—*Woodwardia*—*Scolopendrium*. The condition of *Matteuccia* is that of a Fern with superficial, non-indusiate, basipetal sori, covered over by the revolute margins of the pinnae (Fig. 226). *Onoclea* is the same, except that a membranous basal indusium is present also (Fig. 236, *O*). In the simplest types of *Blechnum* (*B. discolor*, *tabulare*, *lanceolatum*) the structure is as in *M. intermedia*, except that the rows of sori are threaded together by vascular commissures, forming the fusion-sori already described for them (Fig. 237, *A*). This is followed in the more advanced types by an outgrowth of the upper surface of the pinna along the line of greatest curvature, so as to form a flange on either side of the midrib. Steps in its origin are suggested by sections of sori of various species, ending in the broad-leaved type, such as *B. brasiliense* (Fig. 237, *A*—*G*). This flange is a new formation, extended so as to add to the photo-synthetic area of the pinna, and each is traversed by its own system of vascular strands linked up with those of the rest of the pinna (Burck, 1874; Bower, *Studies*, IV). Meanwhile, however, the fusion-sorus is enveloped by the original leaf-margin, which is often erroneously described as an indusium. It will be noted that it opens in the reverse direction from that of the Pterideae and Davallieae. In them the edge of the flap is directed outwards: in *Blechnum*, and its derivatives *Woodwardia* and *Doodya*, it is directed inwards. The fact is that the so-called indusium of *Blechnum* is by origin a different thing from the indusium of *Pteris*. In the former it is a real leaf-margin which has slid to a superficial position: in the latter it is a superficial indusium which has in the course of evolution passed to a marginal position.

The Acrostichoid condition, which is shown to have arisen along many distinct phyletic lines, has points of special interest as regards the protection of its sporangia. All the examples are ex-indusiate, though there is reason to believe that many of them have sprung from an indusiate ancestry. All have a mixed sorus, so that young sporangia are present after the leaf has unfolded. And yet many of them grow in exposed situations. A prominent example of this is seen in the world-wide *Acrostichum aureum*. It appears as though the adult sporangia of such plants had become less susceptible to exposure than their predecessors. Nevertheless the young sporangia are often well protected. The soral area is frequently very closely packed, as in *Leptochilus tricuspis*, so that the younger and shorter sporangia are protected by the older and higher, while these as they grow up expose their

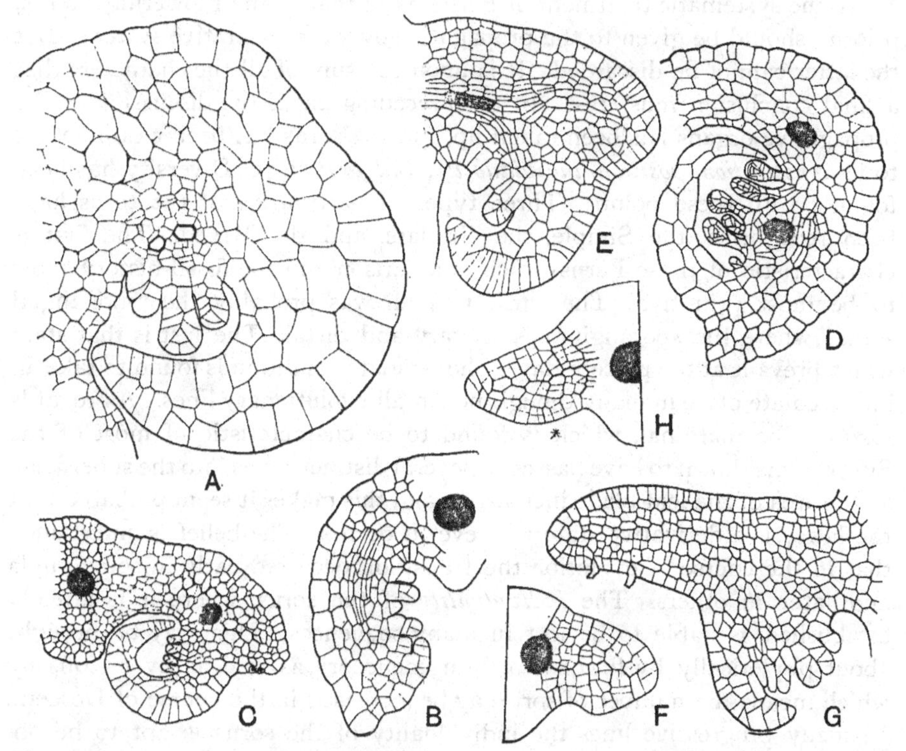

Fig. 237. Vertical sections through the margins of fertile pinna of various species of *Blechnum*, showing how the "flange" has arisen as a new formation. *A*. *Blechnum discolor*, there is no flange. (× 125.) *B*. *B. penna-marina*, flange slightly indicated. (× 125.) *C*. *B. Fraseri*, flange of considerable size bearing a stoma. (× 50.) *D*. *B. L'Herminieri*, flange still minute. (× 50.) *E*, *F*. *B. procera*, the flange has distinct marginal segmentation; the "indusium" which is the real phyletic margin arises superficially and is delayed. (× 125.) *G*. *B. brasiliense*, this state is still more pronounced than in *B. procera*. (× 125.) *H*. *B. occidentale*, shows the early segmentation of the flange, while the receptacle and "indusium" would arise later at point of star (*).

heads freely for dehiscence. But besides this such sori are often provided with numerous hairs, which branch, and are sometimes peltate, sometimes glandular. These means of protection are shared by many Polypodiaceous Ferns that have definite sori. Such protections by branched hairs are seen in *Metaxya*, *Niphobolus*, and *Platycerium*: by glandular hairs in *Vittaria*, and *Acrostichum aureum*: by scales in *Polypodium lineare*, *verrucosum*, and *piloselloides*, and in *Drymoglossum*. Occasionally hairs are borne upon the sporangia themselves, as in *Polypodium crassifolium* and *Dryopteris Poiteana*. Such devices, which are well illustrated in Hooker's *Genera Filicum* (VI), may be held to serve functionally as a sort of diffused indusium.

General conclusions regarding the Sorus, and its place in Phyletic Morphology

In the systematic treatment of Ferns, as in that of the Flowering Plants, priority should be given to the propagative over the vegetative system. But the latter cannot be discarded. It is upon the sum of all the characters that a final conclusion must rest. In the preceding pages the discussion of the propagative organs has been taken up, and the sorus has been examined as to its *constitution*, *position*, *individuality*, *and protection.* Diversity has been found on all these points. Three types of *constitution* of the sorus have been recognised, the Simple, the Gradate, and the Mixed. The first is characteristic of those Ferns which comparison and the fossil history show to be most primitive. The earliest of all was probably the "monangial sorus," where the sporangium is solitary and distal. The last is that state which prevails at the present day. The Gradate condition is found to take an intermediate place in many though not in all evolutionary lines. As regards *position*, the marginal, which is found to be characteristic of most of the Simplices, is shown to have passed in several distinct series into the superficial, as the area of the leaf-blade increased: and this makes it seem probable that the latter is a derivative state wherever it occurs. The belief is entertained that in the course of evolution the transition occurred early in some phyla and later in others. The *individuality* of the sorus, where it is already established, is liable to be lost in many sequences. This may be brought about phyletically by fusions to form coenosori, as well as by fissions by which means the number of sori may be increased in the course of Descent. In many progressive lines the individuality of the sorus is apt to be obliterated by spreading of the sporangia generally over the surface of the sporophyll. In Ferns at large the *protection* of the sporangia while young is effected in various ways. Primarily it is by the circinate vernation, aided it may be by hairs and scales, and incidentally by the massing of numerous sporangia in the crowded sori. The most primitive Ferns relied mainly on these simple devices for protection of the young sporangia. Where special protective growths, designated by the general term "indusium," are present they may be traced to very various origins. For instance, they may be special outgrowths from the leaf-surface, independently originated in a plurality of phyletic lines: or they may be products of the actual margin of the leaf or pinna, which curls over the sori. Such developments, which are plainly not homogenetic throughout, are specially characteristic of Ferns occupying a middle position phyletically. Many of the most advanced types, belonging to some half a dozen or more distinct phyletic lines, are without special protective indusia; but steps of abortion of them may be seen in some of these, and often there is at the same time an advance to the Acrostichoid condition. Frequently, however, in these Ferns hairs grow up together with the massed sporangia, thus helping to protect them while young. In all of these features there is material which may be used in phyletic seriation. None of them can of itself be held rigidly as ground sufficient for a final decision, but all must be taken into account in the general sum of characters upon which such decisions should rest.

As contributing to this end the characters of the sorus are not the most trustworthy or the most important of those derived from the propagative system. The sorus itself is not a constant entity; the term "monangial sorus" is almost a contradiction in terms, while in the Acrostichoid state no definite sori can be recognised at all. It was pointed out at the opening of this Chapter that the sporangium is the essential body in spore-production. It is constant in its occurrence throughout the whole series of Ferns, and no life-cycle can be completed without it. It will be found that when treated comparatively its characters give a consistent basis for phyletic treatment far more reliable than those of the sorus which it constitutes. Nevertheless the two are intimately related biologically. In definitely soral Ferns the structure of the sporangium is closely related to the constitution of the sorus in which it is borne; both take their part in the nutrition, protection, and final distribution of the spores. Parts which are so mutually dependent will be used in cooperation in the arguments which are to follow; and it will be found that the conclusions drawn from them will yield mutual support.

BIBLIOGRAPHY FOR CHAPTER XII

196. BAUER & HOOKER. Genera Filicum. London. 1842.
197. BURCK. Indusium der Varens. Haarlem. 1874.
198. PRANTL. Schizaeaceen. Leipzig. 1881.
199. ENGLER & PRANTL. Natürl. Pflanzenfam. i, 4, p. 146, etc. 1902.
200. CHRIST. Die Farnkräuter. Jena. 1897.
201. BOWER. Studies. III, IV. Phil. Trans. Vol. 189 (1897), 192 (1899).
202. BOWER. Studies. I—VII. Ann. of Bot. 1910—1918.
203. ARMOUR. Sorus of *Dipteris*. New Phyt. 1907, p. 238.
204. DAVIE. *Peranema* and *Diacalpe*. Ann. of Bot. xxvi, 1912, p. 245.
205. SCHUMANN. Flora. 1915, p. 201.
206. THOMPSON. *Deparia Moorei*. Trans. Roy. Soc. Edin. 1915. Vol. l, p. 837.
207. VON GOEBEL. Organographie. II Aufl. 1918. Teil ii, 2.
208. SCOTT. Studies in Fossil Botany. 3rd Edn. 1920. Vol. i, pp. 250, 325.
209. SVEDELIUS. Einige Bemerkungen über Generationswechsel. Ber. d. D. Bot. Ges. xxiv, 1921, p. 178.
210. BOWER. Origin of a Land Flora. 1908. Chaps. XXXII—XL.

CHAPTER XIII

THE SPORE-PRODUCING ORGANS (*continued*)

(B) THE SPORANGIUM

THE sporangium of *Dryopteris* has already been described as a fair example of a type common in Leptosporangiate Ferns. Its parts when mature are the *stalk* of attachment, the distal *capsule* with thin lateral areas of wall, and the indurated *annulus*, and *stomium*. Within are the *spores*, dry and dusty when mature, but surrounded during development by the nutritive *tapetum*. Each whole sporangium of *Dryopteris* springs from a single superficial cell, which after segmentation according to a regular plan gives rise to the several parts. At the centre of the young sporangium a single cell may be recognised at an early stage as the *archesporium*, from which, by further divisions, the *tapetum* and, finally, the *spore-mother-cells* are derived. The latter are twelve in number and, undergoing the tetrad-division, they form the definite number of *forty-eight spores* (Chapter I).

The characters thus seen constantly in the sporangia of this Fern are nevertheless very variable within the Class of the Filicales. The sporangium may originate not from one cell but from a group of cells (Eusporangiate Ferns). The stalk may be absent, and the sporangium may then be deeply immersed in the tissue of the sporophyll (*Ophioglossum*). The sporangial wall may consist not of a single layer of cells, as it is in *Dryopteris*, but of many layers, as in *Botrychium* or *Angiopteris*. The opening mechanism may appear much more massive and complicated than in *Dryopteris*: in some cases there may appear to be no opening mechanism at all (Hydropterideae). The tapetum may not be represented by any definite band of cells (*Ophioglossum*). Finally, the spores may be of a number largely exceeding, but sometimes less than the number forty-eight seen in *Dryopteris*. Frequently the individuality of the sporangia is not maintained: for, as will be shown, fission may result in an increase of their number (*Danaea*), while in other cases many of them may be coherent so as to form a massive synangium (*Marattia*, *Ophioglossum*). It may well be expected that a body of the essential importance of the sporangium, but showing such a high degree of variation, should provide characters of the greatest value for comparison and for the phyletic treatment of the Class. And the fact that their features are relatively constant for any given species, genus, or even larger group, makes them all the more valuable for this end.

If the body in question be so variable in the Class, as it is thus seen to be, it is important in the first instance to have a clear definition of what is meant by the word "sporangium." When all the variable features are put aside there remain simply the spore-mother-cells, which in Ferns are usually numerous: and there is also the protective wall, for the spore-mother-cells never arise superficially. The definition of a sporangium will then be this: "Wherever there is found an isolated spore-mother-cell, or a connected group of them, or their products, this, together with the tissues that protect it, constitutes the essential feature of an individual sporangium." This definition gets rid of the variable accessories, however useful these may be in detailed comparison, or biologically important to the plants in which they are found. It fixes attention upon what is really essential, viz. the spore-mother-cells, and the tissues that protect them.

THE ONTOGENETIC ORIGIN OF THE SPORANGIUM

The customary distinction of the Leptosporangiate from the Eusporangiate type of Ferns is based on the origin of their sporangia respectively from one parent-cell or from a group of them. The one type is relatively delicate in construction from the first; the other is relatively massive. Comparative observation shows, however, that this distinction does not depend on any essential difference in kind of the two organs, but only of degree. Transitional types may be found which bridge over the distinction, and suggest that the one may have been derivative in Descent from the other. It has already been concluded, partly from comparison but largely from palaeontological evidence, that the Eusporangiate was the relatively primitive and the Leptosporangiate the derivative state (p. 118). The series of diagrams shown in Fig. 238 represents young sporangia from the several Ferns named in the rubric below it. They are all based upon published drawings by authors of repute. A glance at them shows that they form a series between two extremes. Fig. 238, *a*, represents a state which may be seen in advanced Polypodiaceous Ferns. It is based upon Kny's diagrams of *Dryopteris Filix-mas* (Fig. 15, 1–3). Fig. 238, *g*, shows the scheme constantly presented in the earliest stages of the sporangium of *Angiopteris* and other Marattiaceae. It will be well to take the series in reverse of the order of lettering, that is, so as to follow presumably the evolutionary sequence. The massive sporangium of the Marattiaceae has its archesporium deeply sunk; the walls all cut one another at right angles, and since the outer surface is but slightly convex the anticlinal walls are almost parallel, and the archesporial cell is approximately cubical (*g*). The segmentation of the Osmundaceae is variable, and it has been observed to be so even in sporangia of the same plant in *Todea barbara* (*e*, *f*). In some cases the archesporium is still square-based (*f*), and appears square also in transverse section; it is in fact cubical, a form according naturally, as in *Angiopteris*, with the gentle convexity of its massive form. But in smaller sporangia, where the initial convexity is greater, the anticlinal walls converge, and the archesporium has the form of a three-sided pyramid (*e*). Here the wall (x, x) is inserted on an inner periclinal; but in (*d*), which represents the segmentation in *Schizaea*, *Trichomanes*, or *Thyrsopteris*, the wall (x, x) cuts another anticlinal. This marks another step in attenuation of the sporangium, though only a slight one, and in other respects the segmentation of (*d*) resembles that of the simplest sporangia of *Osmunda* or *Todea*. The segmentation of the young sporangium common for the Gradatae is shown in Fig. 238, *b*, *c*. The latter is seen in the Cyatheaceae; the former is a slight variant also seen in them, and in *Ceratopteris*. In the more advanced Polypodiaceae, however, where the sporangium is often long-stalked, the wall cut by the wall (x, x) may be no longer inclined

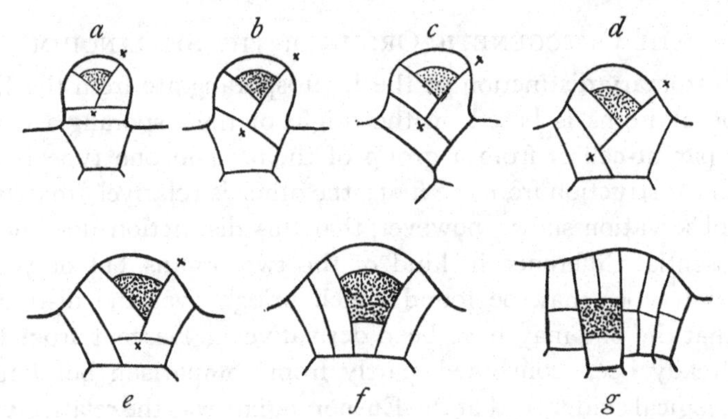

Fig. 238. Diagrams illustrating the segmentation of sporangia of various Ferns. *a* = Polypodiaceae (compare Kny, Wandtafeln XCIV); *b* = *Ceratopteris* (compare Kny, *Parkeriaceen*, Taf. XXV, Fig. 3); *c* = *Alsophila*; *d* = *Schizaea*, or *Thyrsopteris*, or *Trichomanes*; *e*, *f* = *Todea*; *g* = *Angiopteris*.

to the axis of the sporangium, but transverse (Fig. 238, *a*). From this series of diagrams it is seen how gradual are the steps from the segmentation typical of Eusporangiate Ferns to that extreme simplicity to be found among the Leptosporangiates. There is in fact no strict phyletic barrier between the two types; the difference is so gently graded over that the unity of the scheme seems clear. The initial segmentation of the sporangium may itself be held as an index of the progressive attenuation of the sporangium in terms of Descent. It will be shown that it goes along with attenuation of the sporangial stalk, simplification of the sporangial head, and finally with the progressive reduction of the individual productiveness of the sporangium, as measured by the spore-output of each of them.

In such cases as those of *Ceratopteris*, *Alsophila*, and *Schizaea* (Fig. 238, *b*, *c*, *d*), a many-rowed stalk is initiated by the first segmentations. But in those sporangia where the first segmentation is transverse, as it is in *Dryopteris*, either of two things may happen. The stalk-cell may undergo divisions by longitudinal cleavages, thus giving rise to the three-rowed stalk so common in Leptosporangiate Ferns. Or it may remain undivided. The former is seen to occur in *Dryopteris*, as described by Müller (Fig. 15, 3). The latter occurs frequently in advanced types, in which, as in *Scolopendrium*, *Asplenium Trichomanes*, or *Phlebodium aureum*, the long stalk consists of a single row of cells (Fig. 239). In such cases the more complicated segmentation of the sporangial head may result in the upper region of the stalk being still three-rowed. The relation of this to the initial segmentation is suggested by what is seen in *Phlebodium aureum* (Fig. 240, *a*—*f*). Here the first cleavage of the primordium is by an oblique wall (*a*, *b*, *c*), which may extend below the level of the epidermal wall (*b*, *c*), or be clearly above it (*a*). In the latter case growth of the stalk without further segmentation, as in (*e*), would give the condition of the sporangium seen in Fig. 239, of *Asplenium*.

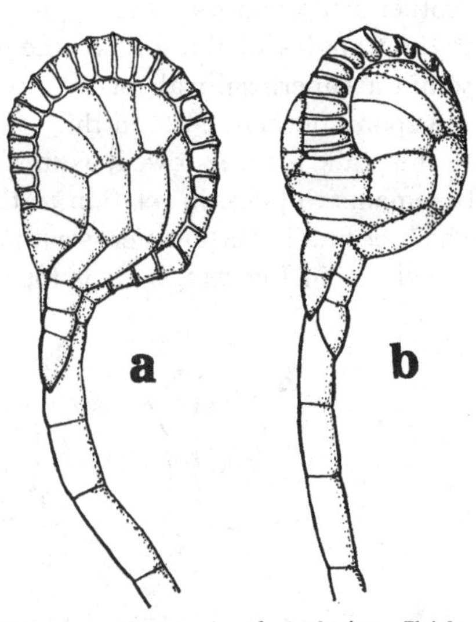

Fig. 239. Sporangia of *Asplenium Trichomanes*, L., after C. Müller, showing stalk of a singe row of cells. (× 140.)

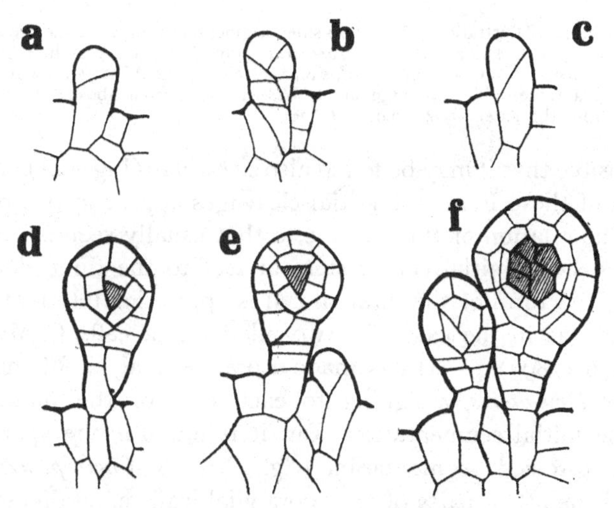

Fig. 240. Stages *a*—*f*, illustrating the development of the sporangium in *Phlebodium aureum*. Compare text. (× 200.)

The further segmentation of the sporangial head has already been described in the case of *Dryopteris*, which is that usual for Leptosporangiate Ferns (Chapter I, p. 13). The inclination of the three oblique walls to one another at an approximate angle of 120 degrees, followed by a fourth wall at right angles to the axis of the sporangium, results in a tetrahedral cell within a superficial wall of four cells. This cell gives rise to the tapetum and spore-mother-cells. In this scheme of segmentation the usual structure of the stalk is three-rowed, as in Fig. 243, *l*, *o*, *p*, or sometimes as in *n*, *q*, *r*. It remains to place in relation to this another scheme hitherto overlooked, which occurs in certain Ferns with a sporangial stalk composed of four rows of cells, as in Fig. 243, *k*. It is found in *Metaxya*, *Dipteris*, and *Cheiropleuria*,

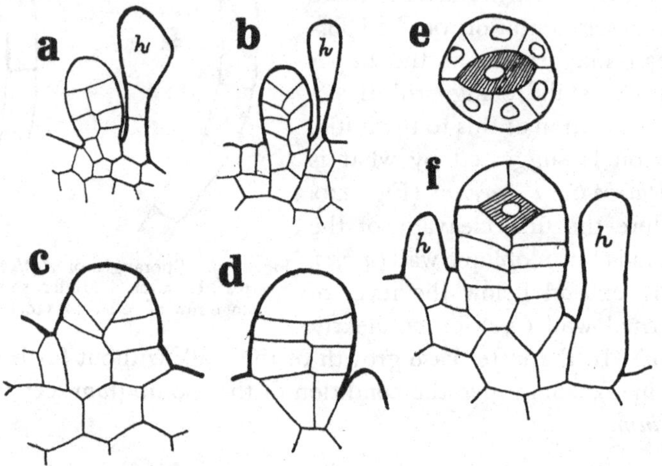

Fig. 241. Illustrations of the two-sided segmentation of Fern sporangia. *a*, *b*, young sporangia of *Cheiropleuria*, showing aspects at right angles to one another. (× 165.) *c*, *d*, similar drawings of *Metaxya*. (× 200.) *e*, a rather older sporangium of *Metaxya* seen from above; *f*, seen from the side. *h*, *h*=hairs. (× 200.)

and it is possible that it may be found also elsewhere (Fig. 241). In the young sporangium of these Ferns the initial cleavages appear in two rows, instead of three. The relation of this scheme to that usually seen is comparable to that of a two-sided initial cell of stem or leaf to the three-sided, or tetrahedral; and the form of the internal cell is approximately that of half of a biconvex lens, as in the case of a two-sided initial cell. C. Müller (*Ber. d. D. Bot. Ges.* ii (1893), p. 54) has made a precise analysis of the sporangial structure in *Dryopteris*, assigning to each part of the mature head its origin in the initial segmentation. But it is found in the sporangia which show the two-rowed segmentation (e.g. *Dipteris*, *Cheiropleuria*) that the annulus and the other parts of the sporangial wall maintain essentially the same relative positions as those of allied Ferns which have a three-rowed segmentation (*Platycerium*). This indicates that the reference of such parts to the initial segmentation in any special case studied has no fundamental significance, for it is not generally applicable. It appears in this as in other examples that there is no general or necessary relation between initial segmentation and the production of mature structures.

THE STALK OF THE SPORANGIUM

In the deeply immersed sporangium of *Ophioglossum* it is impossible to recognise any stalk at all. But in *Botrychium*, in which the sporangium projects, the stalk is a massive column. In *B. daucifolium* it is about six layers of cells in thickness (Fig. 242, *c*). A vascular strand may curve into each

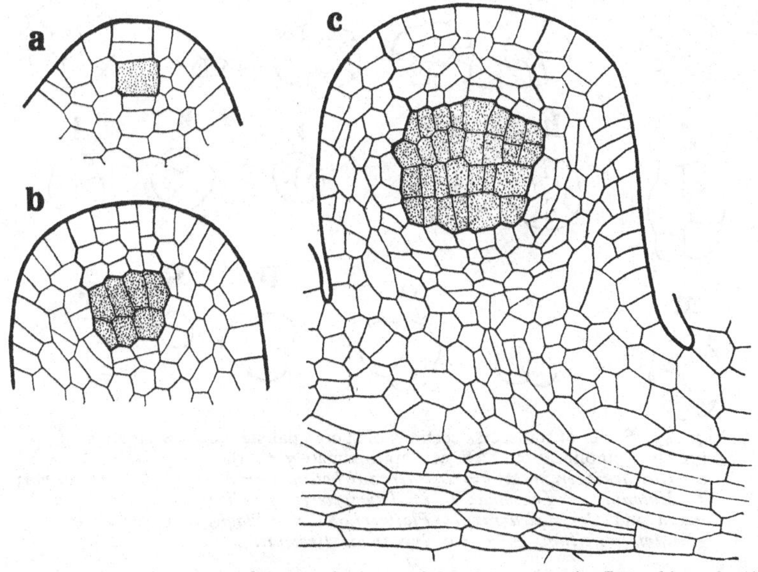

Fig. 242. *a*, *b*, *c*, successive stages of the development of the sporangium in *Botrychium daucifolium*. (× 250.) Compare the massive stalk with those shown in transverse section in Fig. 243.

sporangium, stopping short below the capsule itself. But it is only in the largest sporangia of Ferns that there is any individual vascular supply: when the stalk is less massive there is none, as in *Angiopteris* where the stalk consists of about three layers of cells (Fig. 244). Passing from these Eusporangiate Ferns to those Leptosporangiates styled the Simplices, the relatively large sporangia have usually short stalks. In some of the largest of them the stalk consists of a central column of cells surrounded by a superficial series, as in *Gleichenia circinata*, and sometimes in *Osmunda* (Fig. 243, *a*, *d*). In others the central column is represented only by a single cell-row (*e*, *i*, *h*), but in most of the Leptosporangiates this is absent. The transverse section of the stalk then shows a radial segmentation, as in Fig. 243, *b*, *c*, *f*, *i*, etc. and this leads by successive steps of simplification to the three-rowed stalk (*l*, *o*, *p*), which is the commonest of all in Leptosporangiate Ferns. But it again may pass into still simpler structure (*n*, *q*), till finally the stalk comes to consist of but one row of cells (*r*), a state only found in the most advanced types (compare Fig. 239, of *Asplenium*). A comparison of the figures shows that structural complexity is not always determined directly by absolute size, though on the average the thicker stalks show most cells in the transverse section. On the other hand it is found that the stalks are roughly proportional in transverse section to the bulk of the sporangia, and to the numerical spore-output. (See Tables on pp. 262—263.)

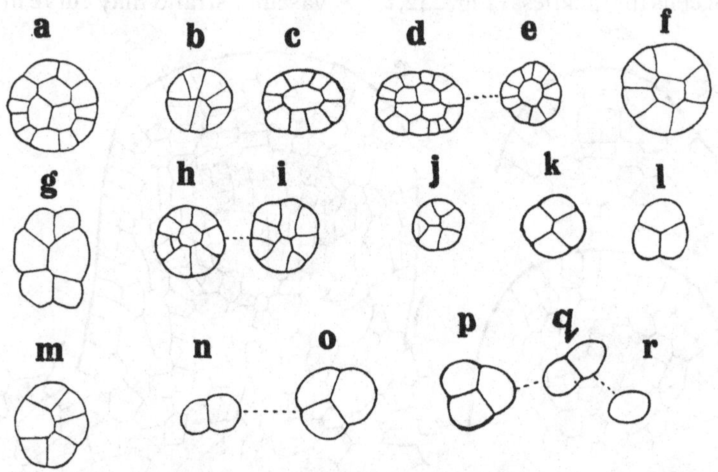

Fig. 243. Series of transverse sections of sporangial stalks, showing steps of progressive simplification. All are approximately to the same scale. (× 150.) *a* = *Gleichenia circinata*; *b* = *Gleichenia dichotoma*; *c* = *Mohria*; *d*, *e* = *Osmunda*; *f* = *Matonia*; *g* = *Loxsoma*; *h*, *i* = *Thyrsopteris*; *j* = *Cibotium culcita*; *k* = *Metaxya* and *Cheiropleuria*; *l* = *Platycerium*; *m* = *Plagiogyria*; *n*, *o* = *Elaphoglossum latifolium*; *p*, *q*, *r* = *Hypoderris Brownii*.

A comparison of (*a*, *i*, *j*) with (*k*, *o*, *p*) shows that the former are more elaborate types, the latter simpler examples of the same tri-radiate segmentation, which corresponds to that usual in the sporangia of Leptosporangiate Ferns. A comparison of (*d*) with (*k*) suggests a four-rowed structure, with possibly a two-rowed initial segmentation, which is unusual in them. It has, however, been shown above that such a segmentation actually exists, though it has long been overlooked (Fig. 241).

The length of the stalk varies greatly. In the Eusporangiatae it is consistently short, or even absent. The Osmundaceae, Gleicheniaceae, Schizaeaceae, Hymenophyllaceae, and Matonineae have spherical or pear-shaped sporangia, with short massive stalks. But the Dicksonioid-Davallioid-Pterid series, as also the Cyatheoid derivatives, illustrate an increase of length of stalk, accompanied by decrease in its thickness; and this is commonly associated with the mixed type of sorus. Finally, many of the advanced Leptosporangiates have very long stalks, often consisting of only a single row of cells (Figs. 239, 240). Here the stalk lengthens rapidly as the sporangium ripens, a condition favourable for mutual protection of the sporangia when young and liberation of the spores when mature in sori where the receptacle is flat and crowded. *Thus speaking generally, massive, short stalks may be held as primitive, elongated and thin stalks as derivative and specialised in character.*

THE WALL OF THE SPORANGIUM, AND THE ANNULUS

The Wall of the sporangium in Ferns varies greatly both in thickness and in mechanical construction. In most of the Simplices it consists of several layers of cells. This is seen in the fossil *Stauropteris* where the adult wall has some four layers, of which the outermost is strongly differentiated. The wall of *Helminthostachys*, and of *Botrychium daucifolium*, is similarly constructed (Fig. 242), as is also that of *Angiopteris*. But in the Leptosporangiate Ferns it consists of only a single layer of cells. Occasional tabular cells may, however, be found lining the outer wall internally in *Osmunda*, and *Lygodium*, which thereby confirm the position of these Ferns as transitional types. We conclude from these facts that a many-layered wall was primitive, and a single-layered wall a derivative state.

The chief interest in the wall, and its special value for comparison, lie in the construction of the mechanism of dehiscence by local induration of the cells. In advanced cases the result appears as the *annulus*, and the *stomium*. But in the simplest no specially differentiated annulus is to be seen. This is so in *Stauropteris*, where dehiscence is by means of a distal pore (compare Fig. 197, Chapter XII). In modern Eusporangiate Ferns, however, and in most of the fossils, the dehiscence is by a slit which runs longitudinally down the side of the sporangium. *Angiopteris* gives a good example (Fig. 244). Here the wall is some three or four layers of cells in thickness: but the superficial layer alone is mechanically developed, and only parts of it are indurated, the rest consisting of thin-walled cells. The annulus appears as a firm vertical arch, broad on either side of the sporangium, but narrow at the distal end (Fig. 244, *D, E, F*). The slit of dehiscence runs down the side next the centre of the sorus, and it is drawn open by the contraction of the peripheral side of the sporangium, the annulus acting like an elastic hoop. This involves a change of form possible only in a free sporangium. In *Marattia*, *Danaea*, and *Christensenia* on the other hand, in which the sporangia are fused into synangia, there is no developed annulus, and the short slit of dehiscence is opened by the drying and contraction of the adjoining cells. The sporangia of many Carboniferous Ferns appear to have behaved in the same way as these living Ferns. *Corynepteris* and *Etapteris* had indurated vertical hoops of tissue, which appear to have acted like those of *Angiopteris* (Fig. 245): the

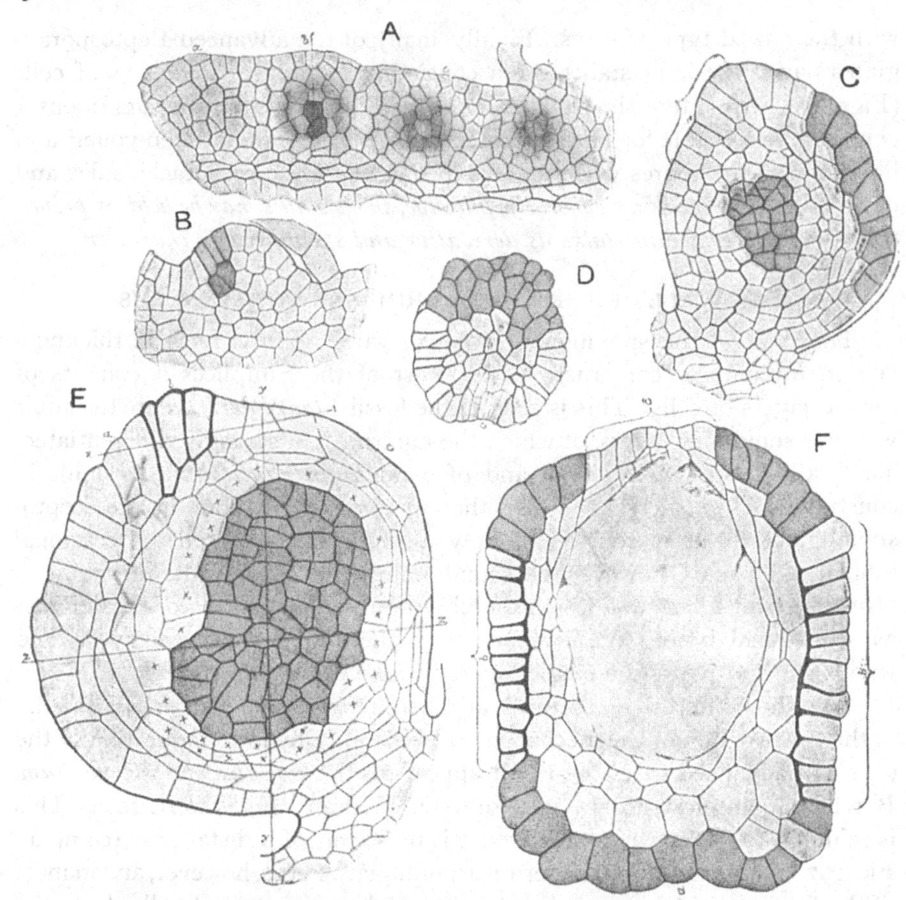

Fig. 244. *Angiopteris evecta*, Hoffm. *A* = part of a young sorus seen in surface view from without. *B* = vertical (radial) section of a sporangium such as would be seen on cutting a sporangium (*b*) in Fig. *A*, along the line indicated. *C* = vertical section of an older sporangium showing genetic grouping of cells. *D* = apex of an almost mature sporangium seen from above: such a section as along the line *x*, *x* in Fig. *E*. *E* = vertical section of a sporangium with spore-mother-cells: the tapetum is marked *x*. *F* = transverse section of an almost mature sporangium. *b*, *b* = annulus; *c* = region of dehiscence; *a* = thin-walled cells which contract. (All × 200.)

Fig. 245. *A* = *Corynepteris Essenghi*, Andrae, from the Carboniferous (Westphalian). Fragment of a fertile pinna. (× 6.) *B* = *C. coralloides*, Gutbier, from the Westphalian. Fragment of a fertile pinna. (× 4.) *B′* = sorus of the same species seen laterally. (× 28.) (After Zeiller.)

sori and sporangia of *Ptychocarpus unitus* are structurally like those of *Christensenia*. In the Ophioglossaceae the outermost layer of the wall is uniformly indurated over the whole surface, excepting along the line of dehiscence, where the cells are smaller than the rest. The slit thus defined runs transversely to the fertile leaf-segment, but longitudinally to the sporangium itself. By drying up of the softer tissue of the wall the lips are drawn apart. It thus appears that the method of dehiscence is not of a high order in any of the Eusporangiate Ferns. The annulus is either absent, or it is of a non-specialised type.

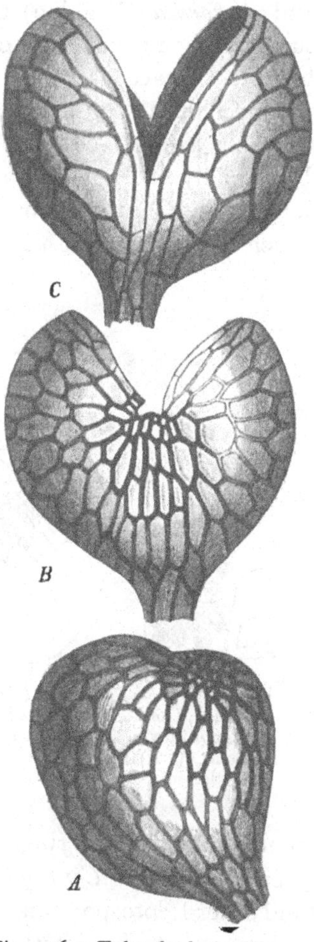

Fig. 246. *Todea barbara*, Moore. Sporangium *A*, in side view, closed; *B*, seen from behind; *C*, from in front, in both cases after dehiscence: the annulus is darkly shaded. (×80.) (After Luerssen.)

A particular interest attaches to the sporangia of the Osmundaceae, with which are to be associated those of the Carboniferous fossils *Kidstonia*, *Boweria*, and probably others also. The sporangia are all separate and pear-shaped, with relatively thick stalks (Fig. 246). The annulus when ripe consists of a group of polygonal thick-walled cells near to its distal end, but rather to one side of it. The slit passes from the centre of the annulus, over the distal end; it extends downwards and gapes widely owing to the action of the mechanical cells. Here is a type of sporangium which has neither the mechanical equipment of the Eusporangiate nor of the Leptosporangiate Ferns. The family is intermediate in so many other characters that this peculiar mechanism of dehiscence claims all the greater attention.

In the Leptosporangiate Ferns the mechanism is precise, and that precision comes with an increasing definiteness of the structure and position of the annulus, and of the stomium. Instead of the outer cell-layer of the sporangial wall being generally indurated, as in the Ophioglossaceae, or a broad band of cells being thickened, as in *Angiopteris* or *Etapteris*, the mechanical thickening is localised in a single row of cells, which forms in all the more advanced types a very efficient means of dispersal. An example has already been described in *Dryopteris* (p. 14). It is possible by comparison of fossils with certain primitive living types to see steps which may have led to this state. The indurated sclerotic cap of the Osmundaceous sporangium corresponds in position though not in detailed structure to that of *Senftenbergia* from the Carboniferous Period, where the cells appear to be arranged in three to four more or less definite ring-like rows (Fig. 247, *A*). In *Klukia* (Jurassic) and *Ruffordia* (Wealden) the cap is replaced by an apparently simple ring surrounding a distal plate of thinner cells (Fig. 248). *Lygodium*, which dates back to Cretaceous times, suggests a transition to this simpler state. Here the mechanical ring may consist of more than one row of cells, and extends down the frontal face of the sporangium on either side of the slit (Fig. 249, *A*). But usually in *Lygodium*, as in *Schizaea* and *Anemia*, it consists of only one row. These comparisons indicate an origin of the single-rowed annulus, which is so constant a feature of the Leptosporangiate Ferns, by simplifica-

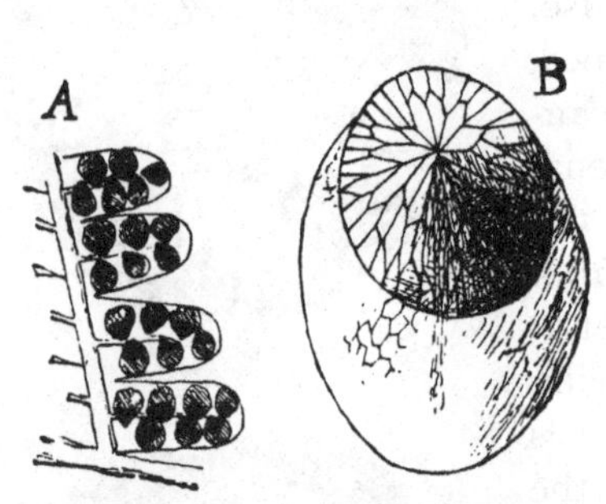

Fig. 247. *Senftenbergia* (*Pecopteris*) *elegans*, Corda. *A* = a small piece of a sporophyll. (× 4.) *B* = a sporangium. (× 35.) (After Zeiller, from Engler and Prantl.)

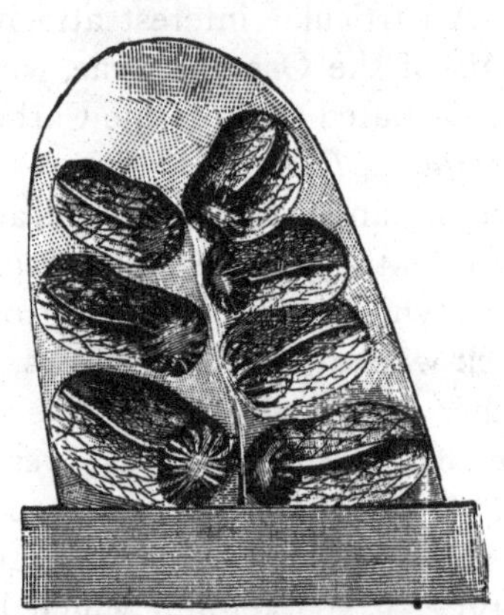

Fig. 248. *Klukia exilis* (Phillips), Raciborski. Fertile pinnule of the last order, seen from below. (× 20.) From the Jurassic of Krakow. (After Raciborski, from Engler and Prantl.)

tion from a complex ring. Irregular doubling of the ring, which is sometimes seen in *Gleichenia*, *Ceratopteris*, *Platyzoma*, *Cryptogramme*, *Acrostichum aureum* and other Leptosporangiate Ferns, even where the annulus is normally single, serves also to bridge the gap between the many-rowed annulus and the simple ring of the Leptosporangiate Ferns.

Along with the simplification of the annulus goes the enlargement of that disc of thin-walled cells which the ring surrounds. It has been styled the "Platte" by Prantl (*Schizaeaceae*, p. 17). In *Senftenbergia* it appears to be absent, though possibly only overlooked owing to faulty preservation. But in the living Schizaeaceae Prantl states that it is always present, though only a single cell represents it in *Lygodium* and *Schizaea* (Fig. 249, *C*). In *Anemia* (*D*) it consists of some 12 cells forming a *distal* or *peripheral region of the wall*. The thin area between the annulus and the stalk may be distinguished as the *proximal* or *basal region of the wall.* The indurated annulus separates these two regions of thin-walled tabular cells from one another. These two parts of the sporangial wall are constantly present throughout the Leptosporangiate Ferns. The external differences which their sporangia show depend chiefly upon their varying proportion, and upon the position of the annulus which separates them relatively to the sporangial stalk.

In the case of *Lygodium* and of *Schizaea* the distal wall is extremely small in area, being represented only by a single cell; but the proximal region is distended with the result that the sporangium appears ovoid, with the annulus as a cap at its apical end (Fig. 250, *a*). The dehiscence is by a median longitudinal slit (Fig. 249, *A*, *C*). But if the distal region were en-

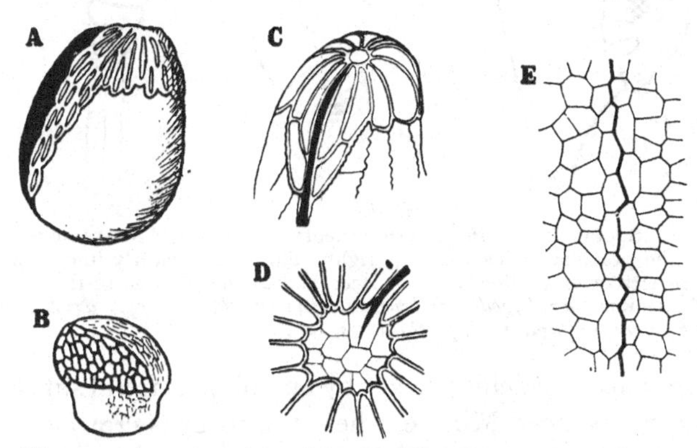

Fig. 249. *A* = sporangium of *Lygodium lanceolatum*, Desv. (× 50), showing annulus of more than a single row of cells; *B* = *Kidstonia heracleensis*, Zeiller, lateral view of sporangium (× 50); *C* = *Schizaea*, apex of sporangium with distal wall ("Platte") of only one cell; *D* = *Anemia*, apex of sporangium with "Platte" of many cells; *E* = line of dehiscence of sporangium in *Ophioglossum reticulatum* (× 100). (*A*, *B*, after Zeiller; *C*, *D*, from Engler and Prantl.)

larged so as to form a considerable convex area, and the annulus were dilated so as to form a zone running obliquely round the body of the sporangium, the type of *Gleichenia* would result, though still with the same constituent parts, and with the dehiscence in a median longitudinal plane (Fig. 250, *b*). A slight difference in proportion, but with only partial induration of the annulus, gives the sporangium of *Loxsoma*, the only known Fern with a gradate sorus which retains the median longitudinal dehiscence (Fig. 250, *d*). In all other Gradatae and in all of the Mixtae the dehiscence is lateral. In the least modified of these the annulus still remains as a complete ring. The lateral dehiscence which they show results from a change in development of certain of its cells. For instance, in *Lophosoria*, *Dicksonia*, *Hymenophyllum*, or *Plagiogyria* the annulus is oblique and complete, and the stomium

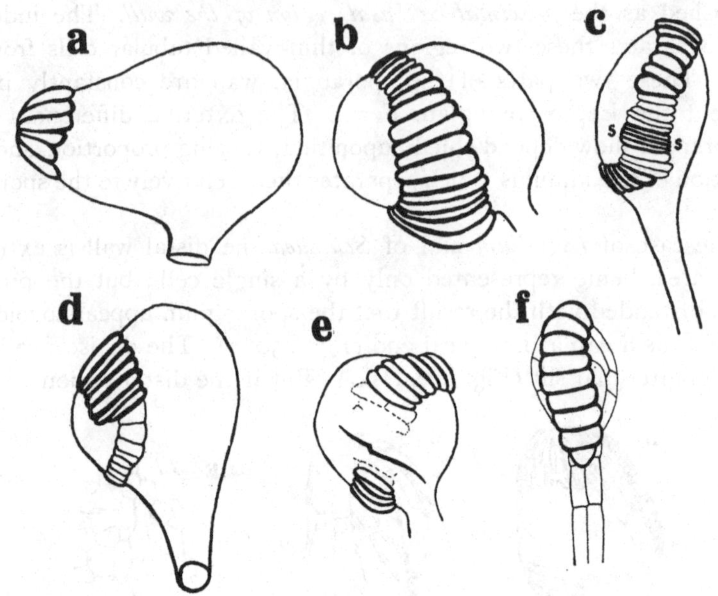

Fig. 250. Sporangia of various Ferns, orientated so that the distal face is to the left, the proximal or basal to the right. This brings clearly into view the differences of proportion of those faces, and of the position of the annulus and stomium. *a* = *Lygodium*; *b* = *Gleichenia*; *c* = *Plagiogyria*; *d* = *Loxsoma*; *e* = *Hymenophyllum*. *f* = *Leptochilus*. *s, s* = stomium.

lies at the side, while the dehiscence is by an oblique lateral slit (Fig. 250, *c*, *e*). But in the more advanced Mixtae, where the sorus is crowded, the annulus tends to a vertical position. The induration of the cells adjoining the stalk is omitted, and the continuity of the cells of the ring is interrupted by its insertion (Fig. 250, *f*). In some of the less altered types, where the sorus is still irregularly gradate, the cells of the ring may be seen to be continuous past the stalk, as in *Cheiropleuria* (Fig. 251, *B*), and in *Platycerium* (*C*). The like is also seen in *Dennstaedtia apiifolia* (Fig. 252, *C*); but in the related *D. rubiginosa* the series is clearly interrupted, as it is in all the more advanced Leptosporangiate Ferns (Fig. 252, *D*). It has been shown in several phyletically distinct groups that such a progression, from the oblique and continuous to the vertical but interrupted annulus, may be traced by

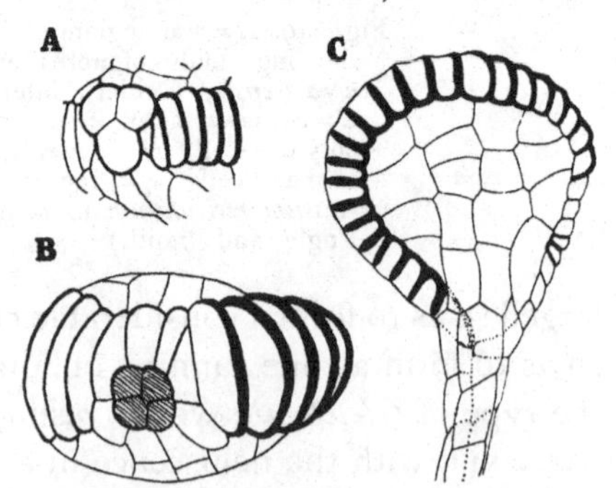

Fig. 251. *A*, base of a sporangium of *Pteris grandifolia*, showing the interruption of the annulus, and the three-rowed stalk; *B*, base of a sporangium of *Cheiropleuria*, showing the ring continuous, and the four-rowed stalk; *C*, sporangium of *Platycerium*, showing cells of the annulus almost interrupted, though still actually continuous. In *A*, *B*, the distal wall is directed upwards, in *C* it is turned away from the observer.

comparison of allied genera and species. The dehiscence consequently becomes not oblique as in the Gradatae (Fig. 250, *e*), but transverse, as it is in *Dryopteris* (Fig. 17). Here, nevertheless, as in other Ferns with a vertical annulus and transverse dehiscence, it is still possible to distinguish the phyletically distal from the phyletically proximal sides of the sporangium by the segmentation. Thus in Fig. 16, 4*a* presents its distal or peripheral face to the observer, while 4*b* presents its proximal or basal face. Professor von Goebel has suggested the distinction of the "longicidal" from "brevicidal" sporangia. These terms may be found useful in a descriptive sense; but the one type merges so frequently into the other that it cannot be held to mark any real distinction of race. The facts which run parallel with those relating to other criteria of comparison suggest that there has been in a plurality of phyletic

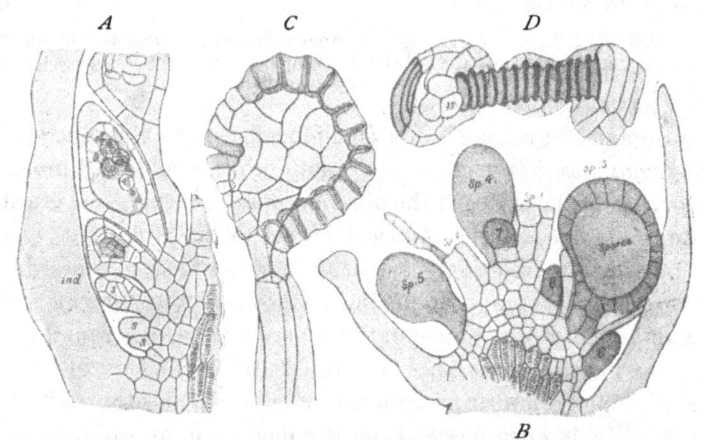

Fig. 252. *A* = *Dennstaedtia apiifolia*, Hook. Sorus showing basipetal succession throughout. *C* = dehiscent sporangium of the same, showing very slightly oblique annulus. *B* = *Dennstaedtia rubiginosa*, Kaulf. Sorus in vertical section showing that it has been at first basipetal, but a mixed character has supervened. *D* = dehiscent sporangium of the same, seen from the base, showing that the annulus stops short on either side of the insertion of the stalk (*st.*). All × 100. *s*, *sp* = sporangia; *ind* = indusium.

lines an actual swing of the annulus from the primitively transverse position to the oblique, and finally to the vertical position, with the natural consequence that the dehiscence has changed from the longitudinal to the transverse plane.

The case of *Loxsoma*, where half of the annulus, though indicated by the cell-arrangement, is not indurated, and hence is mechanically ineffective, suggests that when circumstances prevent its mechanical function being performed the annulus may be partially abortive (Fig. 250, *d*). In the Hydropterideae, which shed their spores in water, it may be wholly abortive. In the Salviniaceae there is no structural evidence of an annulus. But in the Marsiliaceae, which in many characters show affinity with the Schizaeaceae, Campbell has noted in *Pilularia americana* a vestigial annulus, with thin cell-walls. The form of the microsporangium and the arrangement of the cells round its apex so clearly resemble those in *Schizaea* or *Anemia* that they may be held as the vestigial equivalent of the annulus seen in these genera (Fig. 253).

As the specialisation of the annulus increases so does that of the *stomium*. In Eusporangiate Ferns the dehiscence is defined by two parallel rows of cells, as in *Ophioglossum* or *Botrychium* (Fig. 249, *E*). In the Schizaeaceae and Gleicheniaceae the condition is but little more advanced (Fig. 249, *C*, *D*), and even in the Hymenophyllaceae and *Loxsoma* there is no constantly organised group of cells which localises the dehiscence (Fig. 206); but in *Dicksonia* and *Plagiogyria*, and many other Gradatae, a more definite group, belonging to the series of the annulus, may be recognised as a stomium. Though the number of the thin-walled cells adjoining that group may often be indefinite in relatively primitive Ferns, the stomium settles down to a structure usually composed of four sister-cells (Fig. 250, *c*), a condition which is maintained through most of the advanced Leptosporangiates, though sometimes the number may be only two. As this definition of the stomium progresses it swings round from a median to a lateral position, with the effect that the slit of dehiscence passes from the median to an oblique, and finally where the annulus is vertical to a transverse plane. This last, which is strictly maintained by the Mixtae, is obviously in relation to the closely packed sorus, in which the apex of the sporangium alone is exposed so as to allow free movement to the distal arch of the annulus. It thus appears from comparison of the opening mechanisms of Ferns that in the Eusporangiate types the annulus was ill-defined and not a precise mechanism; that in the Leptosporangiate Ferns the annulus became definite, as a single ring of cells: that it changed its position from being almost transverse to the axis of the sporangium (*Schizaea*), through an oblique position (Gradatae) to vertical (Mixtae). The stomium, originally ill-defined, became a definite group of cells, while the slit of dehiscence shifted from a median to an obliquely-lateral, and finally to a transverse position.

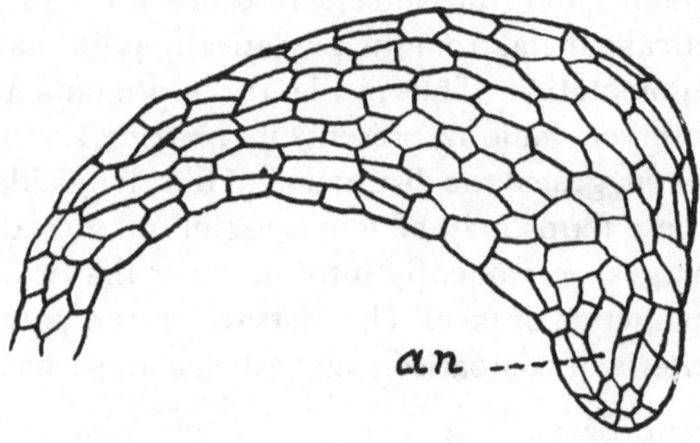

Fig. 253. Sporangium of *Pilularia americana*, after Campbell. It is of the *Schizaea* type, but with the annulus (*an*) not indurated.

Professor von Goebel (*Organographie*, II. Aufl. II. Teil, p. 1180) has discussed the question whether or not a change of position of the annulus has actually occurred in individual groups of Ferns. He admits that such a change might appear where form is already differentiated, and a change of function has occurred. But he points out that often "developmental potentialities" may be variously realised under the influence of external and internal stimuli. He figures to himself a primitive sporangium with the capacity for the cells of its wall to develop an annulus according to the position of the sporangium; and he suggests that the ring, for instance of *Hymenophyllum*, never had any other position than that which we now see. He would only admit that a swing of the annulus had taken place if it were proved that the position and form of the sporangium had previously been other than that now seen.

This somewhat drastic demand for evidence of an actual phyletic swing of the annulus seems to be adequately met by two lines of rejoinder, one of which von Goebel supplies himself by the facts detailed on the next page of his book (p. 1181). The existence of a non-functional side of the annulus of *Loxsoma*, clearly traceable by its cell-divisions, is in itself evidence of greater stability of development of the annulus than von Goebel contemplates in his primitive sporangium. So also is the presence of a non-functional annulus recognised by Campbell in *Pilularia*, as above described.

On the other hand the very natural *Dicksonia-Dennstaedtia* series appears to supply very cogent evidence of a swing having taken place. In *Dicksonia* the annulus appears as a complete oblique ring. But in *Dennstaedtia apiifolia*, notwithstanding the basipetal sequence of the sporangia, the annulus is nearly vertical, though still the chain of its cells is completed by contact close to the insertion of the stalk (Fig. 252, *C*). But in *D. rubiginosa*, in which the mixed character of the sorus has been assumed (Fig. 252, *B*), the annulus is actually interrupted at the insertion of the stalk (Fig. 252, *D*). When it is remembered that *Dennstaedtia* was for long included in the genus *Dicksonia* this series appears to illustrate within very near affinity a swing of the annulus, and a final interruption of it which runs parallel with other features in this progressive series. A second instance is seen in the Dipterid series. The oblique annulus of *Dipteris* appears to be shifted by gradual steps to the vertical in the more advanced members of the series, such as *Cheiropleuria* and *Platycerium* (Fig. 251, *B*, *C*). These and other sequences are held to give support to the view that a swing of the annulus from the transverse to the oblique, and from the oblique to the vertical position, with final interruption at the insertion of the stalk, has actually taken place. In particular the occurrence of rudimentary, non-functional cells of the annulus appears to indicate that the annulus is something more than an immediate response to the influence of external and internal stimuli; probably it is a thing impressed upon the organism, so that its features are subject to hereditary transmission, but liable to modifications such as those involved in a phyletic swing from a transverse to an oblique, and finally to a vertical position.

Origin of the Sporogenous Cells

The tetrahedral cell within the young sporangium, from which by further segmentation the tapetum and the spore-mother-cells are formed, as seen in *Dryopteris* (Fig. 15, 3), has sometimes been called the *archesporium*. In most Ferns all the spores, and frequently the tapetum also, may be referred in origin to a single archesporial cell. But this is not always possible. Exceptions have been found to occur, for instance in *Marattia* (Fig. 254), and the case seems doubtful also in *Ophioglossum*, though many preparations of it would accord with such an origin. It does not appear that any general principle is disclosed by tracing the sporogenous tissue to its source, through "a last analysis" of the segmentations which give rise to the sporogenous cells, or to the "archesporium" itself (see *Land Flora*, p. 108). The constant feature for the Ferns, and indeed for Pteridophytes at large, does not lie in the sequence or number of the segmentations, but in the fact that the sporogenous cells originate ultimately from the division of superficial cells of the plant-body. These divide periclinally: the outer product then forms part of the sporangial wall; the inner gives rise by further divisions to sporogenous cells. The final result of those divisions is a number of highly protoplasmic, thin-walled cells. These round themselves off as *spore-mother-cells*, separating more or less completely, and floating in a semi-fluid medium which fills the cavity of the rapidly enlarging sporangium. Here they undergo the tetrad-division, as already described (Chapter I, p. 22, Fig. 30). The chief points for further discussion will be the development, nutrition, and ripening of the spores: also the number of the spore-mother-cells, and the consequent spore-output of the sporangium. An accessory interest also attaches to the *tapetum*, which is variable in its relation to the sporogenous tissues, as well as in its behaviour in the ripening of the spores.

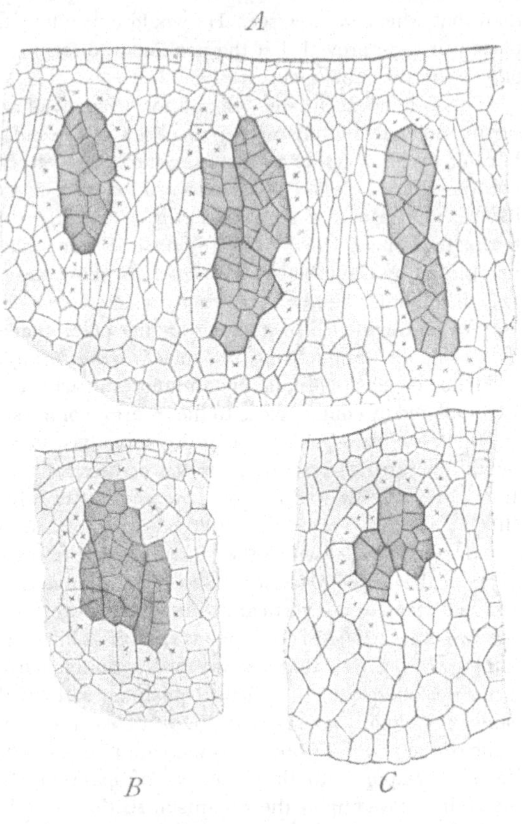

Fig. 254. *Marattia fraxinea*, Smith. A=section transversely through a sorus: the sporogenous cells shaded, the tapetum marked (x, x): the left-hand sporangium shows the most usual arrangement of the sporogenous tissue, the other two less frequent. B, C show in similar section irregular groupings not referable to a single parent cell. (× 200.)

THE TAPETUM

The tapetum is not a morphological constant. In the Bryophyta it is not represented. In the Pteridophyta, and even within the Filicales, it is not uniformly of the same origin. In the Ophioglossaceae and Marattiaceae it appears outside the definitive sporogenous group as an ill-defined band consisting of several layers of tabular cells. But in the smaller sporangia of the Leptosporangiate Ferns the cells which go to form the more definite tapetum are cut off from the sporogenous cell or cell-group. There is good reason to believe that there has been a progressive change of origin of the tapetum within certain circles of affinity. Indefinite and non-specialised nutritive arrangements are characteristic of larger and probably primitive sporangia: but more definite tapetal layers are found in the smaller and are probably derivative. Thus in the large sporangia of *Lycopodium* the tapetum originates outside the sporogenous group: in the smaller *Selaginella* it is cut off from the superficial cells of the group. The same difference is seen in Ferns on comparing the Marattiaceae and Ophioglossaceae with the Leptosporangiate Ferns. In the former indefinite layers of slab-like cells outside the sporogenous group become disorganised, and act as a tapetum. In the latter the tapetum appears as two definite layers of tabular cells, which spring from the "archesporium" itself. The difference of origin seems to depend on size.

As the spore-mother-cells round off prior to the tetrad-division, the tapetal cells fuse together to form a *plasmodium*, their nuclei retaining their identity, and even multiplying by fragmentation. As the spore-mother-cells separate this nucleated plasmodium intrudes between them, so that they are suspended, usually isolated and rounded off, in a rich nutritive medium (Fig. 255). In all the more primitive Ferns the material of the plasmodium is absorbed into the developing spores: but in certain advanced types much of it may remain as a deposit on the outer wall of the spore, and is then called the "*perispore.*"

The tetrad-division itself has been described for *Dryopteris* (p. 22): it is essentially the same for Ferns generally, excepting that the grouping of the spores in the tetrad may follow either of two plans. In the first the division of the spherical spore-mother-cell is by six walls, giving the spores the tetrahedral form (Fig. 256, *B*): in the second the spore-mother-cell divides by three walls to form spores of wedge-shape, with two flattened sides (Fig. 256, *A*). Though these seem so different they may both occur within nearly related circles. The Marattiaceae and the Schizaeaceae include both. The spore-form is thus an uncertain guide to affinity.

The spore-wall is often marked by characteristic sculpturing (Fig. 256). But more important than this is the existence or absence of a deposit of the *perispore* from the tapetum. Ferns may in fact be divided into two groups according to the presence or absence of a perispore. None is seen in the Eusporangiate Ferns, nor in the Osmundaceae, Schizaeaceae, Gleicheniaceae, Hymenophyllaceae, Cyatheaceae, Davallieae, or in *Ceratopteris*. In fact it is absent from the relatively primitive Ferns. Of the more advanced Leptosporangiate Ferns it is wanting in the Vittarieae, Gymnogrammeae, Polypodieae, and Pterideae, but present in the Asplenieae (with *Athyrium*), and in the Aspidieae (Fig. 257). It is interesting to note that it is absent in *Acrostichum*

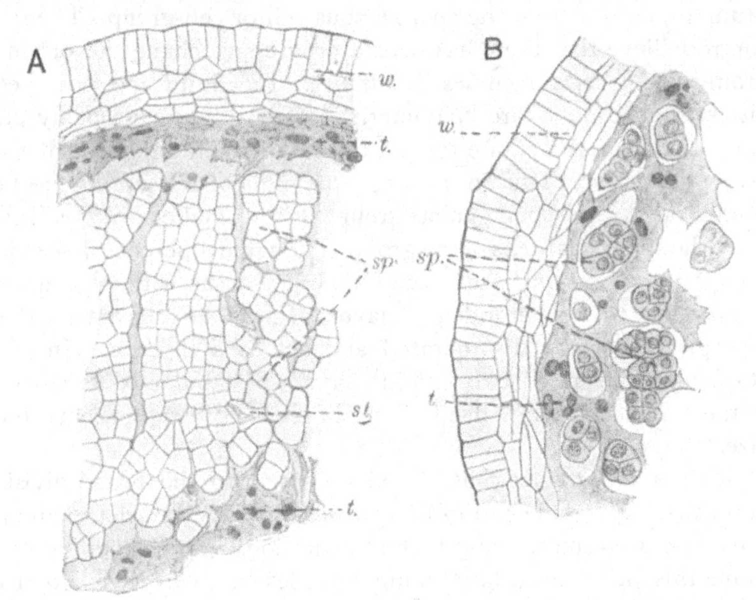

Fig. 255. *Ophioglossum vulgatum*, L. Portions of sporangia showing the sporogenous tissue in two stages of disintegration. In *A* the tapetum (*t.*), evidently derived from more than a single layer of cells, has formed a plasmodium with many nuclei, which is beginning to penetrate the sporogenous tissue, in which an occasional cell (*st.*) is seen disorganised. *B* shows a more advanced state, where the sporogenous cells (*sp.*) appear in small clusters, or isolated, embedded in the tapetal plasmodium (*t.*): *w*=sporangial wall. (× 100.)

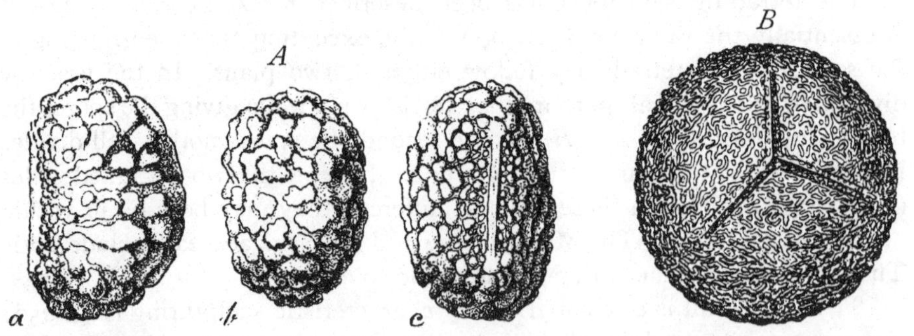

Fig. 256. *A*. Spores of *Polypodium vulgare*, L., showing the wedge-shaped, two-sided form of spore from three aspects. *B*. Spore of *Osmunda regalis*, L., of tetrahedral form. (After Luerssen.) (× 444.)

aureum, a detailed fact supporting its affinity with the Pterideae suggested by Sir W. Hooker 50 years ago. It is absent also in *Cystopteris* and *Onoclea*. The perispore thus possesses a certain value for comparison, but confidence in it as a safe criterion is shaken by the fact that while it is present in *Blechnum* and *Woodwardia* it is absent in the closely related *Sadleria*, *Brainea*, and *Doodya*. It is clearly a feature adopted late in Descent, and apparently restricted to certain circles of affinity.

In *Polypodium imbricatum* the perispore takes the form of fine hygroscopic fibres which lie among the spores in the sporangia. They may be compared to the elaters of *Equisetum*, both being specialised products of the tapetum. It is suggested that their biological use in this epiphytic Fern lies in fixing the large spores upon the surface of the tree-stems (G. Karsten, *Flora*, Vol. 79, 1894, p. 86). The perispore finds its most striking development in the Salviniaceae. In *Azolla* the limitation of the "massulae" in the microsporangium, and the formation of the "glochidia," and also of the swimming organ in the megasporangium, are highly specialised types of perispore (Hannig, *Flora*, Vol. 102, 1911, p. 243; also Vol. 103, p. 321).

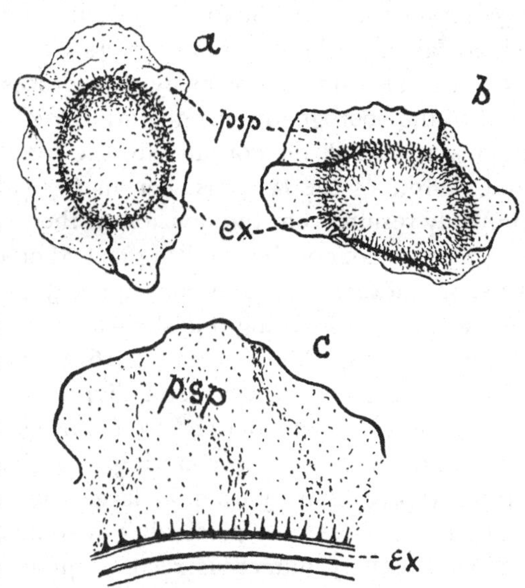

Fig. 257. Spores of *Aspidium trifoliatum*, after Hannig. *a*, *b*=ripe spores with prickly exospore (*ex.*) and transparent perispore (*psp.*), appearing like a loose sac. (× 500.) *c*=part of the exospore and perispore more highly magnified.

SPORE-OUTPUT

By the definition of the sporangium given above the spore-mother-cells form its essential feature. The structural characteristics of the sporangial stalk and wall are developed in accordance with their number: and since each divides to form four spores, all of which grow to maturity in normal sporangia, the numerical output of spores per sporangium will give an approximate basis for their comparison. This was first suggested by Russow (*Vergl. Unters.* p. 86). An estimate of the spores of a single sporangium may be made either by actual counting, which may be easy for small sporangia where the numbers are relatively low: but this is a tedious method where the numbers are large. In the latter case an approximate estimate may be formed by the examination of transverse and longitudinal sections in the stage of the spore-mother-cell. Subject to certain exceptions and limitations the numbers thus obtained are approximately constant for the individual or species, and accordingly they give a consistent basis for comparison of Ferns at large. The numbers range from thousands of spores in each Eusporangiate sporangium down to a single matured spore in the megasporangia of the Hydropterideae. In the great majority of cases they centre round numbers which fall into the series of powers of two: 8, 16, (24), 32, (48), 64, 128, 256, etc. This points obviously to a process of repeated bipartition of the cells of the sporogenous group, terminated by the usual tetrad-division. The figures 24 and 48 do not fall into the series; they are not uncommon, and an example of the latter is seen in *Dryopteris Filix-mas*. The number 48 probably results from the division of the primary sporogenous cell into 2, 4, and 8; only four of the resulting cells then undergo the next division, giving 12 spore-mother-cells, and consequently 48 spores (see Fig. 16, 2, 5, 6, 7). A similar process stopping one step earlier in the series of divisions would give the number 24. The figures 32, 16, 8 will follow from similar, but complete omissions of the successive divisions in the sporogenous group of cells.

Numerical results derived from various Ferns, either based on estimates from sections or on actual countings, are given in the subjoined table, which is arranged progressively from the largest numbers to those which are smaller, while at the same time preserving in some degree the affinities of the Ferns cited. The very numerous records quoted will amply suffice for giving a general conspectus of spore-numbers for living Ferns at large. In the last column the name of the observer is given; where no name is quoted the observation or estimate was made by the author.

Name	Result of Countings	Estimated Typical No.	Authority
Ophioglossum pendulum	—	15000	
Botrychium Lunaria	—	1500–2000	
Christensenia	—	7850	
Marattia	—	2500	
Danaea	—	1750	
Angiopteris	—	1450	
Gleichenia flabellata	838, 794, 695, 684	512–1024	
,, pectinata	—	256–512	
,, dichotoma	319, 251	256 or more	
,, hecistophylla	265, 272	256 or more	
,, circinata	241, 242	256	
,, rupestris	244, 232, 220	256	
,, linearis	—	256 or less	
Stromatopteris	480, 415	256–512	Thompson
Osmunda regalis	476, 462, 396, 373	256–512	
Todea barbara	478, 445, 442, 225, 238	256–512	
,, superba	206, 306, 342	256 or more	
,, hymenophylloides	112, 115, 120, 124, 205	128–256	
Lygodium dichotomum	232, 246	256	
,, javanicum	236, 238, 245	256	
,, pinnatifidum	128, 127	128	
Anemia phyllitidis	114, 111, 104	128	
Mohria caffrorum	107, 107, 101	128	
Schizaea malaccana	90, 110, 112, 115	128	Tansley
Hymenophyllum tunbridgense ...	413, 416, 421	256–512	
,, sericeum	216, 239	256	
,, dilatatum ...	121, 127, 127, 127	128	
,, Wilsoni	119, 121	128	

Name	Result of Countings	Estimated Typical No.	Authority
Trichomanes reniforme	247, 243	256	
" crispum	51, 52, 59	64	
" rigidum	32, 48, 56	32–64	
" radicans	46, 58, 62	48–64	
" javanicum	38, 42, 48	32–48	
" spicatum	48	48	
" pinnatum	32, 48, 32	32–48	
Loxsoma Cunninghamii	64, 62, 63	64	
Thyrsopteris elegans	—	48–64	
Dicksonia antarctica	64, 62, 63	64	
" Menziesii	64	64	
Dennstaedtia apiifolia	61, 62	64	
" punctilobula... ...	—	64	Conard
Pteris tremula	—	64	Russow
" cretica	—	32	"
Davallia speluncae	64, 64	64	
Matonia pectinata	—	48–64	
Dipteris conjugata (D. Horsfieldii)	62, 60, 61	64	Thomas
" Lobbiana	50, 42, 47	64	"
Cheiropleuria bicuspis	124, 123, 108	128	"
Platycerium alcicorne	60, 62	64	"
Leptochilus tricuspis	60, 61, 63	64	"
Metaxya	—	64	
Lophosoria quadripinnata	—	64	
Alsophila excelsa	64, 60	64	
" aspera	—	64	Russow
" atrovirens	57, 62	64	
Cyathea medullaris	57, 61	64	
" dealbata	16, 8, 8, 16	8–16	
Peranema cyatheoides	—	64	Davie
Diacalpe aspidioides	—	48	"
Dryopteris Filix-mas	—	48	Russow
" cristatum	—	64	"
" spinulosum	—	64	"
Sadleria cyatheoides	16	16	
Jamesonia scalaris	58, 63, 64, 68, 69, 71, 72	64–72	Thompson
Llavea cordifolia	52, 46	48–64	"
Cryptogramme crispa	45, 46, 48, 50, 51, 52	48–64	"
Gymnogramme trifoliata	49, 48, 46, 44, 43, 39, 35	48–64	"
Pellaea falcata	60, 56	64	Marsh
" intramarginalis	32	32	Thompson
" andromedifolia	30, 31	32	Marsh
" hastata	24, 16	24	"
Cheilanthes Fendleri	53, 57	64	"
" gracillimum	30, 31	32	"
" lanuginosum	32, 30	32	"
" tenuifolia	32, 28, 27	32	Thompson
" microphylla	32, 30	32	"
Notholaena trichomanoides ...	—	48	"
" sinuata	24, 32, 32	24–32	
" "	—	16–32	"
" affinis {small spores	64, 56, 55, 54	64	"
" affinis {large spores	18, 16, 14, 12	12–24	"
Platyzoma {small spores	30, 29, 28, 26	24–32	"
Platyzoma {large spores	16, 14, 12	16	"
Ceratopteris thalictroides	32 (16 Kny)	16–32	
" deltoides	—	16	Benedict
Marsilia {microsporangia	—	64	Russow
Marsilia {megasporangia	—	1	
Azolla {microsporangia	—	64	Campbell
Azolla {megasporangia	—	1	

Comparing the numbers in this table we see first how great are the differences in spore-output per sporangium within the Filicales: secondly, that in many examples where the figures are moderate the actual number of spores is approximately constant for a given species, and often for a genus: but that where deviations occur in the same species, as in *Cyathea dealbata*, *Platyzoma*, or *Ceratopteris*, or within genera, as in *Todea*, *Trichomanes*, *Pellaea* or *Cheilanthes*, the differences suggest the omission of one of the synchronous divisions prior to the tetrad-division. The larger numbers of spores are associated with those Ferns which on other grounds are held to be relatively primitive. Comparison with early fossils shows that high figures ruled in them also: for instance in *Zygopteris* there was a spore-output that cannot have been less than 500 to 1000 (Fig. 199), while *Stauropteris* and *Pteridotheca* also show evidence of large figures. These plants are on general grounds of comparison related to such living Ferns as are placed at the head of the list. It may then be held that in the more primitive Ferns, and especially in those types which are represented in the Primary Rocks, the spore-output was relatively large.

Caution should, however, be observed in the use of these estimates as a strictly arithmetical guide. It cannot be truly stated that those Ferns which have the largest spore-numbers per sporangium are the most primitive of all. *Ophioglossum pendulum* which stands first on the list with the high estimate of about 15,000 is not on that account alone to be held as the most primitive type named. Reasons will be advanced later for the view that *Botrychium* is a more primitive type, though its estimated output is only 1500 to 2000. But both belong to a very primitive family, as their relatively high spore-output would indicate. It is, in fact, as an indication of general relationship that the numbers can be correctly used, rather than as an exact guide. In this way the fact that "Polypodiaceous" Ferns, which constitute the vast bulk of present-day species, have numbers ranging from 64 downwards is impressive by its constancy. The only exceptions which have so far come to light are *Jamesonia scalaris*, where the number may reach 72, and *Cheiropleuria*, where the number appears constantly to approximate to 128. On the other hand some Ferns show marked instability as regards the lower figures: for instance, *Cyathea dealbata*, 8—16; *Notholaena sinuata*, 16—32; *Ceratopteris*, 16—32. In all of these the variation seems to turn upon one division more or less in the sporogenous tissue. The most remarkable case of all is *Notholaena affinis*, which has sporangia with small spores ranging to 64 in number, and others with large spores ranging from 12 to 24. The nearest parallel is found in *Platyzoma*, which has large sporangia with few large spores ranging up to 16 in number, and smaller sporangia with 32 spores of smaller size (Fig. 258). These facts suggest initial heterospory, but they have not yet been tested by cultivation of the prothalli.

Sections of the sporangia of Ferns transitional between the Eusporangiate and the Leptosporangiate types support the spore-counts which are transitional also. For instance, a vertical section of a sporangium of *Osmunda* (Fig. 259) traverses 30 to 32 spore-mother-cells. The sporogenous group is

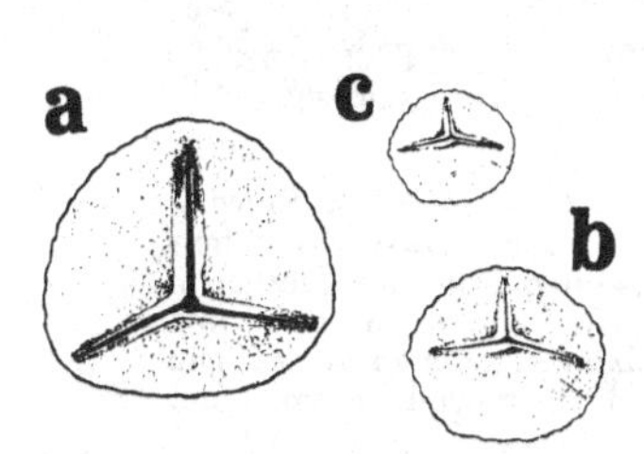

Fig. 258. Spores of *Platyzoma*, after McLean Thompson, all drawn to the same scale. *a*, one of the largest; *b*, of medium size; *c*, one of the smallest. (× 26.)

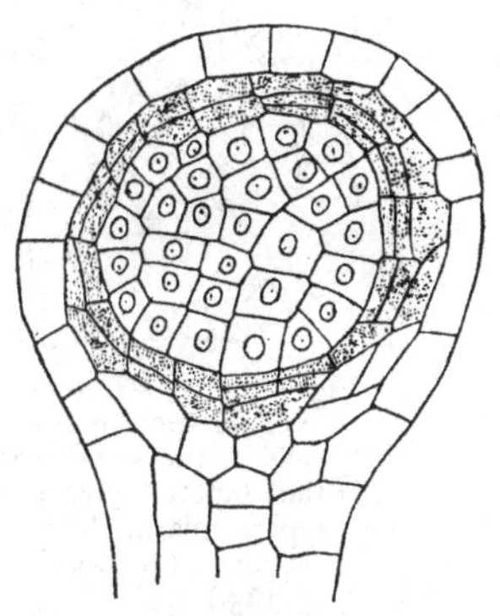

Fig. 259. Sporangium of *Osmunda regalis*, L., containing a large sporogenous tissue, surrounded by a tapetum consisting in parts of three layers of cells. (After von Goebel.)

spherical, and the diameter of each cell is about one-sixth of the whole sphere. Accordingly their total number would approximate to 128, which accords with an estimate of 512 spores. It has moreover a tapetum widened at places to three layers, and a massive stalk (Fig. 243, *d*, *e*). All these points are transitional between the Eusporangiate and the Leptosporangiate types. Again in *Lygodium*, with a spore-output of 128—256, the section of a sporangium shows about 20 spore-mother-cells. Here also there are irregularly three layers of cells between the sporangial wall and the sporogenous group, while the stalk is relatively massive (Fig. 260). In the genus *Gleichenia*, including *G. flabellata* with relatively large spore-output (512—1024), and *G. dichotoma* with relatively small output (256 or more), sections again bear out the spore-counts (Fig. 261, *A*—*D*), while the thickness of the stalk follows suit (Fig. 243, *a*, *b*). In those Ferns with numbers below 64 the stalk is always thin and, as has been seen, may fall to a single row of cells (Fig. 239). Lastly, the mechanism of the annulus is essentially an opening mechanism in the larger types, and nothing more. But in the sporangia with

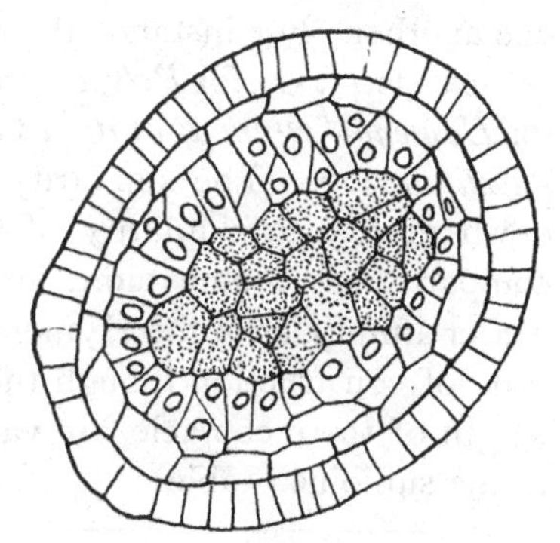

Fig. 260. Section through a sporangium of *Lygodium circinatum*, after Binford. 20 spore-mother-cells are cut through: the tapetum is more than doubled. (× 480.)

smaller output it becomes a mechanism highly specialised for ejection of the spores. All of these features indicate evolutionary progression with decrease

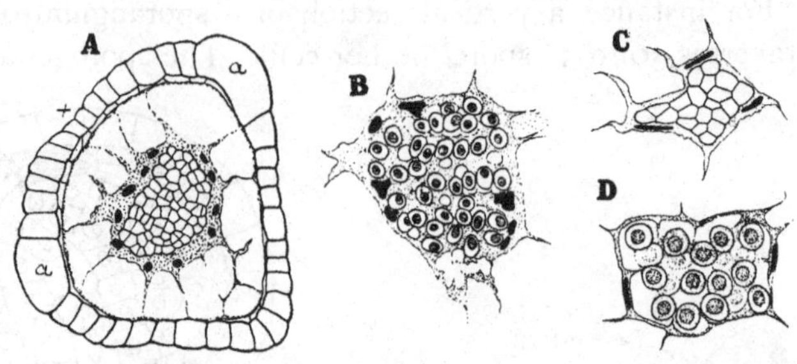

Fig. 261. *A*, section of sporangium of *Gleichenia flabellata*, showing over 60 spore-mother-cells in section. (× 100.) *B*, spore-mother-cells of the same plant, older, and separated from one another in the tapetal plasmodium. (× 165.) *C*, spore-mother-cells and tapetal plasmodium of *Gleichenia dichotoma*; only 20 are seen in section. (× 100.) *D*, cells separated in tapetal plasmodium. (× 165.)

in size and simplification of structure of the sporangium, and a fall in the individual output of spores, but a perfecting of the mechanism of their distribution. This is seen as we pass from the relatively primitive and geologically early Ferns to those which Palaeontology and comparison both indicate as later in evolution.

Notwithstanding the progressive fall in the spore-production of the individual sporangium, the output per sorus may remain approximately the same. This is shown by comparison of Ferns systematically remote from one another. For instance the estimated production per sorus for *Marattia fraxinea* is 45,000, for *Polypodium aureum* 57,000, for *Angiopteris evecta* 14,500, for *Hymenophyllum dilatatum* 11,500, for *Alsophila excelsa* 3200, for *Gleichenia flabellata* 3000. The similarity in output in such pairs of cases may be set down merely to similarity of the underlying nutritive mechanism. Such comparisons become more interesting when the plants compared are of nearer affinity, as in the Hymenophyllaceae, in which the sorus has a uniform type of construction though the size and number of the sporangia, and the length of the receptacle, are variable. The results of comparison are given in the subjoined table:

Name	Sporangia per sorus	Spores per sporangium	Output per sorus
Hymenophyllum tunbridgense ...	20	420	8,400
Trichomanes reniforme	40	265	10,240
Hymenophyllum dilatatum... ...	90	128	11,500
Trichomanes radicans	140	64	8,960

It thus appears that notwithstanding the great variations of sporangial output, the result per sorus is approximately uniform for the cases quoted from a very natural family of Ferns. This table illustrates a principle of very wide application in Ferns, viz. that diminution of productivity of the individual sporangium does not necessarily lower the productivity of the sorus: for it is often compensated by the larger number of the sporangia. Moreover, their appearance is liable in gradate and mixed sori to be spread over a long period of time, with advantageous results as regards nutrition.

HETEROSPORY

The essential feature of Heterospory as seen in Ferns is that, while the spores which on germination produce antheridia and spermatozoids (microspores) remain of the approximate size and number of the homosporous spores, others which on germination produce one or more archegonia or ova, grow to a relatively large size, but are fewer in number (megaspores). The two types of spore are produced in different sporangia. Nevertheless the mode of development of those sporangia remains the same as in related homosporous plants, of which they may be held to be specialised examples. The changes involved in heterospory are thus intrasporangial, and do not appear to have brought far-reaching effects upon other parts than the contents of those sporangia.

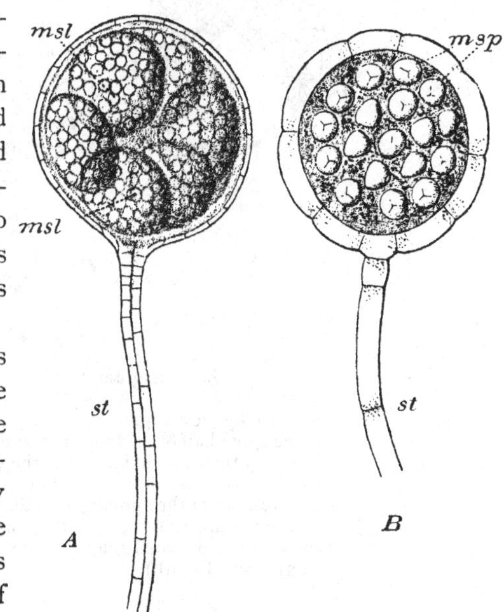

Fig. 262. Microsporangia of the Salviniaceae. *A* = *Azolla*: *msl.* = the massulae; *st.* = the stalk consisting of two rows of cells. *B* = *Salvinia*: *msp.* = microspores; *st.* = stalk. (After Strasburger, from Engler and Prantl.) (× 80.)

In the vast majority of Ferns the spores are all alike. Some few exceptions occur of which the most remarkable are those of *Platyzoma* and *Notholaena* already noted. Cultivation must decide whether or not the differences there seen are merely results of varying nutrition, or are initial steps towards a heterosporous sex-difference. This state is characteristic of the Marsiliaceae and Salviniaceae, families which differ so widely that it seems inevitable that their heterospory must have been separately acquired. The most highly specialised example is that of *Azolla*, which may serve as an example though an extreme one. Both types of sporangia here arise by the growth and segmentation usual for Leptosporangiate Ferns. The number of the spore-mother-cells in the microsporangia is 16, and accordingly the spore-number is 64, all of which are normally developed as microspores. In the mega-sporangia there are only 8 spore-mother-cells, the last division having been omitted. The potential number of megaspores is therefore 32, but only one of these is matured. A double layer of tapetum is laid down, which merges into a nucleated plasmodium, from which extraordinary developments arise in relation to the megaspore. The two types of sporangia which result are seen in Figs. 262, 263. Fig. 262, *A* represents a single ripe microsporangium

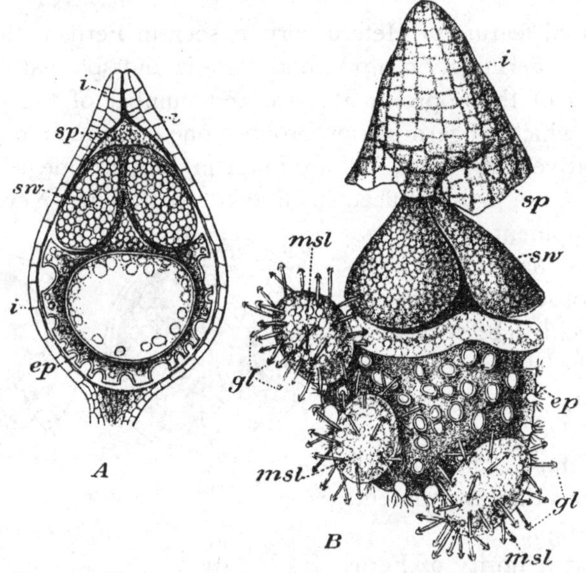

Fig. 263. Megasporangium and Megaspore of *Azolla filiculoides*, Lam. *A* shows a megaspore in longitudinal section surrounded by the complete indusium. *B* = a megaspore suspended in the upper half of the indusium: the massulae (*msl*) are attached firmly by their glochidia (*gl*) to the epispore with its many fine hairs. *i.* = indusium; *ep.* = epispore; *sw.* = swimming apparatus; *sp.* = residue of the sporangial wall, which in *A* still covers the swimming apparatus within the apex of the indusium: in *B* it has assumed a funnel-shape. (× 75.) (After Strasburger, from Engler and Prantl.)

of *Azolla* containing a number of "massulae," each of which is a clump of periplasmic material including many microspores. These massulae may be set free separately into water, and being provided with anchor-like processes, or "glochidia," they attach themselves to the rugged surface of the mega-spore (Fig. 263). The latter is of large size, and like the rest of the spores it is tetrahedral in form, and it lies in the sporangium with its apex directed upwards, that is away from the stalk. Each megasporangium is surrounded by a flask-shaped indusium, which envelopes it completely except for a narrow channel like a micropyle, towards which the apex of the megaspore points. The distal end of the sporangium is occupied by three vacuolated masses, which at maturity are stated to be filled with air, and to serve together as floats for the large megaspore in the water into which it escapes.

The whole of these remarkable structures may be referred in origin to the spore-mother-cells and to the tapetum. In the microsporangium after the 64 spores are formed they pass out towards the periphery of the nucleated plasmodium formed by disorganisation of the tapetum. This then forms a number of vacuoles, into which the spores are grouped. Presently, starting from its centre, each vacuole becomes so partitioned off by irregular septa that it forms an alveolated "massula," in which the spores are enmeshed (Fig. 264, *c*). The origin of the glochidia precedes this partitioning of the vacuole. Out-

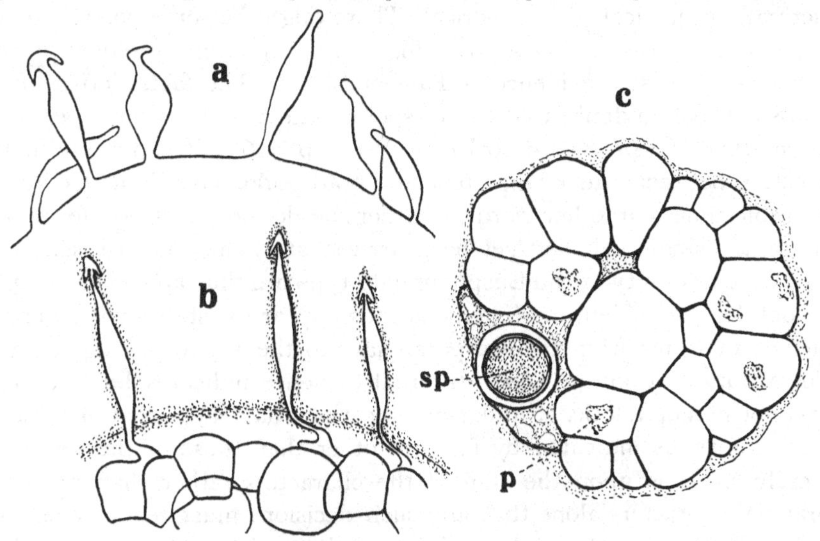

Fig. 264. *a* = vacuolar membrane of a "massula" of *Azolla* in a young state, showing glochidia as extensions of its surface; *b* = the glochidia more advanced, showing their insertion upon alveolae of the massula which have now appeared; *c* = transverse section of a mature massula. *sp.* = a microspore; *p.* = plasmatic residue shrunken in one of the alveolae. (After Hannig.)

growths appear like glove-fingers from the wall of the original vacuole, and they become barbed at the apex (Fig. 264, *a*). Later they have the appearance of being seated each on a single alveola, though this is not their actual origin (Fig. 264, *b*). Each such massula escapes at maturity separately into the water in which the plant grows, with its stiff glochidia radiating outwards.

In the megasporangium the single large megaspore is covered by the true spore-wall, outside of which is a rugged perispore, variously sculptured in different species, and sometimes provided with superficial whip-like threads. The whole may be held as the correlative of one massula of the microsporangium, which is almost wholly occupied by the megaspore. The three vacuolated bodies above it represent three other massulae, in which the remains of the 31 aborted spores and also the tapetal nuclei may still be recognised. This example will serve as illustrating the climax of development of the tapetal plasmodium. It may be seen to contain chloroplasts and starch-grains, and to undergo growth and development after the identity of its constituent cells has been lost.

From the description of details and the discussion of them given above it appears that the sporangium of the Filicales provides much material for their comparative treatment. Their sporangia range from structures relatively massive in their several parts and short-stalked, with large spore-output from the individual sporangium, and with non-specialised mechanism for their dispersal, to relatively small structures, more delicate in their several parts, often long-stalked, with small spore-output from the individual sporangium, and with highly specialised mechanism for their dispersal. The two ends of the series are in strong antithesis to one another, though the intermediate gradations between them show a range of most gentle steps, structural, mechanical, and numerical. There might be some possible doubt of the reality of these steps as marking a true evolutionary sequence were it not for the decisive evidence of Palaeontology. The facts derived from the fossils leave no doubt that the Eusporangiate was the prior type. Scott has remarked (*Fossil Botany*, 3rd Edn. Vol. i, p. 366), "It is doubtful if the distinction between Eusporangiate and Leptosporangiate Ferns existed in Palaeozoic times—in other words, whether the development of the sporangium from a single cell had yet been arrived at....The conclusions...as to the relative antiquity of the Eusporangiate type are thus amply justified by palaeontological evidence." The facts thus appear to establish the general sequence of forms from the Eusporangiate to the Leptosporangiate, as a valid evolutionary progression. Internal evidence indicates, however, that it was not monophyletic. Similar modifications have appeared in Ferns of divers affinity, as indicated by features other than those of the sporangia themselves. It is upon the sum of the characters rather than upon the sporangial characters alone that any such decisions must rest. When this is grasped, together with the biological probability that underlies the various details of soral and sporangial modification, it appears that there has been a general phyletic drift in the evolution of the Filicales: that it has involved progressions worked out in a number of distinct evolutionary lines: and that these have led from a relatively massive construction, typified in the more primitive forms by the Eusporangiate sporangium, to a relatively delicate and precise construction, typified by the later and derivative Leptosporangiate sporangium.

Nevertheless the actual spore-output as a whole does not at the same time fall in any marked degree. In some living Ferns, such as the larger Tree Ferns, it is probably as high as in any of their predecessors. The important advantage of high propagative capacity is secured in them by the compensating elaboration of the sorus. In fact the evolution of the sorus and of the sporangium have marched along complementary lines. As the individual sporangium diminished, and its output fell though its mechanism became more precise, the sorus held its own in productive capacity. The most effective feature introduced in its later stages lies in the spread of the physiological drain of spore-production over a lengthened period of time, by the introduction of the gradate and mixed conditions. The interdependence of soral and sporangial progress thus led to this final result; that of all the homosporus Pteridophytes, the Filicales, whether Eusporangiate or Leptosporangiate, include the most successful exponents of the method of survival and spread of the race that depends primarily on spore-number. The high multiplication of individual chances of success in the Ferns may be held as an offset against the risks of a very vulnerable life-history. Their prevalence at the present day shows that the balance has been in their favour.

It may be that increase in number and in spread of individuals may not have been the sole, or even the initial reason for the high output of spores seen in Ferns. It has been suggested by Svedelius (*Ber. d. D. Bot. Ges.* 1921, Bd. xxix, p. 178) that the most important feature of the reduction-division lies in the possibility of producing new combinations of chromosomes in the daughter-nuclei: and that the more numerous the spore-mother-cells the more frequently acts of reduction will be effected, and the possible number of combinations increased. In this he sees a possible explanation of the origin of the diploid "soma," or sporophyte: for thereby numerous reduction-divisions are secured. The result should be, especially where the number of chromosomes is large, a high degree of variability in the progeny, which is certainly the case in Ferns. Though this theory may offer a new aspect of the origin of a somatic sporophyte, increased variability cannot be regarded as the sole factor which has led to its extension, and finally to its high spore-output, as in Ferns. It seems impossible to ignore for them the effect of their high fecundity as in itself a positive advantage, specially fitting them for survival under competition, and for spread to new and untenanted areas.

BIBLIOGRAPHY FOR CHAPTER XIII

211. BOWER. Studies in the morphology of Spore-Producing Members. II. Dulau. London, 1896. III and IV. Phil. Trans. Vol. 189 (1897), Vol. 192 (1899).
212. BOWER. Studies in the Phylogeny of the Filicales. I—VII. Ann. of Bot. 1910—1918.
213. BOWER. Origin of a Land Flora. 1908, pp. 637—646.
214. ENGLER & PRANTL. Natürl. Pflanzenfam. i, 4, p. 79, etc. Here the literature up to 1902 is fully quoted.
215. KNY. Botanische Wandtafeln. Text, p. 418, etc. 1895.
216. CAMPBELL. Mosses and Ferns. 2nd Edn. (1905), where full references to the literature are given.
217. VON GOEBEL. Organographie. 2nd Edn. (1918), p. 1159, etc.

218. SCOTT. Studies in Fossil Botany. 3rd Edn. Vol. i, p. 366, etc. (1920).
219. RUSSOW. Vergleichende Untersuchungen. St Petersburg, 1872, p. 86, etc. Here the first comparison of spore-numbers is to be found.
220. HANNIG. Ueber die Bedeutung der Periplasmodien. Flora. Vol. 102, p. 243.
221. HANNIG. Ueber das Vorkommen der Periplasmodien bei den Filicineen. Flora. Vol. 103, p. 321 (with full references).
222. STRASBURGER. Ueber *Azolla*. Leipzig. 1875.
223. PRANTL. Die Hymenophyllaceen. Leipzig. 1875.
224. PRANTL. Die Schizaeaceen. Leipzig. 1881.
225. THOMPSON. *Stromatopteris*. Trans. Roy. Soc. Edin. Vol. lii, p. 133. 1917.
226. THOMPSON. *Platyzoma*. Trans. Roy. Soc. Edin. Vol. li, p. 631. 1916.
227. THOMPSON. Further...on *Platyzoma*. Trans. Roy. Soc. Edin. Vol. lii, p. 157. 1917.
228. YAMANOUCHI. Sporogenesis in *Nephrodium*. Bot. Gaz. xlx, p. 1. Here full references are given to cytological literature.
229. LUERSSEN. Rab. Krypt. Flora, Vol. iii. Farnpflanzen, *passim*.

CHAPTER XIV

THE GAMETOPHYTE, AND SEXUAL ORGANS

THE germination of the spore produces in every normal cycle of Ferns that alternate somatic phase which is called the Gametophyte. It might be anticipated that, as in the sporophyte, it also would provide features of value for comparison, in respect of external form and of internal structure. It has been seen how the sporophyte has definite form, with axis and appendicular organs produced in regular sequence. Also that it shows a high degree of differentiation of tissues. It has been remarked that the comparative value of the vascular system of the sporophyte depends upon its "phyletic inertia": that is, upon the fact that it shows so high a degree of conservatism in its construction, that comparisons can be effectively drawn not only between plants closely related, but also between those of wider affinity. But the constituent tissues of the gametophyte in the Filicales show a much lower scale of differentiation, while its external form, which is without any regular sequence or relation of parts, is frequently very simple, and is liable to be directly impressed in high degree by the external circumstances under which it develops. These facts detract heavily from the value of the gametophyte for comparative purposes, so far as its vegetative characters are concerned. Comparison is therefore thrown back upon the *sexual organs*, or *gametangia*, as its most reliable features. Nevertheless certain somatic characters of the gametophyte may be made use of, and these will be considered first. It should, however, be borne in mind that knowledge of the prothalli of Ferns is far less complete than that of the sporophyte-plant, and present views may be subject to drastic revision as the facts are more fully disclosed by further investigation.

THE CORDATE TYPE

The commonest type of prothallus for the Leptosporangiate Ferns is that seen in *Dryopteris* (Fig. 265). It is characterised by the presence of a single growing point, which is deeply set between two more strongly growing lateral lobes. Each of these consists of a single layer of cells, while the central region is more massive, and it is attached to the soil by numerous rhizoids. In adult prothalli an acropetal succession of archegonia borne on the lower face of this cushion leads to the apex itself: while the antheridia, which are commonly borne on the same prothallus, are scattered irregularly over the basal region of the cushion, and of the lateral lobes. This is the customary type of prothallus described in textbooks, and in them it is often the only one mentioned for Ferns. But variants from this type are common in *Dryopteris* itself, as well as in other Ferns. If the prothalli are crowded during development, and consequently half-starved, they assume an attenuated filamentous form, with frequent branching, and they bear antheridia only (compare Fig. 20, 1; also Fig. 266). A partial separation of the sexes is the result. Prothalli of *Dryopteris* grown in deficient light may persist for years, with a very attenuated form, and bear antheridia only. But if transferred to normal conditions they will then develop normally,

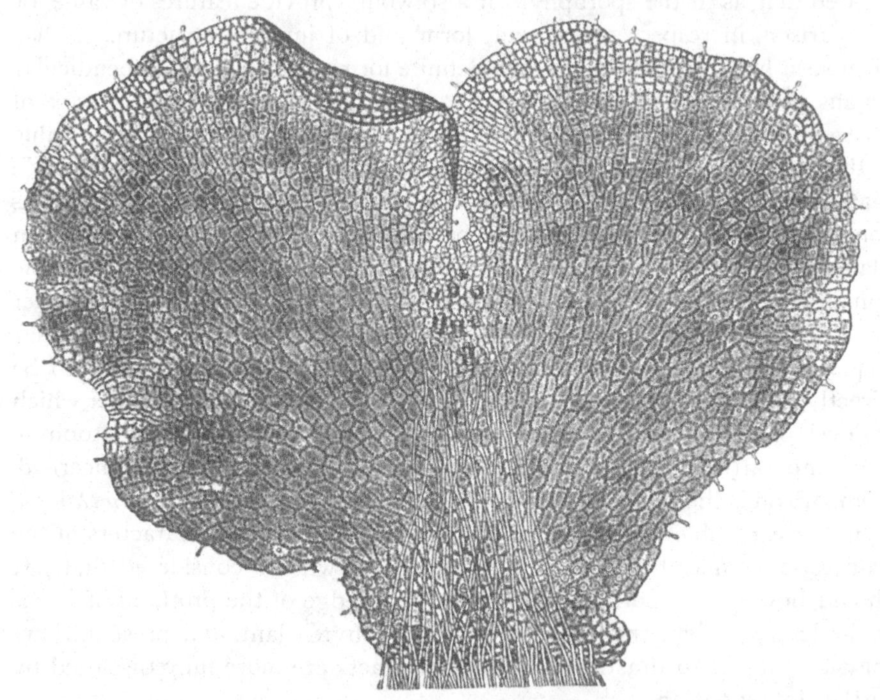

Fig. 265. Mature prothallus of *Dryopteris Filix-mas*, as seen from below, bearing antheridia among its rhizoids, and archegonia near to the apical indentation. (After Kny.)

bearing functional archegonia (Phillips, *Ann. of Bot.* 1919, p. 265). Such facts suggest a high degree of plasticity of ordinary Leptosporangiate prothalli. They appear to be more directly the creatures of circumstance than the sporophyte generation. The behaviour on germination of the spore may be held to give facts of value. Usually a filament composed of one or more cells is formed first, the length of which depends upon the conditions. Under insufficient light the filamentous stage is prolonged. On the other hand, occasionally in the Polypodiaceous Ferns, and normally in some Marattiaceae, the filamentous stage may be omitted where the light is specially favourable. It may be held that a natural juvenile form for Ferns is that of the filament. This filamentous state may be prolonged, or even its repetition induced by cultivation under suitable conditions, and especially so in prothalli that are still young. But in most cases it is only on some massive development of tissue, such as the cushion normally supplies, that the archegonia themselves are borne.

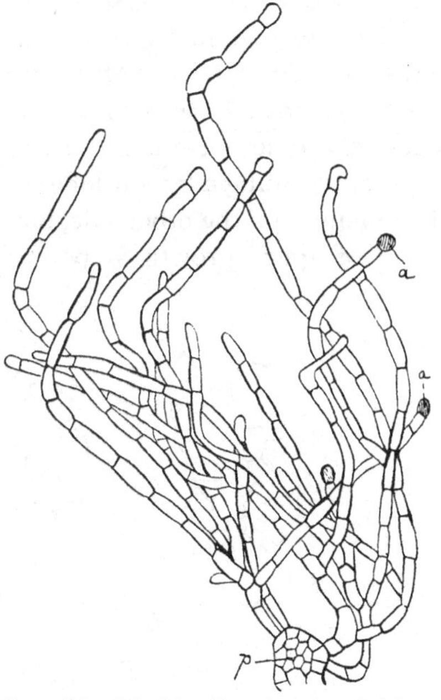

Fig. 266. *Woodsia ilvensis*, after Schlumberger. A separated fragment of a prothallus (p), from which, under feeble light, branched filaments have sprung, bearing antheridia (a).

Fertilisation of an archegonium habitually arrests the growth of the prothallus that bears it, the available material being used in the nursing of the embryo, and the prothallus is as a rule exhausted. But if fertilisation be prevented the prothallus may attain large size, and continue its growth for even five or six years. The absence of external water, without which fertilisation cannot take place, or growth under light of insufficient intensity for the formation of the archegonia, may bring this result. But even then the actual size rarely exceeds one or two inches, as seen in Ferns of robust habit such as the Marattiaceae and Osmundaceae. Von Goebel figures a prothallus of *Osmunda regalis* about two inches in length, and such prothalli occasionally branch (von Goebel, *l.c.* Fig. 918). They resemble in their size, fleshy texture, dark green colour, and method of apical growth and of branching some of the thalloid Liverworts, such as *Pellia* or *Aneura*. Differences of character may arise according to the degree of branching which prothalli show. In filamentous forms branching is common, and may appear to follow an almost regular scheme, as in *Trichomanes*, and *Schizaea* (see below, Fig. 269, *A*). In flattened types it is less common. *Gleichenia pectinata* is one of the best examples, in which adventitious shoots are formed in numbers upon the margin or from the ventral surface, and may develop into perfectly normal prothalli (Campbell, *Mosses and Ferns*, 2nd Edn. p. 367, Fig. 208). It has been suggested that since certain characteristic features, usually absent in normally fertilised prothalli, appear on those which have not been thus arrested in their development by embryo-formation, the present-day prothallus may be a structure the ancestors of which showed a greater complexity. An example in point is seen in those scales and bristles which are borne on old prothalli of *Loxsoma* and *Dicksonia* (Fig. 268). But speculation on this point cannot be fruitfully pursued until facts are available more numerous and more cogent than at present (see von Goebel, *l.c.* p. 950, Figs. 939—941).

Though commonly the apical growth arises from the distal end of the primary germinal filament this is not always so. Von Goebel cites the case of *Pteris longifolia* in which a spathula-shaped thallus is first formed (Fig. 267, *A*): presently a meristem is established in one of the wings, often with an initial cell, and forming a lateral lobe (Fig. 267, *B*). Later another lobe is formed in like manner on the other side, but smaller (*C*). This illustrates the establishment of apical growth at points on the margin other than the originally

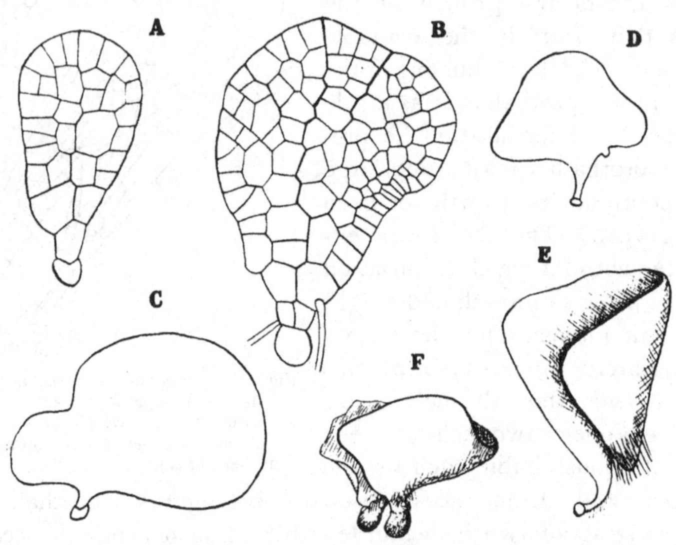

Fig. 267. *A*, *B* = young prothalli of *Pteris longifolia*, showing the establishment of a lateral growing point; *C* = older prothallus showing unequal lobes, the larger was first formed, the smaller is the later lobe; *D* = young prothallus of *Anogramme chaerophylla*, with a young archegoniophore projecting from the side; *E* = an older prothallus seen from the side; *F* = a prothallus derived from the tuberous archegoniophore of the previous year, having already formed a new archegoniophore. All after von Goebel.

distal end, a process which serves to explain the lop-sided form of many prothalli. In particular it is in this way that the peculiar arrangement for perennation in *Anogramme chaerophylla* finds its elucidation. A lateral meristem is here established on one side, and a lop-sided prothallus results (Fig. 267, *D*). Just behind the lateral meristem a tuberous archegoniophore is produced, which burrows into the soil, and becomes stored with starch, etc. (Fig. 267, *E*). In this state it can resist drying out, and in case an embryo is already established, it can advance quickly when external conditions are favourable. If there is no embryo a new prothallial flap may grow out from it, which later produces a second archegoniophore (*F*), the perennation being after the model of *Phylloglossum*, and of *Orchis*; but here it is carried out by the gametophyte. The case is still more complicated in *Anogramme leptophylla*, in which the sporophyte is annual, the species depending entirely upon the gametophyte for perennation. The tuberous archegoniophore is larger than in *A. chaerophylla*, and the prothallial surfaces that nourish it are more complicated, exposing a funnel-shaped expanse of photo-synthetic tissue (see von Goebel, *l.c.* p. 967, Fig. 963).

In addition to the rhizoids already mentioned, hairs are frequently found on prothalli, usually borne in a marginal position, but sometimes also superficially. They may be either glandular, as in *Dryopteris* (Fig. 265); or less commonly they are stiff bristles, as in *Polypodium obliquatum* (Fig. 268, *a*). Sometimes they resemble very closely the hairs borne upon the sporophyte of the same plant. This is particularly noticeable in the glandular hairs of *Notholaena* (Fig. 186, p. 199). Occasionally more massive appendages appear on prothalli as they grow old. This has been noted by von Goebel in Ferns related to the Dicksonieae and Cyatheae (*Flora*, Bd. x, 1912, p. 36). For instance in *Loxsoma* they arise on the under side of the old prothalli, right and left of the region where the archegonia are borne. They consist of cell-rows with transverse septa, but with longitudinal walls added in the basal region, and they correspond in this structure to the bristles of the sporophyte in *Loxsoma* (Fig. 268, *b*, *c*). Similar bristles have been found on prothalli of *Hemitelia capensis*, but here they ultimately become flattened as scales (ramenta). As von Goebel himself points out (*l.c.* p. 38), "*Loxsoma* stops short at the stage of bristle-formation, while this is only an intermediate stage in *Hemitelia capensis*." Thus the prothallus of *Loxsoma* shares in its appendages the characteristic feature of the sporophyte of Dicksonioid Ferns, viz. that hairs or bristles, but no

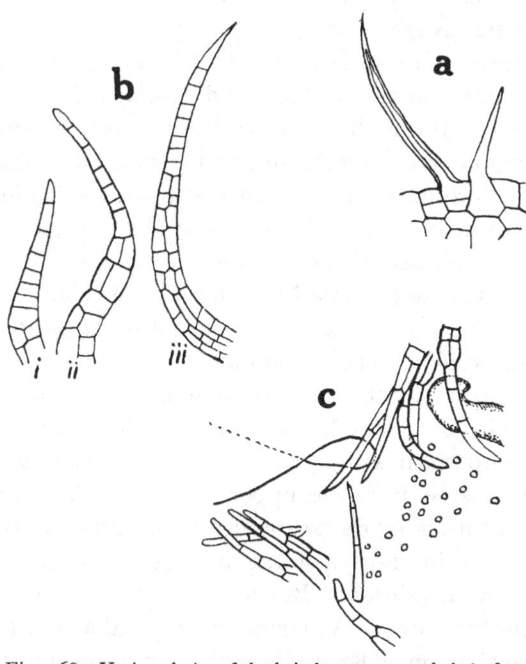

Fig. 268. Various hairs of the bristle-type. *a*, a bristle from the margin of a prothallus of *Polypodium obliquatum*, after von Goebel; *b*, hairs of *Loxsoma*, i, ii, on sporeling; iii, on an old plant; *c*, apical region of a prothallus of *Loxsoma* with bristles: showing also archegonia and the young sporophyte. After von Goebel.

flattened scales, are present: the prothalli of *Hemitelia* share the feature common for the Cyatheoids, viz. the presence of flattened scales. These facts again illustrate the parallelism between the two generations in respect of their appendages already noted for *Notholaena*. Nor is it surprising that the similarity should exist, for both are mere stages in the completed life-cycle.

The arrangement of the gametangia in Ferns appears to be in high degree fortuitous, and dependent on circumstances. In the case of green prothalli light of sufficient intensity is necessary for the formation of both antheridia and archegonia, but the former require less intensity than the latter (Nagai, *Flora*, 1914, p. 326). The position of the antheridia on the *Nephrodium*-type of prothallus may be marginal or superficial: they are usually on the surface, and directed downwards. They appear as a rule before the archegonia, and consequently are most numerous in the basal region of the prothallus, and later their formation may cease altogether on the prothalli where archegonia are being formed. The consequence is a partial separation of the sexes, which is accentuated by the fact that starved prothalli usually bear antheridia only.

The archegonia being borne as a rule on the lower surface of the massive cushion, and in acropetal succession, are more regularly arranged than the antheridia. Their position appears, however, to be determined by the incidence of light: for if both sides of a prothallus are about equally lighted archegonia may be formed on both surfaces. This result has its interest for comparison with what is seen in *Polystichum anomalum*, in which the sori may arise from the upper instead of the lower surface of the leaf. In both a transfer of the stimulus of formation of the part is held to have taken place: in respect of the archegonia the transfer can be correlated with external conditions. But it does not seem probable that the same holds with the sori, for Sir Wm. Hooker states that the Fern retained its anomaly under cultivation at Kew, where the conditions must have been very different from those of the uplands of Ceylon (*Species Filicum*, iv, p. 27). In this instance, as in its characters at large, the sporophyte seems less directly susceptible to the impress of circumstances than the gametophyte (see p. 217).

The Filamentous Type

There is another type of prothallus characterised by a more persistent filamentous structure than that seen as a consequence of growth under diminished light in the instances above described. It is found in the Hymenophyllaceae, and in *Schizaea*. The prothallus of *S. pusilla* is composed entirely of branched filaments: the same appears to hold also for *S. bifida*: flattened expansions have not been recorded in them (Britton and Taylor, *Torrey Bot. Club*, Vol. xxviii, 1901), nor in *S. rupestris* (von

Goebel, *l.c.* p. 957). Nevertheless the allied genera *Anemia*, *Mohria*, and *Lygodium* possess essentially typical flattened prothalli. The filaments consist of rows of chlorophyll-containing cells, having the general habit of a *Cladophora*. Some of these run horizontally along the surface of the substratum, and are attached by brown rhizoids formed in regular relation to distended spherical cells (Fig. 269, *A* (*S*)). The latter are inhabited by fungal filaments, which extend onwards into the rhizoids, the whole structure being clearly a mycorhizic coalition. From their position it seems probable that the spherical cells represent branches of the filamentous thallus specialised for this method of nutrition. Other branches roughly alternating with them may grow upwards from the substratum, branching repeatedly, and func-

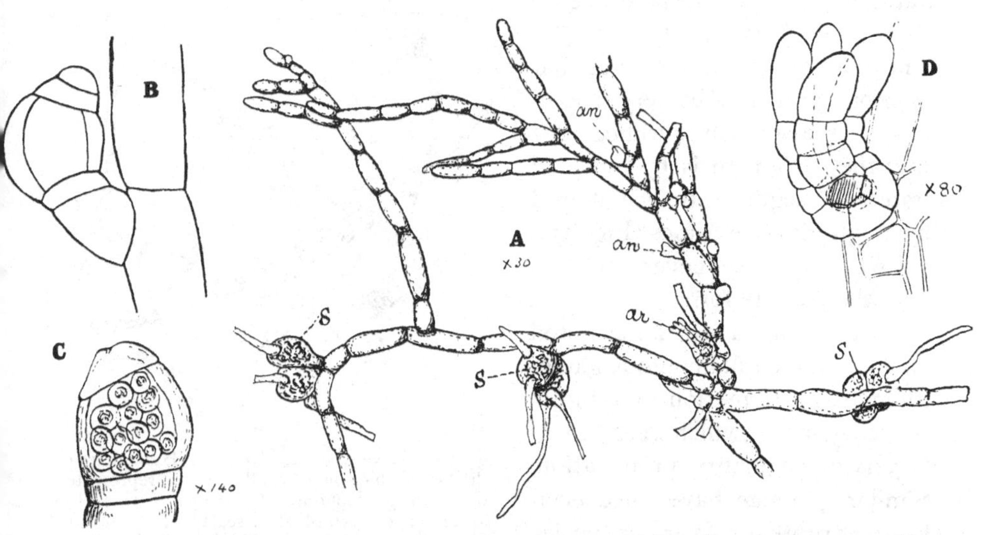

Fig. 269. *A* = prothallus of *Schizaea pusilla*, bearing spherical cells (*S*) with endophytic fungal filaments. *ar* = archegonia; *an* = antheridia. *B* = segmentation of an antheridium of *Schizaea rupestris*: the lid is not always divided. *C*, *D* = mature antheridium and archegonium of *Schizaea pusilla*. *A*, *C*, *D*, after Britton and Taylor. *B*, after von Goebel.

tionally photo-synthetic. They bear also the sexual organs, both of which may be present on the same thallus. They occur singly. The antheridia (*an*) correspond in position to branches of the filament, springing like them from the upper end of the cell that bears them. Each antheridium represents the ending of a dwarf-branch. The isolated archegonia occupy a similar position, but are seated as a rule on small masses of cells resulting from longitudinal division of the cells at the base of a filament (*D*). As von Goebel remarks, "In such free archegonia we see undoubtedly the simplest state in any of the Pteridophyta." The archegoniophore is here in its simplest form. It is a tenable position that *Schizaea* illustrates a really primitive state of the gametophyte.

A similarly filamentous and profusely branched gametophyte has been observed in a number of species of *Trichomanes* (Fig. 270). It is composed of oblong chlorophyll-containing cells, and the branches arise from the distal end of each, while dark brown rhizoids attach it to the substratum. Not uncommonly fungal hyphae are associated with it externally, but no distinctive organs like the spherical cells of *Schizaea* have been noted. In some species the filaments may widen out into broad flattened expanses one layer of cells thick, as in *T. alatum* (Fig. 271, *C*). The relation of these to the filamentous part may be very irregular, as is also their outline. The way they originate is by repeated transverse segmentations followed by longitudinal divisions (Fig. 271, *A*, *B*). But they never seem to take the cordate form.

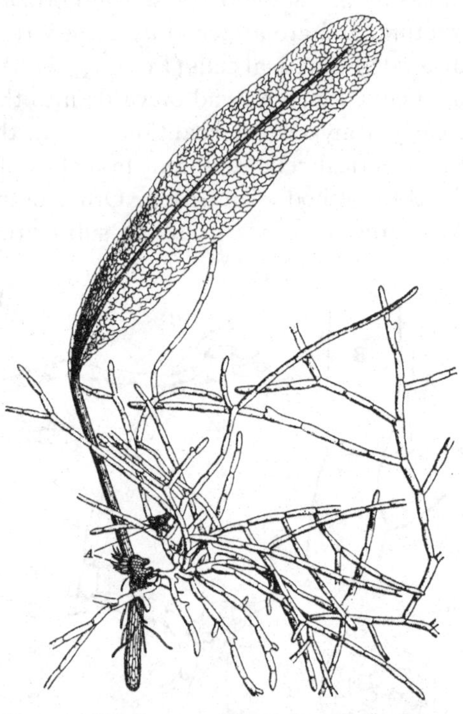

Fig. 270. *Trichomanes rigidum*, Sw. Habit of a prothallus, of which only a small portion is represented, with archegoniophores (*A*), on one of which (the lower) an embryo-plant is seated. (× about 50.) (After von Goebel, from Engler and Prantl.)

Many years ago Cramer noted on an unknown filamentous gametophyte the formation of spindle-shaped gemmae, which were clearly organs of vegetative propagation. Similar gemmae have since been found on prothalli of various species of *Trichomanes* (*T. alatum*, *venosum*, and *pyxidiferum*, etc.). They are borne distally, sometimes upon simple filaments but more frequently at the ends of the flattened expanses (Fig. 271, *C*, *D*, *E*). Distal cells grow out forming sterigmata, and their ends become distended transversely, taking a spindle-shape, and undergoing segmentations. These bodies consisting of about six cells are stored with nutriment, and are easily detached from the brittle neck of the flask-shaped sterigmata, as gemmae. Their germination reproduces the filamentous prothallus (Fig. 271, *D*, *E*, *F*).

The sexual organs of *Trichomanes* are borne very much as in *Schizaea*, the antheridia corresponding in position to dwarf branches (Fig. 272, *a*—*d*). The archegoniophores are, however, more complicated, and of larger size, bearing a number of archegonia, their ventral region being immersed in a mass of parenchyma (Fig. 272, *e*, *f*, *g*).

The allied genus *Hymenophyllum* has branched prothalli consisting of strap-shaped expanses one layer of cells in thickness, excepting at certain marginal regions where they are thickened to several layers, thus forming the archegoniophores. An extreme development after the manner seen in Fig. 272, *a*, *b*, with marginal archegoniophores would account for what is

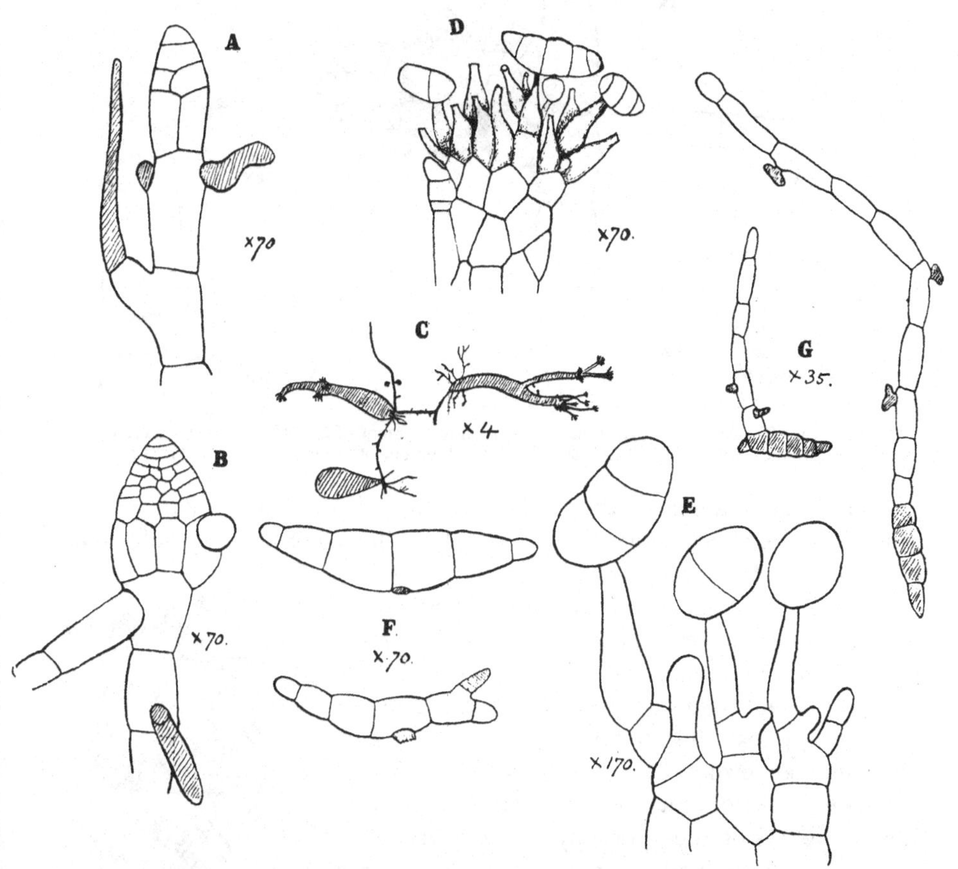

Fig. 271. *A—F*, *Trichomanes alatum*. *A*, *B* show the formation of a flattened prothallus from a filament. *C*, relation of filaments to flattened expansions which bear sterigmata distally. *D*, sterigmata bearing gemmae. *E*, very young gemmae. *F*, mature gemmae detached, the lower beginning to germinate. *G*, result of germination of gemmae of *Trichomanes Kaulfussii*.

actually seen. The antheridia and archegonia are, however, borne together, and they are directed downwards (Fig. 273).

Von Goebel places in relation to these, which may well be held to be flattened derivatives of a filamentous type of gametophyte, the irregularly lobed prothalli of *Vittaria* (*l.c.* p. 958). They also consist of a single layer of cells with marginal growth. At certain points that growth ceases, but the rest of the margin continuing to grow produces very irregular lobes, suggesting

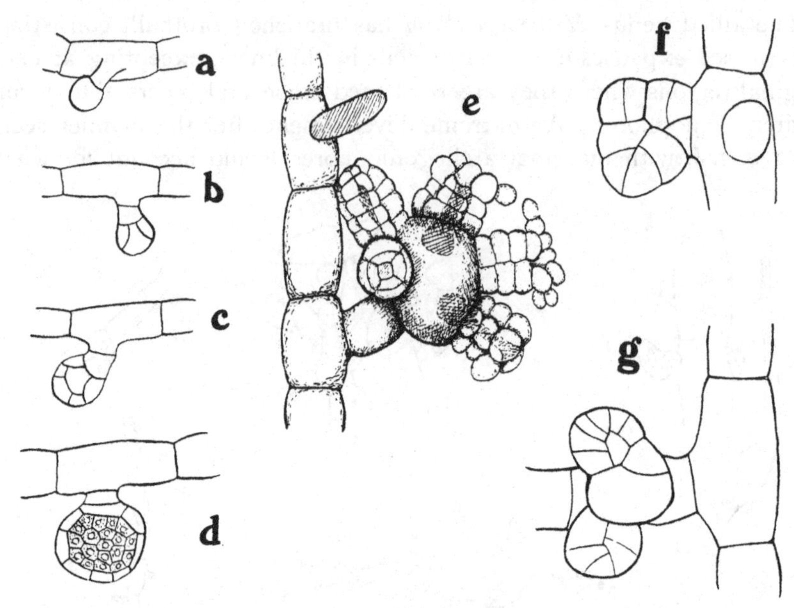

Fig. 272. *Trichomanes pyxidiferum*. *a—d*=development of an antheridium; *e*=archegoniophore with five archegonia; *f*, *g*=development of archegoniophore. (× 150.)

as regards the manner of their origin a comparison with the flattened protonema of *Sphagnum*. These epiphytic prothalli also produce spindle-shaped gemmae, which are attached by one end to their sterigmata. The groups of archegonia originate from the marginal meristem, but are later separated from it by regions which have passed over into permanent tissue. (See von Goebel, *Organographie*, ii, p. 957, Fig. 951.)

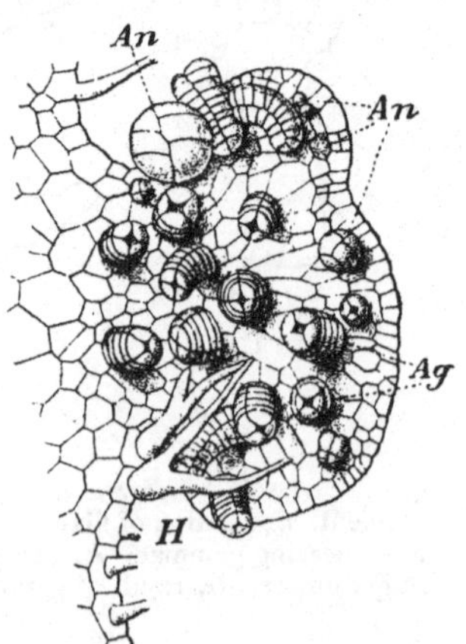

Fig. 273. *Hymenophyllum dilatatum*, Sw. Lobe of a prothallus with a group of archegonia. *Ag*=the individual archegonia; *An* = antheridia; *H* = hairs. (After von Goebel, from Engler and Prantl.)

These somewhat similar forms of prothallus are seen in Ferns not closely allied as regards the characters of the sporophyte, and therefore they cannot be held as phyletically of near relation to one another. Nevertheless they may perhaps illustrate steps of progression from a simple state, reminiscent of certain filamentous Algae. It is easy to figure how from the simple state shown by *Schizaea* a like state with occasional flattened expansions, as seen in *Trichomanes*, might lead to a generally flattened form as seen in *Hymenophyllum* or *Vittaria*. A strict localisation of apical growth at the distal end, instead of its being spread generally along the margin as in *Vittaria*, would then give the ordinary cordate type. Whether this is anything more than a fanciful comparison must for the present remain an open question. Against it is to be placed the fact that all the Ferns quoted are epiphytic except *Schizaea*, which is a ground dweller, growing among *Sphagnum*, etc. It is a possibility that the circumstances of epiphytic growth may favour, in Ferns not related closely to one another, a similar form of prothallus, together with vegetative propagation by gemmae.

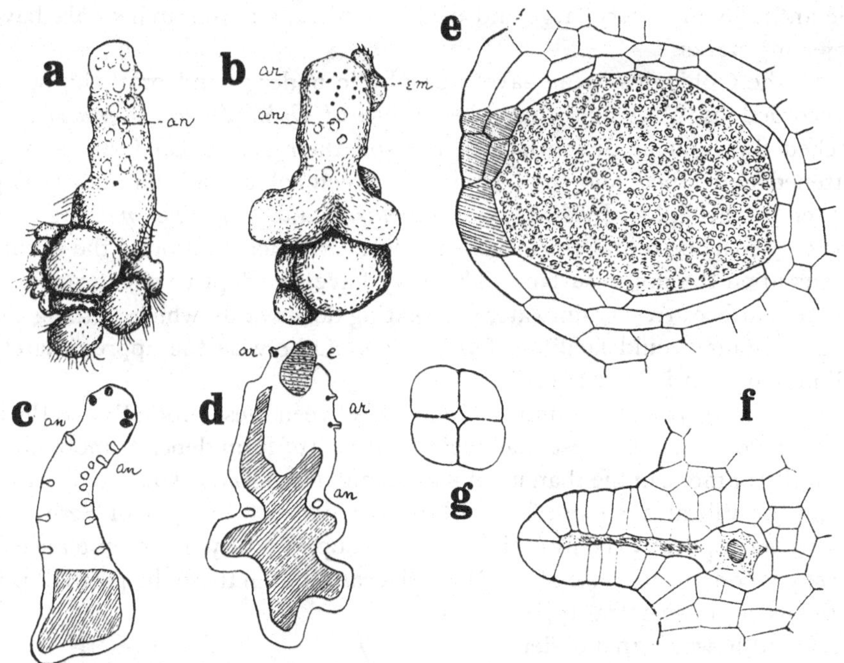

Fig. 274. Prothalli and sexual organs of *Helminthostachys zeylanica*, after Lang. *a*, *b*=prothalli seen from without; *c*, *d*, in section, with mycorhizic regions shaded; *e*=antheridium in longitudinal section; *f*, *g*=archegonia. (*a*, *b*, *c*, *d*, × 7; *e*, *f*, × 200.) In *a*, *b*, *c*, *d*, *an*=antheridium; *ar*=archegonium; *e* or *em*=embryo.

THE MYCORHIZIC TYPE

A third type of prothallus, differing widely from those described, is seen in the Ophioglossaceae. It is as a rule colourless and develops beneath the level of the soil in relation to mycorhizic nutrition. The wholly saprophytic prothallus of *Helminthostachys* investigated by Lang (*Ann. of Bot.* xvi, 1902, p. 32) provides an example of these massive underground types, which offer interesting analogies with those of the Lycopodiaceae and Psilotaceae. Each prothallus consists of a lobed basal portion, or vegetative region, attached to the soil by rhizoids. From this the cylindrical sexual region arises later, usually growing vertically upwards, with a four-sided initial cell at the apex (Fig. 274, *a*, *b*). But since the prothalli occur at a depth of about two inches they do not reach the surface of the soil. The vegetative region is occupied by a symbiotic fungus (Fig. 274, parts shaded in *c* and *d*), but it dies out about the time that the sexual region begins to elongate, the growth being completed at the expense of the starch accumulated internally as a consequence of its activity. There is a partial sexual differentiation; for instance, Fig. 274, *a* and *c*, are exclusively male. But both antheridia and archegonia may be present on the same prothallus, the former appearing first, as in Fig. 274, *b*, *d*. The antheridia are very large and deeply sunken, but the archegonia have projecting necks.

All the Ophioglossaceae have saprophytic underground prothalli, but it is recorded by Mettenius that when those of *Ophioglossum pedunculosum* reached the light at the surface of the soil their cylindrical form became flattened and lobed, and they assumed a green colour, but did not develop further in the light. In *Botrychium Lunaria* and *virginianum* there is a tendency to a horizontal, thick, flattened form with localisation of the sexual organs upon the upper surface. This may have an adaptive significance, so that the surface arrests rain-water percolating downwards, which, bathing the sexual organs, would facilitate fertilisation. Otherwise the approximately cylindrical form is maintained.

There are certain important differences between these prothalli and those of Lycopods which suggest that their similarity of form depends more upon similarity of mode of life than upon real affinity (Lang, *l.c.* p. 50). Comparison within the Filicales is more fruitful, however different this type of Fern-prothallus may appear to be from the rest. It is supported by the resemblance of the sexual organs to those of the Marattiaceae. But in these the prothallus is of the usual dorsiventral type. It is found, however, that under special conditions of cultivation the flattened form may be lost even in Leptosporangiate Ferns, the prothallus continuing its growth as a cylindrical process, with structure, apical meristem, and sexual organs like those of the sexual region of the Ophioglossaceae (Fig. 275). This suggests that if it were possible to cultivate ordinary Fern-prothalli, of a fleshy type like those of the Marattiaceae, below the surface of the soil the cylindrical form might be attained, as in the Ophioglossaceae.

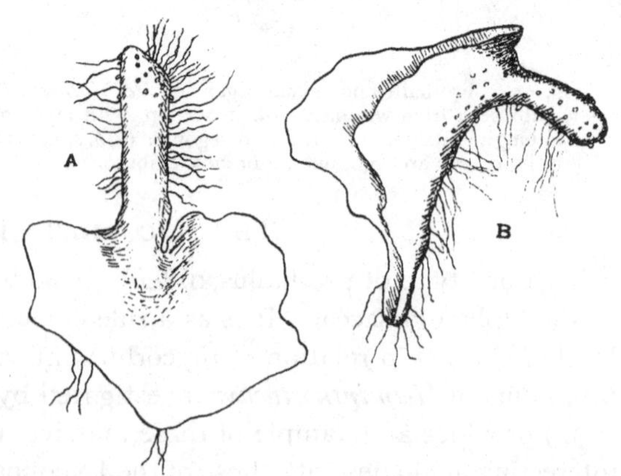

Fig. 275. Prothalli of *Scolopendrium*, after Lang. They have grown out unfertilised into "cylindrical processes," arising (*A*) from the apex, (*B*) from the under surface of the prothallus, and bearing sexual organs all round. (× 6.)

For the latter the adoption of the mycorhizic habit has made such life possible. These considerations indicate that the mycorhizic type of prothallus is not so far removed from that of other Filicales as it would appear to be at first sight. (See Lang, *Phil. Trans.* Vol. 190, 1898, p. 187.)

THE GAMETANGIA

The male and female gametangia resemble the sporangia in the fact that the gametes like the spores are borne internally, and are covered over till maturity by a wall composed of one or more layers of cells. Their structure may be held to be a general consequence of the fact that the plants which bear them are normally exposed to the air during development, and since the gametes are primordial cells they must be protected. Like the sporangia, the gametangia may in some cases be deeply sunk in the parent tissue, in others they may project in more or less marked degree. It will be seen later that the analogies between gametangia and sporangia may be followed into still further detail; but first the gametangia themselves must be described and compared.

The antheridia and archegonia of *Matteuccia Struthiopteris*, which are essentially similar to those of *Dryopteris*, will serve as average examples for ordinary Leptosporangiate Ferns (see Chapter I, Fig. 20). The antheridium originates as a hemispherical protrusion from a single prothallial cell, which is then partitioned off by a transverse or somewhat oblique wall. It contains cytoplasm with chlorophyll, and a large central nucleus. Its first segmentation is by a funnel-shaped wall, the base of which strikes the basal wall of the antheridium (Fig. 276, *A*); sometimes, however, this wall does not reach down to the basal wall, but taking a more or less convex course it cuts off a discoid stalk from the distal head (*B*). The latter then divides by an hemispherical wall nearly concentric with the outer wall. This cuts off an external cell, forming the antheridial wall into which all the chloroplasts pass, from an inner colourless cell. The latter is the mother-cell of the spermatozoids. The cell of the antheridial wall undergoes a further ring-like division, cutting off a cap-cell at the top. The young antheridium then consists of four cells, viz. the central mother-cell, two ring-shaped cells, and a discoid terminal cell. The cells of the wall have meagre cell-contents except the chloroplasts, which

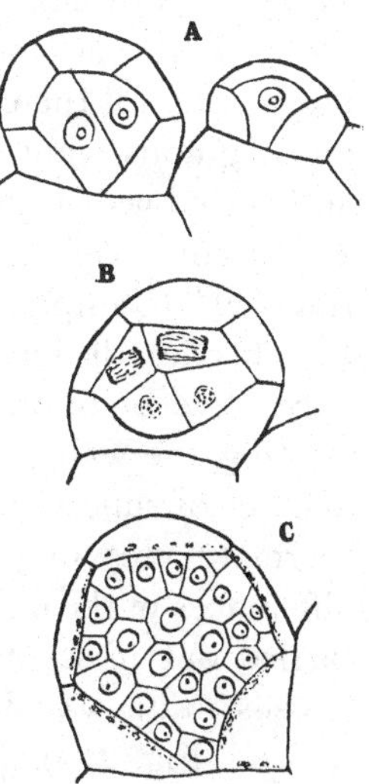

Fig. 276. Antheridium of *Matteuccia Struthiopteris*, after Campbell. *A*, *B* = young antheridia in vertical section, to show segmentation. *C* = adult antheridium with 21 spermatocytes in section. (× 200.)

approach the inner cell-wall. The mother-cell of the spermatozoids has more dense granular cytoplasm. It divides at first regularly, by a vertical wall followed by transverse walls, and again by walls at right angles to these. Bi-partitions then follow in all the cells "until the number may be a hundred or more, but the number is usually much less, about 32 being the commonest" (Campbell, *l.c.* p. 316). They form a mass of polyhedral sperm-cells, or spermatocytes, with dense granular cytoplasm, and large nuclei. Thus the antheridia are not strictly standardised as regards number of spermatocytes. Each nucleus is then transformed into the body of a spermatozoid, of which, however, the forward end is partly of cytoplasmic origin. The fully developed spermatozoid shows about three complete coils in a tapering spiral (Fig. 20, p. 17). The very numerous cilia are attached at a point a short distance back from the apex, and are found to arise from an elongated "blepharoplast." As the spermatozoids develop the sperm-cells separate, their walls appear thick and silvery, and give the staining reactions of mucilage, but each inner wall remains intact so that on ejection each spermatozoid is still enclosed in a delicate membrane, which swells in water, and finally dissolves. A vesicle representing the remains of the cytoplasm is conspicuous at first, but is soon lost from the escaping spermatozoid. Prior to rupture of the antheridium the cells of the wall are compressed by the crowded sperm-cells; but after it is burst they become so distended that they nearly fill the cavity. This behaviour, together with the swelling of the mucilaginous walls of the sperm-cells, is the cause of the rupture of the antheridial wall, which in this case is carried out "either by a central rupture of the cover-cell or less commonly by a separation of this from the upper ring-cell" (Campbell, *l.c.* p. 318).

This description will apply generally for Leptosporangiate Ferns. There may, however, be differences in the number and position of the segmentations forming the wall, but the wall always remains a single layer of cells. The number of subdivisions of the internal mother-cell, and consequently of the spermatocytes, may also vary. It has been stated also that there may be differences in the detail of dehiscence. But von Goebel holds that it is always carried out in Leptosporangiate Ferns by extrusion of the distal cap-cell or cells, in the way described by Schlumberger for the antheridia of *Woodsia ilvensis* (Fig. 277). The mature antheridium (*a*) being tense by swelling of the mucilage within, the cuticle bursts above the cap-cell (*b*). The latter then separates as a whole, and is extruded, the adjoining cells of the wall becoming distended inwards (*b*—*d*). The contents of the antheridium are then thrust out by their pressure. Examination of an empty antheridium from above often shows apparently an irregular hole (*e*), and it used to be held that in certain Ferns the cap-cell itself ruptured. But in point of fact it is extruded bodily, and the ragged appearance is due to the outlines of the intruding cells of the wall projecting inwards further than the ring of attachment of the cap-cell (*r*, Fig. 277, *d*, *e*).

The antheridia of the Eusporangiate Ferns differ widely from those of the Leptosporangiate type. They are more massive, and are deeply sunk in the tissue of the prothallus. The very numerous spermatocytes which they contain are covered in by a wall of which certain of the cells divide periclinally: but one, or sometimes more, may remain undivided, and act as an opercular cell (or cells). (Figs. 278, 283, *A*.) A layer of slab-like cells cut off from the adjoining tissue of the prothallus sometimes surrounds the mass of spermatocytes, and is nutritive in function like a tapetum. Later these cells assist in the extrusion of the spermatocytes by swelling into the antheridial cavity.

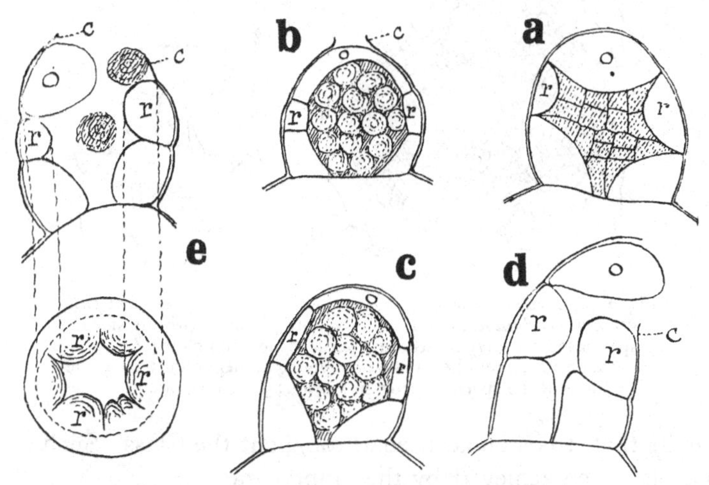

Fig. 277. Antheridia of *Woodsia ilvensis*, after Schlumberger. *a* = antheridium with spermatocytes; *b* = ripe antheridium with cuticle ruptured; *c*, *d* = same antheridium before and after dehiscence; *e* = lateral and vertical aspects of ruptured antheridium. *c* = cuticle; *o* = opercular cell; *r* = ring-cells.

The archegonium of Ferns has its ventral region sunk in the tissue of the prothallus. Sometimes even the neck may also be so far embedded that it projects only slightly beyond the surface, as in *Marattia Douglasii*, or *Ophioglossum pendulum* (Fig. 278, *c*, *d*, *f*). But usually the neck projects as a cylindrical chimney composed of four rows of cells, the number of cells in each row being variable in different Ferns. The central series as seen in *Dryopteris* holds with remarkable constancy for Ferns at large (Fig. 279). It consists of the canal-cell with two nuclei, the ventral-canal-cell, and the ovum. There may be some slight difference whether or not the division of the canal-cell into two is completed, or the division be confined to the nucleus only. Otherwise the central series of cells of the archegonium in Ferns seems to have settled down to a structure which so fully meets the

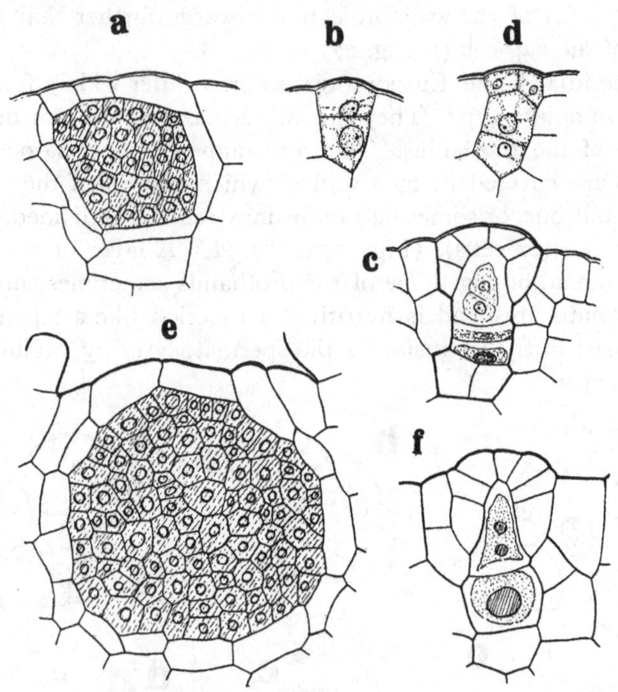

Fig. 278. *a—d* = *Marattia Douglasii*, after Campbell. *e—f* = *Ophioglossum pendulum*, after Lang. *a* = antheridium, with divisions of spermatocytes (32) in section, perhaps not complete; *b* = young antheridium; *c* = archegonium; *d* = young archegonium; *e* = antheridium with 88 spermatocytes in section; *f* = archegonium.

requirements that it is standardised throughout the Class. Such standardisation has also been achieved by the embryo-sac of Angiosperms, which takes an equally essential part in propagation, and shows uniformity of its contents comparable with the uniformity of the contents of the archegonium in Ferns. The only variable characters of the archegonium available for comparison will therefore be the number of cells of the neck, and the degree of its projection beyond the general level of the prothallial tissue. A comparison of the sunken archegonia of *Marattia* and *Ophioglossum* (Fig. 278, *c*, *f*) with that of *Matteuccia* (Fig. 279) will show the range of variability in this respect. Naturally the development of the archegonium is also uniform. The whole archegonium originates like the antheridium from a superficial mother-cell (Fig. 278, *d*). This first

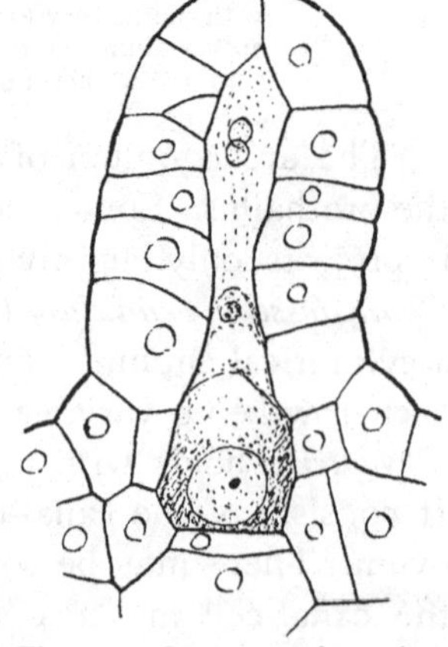

Fig. 279. Mature archegonium of *Matteuccia Struthiopteris*, after Campbell. (× 250.)

divides by a periclinal wall, the inner dividing again, so that three superimposed cells result. Of these the outermost forms the neck, dividing by crossed cleavages into four cells: each of these by subsequent divisions, which vary in number in different Ferns, gives rise to one row of cells of the neck. The innermost or basal cell takes no further direct part, and may perhaps represent a sterilised part at the base of the central series (see below). The middle cell of the series divides again periclinally to form the canal-cell, and the central cell. These each divide again periclinally: in the former the division is often incomplete, and confined to the nucleus: in the central cell the last division gives origin to the ventral-canal-cell and to the ovum (Fig. 278, *c*).

Comparison of Gametangia with Sporangia

As seen in relatively advanced types of Ferns such as *Dryopteris* the antheridia and archegonia appear to differ clearly from one another, both in form and in their contents (Figs. 20, 21). But in certain of those Ferns which on the sum of their characters may be held as relatively primitive their form is less distinct, though still the contents differ. For instance in *Marattia*, where both are sunk in the tissue of the prothallus, not only does each spring from a single deeply sunk superficial cell of cubical form, and maintain that form to maturity, but the segmentation of the parent cell by a periclinal wall gives rise to an inner and an outer cell (Fig. 278, *b*, *d*). The latter produces on the one hand the protective antheridial wall, and on the other the neck of the archegonium: the former is the parent cell of the gametes, giving rise respectively to the spermatocytes, or to the ovum with its attendant canal-cell and ventral-canal-cell, together with the basal cell. Such a degree of similarity raises the question of homology of the gametangia, in the sense that the two types may have originated from a single type of gametangium. Organs intermediate in character between antheridia and archegonia have not been recorded for the Pteridophyta, and abnormalities in their archegonia are rare. But in the Bryophyta examples have been described and figured (von Goebel, *Organographie*, i, p. 129, footnote). The most notable are those archegonia of *Mnium* which have been seen to contain not only an egg and ventral canal-cell, but sometimes several eggs, while others show masses of spermatocytes, and these may be both above and below the ovum (Fig. 280). Such a structure appears to be in fact a bisexual organ, having both antheridial and archegonial characters. On these facts the view is based that the antheridium and archegonium of Bryophytes are probably homologous organs, derived from a primitively uniform gametangium. In the case of the archegonium extensive sterilisation of the reproductive cells has led finally to the survival of only one functionally perfect cell, viz. the ovum. The essential similarity of the archegonia of the Bryophyta to those of the Pteridophyta, and the close analogies which the antheridia and archegonia of the latter show in their development and structure, justify a similar conclusion for them also.

The analogy of the sexual differentiation of gametangia thus contemplated with that of the differentiation of megasporangia and microsporangia from a primitively homosporous sporangium, so fully illustrated in the Pteridophytes, cannot be mistaken. These two progressions may be held to involve the two most important evolutionary steps which have followed upon the initial differentiation of sex in the gametes themselves. Naturally in Descent the differentiation of the gametangia must have preceded that in the sporangia. For purposes of comparison the microsporangium is regularly found to be the more conservative in its characters, continuing to represent the primitive state. Thus it offers features of greater value for comparison *downwards*: but the megasporangium is more prone to initiate new features in accordance with the promotion of the megaspore, consequently its special value is in comparisons *upwards* in Descent. Similarly with the gametangia: the antheridium will reflect more nearly the archaic type of gametangium, the archegonium being a specialised advance upon it. Accordingly the former will present the greater interest in relation to questions of phyletic origin, the latter will be suggestive rather of later and derivative states. Prof. von Goebel has suggested the propriety of establishing a conformable terminology for both sporangia and gametangia. If the former are segregated by sexual differentiation as *mega-sporangia* and *micro-sporangia*, so the latter may be distinguished as *mega-gametangia* and *micro-gametangia*.

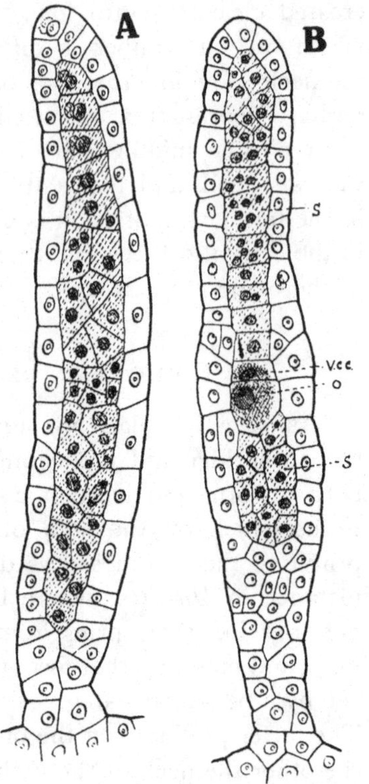

Fig. 280. Bisexual archegonia of *Mnium cuspidatum*. *s* = spermatocytes; *v.c.c.* = ventral-canal-cell; *o* = ovum. (After Holferty.) (× 300.)

It is not only in point of sexual differentiation that gametangia may be compared with sporangia. The comparisons can also be drawn as regards the relative size of the gametangia and sporangia in different types of plants, and the Filicales offer good examples of this. Still it is necessary to remember that these organs are essentially distinct in origin and in nature; such similarity as they show may be held as an indication of the general organisation of the plants that bear them, modified it may be in some measure by the biological conditions under which they live. The comparison is best drawn between antheridia and homosporous sporangia, both of which, as already shown, are conservative in their character. A rough but by no means exact numerical estimate may be made by counting the spermatocytes traversed in the median vertical section of an antheridium, and comparing it with the number of spore-mother-cells in a similar section of the corresponding sporangia (Fig. 281). Much depends upon the section being really median, a tangential section being liable to give too low a figure. Other sources of inaccuracy of the comparison may arise from the shape, and the degree of standardisation of the parts. In most of the advanced Ferns the sporangium is strictly

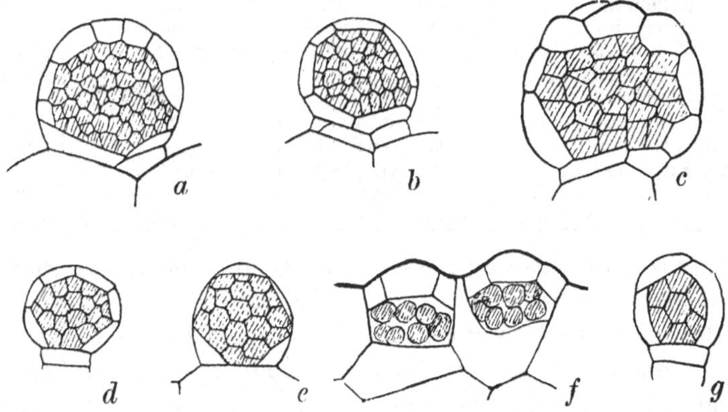

Fig. 281. Antheridia of various Ferns, to show differences in number of spermatocytes. *a* = *Gleichenia pectinata*, after Campbell (231, Fig. 209, *B*), 46 cells in section; *b* = *Osmunda cinnamomea*, after Campbell (231, Fig. 195, *D*), 32 cells; *c* = *Hymenophyllum*, after Campbell (231, Fig. 217), 26 cells; *d* = *Trichomanes venosum*, after von Goebel (*Organographie*, Fig. 912, *v*), 18 cells; *e* = *Nephrodium Filix-mas*, after Kny (Wandtafeln), 19 cells; *f* = *Ceratopteris thalictroides*, after Kny (*Parkeriaceen*, Taf. XVIII, Fig. 11), 7 cells; *g* = *Schizaea pusilla*, after Britton and Taylor (Pl. 3, Fig. 52), 7 cells.

standardised in size and shape, and the spore-output can therefore be fairly arrived at from the number of spore-mother-cells traversed in median section. In the Eusporangiate Ferns they are more apt to be variable even in the same species or individual, as is seen in *Danaea* (*Land Flora*, Fig. 286), or *Marattia* (*l.c.* Fig. 285). The antheridia may also be variable in the species or even in the individual, as is clearly shown by Campbell's drawings of *Gleichenia laevigata* (*Ann. Jard. Bot. Buit.* VIII, Pl. XI), which give spermatocyte-counts as divergent as 95, 76, 35. Notwithstanding such sources of error the following table shows a substantial parallelism of the figures for the spermatocytes and the spore-mother-cells as seen in vertical sections of the antheridia and the sporangia of the Ferns named. In the fourth column the spore-output is given, as estimated in the larger sporangia from sections, or based in the smaller upon actual spore-counts. When the parallel results shown in these columns are compared with the general facts of structure of the Ferns in question, and especially with the details of their apical meristems as set forth in Chapter VI, they will be seen to give consistent support to the general progression there traced.

Name	Number of spermatocytes in vertical section of the antheridium	Number of spore-mother-cells in vertical section of the sporangium	Estimated spore-output
Ophioglossum pendulum...	88 (Lang)	500	15000
Christensenia	74, 60 (Campbell)	74	7850
Angiopteris	55 (Campbell)	67	1450
Gleichenia flabellata ...	—	66	512–1024
" laevigata ...	95, 76, 35 (Campbell)	—	—
" dichotoma ...	42, 30 (Campbell)	26, 25	256 or more
Osmunda cinnamomea ...	32 (Campbell)	—	128
Hymenophyllum sp. ...	25 (Campbell)	—	128
Dennstaedtia punctilobula	24 (Conard)	8	64
Trichomanes venosum ...	18 (von Goebel)	—	64 or less
Dryopteris Filix-mas ...	16 (Kny)	8 or less	48
Ceratopteris	7 (Kny)	–	32 or 16

Such comparisons may be pursued further into the early segmentations of the young antheridia. It has been found possible to trace a sequence of steps in the initial segmentation of the sporangia of Ferns, leading from those of the massive Eusporangiatae to those of the delicate Leptosporangiatae (Fig. 238, Chapter XIII). The necessary data are not yet to hand for an exhaustive comparison of this nature in the case of the antheridia. But three diagrams based upon drawings from reliable sources may be placed in sequence, as a step towards a more complete demonstration (Fig. 282, *a*, *b*, *c*). Their correspondence with the diagrams *a*, *d*, and *g*, *e*, of Fig. 238, Chapter XIII, shows the essential similarity of the initial segmentation of the two quite distinct organs, viz. the sporangia and the antheridia. It proves again, if further proof were wanted, that the progression in organisation from the relatively massive Eusporangiate to the delicate Leptosporangiate state is quite general for the various parts.

A very natural concomitant of the diminishing output of spermatocytes is a progressive simplification in structure of the antheridial wall. In the large sunken antheridia of Eusporangiate Ferns the antheridial wall may consist, partially, of two layers of cells, but usually of only one (Fig. 274, *e*). There may be several opercular cells, as in *Botrychium Lunaria*, or only one (Fig. 283, *A*). In either case the cell is cut out from the distal cap-cell of the antheridium by successive segmentations (Fig. 283, *E*, *F*). Similar segmentations appear also in the Cyatheoid Ferns, dividing the cap-cell into two or three, of which the innermost is the single operculum. These divisions tend, however, to be irregular. In *Diacalpe* sometimes two divisions exist, but usually only one (Fig. 283, *H*, *I*): in *Woodsia obtusa* one is usually present (Fig. 283, *G*), but undivided cap-cells have been seen to occur; a condition which is described as usual for *W. ilvensis*. This is the state which

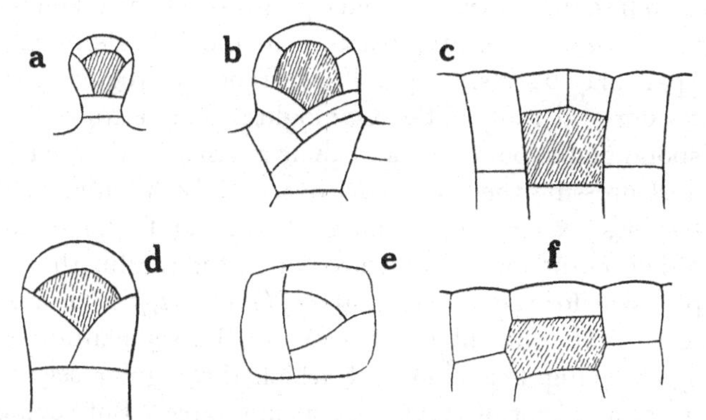

Fig. 282. Segmentation of antheridia in Ferns and in *Equisetum*. *a* = type of *Trichomanes*, and of Leptosporangiates generally; *b* = type of *Osmunda* (Campbell, *Mosses and Ferns*, Fig. 195, *C*); *c* = type of Marattiaceae and Ophioglossaceae; *d*, *e*, *f* = *Equisetum*, after von Goebel; *d* = antheridium on end of filament; *e* = view of the same from above; *f* = the antheridium sunk in the massive thallus. (See von Goebel, *Organographie*, 2nd Ed. Fig. 910.)

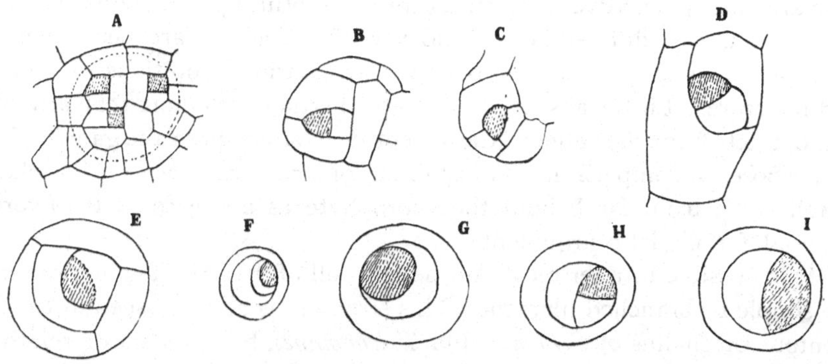

Fig. 283. Antheridia of various Ferns seen in surface view: the operculum is shaded. The scale is not uniform. They show a progressive simplification of the segmentations of the cap-cell. In *F*—*I*, the inner circle represents the cap-cell. In the Polypodiaceae this is not divided at all, but comes away bodily as the operculum. Compare Fig. 277. *A* = *Botrychium Lunaria*, after Bruchmann; *B* = *Angiopteris*, after Campbell; *C* = *Kaulfussia*, after Campbell; *D* = *Danaea*, after Campbell; *E* = *Osmunda*, after Heine; *F* = *Trichomanes*, after von Goebel; *G* = *Woodsia obtusa*, after Schlumberger; *H*, *I* = *Diacalpe*, after Schlumberger.

is general for the "Polypodiaceous" Ferns, in which the cap-cell develops without segmentation into the single opercular cell. These steps of progressive simplification express in details of wall-structure a progression roughly parallel with the diminution in number of the spermatocytes, as shown in the table. In all cases it appears that the opercular cells are extruded bodily when the antheridium opens.

It is in the persistence of these parallels in so many cases, in respect of the various parts of the sporophyte and of the gametophyte, that the cogency of the conclusion lies, that an evolutionary progression has taken place from a more massive to a more delicate structure in the Filicales. But the facts from a single part may be misleading. For instance the regularly segmenting apical cell in stem and root of the Ophioglossaceae accords ill with their typically eusporangiate sporangia, and their massive antheridia. A further warning against pressing these comparisons unduly is found in *Equisetum*. Von Goebel shows extreme divergence of form and segmentation of its antheridia. When borne on a massive region of the thallus the antheridium is of the type usual for the Eusporangiates (Fig. 282, *f*). But when borne on the end of a prothallial filament it shows the segmentation typical of Leptosporangiates (Fig. 282, *d*, *e*), with which the regular segmentation at the apex of stem and root of *Equisetum* would agree. But such exceptions do not vitiate the comparisons where, as in the great majority of cases, the correspondence in structure actually exists.

It remains to estimate the value of the characters of the gametophyte for purposes of phyletic argument. Taking first the vegetative features: they are chiefly negative. The structure is uniformly parenchymatous, with the minimum of differentiation, and very few features are permanent by inheritance. The form is very plastic under varied conditions of lighting and moisture. It is the absence of phyletic inertia which more than anything else detracts from the value of the vegetative characters of the gametophyte for purposes of comparison. Its instability of structure, as well as its relative simplicity, place it far behind the sporophyte as a source of trustworthy material for phyletic argument.

It is possible to refer most and perhaps all of its types to an origin from the simple or branched filament. This structure is actually seen in the filamentous prothallus of *Schizaea* and *Trichomanes*, both genera of relatively primitive position. In both of them the gametangia may take the place of the end of a filament, a point which is suggestive of comparison with certain Algae. *Trichomanes* and *Hymenophyllum* both suggest progressive steps to a flattened vegetative expanse a single layer of cells in thickness. Localisation of segmentation gives rise to definite apical activity, and the establishment of this may be seen in the ontogeny of many prothalli. In the great majority both of Eusporangiate and Leptosporangiate Ferns that localisation is distal, and the result is the symmetrical prothallus of the textbooks. But such cases as *Anemia*, *Vittaria*, *Pteris longifolia* (Fig. 267), and many others, show that the localisation may be at some indefinite point at the margin of the flattened thallus, giving it an irregular or lopsided form (Fig. 267, *B*, *D*). The addition of periclinal divisions in the cells lying behind the apical region, giving the massive cushion, is all that is then required to complete the ordinary cordate type where the apex is distal. In these peculiar lopsided forms, such as are seen in *Anemia*, or in *Anogramme*, which show that peculiar method of perennation already described, the form which is adopted results from the establishment of a lateral apex, followed as before by periclinal divisions. The initiation of such divisions is exemplified in the ontogeny of every prothallus which has a fleshy cushion. All these forms, together with the large and fleshy prothalli of the Marattiaceae and Osmundaceae, can without difficulty be referred back to the simple filament as their ultimate phyletic source. It is indeed, as a rule, from that source that they originate ontogenetically, and to it they may frequently return as a consequence of growth under special conditions (Fig. 266).

The massive mycorhizic prothalli of the primitive Ophioglossaceae appear to have diverged far from the simple filamentous structure. There is, however, no need to assume that their present state is itself primitive. The key to their origin is probably to be found in the fact that certain well-known Ferns under special conditions of culture produce cylindrical processes, which bear gametangia (*Scolopendrium*, etc. Lang). If the massive prothalli of primitive Ferns, such as the Marattiaceae and Osmundaceae, living in the absence of light were to do the like, assuming at the same time a mycorhizic habit, the result would be something of the same nature as the saprophytic prothalli of the Ophioglossaceae: in which case they also would be referable back ultimately to a filamentous origin, which has probably been the source for the gametophyte of all of the Filicales.

It has already been shown that the two types of gametangia have many points in common, and that it is reasonable to suppose that they diverged as a consequence of sexual differentiation from a single type of gametangium. In the filamentous prothalli of *Schizaea* and *Trichomanes* their position is commonly terminal on lateral branches. This brings them into line with the gametangia of Algae, which frequently have a like position. Comparison of them within the Filicales shows that the archegonium was standardised early, assuming that degree of constancy of structure which is seen even in primitive Ferns. The venter is always sunk in the tissue of the thallus, except in the simplest examples of filamentous prothalli, such as *Schizaea*. But there is some variation in the protrusion of the neck. In Eusporangiate types it may be almost wholly immersed, as in *Marattia* and *Ophioglossum*; but in Leptosporangiate types the neck projects freely. The same is the condition of the antheridia. As shown in the table above there may be differences in number of the spermatocytes, and where the number is large the antheridium is massive and thick-stalked, or even wholly immersed in the prothallial tissue. These characters run parallel with those of the sporangia. The largest sperm-numbers and the most massive and sunken antheridia are found in the Eusporangiate Ferns, where the sporangia also are massive and often deeply sunk, and the spore-output the largest. This parallelism, which however cannot be pursued into strict numerical detail, provides useful material for phyletic comparison.

The general conclusion which follows from a study of the gametophyte generation in Ferns is that its vegetative characters are deficient in stability and in variety of detail, and consequently they are only of minor importance for phyletic comparison. Of the gametangia the archegonium is so fully standardised that it can only yield material as regards its position relatively to the surface. Consequently it is upon the antheridium that the chief weight of the comparisons must fall. This use of it finds its justification in the parallel which has been traced between the antheridia and the sporangia. The facts and conclusions derived with greater certainty and profusion from the study of the sporangia, as set forth in Chapter XIII, gain collateral support from those relating to the antheridia, organs quite distinct from them in immediate origin and in function. What thus applies to the antheridia is found to hold also in general terms for the whole gametophyte that bears them. But at best it is only as ancillary to the comparisons based on the sporophyte that the observations on the gametophyte take their natural place in the study of the Filicales.

BIBLIOGRAPHY FOR CHAPTER XIV

GENERAL

230. ENGLER & PRANTL. Natürl. Pflanzenfam. i, 4, p. 15.
231. CAMPBELL. Mosses and Ferns. 2nd Edn. 1905.
232. CAMPBELL. Eusporangiate Ferns. 1911.
233. VON GOEBEL. Organographie. II Aufl. Teil ii, 1918, p. 947.

In these the chief literature is referred to.

SPECIAL MEMOIRS

234. BAUKE. Keimungsgeschichte d. Schizaeaceen. Pringsh. Jahrb. xi, 1876.
235. BAUKE. Aus dem Bot. Nachlass. Bot. Zeit. 1880.
236. CRAMER. Gemmae in *Trichomanes*. Denkschr. Schw. Nat. Ges. Bd. xxviii, 1880.
237. BOWER. Oophyte in *Trichomanes*. Ann. of Bot. i, 1888.
238. BOWER. Gemmae in *Trichomanes*. Ann. of Bot. viii, 1894, p. 465.
239. HEIM. Unters. über Farnprothallien. Flora. 1896, p. 329.
240. JEFFREY. Gametophyte in *Botrychium virginianum*. Proc. Canad. Inst. Vol. v, 1898.
241. BRITTON & TAYLOR. Life History of *Schizaea pusilla*. Torrey Bot. Club. 28 Jan. 1901.

242. VON GOEBEL. Über Homologien männlicher u. weiblicher Geschlechtsorgane. Flora. 1902. Bd. 90, p. 279.
243. VON GOEBEL. *Loxsoma* und das System der Farne. Flora. Bd. 105, p. 33.
244. LANG. Phil. Trans. Vol. 190, 1898, p. 187.
245. LANG. Prothalli of *Helminthostachys zeylanica*. Ann. of Bot. xvi, 1902, p. 23.
246. DAVIS. Origin of the Archegonium. Ann. of Bot. xvii, 1903, p. 477.
247. HOLFERTY. Bot. Gaz. Vol. 37, 1904, p. 106.
248. SCHENCK. Phylogenie der Archegoniaten. Engler's Bot. Jahrb. xlii, 1908, p. 1.
249. CAMPBELL. Prothallium of *Kaulfussia* and *Gleichenia*. Ann. Jard. Buit. Vol. viii, p. 69. 1908.
250. SCHLUMBERGER. Flora. 1911. Bd. 102, p. 265.
251. NAGAI. Physiologische Untersuchungen über Farnprothallien. Flora. 1914. Bd. 106, p. 281.
252. PHILLIPS. Ann. of Bot. 1919, p. 265.

CHAPTER XV

THE EMBRYO

THE act of Syngamy, consisting in the fusion of the spermatozoid with the ovum, produces the zygote, which is the starting-point for the diploid generation or sporophyte. So far as is known for Ferns at large the process is uniform, and the characters of the gametes involved in it are so similar that no comparative arguments have hitherto been based upon them. It is true that the spermatozoid of *Marsilia* is a body with many coils of its spiral form, while that in most Ferns has fewer coils. But at present the detailed facts are deficient for the Filicales as a whole, and no satisfactory use can be made of them till a better position is reached. The prevailing uniformity of the ovum is equally a barren source for comparative data. At present it is not from microscopic observation of the gametes themselves that assistance in phyletic argument can come. The zygote which results from their fusion is at first a primordial cell, which at once secretes a pellicle of cell-wall; and the roughly spherical cell thus produced, sunk in the venter of the archegonium, presents a very uniform starting-point for Fern embryology. It has been seen how uniformly standardised is the archegonium in Ferns. It seems a natural consequence that at first the zygote which it contains should also appear thus standardised. But the surroundings of the zygote as it develops into the embryo are by no means uniform in Ferns at large. The form of the gametophyte which bears the archegonium, its relation to external conditions of aeration and of light, the position of the sources of food relative to the archegonium, and in particular the orientation of the archegonium itself, are so diverse in the different types, that it might well be expected that divergent details should appear in embryos so differently placed.

In the embryo, as is the case of any shoot or part of the adult plant, two factors influencing development must be considered. First the qualities inherited from the ancestry, and secondly the features resulting from the impress of external circumstance. The theoretical problem in all embryology of plants will be to realise in the examination of the embryo as it is, how the balance has been struck between the first and the second of these factors, which, working together, may be held to have produced the effect which we see. In earlier days the tendency was to lay the greater stress upon the inherited factor; and indeed to translate the observed details directly into terms of Descent. More recently a tendency has arisen to refer the form of the embryo largely to the biological conditions under which it develops.

Probably a middle position will ultimately be found to give conclusions nearest to the truth. We shall therefore be prepared in studying the embryology of the Filicales to recognise underlying features, inherited possibly from a remote ancestry; and to see how the working out of the ancestral type may have been modified in any given case by biological conditions affecting the embryo during its development. Doubtless opinion will vary according as greater weight is attached to the first or the second factor. This is indeed a legitimate field for discussion as part of the embryological problem, when once the facts have been ascertained. In view of the many recent additions to knowledge it may be held that, as regards the embryology of the Filicales, the detailed facts are so far before us that a reasonable basis for forming such opinions already exists.

Embryological argument will necessarily turn in great measure on the segmental cleavages in early states of the embryo. To-day these are taken as indicators of the direction or localisation of growth. Such growth being unequal at different points and in different directions, it will result in the adoption of certain features of form by the embryo; and these should be considered in close relation to the cell-cleavages. In the nature of the case the study of embryology thus becomes a highly technical branch of the subject, and one in which it is specially necessary to discriminate between what is essential and recurrent in many types or in all, and what is of inconstant occurrence, and on that account to be held as accessory.

The embryo of an ordinary Leptosporangiate Fern has been described in Chapter I, and stages in its development are shown in median plane of section in Figs. 24 and 25, for the case of *Adiantum*. Like other embryonic tissues of the more advanced Ferns the embryo shows very regular segmentation, and it is possible from an early stage of development to refer the several parts of the young plant, viz. the stem, leaf, root, and foot, to definite segments cut off by the first cell-cleavages in the zygote. Such reference gave the opportunity for the facile conclusion that there is a causal connection between cleavages and the initiation of the several organs. The facts of segmentation seen in the Fern-embryo and in the young sporogonium of a Moss have been used, more than perhaps any other examples, as a foundation for an elaborate theory of embryology based upon cell-cleavages, or as von Goebel has styled it, a sort of "Theory of Mosaics" (*l.c.* p. 978). The elucidation of the facts coincided in time and in tendency with advances in the knowledge of the genesis of tissues from apical meristems, as demonstrated in particular cases by Hanstein. These appeared also to accord with the development of the science of animal embryology, in which the theory of germinal layers took a firm hold. Botanists allowed themselves to be influenced by these results, and many of them accepted the view that the segmentation at the apex of the stem determined the disposition of tissues, as well as the origination of organs. It was natural to trace this method back to the embryo, especially in cases where the cleavages are as clearly marked as they are in the Leptosporangiatae, the Ferns which naturally were examined first. Thus for a time the study of cell-cleavages dominated genetic morphology, finding a favourable centre in the embryology of such Ferns as had happened to be the best known.

In Chapter XIV of the *Land Flora* this position has been critically examined, and the conclusion stated (p. 180) that instead of accepting a general embryology as based upon cell-cleavages, it would be more natural to regard the embryo as a living whole: to hold that it is liable to be segmented according to certain rules at present little understood: that its parts are initiated according to principles also as yet only dimly grasped: and that there may be, and sometimes is, coincidence between the cleavages and the origin of the parts, but that the two processes do not stand in any obligatory relation one to the other. In particular this conclusion was found to apply to the embryonic organ called the "foot." The inconstancy of its position, and even of its existence in various types, suggests that it is an organ formed only where it is required for the first stages of development and nutrition of the embryo. This idea has found support in the embryological facts disclosed in the Lycopods and Ophioglossaceae. Later von Goebel formulated a general opportunist position as applied to the embryo: viz. that root, shoot, and haustorium are laid down in the positions that are most beneficial for their function: that external forces do not come into consideration in the arrangement in space of the parts of the embryo; and accordingly that we have only to consider internal factors. (Von Goebel, *Organographie*, 1898, p. 452, English Edn. p. 246.)

A revision of the embryology of the whole series of Pteridophytes leads, however, to the conclusion that the form of the embryo is not so plastic as this would imply (see *Land Flora*, Chapter XLII). Comparison shows that the polarity of the embryo is indicated by the first segmentation of the zygote. Of this segmentation there are two types according as a suspensor is present or absent: otherwise it shows remarkable constancy. Where a suspensor is formed in Ferns, the first segment-wall divides the zygote approximately at right angles to the axis of the archegonium: the cell nearer to the neck forms the *suspensor*, the other is the *embryonic cell*. Where there is no suspensor the zygote is itself the embryonic cell, and the first cleavage is not necessarily transverse to the axis of the archegonium. In either case the embryonic cell is ultimately divided into octants by cleavages at right angles to one another (Fig. 284). The two types, with and without a suspensor, are represented both in the Lycopods and in the Filicales. In all fully investigated cases the octant-division takes place, though the sequence of the cleavages is not uniform: and the embryo is accordingly composed of *epibasal and hypobasal tiers*, each consisting of four cells and separated by a septum which is called the *basal wall.* It appears to be the rule that *for all fully investigated embryos of Pteridophytes the position of the apex of the axis is constant in relation to this basal wall.* Whatever may be the other fluctuations of form of the embryo, or of the relations of the other parts, *the apex of the axis originates as nearly as possible to the centre of the epibasal hemisphere of the embryo*, that is, in close relation to the intersection of its octant walls. This being so, it is clear that *the polarity of the embryo is determined structurally at the time of the first segmentation of the zygote: or objectively prefigured by the position of the nuclear spindle in that first segmentation.* This generalisation was illustrated in Chapter XLII of the *Land Flora* by reference to the various types of embryogeny then known, and all those which have been fully described since conform to it.

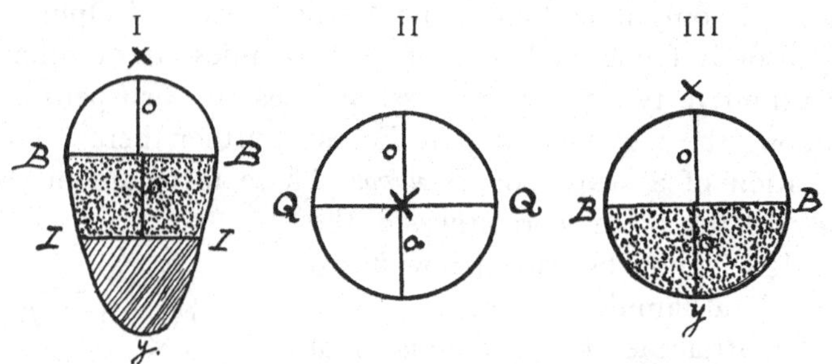

Fig. 284. Diagrams illustrating the segmentation of embryos. I = where a suspensor is formed, which is cut off by the first wall, *I, I*: the suspensor is cross-hatched; *B, B* = the basal wall, separating the hypobasal hemisphere (dotted) from the epibasal (clear). II = the same seen from above, *x* marking the pole. III = an embryo where no suspensor is formed, and the segmentation resembles that in the embryonic cell where the suspensor is present; the lettering corresponds: *x, y* indicate the polarity. Each hemisphere divides into four by quadrant walls (*QQ* in II), and octant walls (*o, o*).

A general conception of the embryo of Pteridophyta which follows from comparison of them all, whether with or without suspensor, is that it is a body possessed of polarity from the very first: that in form it is at first commonly a more or less spindle-shaped body, composed of two or more component tiers. It is in fact based on the ultimate type of a transversely septate filament. This applies also for Seed-Plants. It is most apparent in those embryos which retain the suspensor. In them the whole product of the zygote consists at an early stage of a simple row of cells, which may be more or less abbreviated; the cells are liable to lateral distension, with or without further cell-divisions. Upon the primordial spindle thus formed appendages may originate, but with latitude of difference in their number and in their proportion. The determination of their dimensions and time of appearance may be correlated with circumstances. For instance, in mycorhizic Ferns, such as *Botrychium Lunaria* and *Ophioglossum vulgatum*, the root is hurried quickly forward, and the shoot correspondingly delayed. On the other hand, in the embryos which spring from the delicate prothalli of Leptosporangiate Ferns the cotyledon is advanced quickly, so as to take up autotrophic nutrition early. In the aquatic *Salvinia*, though the origin of the shoot is as in other Leptosporangiate Ferns, the root is absent, a fact which accords with the habit of the plant, and with the rootless state of the adult.

Embryos having a suspensor, and others without one, are included within the Filicales; and the Lycopodiales are in a similar position. In both the hypothesis may be put forward that the suspensorless type has been derived by elimination of the suspensor in the course of Descent: and the change appears to have been related to the bulk of the nutritive prothallus. If this were so, then embryos with a suspensor would be held as relatively primitive in that respect, and those without it as relatively advanced (Lang, 256). Among the Filicales it is only in certain of the Marattiaceae and Ophioglossaceae that a suspensor is found, and both of these families are on other grounds held to be primitive types of the Class. No case of a suspensor has been recorded among the Leptosporangiate Ferns. Further there is no evidence of the formation of a suspensor *de novo*. These facts are in themselves *prima facie* evidence of the correctness of the hypothesis. It will now be shown how it applies to the several families of Ferns.

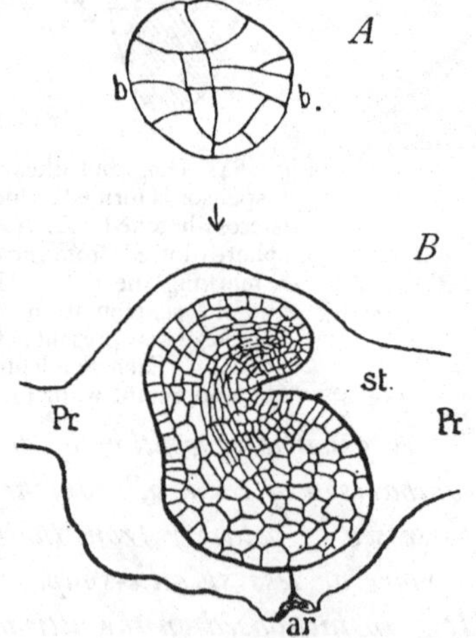

Fig. 285. *Marattia Douglasii*. A = longitudinal section of a young embryo (× 225); b, b = the basal wall: the arrow points to the neck of the archegonium. B = a similar section of an older embryo, showing its position in the prothallus; st = stem; Pr = prothallus; ar = neck of the archegonium (× 72). (After Campbell.)

In the Marattiaceae the prothallus is of the cordate type, but unusually fleshy, and the archegonia are directed downwards. The embryo develops with its basal wall cutting the axis of the archegonium transversely, and its polarity is consequently vertical from the first, so that the apex of the axis points upwards: the result is that the upper surface of the prothallus is ruptured, and the sporeling emerges with its cotyledon and apical bud erect, while its root projects downwards into the soil (Fig. 285). If a series of embryos of the Marattiaceae be all orientated as they would be in nature, with their basal wall (*b*, *b*) horizontal, they would appear as in Fig. 286. In *Angiopteris* (*a*), *Christensenia* (*b*), and *Marattia* (*c*) there is no suspensor, and the segmentation, though less regular, follows in essentials that seen in the simpler Leptosporangiate Ferns (cf. Fig. 294). But in *Danaea jamaicensis* (*d*, *e*) a short suspensor has been found, and a similar organ appears also in *D. elliptica*. This goes with a difference in form of the embryo, and Campbell contrasts the pear-shaped embryo of *Danaea* with the broadly elliptical and much depressed embryos of corresponding stages in the other Marattiaceae. Recently he

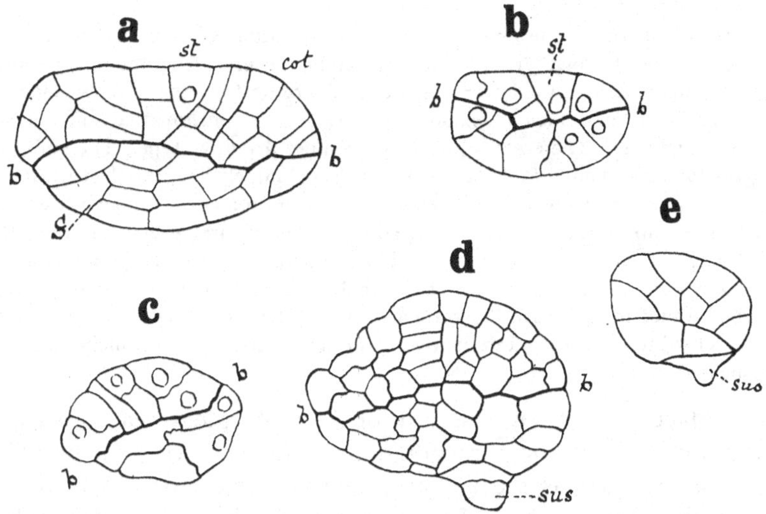

Fig. 286. Embryos of Marattiaceae, all orientated with the archegonial neck downwards as in nature. *a*=*Angiopteris*; *b*=*Kaulfussia*; *c*=*Marattia*; *d*, *e*=*Danaea jamaicensis*. (All from Campbell's *Eusporangiatae*.)

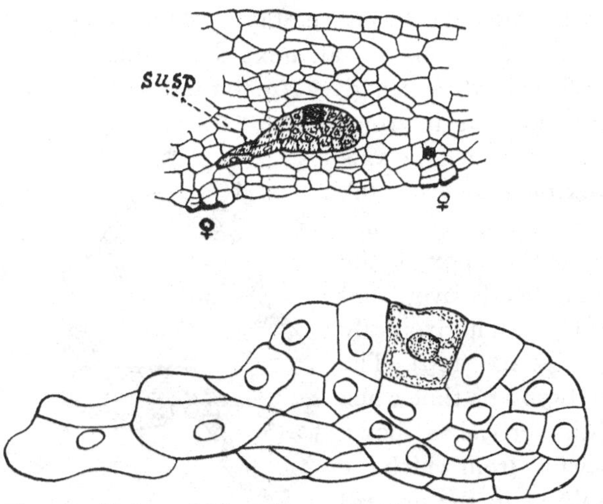

Fig. 287. Embryo of *Macroglossum*, after Campbell, showing the natural orientation, and the relation of the suspensor to the archegonium.

has found a similar organ present in the embryo of the new genus *Macroglossum* (Fig. 287). Here the long pear-shaped embryo takes an oblique course in its development, as though accommodating itself to the narrow bounds of the flattened prothallus, while the stalk which points towards the archegonium is clearly a pluricellular suspensor. In this respect *Danaea* and *Macroglossum* may be held as retaining an archaic feature, which appears to be wanting in the rest of the Marattiaceae.

The difference between the embryology of the Marattiaceae and that of the Leptosporangiate Ferns is striking. In the former the embryo is erect from the first, perforating the prothallus upwards. In the latter it is prone, emerging from the lower surface of the prothallus. Hitherto no sufficient explanation has been given how that difference may have been bridged over. The erect is probably the more primitive type. The embryology of *Macroglossum* gives a clue: for here the embryo lies obliquely between the surfaces of the prothallus. Its position suggests some obstacle to perforation of the upper surface, while the body of the embryo is more nearly related to the lower surface than in other Marattiaceae. A slight further modification would then suffice for its emergence with a still prone position from the lower surface of the thallus, as it does in the Leptosporangiate Ferns. They would in that case illustrate a later and derivative type of embryology, as they are also held to be later and derivative in so many other features, including the loss of the suspensor.

Similar facts have been observed for the Ophioglossaceae. In *Helminthostachys* the axis of the archegonium is usually horizontal or oblique. Lang has shown that as the zygote develops, it extends obliquely downwards from the venter of the archegonium into the massive prothallus before segmentation. Then follow two transverse walls, so that a row of three cells is formed. The two cells next the venter supply the first and second tiers of the suspensor: the distal cell is the embryo proper. This is at first straight; and it divides into a hypobasal half next the suspensor, which forms the foot; and an epibasal half which gives rise to the stem and leaf, and probably also to the first root. As it grows larger the embryo, of which the axis is at first approximately horizontal, curves so that its apex points upwards; but later the apex of the adolescent plant bends over, and growth proceeds again horizontally (*Land Flora*, Figs. 108, 109). The form of the embryo at a time when the apex is still vertical is shown by Lang's drawing, from which the curvature resulting in the upward turn of the apex is clearly seen (Fig. 288). This relation of the parts of the embryo is very similar to that shown by the photograph of *Botrychium obliquum* of similar age taken by Dr Lyon. Here also the first direction of growth was found to be inwards

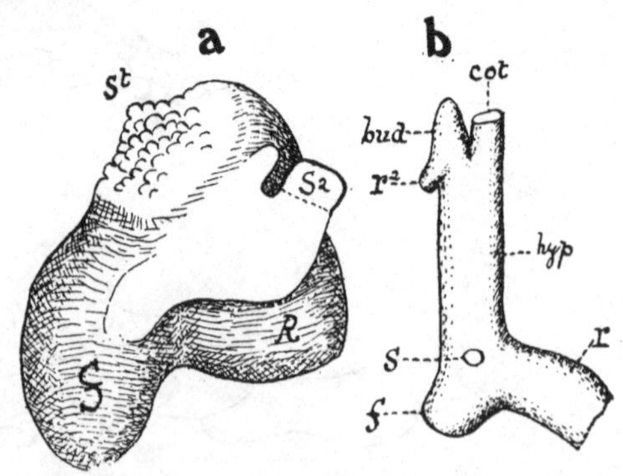

Fig. 288. Form of embryo of *Helminthostachys*. *a*, in a younger; *b*, in a more advanced state; *s* = suspensor; *f* = foot; *r*, r^2 = roots; *st* = stem; *cot* = base of cotyledon; *hyp* = hypocotyl. (After Lang.)

from the archegonial neck, and later a change in direction of growth results in the shoot pointing vertically upwards, and the root downwards (Fig. 289).

The development of the embryo is now well known in *Botrychium* (*Sceptridium*) *obliquum* (Campbell, *Ann. of Bot.* xxxv (1921), p. 141, where references are given). The archegonial neck is directed upwards. The zygote while still undivided grows into an elongated tube, which penetrates by an irregular course into the tissues of the prothallus (Fig. 290). Its nucleus settles down to the distal end as the growth proceeds. A transverse wall apparently in a horizontal plane separates the embryo proper from the suspensor; it is followed by a second division, by a "basal" wall, in the terminal cell or embryo, which then consists of epibasal and hypobasal tiers. The embryo thus constituted develops further along lines corresponding to those of the Marattiaceae. The epibasal hemisphere gives rise to the stem-apex and the cotyledon, the latter being the more marked feature. The hypobasal hemisphere forms the foot, and the first root originates "near the centre of the embryo, very near the basal wall" (Campbell). As in *Helminthostachys* the original direction of growth alters by strong curvature of the embryo. This has the effect of turning the apex of the stem and leaf upwards, while the root projects vertically downwards, penetrating the tissue of the foot (Fig. 289). In both cases a strong curvature away from the original line of growth of the embryo was necessary to produce this result; and the degree of that curvature is brought into prominence in the case of *Botrychium* by examination of the section taken in a horizontal plane through the developed embryo of the age shown in Fig. 289. Here, though the suspensor is traversed, the apex of the embryo is still seen in surface view (Fig. 291). No suspensor has been found in any other species of *Botrychium* or of *Ophioglossum* so far examined. The orientation of their embryos within the archegonium is, however, inverted as compared with those of

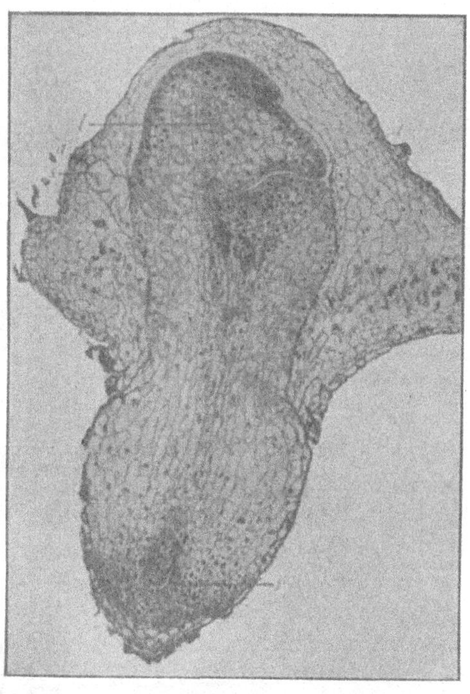

Fig. 289. *Botrychium* (*Sceptridium*) *obliquum*, Muhl. Photo-micrograph of a section through a gametophyte and young sporophyte. The root has already protruded from the under side of the gametophyte. a = archegonium; s = suspensor; t = stem-tip; l = first leaf; r = root. (×60.) (After H. L. Lyon.)

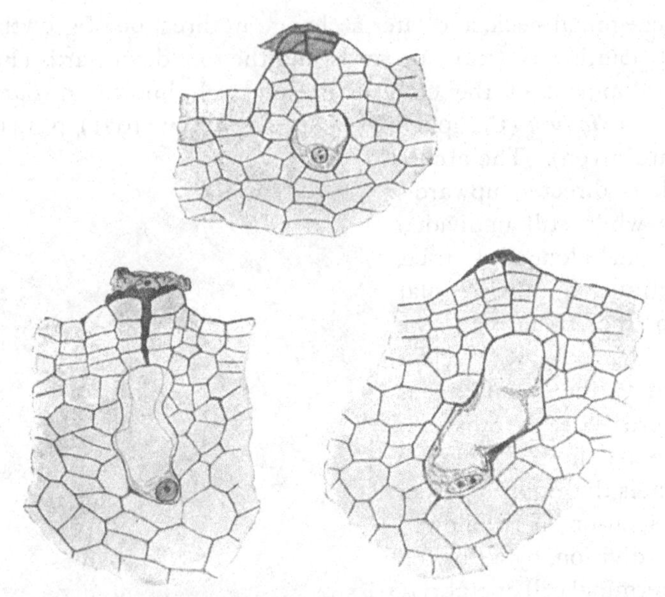

Fig. 290. *Botrychium obliquum.* First stages in the embryogeny. Before the first segmentation the zygote grows into an elongated tube, cut off later as the suspensor, which burrows its way irregularly into the tissue of the prothallus. (× 150.) From sections lent by H. L. Lyon.

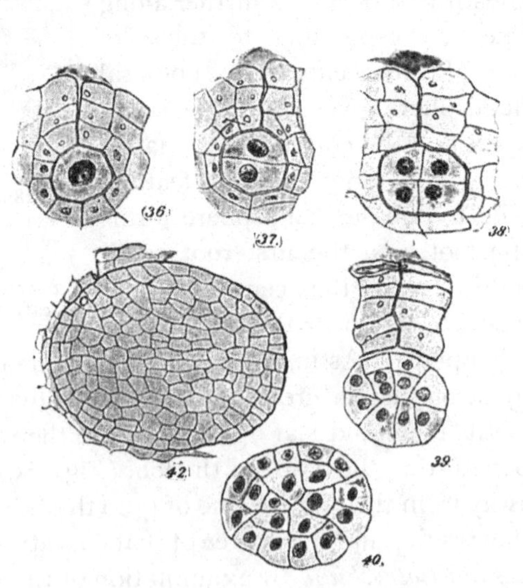

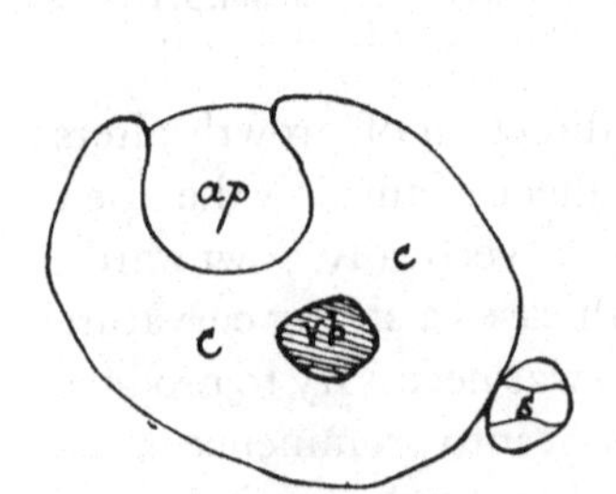

Fig. 291. Embryo of *Botrychium obliquum* in transverse section at the level of the stem-apex (*ap*). *c*, *c*=cotyledon; *s*=suspensor; *vb*=vascular bundle. (From a preparation lent by H. L. Lyon.)

Fig. 292. *Botrychium Lunaria*, L. 36=a fertilised archegonium; 37=zygote, showing the first segmentation; 38=embryo of four cells; 39, 40=embryos cut in direction of the axis of the archegonium; 42=an embryo breaking out of the prothallus; 36—40 × 225; 42 × 150. (After Bruchmann.)

Helminthostachys or *Botrychium obliquum.* In them the apex of the shoot points from the time of its first appearance directly towards the neck of the archegonium, and there is no curvature seen in the further development. This follows from a comparison of the drawings of Bruchmann from *Ophioglossum vulgatum* and *Botrychium Lunaria* (Figs. 292, 293). The apex as before originates from the epibasal hemisphere, an arrangement closely resembling what is seen in those Marattiaceae which have no suspensor. Essentially the

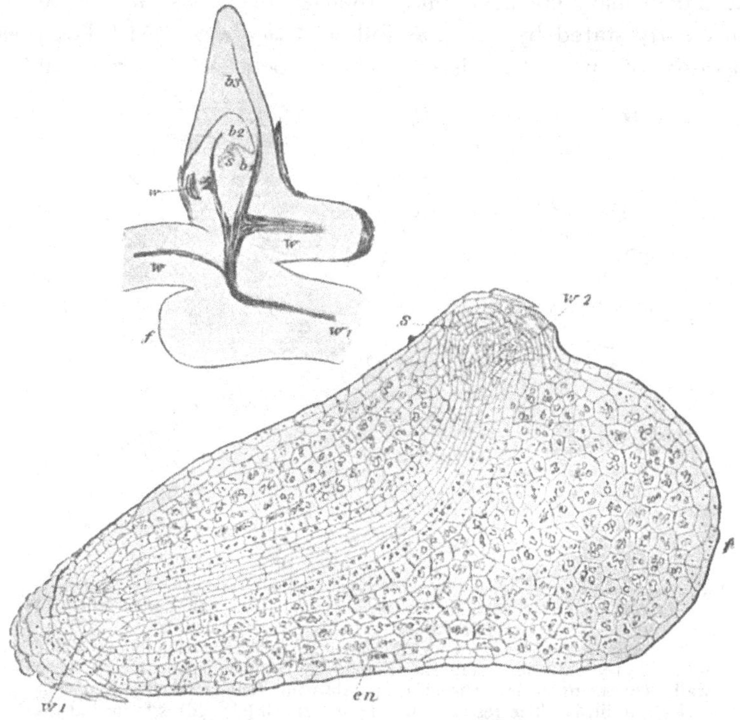

Fig. 293. *Botrychium Lunaria*, L. The lower figure represents an old embryo with well-developed foot (*f*); w_1=apex of first root; *s*=apex of the rhizome with the second root, w_2. The endophyte (*en*) is already in the cells. (× 52.) The upper figure is a diagrammatic section of a seedling, with six to eight roots, of which three are in the plane of section. *f*=foot; w_1=first root; *w*=other roots; *s*=apex of rhizome; b_1—b_3=developing leaves. (× 6.) (After Bruchmann.)

same plan of construction holds for all Leptosporangiate Ferns, but with greater precision in the segmentation, which accords with that greater precision seen in all their growing points in the adult state. There is, however, a difference in the orientation of their embryos relatively to the axis of the archegonium, their polarity being defined by the first cleavage of the zygote (Fig. 294). For here the first wall is approximately in a plane which includes the axis of the archegonium, to which consequently the axis of the embryo is approximately at right angles. This goes along with the fact that the sporeling emerges on the lower side of the prothallus, its axis being prone, and without any curvature within the thallus. Curvatures may, however, appear later in the adolescent state, varying according to the adjustment of the shoot for adult life.

These facts relating to the primary embryology of the Filicales provide material for a general opinion on the polarity and orientation of the embryo. It appears probable that a suspensor was characteristic of primitive Ferns, and that the primary construction of their embryo was filamentous. This has been clearly stated by Lang, as follows (256 *bis*, p. 35): "The presence of a suspensor of one or two tiers appears to be a fact of organisation in a

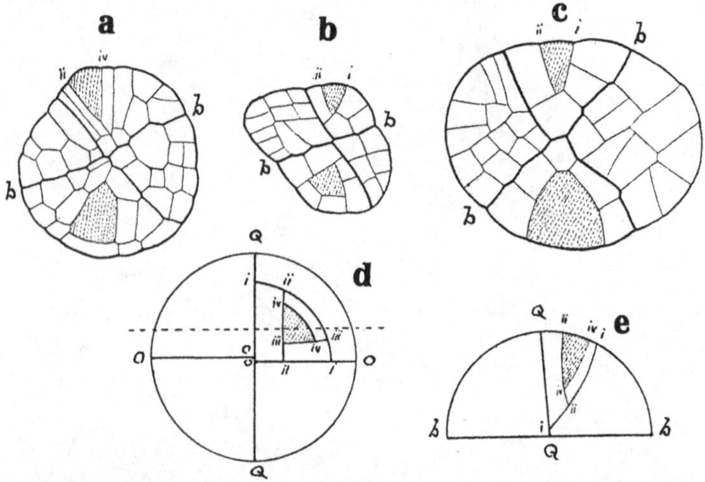

Fig. 294. *a* = embryo of *Equisetum*, after Sadebeck; *b* = *Marsilia*, after Hanstein; *c* = *Adiantum*, after Atkinson: all are orientated with the axis of stem and root vertical, to which line the basal wall is variously inclined.
d, *e* are diagrams to show in view from above (*d*) and in section (*e*) how a single tetrahedral initial cell is established in the epibasal hemisphere. *Q* = quadrant wall; *O* = octant walls. The cleavages thus initiated are continued as the series i, ii, iii, iv. The result is that the initial cell (shaded) is formed at the nearest possible point to the centre, consistent with the sequence of the segmentations.

number of forms which are relatively primitive. Its presence may be looked upon as the last indication of the construction of the plant-body from a filament or row of cells, i.e. as a juvenile stage in the development rapidly passed over, and often suppressed." That it was liable to be suppressed both in the Filicales and the Lycopodiales (and possibly also in the Equisetales) accords well with the facts disclosed for these classes. Its survival may perhaps have been determined by the advantage of thrusting the embryo into the massive nutritive prothallus. But on the other hand the presence of a suspensor appears to have fixed the orientation of the spindle-like embryo so that where it is present the apex of the shoot is always directed away from the neck of the archegonium; it might be described as being *endoscopic*. Consequently the varying inclination of the archegonium to the vertical necessitated in certain cases a subsequent curvature of the embryo, so as to secure the upward direction of the shoot. The embryos of *Botrychium obliquum* and *Helminthostachys* (Figs. 288. 289), as well as those of various species of *Lycopodium* and *Selaginella*, illustrate the inconvenient shifts that followed from a polarity controlled, in its relation to the archegonium, by the inherited suspensor. Those of *Ophioglossum vulgatum*, and *Botrychium Lunaria* (Figs. 292, 293), as well as of *Isoetes*, show how the absence of the suspensor has removed the restricting tie, and thus simplified the problem of establishment of the sporeling without such contortions. In the case of the Marattiaceae no curvature is necessary, but "the suspensor seems to be of no particular use" (Lang). It certainly helps to embed the embryo in the prothallus: but though relatively fleshy the prothallus is a flattened body, and the embryo of *Macroglossum* shows how the elongation of the suspensor runs parallel to its surfaces, not vertically inwards (Fig. 287). Its elimination in most of the Marattiaceae seems to have removed a useless vestige. The facts for the Eusporangiate Ferns, and for the Lycopodiales as well, indicate that the suspensor is vestigial: that its survival is intelligible in certain cases, but that its removal has set the embryo free from an unnecessary and inconvenient tie[1].

As in *Isoetes* among the Lycopodiales, or *Ophioglossum* and *Botrychium Lunaria* among the Ophioglossaceae, so the Leptosporangiate Ferns, having no suspensor, are not restricted in the orientation of their embryos relatively to the axis of the archegonium. They also are free to adjust the polarity of the embryo, and they have assumed an orientation peculiar to themselves. Throughout the Leptosporangiate Ferns the first segment-wall lies approximately in a plane that includes the axis of the archegonium, instead of transversely to it. This provides for the prone position of the axis. The adjustments of the cotyledon and root are such as to allow their ready protrusion, while the enlarged foot maintains communication with the parent prothallus. The proof of the fitness of these arrangements which follow upon the prone position appears in their constancy in the great series of the Leptosporangiate Ferns. Notwithstanding the loss of the suspensor, which consequently gives a less obviously filamentous form to the young sporophyte, it is still correct to preserve the view of it as a spindle-like body theoretically and by Descent, with polarity marked by the position of the

[1] No suspensor has been described for *Tmesipteris*. Lawson and Holloway both agree in the statement that the shoot-region of the embryo is directed towards the archegonial neck, while a suctorial organ with filamentous outgrowths penetrates into the prothallus. This orientation resembles that in *Isoetes*, *Equisetum*, *Ophioglossum vulgatum*, and *Botrychium Lunaria*, which are all types without a suspensor, having the apex directed towards the neck of the archegonium (*exoscopic*). But the absence of a suspensor is remarkable and rather unexpected in an organism so primitive as *Tmesipteris*, and especially since its prothallus is relatively massive.

growing apex of the axis. The actual pole or centre of the stem-apex is indicated by the cell-cleavages. It may be located accurately where there is a single apical cell, as there is in the Leptosporangiate Ferns, or in *Equisetum.* After the epibasal hemisphere is defined by the basal wall (*b*, *b*), the segments which follow it establish at once the initial cell *at the point nearest to its centre that is possible by the method of successive segmentation* (Figs. 294, *a–c*). In these cases it is not going too far to say that the epibasal hemisphere itself is the original initial cell for the shoot. In the Marattiaceae, and others where the apical point is less exactly defined by segmentation, its position is still the same. Accordingly the conception of the young sporophyte as a primordial shoot of spindle-construction may be accepted as general for the Filicales, and it applies also for other Pteridophytes (267).

The next step will be to trace the relations of the other parts of the embryo to the primitive spindle thus defined. They are, the first leaf, the first root, and the haustorial foot. Of these parts, the root and foot appear to be related to the central spindle according to convenience. Neither time of origin, position, nor even their existence is fixed or immutable, but variable; and especially in those forms which are recognised as relatively primitive. This is the general position for Vascular Plants at large, and within limits it is illustrated among the Filicales.

Taking first the *haustorium* or *foot*, it can hardly be defined as a morphological entity. It originates from the hypobasal hemisphere of the embryo. Where a suspensor is present, and the embryo is curved in its development, the hypobasal hemisphere enlarges on its convex side. This is seen especially in Lycopods; and it appears also in *Helminthostachys*, and the swelling is described as a "foot." In *Botrychium obliquum*, however, no definite foot is found. Where the suspensor is small or absent, the greater part or even the whole of the product of the hypobasal hemisphere may develop as a haustorial organ. According to circumstances it may be symmetrical or lop-sided; as for instance, in *Ophioglossum vulgatum*, and *Botrychium Lunaria* and *virginianum* (and also in *Isoetes*). In these plants the first root appears to be epibasal in origin, and the whole hypobasal region is recognised as "foot." In *Salvinia* where no root is present it is the same. On the other hand, a definite foot is not recognisable in the Marattiaceae (Fig. 286). Here the root, which according to Campbell originates late, and apparently from about the limit between the epibasal and hypobasal regions, grows directly downwards and approximately through the centre of the hypobasal hemisphere. A foot as such is never organised. According to von Goebel (254, p. 992), the cotyledon in these Ferns acts at first as an haustorium, but it finally breaks through the upper surface of the prothallus, and this he regards as an explanation of the absence of a definitely organised foot. As von Goebel justly remarks (254, p. 979), "We can only speak of a special suctorial organ when it appears externally as such. That is not always the case, and there is no ground for assuming the presence of an haustorium if it is not clearly recognisable as a special organ." The fact appears to be that tissues derived from the hypobasal tract are liable to distension in various ways in relation to the transfer of nutriment from the prothallus to the embryo, taking a rounded form, and projecting towards the source of supply. When such a suctorial organ is clearly distinguishable it may be designated a "foot." But it is not constant in occurrence, size, or position for Ferns. The relative constancy of a foot in the highly standardised Leptosporangiate embryo has given such swellings a morphological recognition they do not deserve even in the Filicales taken as a whole, and much less in the Pteridophyta at large.

The most important feature of the embryogeny, next to the definition of polarity and consequent establishment of the axis, is the initiation of the *first leaf*, or *cotyledon*. It always arises from the epibasal hemisphere, and is orientated definitely in relation to the axis: but its time of appearance may vary, and this is closely related to the nutrition of the embryo, whether autotrophic or by mycorhizic saprophytism. No Family of Ferns shows greater variety in this than the Ophioglossaceae. In *Helminthostachys*, or in *Botrychium obliquum* and *virginianum*, we probably see a relatively primitive state (Figs. 288, 289). The cotyledon arises from the epibasal hemisphere side by side with the apex of the axis and, soon appearing above ground, expands as the first photo-synthetic leaf. Campbell and Lang have, however, observed that in *Helminthostachys* the cotyledon itself may sometimes remain rudimentary, as are several of the early leaves of *Botrychium Lunaria*: this is probably a derivative state following on mycorhizic nutrition, and bringing with it a delay in their initiation. A curiously contrasted modification has been described for *Ophioglossum pedunculosum* and *moluccanum*(255). Here the cotyledon appears early and rises at once above ground as a photosynthetic leaf, but it extends downwards directly into the first root, while the stem-apex is arrested, and appears to develop no further. It is stated that it is replaced later by an adventitious bud originating from the root. This also is probably a derivative condition from that seen in *Helminthostachys*. Such peculiarities may affect the time of appearance and the proportions of the cotyledon, but not its position relatively to the axis and other parts.

Von Goebel has, however, pointed out an apparent difference in relative position of these organs between the embryos of the Marattiaceae and the Leptosporangiate Ferns (254, p. 994). In the Marattiaceae the position is as in the Ophioglossaceae, with the axis and cotyledon directed upwards, that is away from the archegonial neck. In the Leptosporangiate Ferns the cotyledon is directed downwards, that is, it is on the side next to the archegonial neck. He ascribes the difference to the early development of the foot, which has brought about the change in position. He denies the fact of an actual rotation of the embryo in the venter as suggested by Conway Macmillan (257 *bis*): and probably rightly. But in my view, all that is necessary to effect the change is a swing of position of the first nuclear spindle in the zygote. This would determine the plane of the basal wall so that it should include the axis of the archegonium, instead of being at right angles to it. In fact the initial polarity of the embryo would be altered, but not the relative position of its parts. After this all would follow in the normal course, and the relation of the cotyledon to the apex of the shoot would remain unchanged. In either case the succession of parts as seen in the median longitudinal section of the embryo reads thus: stem, leaf, root, foot, but with the latter ill defined in the Marattiaceae and some others. The Fig. 288 of *Helminthostachys* shows this relation of parts, as also do the Figs. 294, *b*, *c* of Leptosporangiate Ferns. In other words, *in the megaphyllous Filicales the first leaf is on the same side of the axis-spindle as the first root*, a relation which is maintained in many Ferns in the adult state (see Fig. 57). This fact bears physiological reasonableness on the face of it, for it gives the shortest course for supplies from the root to the precocious cotyledon. In microphyllous embryos, where its importance is less, this relation is less constant. It may be noted that this relation accords with Chauveaud's theory of the "Phyllorhize."

The *first root* shows some variety not only in its point of origin, but also in time: in *Salvinia* it may be absent altogether. But in point of orientation its relation to the cotyledon is that above stated. In respect of time it is usually complementary to the cotyledon: that is to say, when for purposes of mycorhizic nutrition it is developed precociously, as in *Botrychium Lunaria* and *Ophioglossum vulgatum*, the cotyledon is usually delayed. But where as in most Leptosporangiate Ferns the embryo is autotrophic, both advance simultaneously. As regards the level of origin, the first root may spring either from the epibasal or the hypobasal hemisphere. In all the Leptosporangiate Ferns it is hypobasal, but in many of the Eusporangiates it is definitely epibasal. Campbell remarks of *Botrychium virginianum* (255, p.48) that "all the organs of the young sporophyte arise, as Jeffrey showed, from the epibasal region, and in this respect *B. virginianum* agrees with the Marattiaceae and with *Ophioglossum*." Still several cases are left uncertain by Campbell: nor is it of material importance that they should all be strictly defined. The important point is that the root is not universally linked with either hemisphere, but may be regarded as an accessory to the spindle-like shoot, of various position and origin, and occasionally absent. In relatively primitive Ferns it arises usually from the epibasal, but in the advanced Leptosporangiate Ferns always from the hypobasal hemisphere.

The facts relating to the Ferns thus appear to be in general accordance with the conception of the embryo as essentially of spindle-like, or even of filamentous construction. The primitive spindle may be abbreviated by elimination of its base, that is the suspensor. It may be further disguised by cell-segmentation, and by lateral distension of the tissues thus formed. The parts which it bears may vary in proportion and in some degree in their apparent relations. But still there remains the fundamental fact of the polarity of the spindle, which is established by the very first segmentation of the zygote. *In point of fact, the embryo of a Fern is a simple leafy shoot from the first, and it may bear accessory suctorial organs, i.e. the foot and the root: but neither of these is always present.*

The question remains how far the characters of the embryo thus constructed can be used in the phyletic treatment of the Filicales. If the biological position of the embryo be constantly kept in mind, that will help towards a just estimate of the value of the features of form and structure which it shows. More especially is it important to remember how nearly the embryo is dependent upon the gametophyte and its capacity for yielding nutrition. The immediate point will be to distinguish between those features which are directly plastic under present circumstances, and those which may be held as inherited features common to the race and to its predecessors. In an earlier phase of the science the majority of the features observed were habitually ranked in the second category: at the present time the tendency is to give full latitude to the former, recognising a high degree of adaptability in the young embryo. With criticism illuminated by such ideas as these, there are four lines of comparison that may be traced from the study of embryos, which may be used in the phyletic treatment of the Filicales. They will be considered in succession.

(1) The most constant character of the embryo of the Filicales is its *polarity*, which is expressed with varying clearness in the spindle-form. But this is liable to be masked by the elimination of the base of the primitive spindle, which we call the suspensor. Those embryos which have a suspensor may be held to retain their primitive state more fully than those which have none. In fact the suspensor is held to be an archaic feature. Nevertheless the recognition of this must not be pressed too far. It may be retained as a vestigium in cases which are not primitive in other features. For instance, because *Danaea* has a vestigial suspensor and *Angiopteris* has none, it does not follow that *Angiopteris* is a more advanced Fern than *Danaea*, though it has lost this archaic feature. The spindle-structure is further obscured by transverse distension, especially seen in the epibasal and hypobasal regions. This is manifest in the embryos of the Marattiaceae (Fig. 286), and in those of *Ophioglossum vulgatum* and *Botrychium Lunaria* (Figs. 292, 293). In all of these the hypobasal and epibasal hemispheres are so dilated laterally that

it is only by comparison, and by their further development, that it becomes apparent that the axis of organisation cuts the basal wall at right angles. These features, combined with the early formation of the appendages, have disguised the spindle-construction of the embryo, so that in many Ferns it is only recognised with difficulty.

(2) A second feature is the *cellular segmentation* of the embryo. In certain Ferns the embryo is relatively massive with less definite segmentation: in others it is less massive with more exact segmentation. This runs naturally parallel with the like differences of segmentation of the several parts as compared in Chapter VI. It was there concluded that the less definite segmentation is characteristic of the more massive and relatively primitive types, and the more definite of the delicate and relatively derivative types. The former is seen in the embryos of the Eusporangiate, the latter in those of the Leptosporangiate Ferns (Figs. 286, 294). This accords with the facts relating to the suspensor, which is retained only by certain Eusporangiatae, and is always absent from the Leptosporangiatae. On both grounds the former are thus indicated as relatively primitive, and the latter more advanced types.

(3) A third feature lies in the *relations of the primordial organs to the basal wall.* In all Ferns the axis and cotyledon arise from the epibasal hemisphere. In many of the Eusporangiates the first root also springs from it, and in others it arises about the border line. But in all the Leptosporangiates it arises from the hypobasal hemisphere. This is held to be a later and derivative state. The foot, or haustorium, where present, arises universally from the hypobasal hemisphere.

(4) Certain embryos are characterised by *delay, or even suppression of the development of certain parts.* In *Botrychium Lunaria* the first leaves appear only as scale-leaves, and it is stated that about the eighth of them is the first leaf to appear above ground. In *Ophioglossum moluccanum* and *pedunculosum* the apex of the axis is arrested; it appears from the descriptions to be replaced later by an adventitious bud on the root. In *Salvinia* no root is formed, but the whole hypobasal region develops as a foot. All of these departures from the normal numbers and relations of parts may be held as secondary; i.e. derivative from that recognised as normal. They may all be explained as biological adaptations.

It may finally be asked what weight is to be accorded in our general comparison of the Filicales to the facts of the primary embryogeny. Its details have been highly estimated in the past, and sometimes they have been made the basis for far-reaching conclusions. A bias towards this may have originated from the success of the recapitulation-theory in Animal Morphology, and this has no doubt influenced botanical opinion. But however convincing the analogies between the two kingdoms may appear to be, strict recapitulation cannot be assumed where, as in plants, continued embryology holds sway, with its successive origination of new organs. In plants the primary steps of foundation of the organism are a less important matter than in animals, where the plan of the whole organism is laid down once for all. For plants a line of inductive reasoning is required quite distinct from that which has led to more definite conclusions for animals.

The facts relating to the suspensor and the initial polarity of the embryo, together with the primary relation of the cotyledon to the apex thus defined, appear on a comparative review of their embryogeny to be the fundamental features for the vascular plants in general, and for the Filicales in particular. The reference of the sporophyte embryo back to a primitive spindle, or filamentous row of cells, which has been shown to accord with the facts of embryology, gives a plan of construction of the embryo which is believed to be fundamental and archaic. To those facts which harmonise most readily with that plan the greatest weight is to be accorded, while those features may be held as of less importance for comparative purposes which have had the effect of disguising that original plan.

In conclusion it may be pointed out that the relation of leaf to axis, as shown in the embryos of Ferns, accords readily with a theory of the ultimate relations of leaf and axis as equivalent branches of an indifferent branch-system, which was possibly dichotomous (Chapter XVII, p. 340). They both originate from the epibasal hemisphere side by side. The one or the other may in concrete cases take precedence in the individual development, a point closely related to the conditions of nutrition. Thus the archaic spindle may have had a power of dichotomous branching at its apex. Whether or not such a primitive sporophyte resembles the actual prototype of the sporophyte of the Filicales, or of the Pteridophyta at large, it is the conclusion to which induction, based upon an organographic study of the adult, combined with the facts of embryology, and the evidence of palaeontology, appears naturally to lead.

POSTSCRIPT

Since this Chapter was written still another view as to the constitution of the leafy shoot as seen in Ferns has come into prominence. The theory of the "phyllorhize" has been propounded by Chauveaud (265), and expounded by Becquerel (266). It claims to be based upon the ontogenetic method, and the central idea is that of a body designated the "phyllorhize," consisting of an upward-directed leaf and a downward-directed root, connected by a middle-region or stalk (caule). It is stated that in a young Fern composed of one or more of such units there is at first no stem (tige). This comes only into existence when a fusion of units has taken place. It then originates laterally upon the "phyllorhize," as a bud which transforms itself gradually into a terminal bud. How then does this suggestion accord with the demonstration of the primitive spindle, which is the result of comparative study of the earliest embryonic stages not of Ferns only, but also of Algae, of Bryophytes, and of Vascular Plants at large? (267).

The two accounts of the embryogeny are entirely out of harmony. The reason for this is that Prof. Chauveaud, while stating that his views are based on ontogeny, does not take sufficiently into account the earliest stages of the embryo; moreover his comparisons are restricted to a limited field. For him the vascular tissue appears to be the criterion of morphological character, and pre-vascular development is not accepted as distinctive (*l.c.* pp. 75, etc.). But if the ontogeny be traced in each case from the first steps of the embryo, and consecutive stages onward are duly followed, it becomes clear that, however retarded or disguised, the apex of the shoot is defined from the very first. The two views diverge on questions of fact as well as of method: and that here advanced is based on the actual ontogeny as recorded in a very wide literature relating to the embryogeny of plants. It is not a morphology of cell-mosaics, nor of vascular anatomy, but of polarity, which is the feature first to be defined in the individual life.

There is a certain similarity between the view of Campbell, as stated in relation to *Ophioglossum pedunculosum* (255), and that of Chauveaud. But the plants which they use as the foundation of their several positions cannot be held as the most primitive even in their own respective alliances. There are good reasons for not holding *Ophioglossum* as the most primitive genus of the Ophioglossaceae: *Ceratopteris* and *Polypodium*, which are used by Chauveaud as illustrations, are both relatively specialised genera of Leptosporangiate Ferns. It would appear undesirable to use either of them as a basis for a far-reaching theory of the constitution of the shoot at large. Any such theory should be founded upon a general comparison. When this is done the primitive spindle, and not the "phyllorhize," appears as the original source of the shoot, while the root when present is accessory to it. But, notably, the root is absent from those early leafless vascular plants, the Psilophytales. What then is to be regarded as the "phyllorhize" in these primitive vascular plants which possess neither leaf nor root, though on Chauveaud's hypothesis these should be its essential constituents? From them we learn that vascular plants antedated the "phyllorhize": and we can only conclude that such a body, even if we concede its apparent existence as a unit in certain instances, is not a fundamental feature for Vascular Plants at large.

BIBLIOGRAPHY FOR CHAPTER XV

253. BOWER. Origin of a Land Flora. Chaps. xiv, xlii; also p. 464.
254. VON GOEBEL. Organographie. II Aufl. Teil ii, pp. 978—996.
255. CAMPBELL. The Eusporangiatae. Washington. 1911.
256. LANG. Embryo of *Helminthostachys*. Ann. of Bot. xxviii, 1914, p. 19.
256 *bis*. LANG. Address to Section K. British Association Report. Manchester. 1915.
257. JEFFREY. Gametophyte of *Botrychium virginianum*. Toronto. 1898.
257 *bis*. CONWAY MACMILLAN. The orientation of the Plant-Egg. Bot. Gaz. 1898, p. 301.
258. BRUCHMANN. *Botrychium*. Flora. 1906. Bd. 96, p. 203.
259. BRUCHMANN. *Ophioglossum vulgatum*. Bot. Zeit. 1904. *Selaginella*. Flora. 1912. Bd. 104, p. 180.
260. LYON. *Botrychium obliquum*. Bot. Gaz. Dec. 1905.
261. CAMPBELL. *Macroglossum*. Ann. of Bot. xxviii, 1914, p. 651.
262. LAWSON. Prothallus of *Tmesipteris*. Trans. R. S. Edin. 1917. Vol. i, p. 785.
263. HOLLOWAY. Prothallus and young plant of *Tmesipteris*. Trans. N. Z. Inst. 1918, Vol. l, p. 1.
264. CAMPBELL. Gametophyte and embryo of *Botrychium obliquum*. Ann. of Bot. xxxv, 1921, p. 141. Also *Botrychium simplex*. Ann. of Bot. xxxvi, 1922.
265. CHAUVEAUD. La constitution des plantes vasculaires révélée par leur Ontogénie. Payot et Cie. Paris. 1921.
266. BECQUEREL. La découverte de la Phyllorhize. Rev. Gén. des Sciences. 28 Feb. 1922, p. 101.
267. BOWER. Presidential Address to the Roy. Soc. of Edinburgh. Oct. 1922.

CHAPTER XVI

ABNORMALITIES OF THE LIFE-CYCLE

THE investigations of the middle years of the nineteenth century made it for the first time possible to give a consecutive account of the various stages in the life-history of the Higher Cryptogams, and in particular of the Ferns. The spores of Ferns were first recognised experimentally as reproductive organs by Morrison, who raised young plants from them (1699). John Lindsay observed the germination of the spores (1792); but it was Brisseau-Mirbel (1802) who was the first to trace the formation from them of the prothallus, a body which had already been described by Ehrhart (1788). Kaulfuss, in 1827, published good drawings of the germinating spore, the prothallus, and the young plant attached to it, and he gave an excellent summary of the literature up to that date. In 1844 Naegeli discovered the antheridia and spermatozoids, while Suminski, in 1848, ascertained the true nature of the archegonium, and its relation to the embryo. But it remained for Hofmeister to put together and to complete the story. In 1849 his description of the germination of *Pilularia* appeared, and two years later, in 1851, he gave to the world his *Vergleichende Untersuchungen*, a work which dealt in the most comprehensive way with the life-histories of archegoniate plants generally. Eight years later Darwin's *Origin of Species* was published, and "the Theory of Descent had only to accept what genetic morphology had actually brought into view" (Sachs). At first no morphological or physiological history was traced in the facts of the individual life. But gradually it became apparent that an interpretation of the life-history in terms of evolutionary history was possible, and in recent years such interpretations have been the subject of frequent discussion.

The life-history of a Fern, as described in Chapter I, is held to be that which is normal for the Filicales, because it is usual for the large majority of them, and also for Archegoniatae generally. It is, however, liable to modifications not only in the form and structure of the plants concerned, but also in the several steps observed. Extra stages may be interpolated in the simple cycle, or phases held to be essential may be entirely omitted in the single life. These irregularities often raise questions of the utmost importance morphologically and physiologically; they may even be held to affect views relating to the Descent of the Filicales, and to influence profoundly the interpretation of their life-history as illuminating the Descent of plants generally. It is therefore necessary to examine and to estimate at

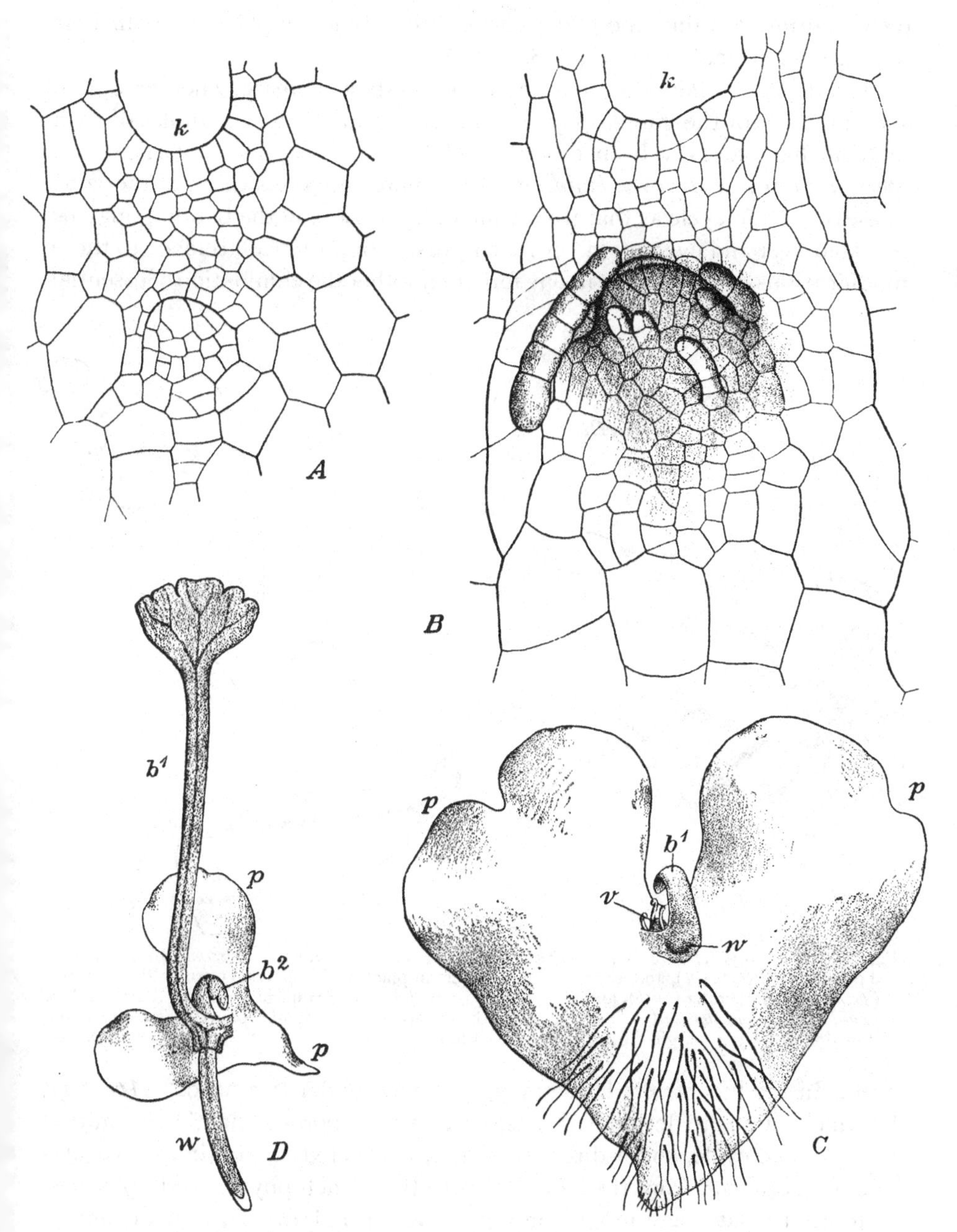

Fig. 295. Apogamy in *Pteris cretica*, L. *A* and *B*=development of the first foliar process close to the emargination (*k*), on the under surface of the prothallus. *C*=a whole prothallus seen from below, showing a young apogamous shoot. *p*=prothallus; b^1=first leaf; *v*=stem apex; *w*=root. *D*=a similar growth more advanced. *A* and *B* × 145. *C* and *D* less highly magnified. (After De Bary, from Engler and Prantl.)

their legitimate value these departures from that simple cycle which is regarded as normal. (See Fig. 28, p. 21.)

It will suffice here to mention the vegetative increase, whether of the sporophyte or of the gametophyte, by buds or gemmae, for examples of both of them have already been described (Figs. 66, p. 72; 271, p. 281). Such *sporophytic or gametophytic budding* results merely in a repetition of the same phase of the life-cycle as that from which they arose, and the budding may be repeated over and over again. A much greater importance attaches to those modifications which involve in one form or another the elimination of essential

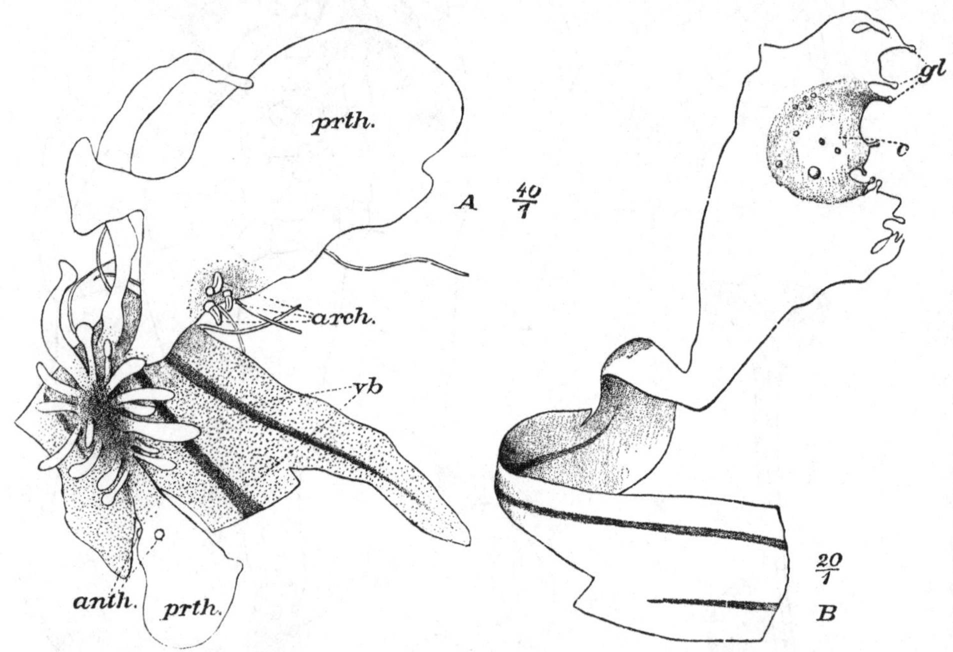

Fig. 296. Apospory. *A* = soral apospory of *Athyrium Filix-foemina*, var. *clarissima*, Jones. Part of a pinnule with veins (*vb*), and a sorus. In the latter, in place of the sporangia prothalli are formed (*prth*), with antheridia (*anth*), and archegonia (*arch*). (× 40.) *B* = apical apospory of *Polystichum angulare*, var. *pulcherrimum*, Padley. A prothallus arises at the tip of a pinnule, as a direct continuation of it. *gl* = marginal glands; *c* = the cushion. (× 20.) *Continued on p.* 321.

events in the life-cycle. These may be ranked under two heads: *Apospory*, by which name are designated those cases where spore-production is omitted from the life-cycle, and a direct transition is effected, by continuous vegetative development, from the sporophyte to the gametophyte; and *Apogamy*, or to use the later and more comprehensive term *Apomixis*, which connotes the omission in one form or another of the act of syngamy, whereby a vegetative transition is effected from the gametophyte to the sporophyte.

The first intimation of a departure from the regular cycle of events in Ferns was made by Farlow (1874), who noted the vegetative production of sporophytic buds from the cushion of the prothallus of *Pteris cretica*, without the intervention of sexual organs. The subject received more general treatment later by De Bary (1878), who introduced the term *apogamy* to include all cases of the elimination of the sexual function (Fig. 296). The converse, viz. the omission of the event of spore-production, was first demonstrated by Druery before the Linnaean Society in June, 1884, and later it was examined

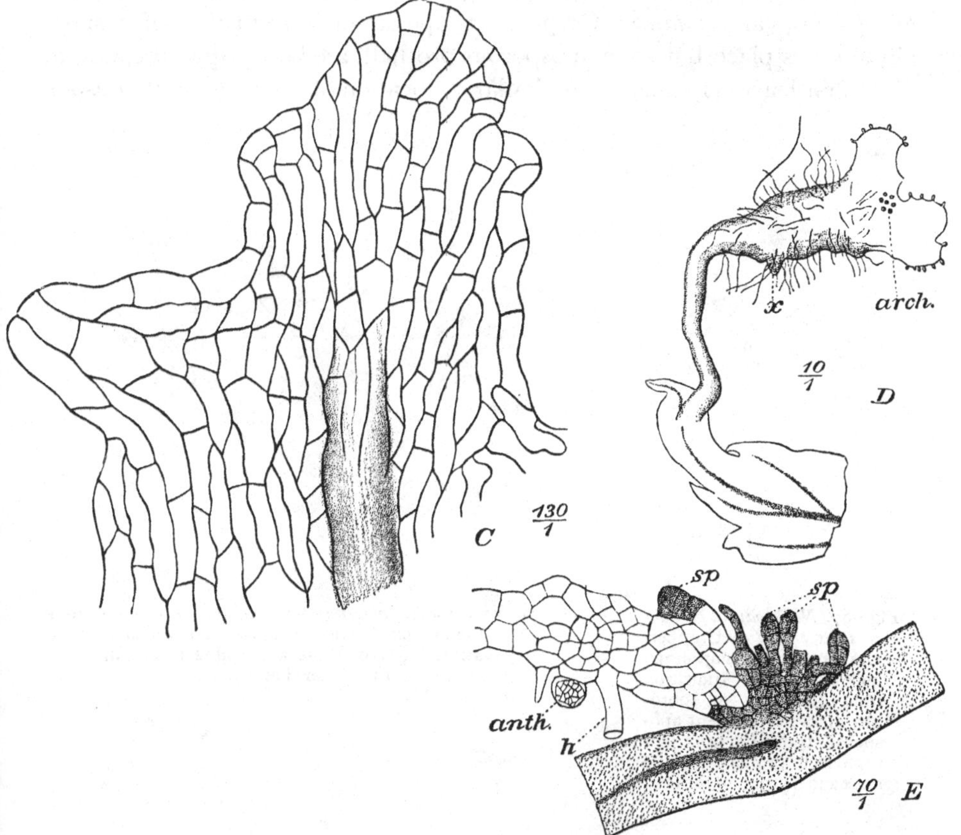

Fig. 296 *continued.* C = the initiation of a prothallus as in B, at the apex of a pinnule. The shading indicates a vein, beyond the tip of which the prothallus arises. (× 130.) D = a similar growth, but borne on an elongated cylindrical process: archegonia (*arch*) are already present. (× 10.) E = soral apospory in *Polystichum angulare*, var. *pulcherrimum*. A prothalloid growth bearing an antheridium (*anth*) and rhizoids (*h*) has arisen from the stalk of a sporangium. (× 70.)

in greater detail by myself in the Proceedings and Transactions of the Society. The term *apospory*, previously introduced by Vines, was adopted to connote all such cases (Fig. 296). Since 1884 very many observations both of apogamy and of apospory have been published, relating to very different families of Ferns, but chiefly to those which are phyletically late and derivative. Thus the phenomena which at first were held to be rare are now seen to be widespread, and it has been shown how in certain cases they can be artificially induced. Not only are the normal limits between the gametophyte and the sporophyte broken down by such observations as these, but the features of the two generations are liable to be mixed up in the most perplexing fashion.

In simple cases a succession of events, in themselves abnormal, may be seen following with some degree of regularity. For instance, in *Nephrodium pseudo-mas*, var. *cristatum*, Cropper, an apogamous production of a sporophyte takes place, but soon aposporous prothalli are borne upon the margins of its first leaves (Fig. 297). A similar collocation of apogamy and apospory

Fig. 297. *Nephrodium pseudo-mas*, var. *cristatum*, Cropper. Drawing by Dr Lang, showing apogamous transition from prothallus to sporophyte, and subsequent aposporous transition from sporophyte to prothallus at the apex and margins of the leaf.

Fig. 298. *Scolopendrium vulgare*. Prothallus from the branched cylindrical process of which ten roots arise: eight of these are visible in the drawing. (× about 6.) (After Lang.)

has been seen in a number of cases. Sometimes the parts of one or of the other generation appear to be formed without any definite sequence. This was made particularly plain by the cultures grown by Lang (1898), who found numerous roots, without any corresponding leafy shoots, borne upon prothalli of *Scolopendrium* (Fig. 298). Sporangia were seen to be borne upon a process growing out from the prothallus of *Nephrodium dilatatum*, while numerous archegonia were seated on its base (Fig. 299). He even observed young sporangia growing out from a transformed archegonium of *Scolopendrium* (Fig. 300). A similar medley of sporophytic and gametophytic features was found by von Goebel (1908) in a regeneration-growth from the primordial leaf of *Ceratopteris thalictroides*, in which a stoma is seated only three cells away from an antheridium (Fig. 301). Such developments appear quite inconsequent. It seems impossible to trace any order in them. They appear so varied that it would be possible by using individual cases to deduce almost any morphological conclusion from them. As M. Henri de Cassini has said of the abnormalities of Flowering Plants, "on verrait en elles tout ce qu'on voudrait y voir."

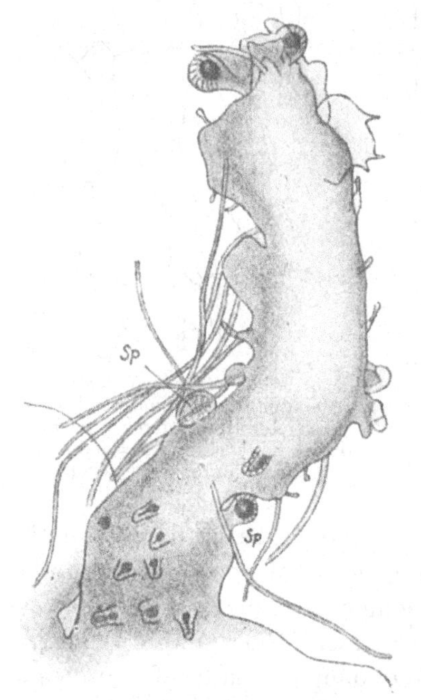

Fig. 299. *Nephrodium dilatatum*, Desv. var. *cristatum gracile*. Prothalloid cylindrical process bearing archegonia near its base. It arises by the side of an imperfect sporangium (*sp*), and bears a similar sporangium (*sp*) on the other side, and on the tip are a number of other sporangia associated with ramenta. (× 35.) (After Lang.)

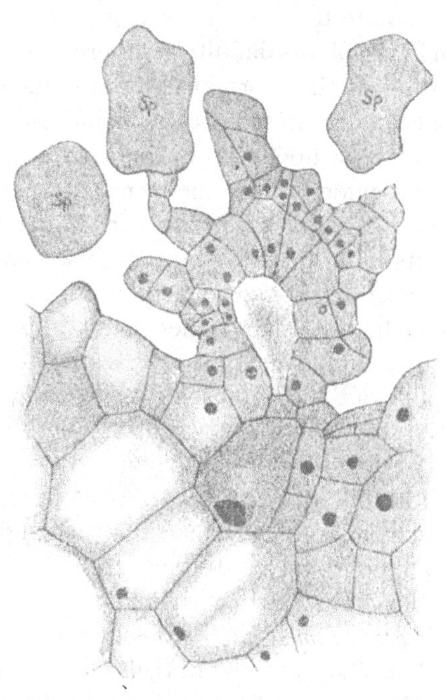

Fig. 300. *Scolopendrium vulgare*. Group of sporangia (*sp*) on a projection, the structure of which indicates its relation to an archegonium. Occasionally two nuclei are present in a single cell. (× 600.) (After Lang.)

In 1894 Strasburger stated generally, for plants showing alternation, that a difference in the number of chromosomes seen on nuclear division showed that the alternating generations differed in nuclear constitution. He recognised the spore-mother-cell in which reduction takes place as the limit between the sporophyte or diploid generation (*2n*), and the gametophyte or haploid generation (*n*). He held that the ovum, which on fertilisation has the number doubled, is the limit between the haploid gametophyte and the diploid sporophyte. The general statement of this distinction, which had already been indicated by Overton, was welcomed as giving a new precision to the facts of alternation. Observation of the details has shown in a vast number of normal life-histories that the chromosome-distinction between the generations is a just one. It applies equally to certain Algae, to the Archegoniatae, and to Flowering Plants, and may be accepted as a state to which a very large proportion of living plants definitely adhere. The recognition of this normal chromosome-cycle at once concentrated attention more critically than ever upon those abnormalities which are included under the terms apogamy and apospory: and the question as to the nuclear facts in such cases was recognised as of the utmost importance in their interpretation. On the other hand, these facts must necessarily influence any true estimate of the value of the chromosome-cycle itself in relation to evolutionary theory.

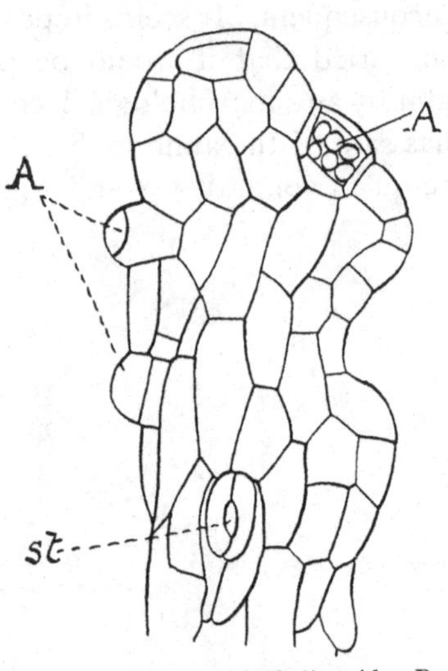

Fig. 301. *Ceratopteris thalictroides*. Regeneration from a primary leaf, showing an intermediate state between prothallus and leaf. *A*=antheridia; *st*=stoma. (After von Goebel.)

Already in 1898 Lang had observed in prothalli of *Scolopendrium* the frequent occurrence of two nuclei in a single cell of the tissue bordering on the change from gametophyte to sporophyte (Fig. 300). More detailed observations have since been made on other apogamous Ferns: for instance on *Nephrodium pseudo-mas*, var. *polydactylum* (Farmer, Moore, and Miss Digby, *Proc. Roy. Soc.* Vol. lxxi, p. 453). Examining young prothalli before any apogamous growths began to manifest themselves, it was found that certain cells contained two nuclei: when that was so a neighbouring cell was seen to be without a nucleus, and cases were found where the passage of the nucleus through a hole in the cell-wall was actually in progress (Fig. 302). Fusion of the two nuclei followed, and the process was regarded as a kind of irregular fertilisation. On their division the nuclei of the apogamous growth thus initiated show evidence of doubling of the number of chromosomes, just as happens in the normal post-sexual stage. On the other hand, meiosis occurs in the spore-mother-cells of this Fern, and the spores germinate to form a haploid prothallus. Thus the chromosome-cycle is essentially the same as in *Dryopteris Filix-mas*, except for the substitution of an irregular fusion in place of the normal syngamy: and the cytological criterion between the two generations holds good.

The apparently simple cycle of *Marsilia Drummondii*, A. Br. was worked out by Strasburger in 1907 (*Flora*, Bd. xcvii, p. 123). It was found to depart

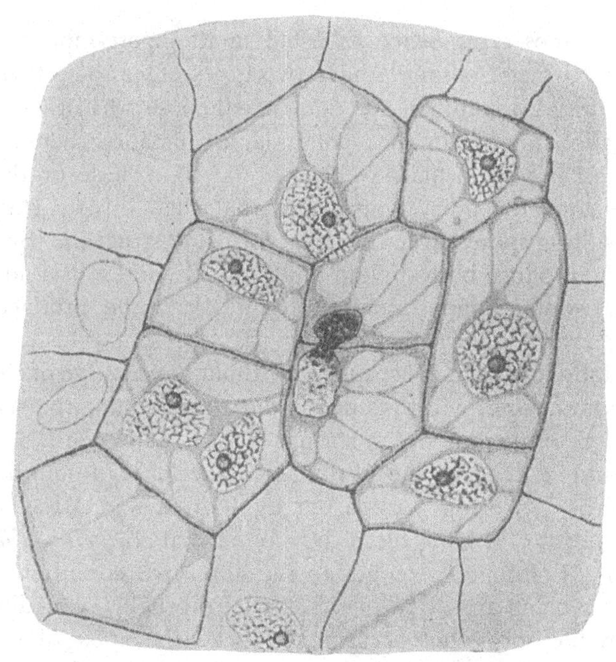

Fig. 302. *Nephrodium pseudo-mas*, var. *polydactylum*. Tissue of prothallus where an apogamous growth is to be found, showing on the left a cell with two nuclei, while an adjoining cell has none. At the centre a nucleus is seen passing through a perforation of the wall, and fusing immediately with that of the cell it enters. (After Farmer, Moore, and Miss Digby.)

further from the normal cycle than the previous case. The chromosome-numbers for the two generations are 16 and 32 respectively, and normal plants show the usual succession of events. But on germination of the megaspores borne by certain plants the gametophyte was found to have the diploid number, and this was seen even in the division to form the ventral-canal-cell of the archegonium; thus the ovum itself was diploid. In such archegonia the neck does not open, so that fertilisation by spermatozoids is impossible (Fig. 303). Nevertheless the unfertilised diploid egg develops apogamously into an embryo, which is naturally diploid also. An examination of the sporangia showed further that while in typical Marsilias the reduction

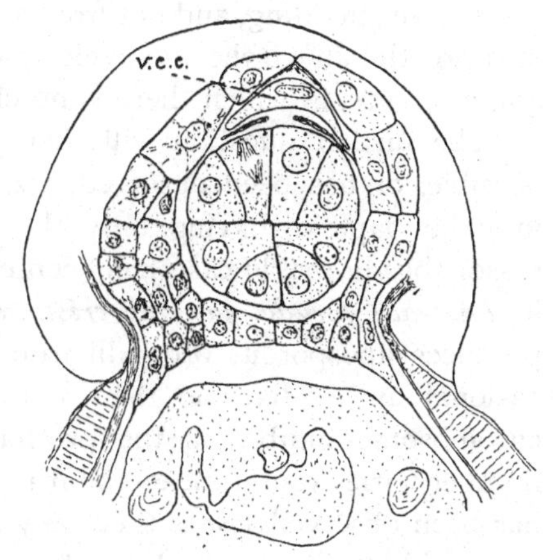

Fig. 303. *Marsilia Drummondii*. Parthenogenetic embryo: the neck of the archegonium has not opened, and the ventral-canal-cell (*v.c.c.*) is still in place. (After Strasburger.)

to 16 chromosomes takes place as usual in the spore-mother-cells, in *M. Drummondii* the megasporangia have two types of spore-mother-cells. The one type is normal in number, and shows reduction: the other type is produced in smaller number in the sporangia, for instance there may be only four spore-mother-cells in place of the normal 16. These on division have diploid nuclei, and the interesting fact is that their diploid state does not divert the resulting spores from the usual form and structure. Since the apogamous plants produce both diploid and haploid spores, it is not surprising that both apogamous and sexual prothalli should be produced on their germination, and it follows that among the representatives of the species there will be individual cycles completed without any change of chromosome-number. Certain cycles will thus be diploid throughout; the spore may be diploid, also the prothallus which springs from it, and even the egg itself.

Almost simultaneously the chromosome-cycle of a number of other abnormal Ferns was being worked out by Farmer and Miss Digby (*Ann. of Bot.* 1907, p. 161). Among these they found that *Athyrium Filix-foemina*, var. *clarissima*, Bolton, corresponds to the abnormal condition of *Marsilia Drummondii* in being diploid throughout. The prothalli are readily produced, usually from the sorus, but occasionally from the apex of the pinnules. They are very fertile, bearing sexual organs freely: but there is no fertilisation, nor any migration of nuclei. There is thus neither reduction nor doubling of chromosomes, and the new sporophyte, so far as observed, "invariably arises from the oosphere." *Scolopendrium vulgare*, var. *crispum Drummondiae* is also like these in all essential details. But *Athyrium Filix-foemina*, var. *clarissima*, Jones, differs sharply even from *A. F.-f.*, var. *clarissima*, Bolton, in the fact that the embryo arises from the prothallus by apogamous budding, and not from a diploid ovum. Apart from these details, in all of the above the life-cycle is diploid throughout, including even the gametophyte, in which there is no change of external character to be seen.

The further question will then present itself whether the converse is possible, or has been observed, viz. that the normally diploid sporophyte may be haploid, having only the reduced number of chromosomes. A reasonably probable case has been established by Farmer and Miss Digby in *Lastraea pseudo-mas*, var. *cristata apospora*, Druery. The detached leaf produces aposporous prothalli from its margin or surface, which bear occasional antheridia, and the sporophyte is produced apogamously. The chromosome-number in the prothallus is about 60: in the embryo the number varies round mean countings of 60 and 78. No migration of nuclei has been observed, nor is there any reduction in the whole cycle. The facts suggest that the gametophyte-character has been impressed upon the sporophyte, the converse in fact of what has been seen in the varieties of *Athyrium*, and in *Marsilia*. A similar condition was very fully made out by Yamanouchi for *Nephrodium molle*, in which normally the chromosome-cycle is based upon the gametophyte-number of 64 or 66. When apogamy occurs a sporophytic bud appears by direct vegetative growth, whose cells on mitosis still give the gametophyte-number of 64 or 66. A further example is recorded with all necessary detail by Steil for *Nephrodium hirtipes*, Hk. in which the prothallus arises by germination of haploid spores: but the gametophyte never produces archegonia: the embryo originates as a vegetative growth from the prothallus. No nuclear migrations or fusions have been observed, and the sporophyte retains the haploid number of 60 to 65 chromosomes.

It seems unnecessary to multiply instances of such irregularities of the chromosome-cycle, or to give an exhaustive abstract of the now voluminous literature. Abundant references are quoted in the memoirs here cited. From what has been stated above it is clear that the gametophyte may be diploid, though normally it is haploid: and that the sporophyte may be haploid, though normally it is diploid: in neither case does the cytological difference appear to affect the form or structure of the gametophyte or sporophyte in question. The general conclusion from these facts may be stated substantially in the terms of Prof. Farmer and Miss Digby (278, p. 197): that no necessary relation exists between periodic reduction in the number of chromosomes and the alternation of generations. Therefore the problem of alternation and its nature must be settled by an appeal to evidence other than that derived from the facts of meiosis. That, so far as the Archegoniatae are concerned, alternation is normally associated with meiosis on the one hand and syngamy on the other no one will dispute. But now that it has been shown that no necessary connection exists between alternation and the usual nuclear cycle, it is scarcely to be wondered at if the presumed correlation should often break down in complex cases. So long as the exceptions were of rare occurrence they have been regarded as monstrosities or abnormalities. But now that records of them have become common, they are to be looked upon as proofs of current physiological instability. Each may be held to result from an individual break-away from the normal course of events. There is no reason to hold that such a break-away represents any state which has had a settled place in the previous history of the individual or of the race. In many cases, or even in most of them, the source may probably lie in hybridisation of gametes in some sense incompatible, so as to prevent or alter the readjustment of the nuclear mechanism involved in meiosis. Such hybridisation is always liable to occur in promiscuous growths of prothalli on moist soil, with casual juxtaposition of those derived from various sources; and a method of attraction determining the movements of the spermatozoids, common to several prothalli, would encourage it.

The instances quoted suggest that within certain limits each plant is a law to itself; but in certain points those limits are binding. Sexuality and meiosis are knit together by the closest physiological ties. The latter is the complement of the former. In like manner there is a relation between apospory and apogamy. So far as our present knowledge goes apospory is always found to imply the absence of meiosis from the life-cycle of the organism. Consequently after apospory apogamy is inevitable. But it does not follow that after apogamy there must be apospory. This is shown by the case of *Nephrodium Filix-mas*, var. *polydactyla*, Dadds. On the other hand it would appear highly undesirable to see in such irregularities of the life-cycle as those described direct evidence suitable for morphological or phyletic argument. Their instability and sporadic occurrence precludes it. It has been seen that apospory may be initiated from the sporangium, or from the vegetative tissue of the leaf. That certain Ferns may sometimes follow the normal cycle, and at others show apogamy (*Marsilia Drummondii*). Further that apogamy may be carried out in different ways: for instance by migration of nuclei (*Nephrodium Filix-mas*, var. *polydactyla*); or by direct vegetative growth from a diploid source a diploid apogamous bud may arise (*Athyrium Filix-foemina*, var. *clarissima*, Bolton); or again from a haploid source a haploid apogamous bud (*Nephrodium molle*, and *hirtipes*). But the most important objection of all to their direct use in comparison with a view to phylesis lies in the fact, that no definite phylum or line of Descent has been permanently established showing any of these aberrant characters as a constant feature. Such considerations will prevent these irregularities of the life-cycle being used in serious morphological or phyletic argument. Their value lies in the fact that they illustrate what is practicable in the chromosome-cycle, and so they tend to relax a too rigid conception of it.

They also provide physiological suggestion. Observations on induced apospory are cases in point. When first apospory was described it was thought that it was specially related to the adult leaf, and particularly to the fertile parts of it. Repeated attempts were accordingly made to induce it in pieces of the sporophylls of many Ferns, but without any result (*Ann. of Bot.* Vol. iv, p. 169). There appears to be a marked disability in the adult leaf for bridging over the limit between the generations in any other way than by spores. But it is different with the juvenile leaf. Von Goebel (*Experimentelle Morphologie*, 1908, p. 197, etc.) has shown how readily the primordial leaves of various Ferns form aposporous prothallial growths (Fig. 304). This clearly points to an essential difference between the juvenile and the adult states. The latter seem to be more limited in their potentialities than the former.

It will naturally be asked, what bearings have these facts of apospory and apogamy as seen in the Filicales upon any general theory of alternation? The reply may be in the form of another question: are we justified in having any "general theory of alternation" applicable to plants at large? It is a fact testified by innumerable instances that syngamy doubles the number of chromosomes, and initiates a diploid state. It is also a fact of experience that the number is halved on meiosis, which heralds spore-formation, and the establishment of a haploid state. Beyond these facts there seems to be nothing fixed for plants at large as regards their life-cycle. Experience shows that where sexuality occurs plants work this chromosome-cycle into their life-history in various ways as regards somatic development, and the consequent establishment of "generations." The facts above detailed for the Filicales show a certain degree of latitude even in this group.

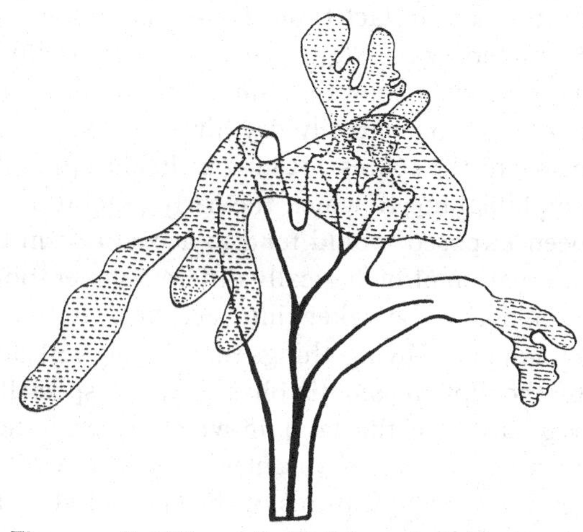

Fig. 304. *Alsophila van Gertii.* Primary leaf with *Aneura*-like prothalloid growths shaded so as to distinguish them from the leaf itself. (After von Goebel.)

But in the Thallophytes the proportions and even the sequence of the so-called "generations" may show a much greater latitude of difference, as illustrated by those relatively few Algae and Fungi in which the details have as yet been thoroughly worked out. It is still an open question whether in any given phylum of Thallophytes the point at which meiosis takes place in the life-cycle has been fixed throughout Descent or has been moveable. This question is especially a moot point for the Rhodophyceae: and when the Phaeophyceae and Chlorophyceae have been further examined it seems not improbable that like questions may arise also for them.

The general impression given by the comparison of Thallophytes so far as their analysis has yet gone, in respect of the fit of the chromosome-cycle upon their somatic development, is that there is a high degree of fluctuation. The two phases do not appear to have settled down to any mutual adjustment that has become general. They illustrate a definite alternation in the making rather than as a present actuality. In the Archegoniatae, however, the life-cycle shows as a normal feature so constant an adjustment of the chromosome-cycle to the somatic development that it justifies the recognition of a definite alternation. No one can doubt in them the normal nature of that succession of events which is demonstrated by all their known life-histories. A biological reason for the constancy of that alternation which they show has been suggested in an accommodation to sub-aerial conditions during their Descent. It is probably this that has stabilised their life-cycle. The cytologically distinct generations differ widely in form and structure. The more delicate gametophyte is well suited to moist conditions, while its ultimate function of sexual reproduction cannot be carried out without the presence of external water. It is in fact typically semi-aquatic in its nature, sharing many of its characters with Algal types. The sporophyte is fitted by its more robust habit as well as by the differentiation of its tissues for successfully enduring exposure to relatively dry air, while this condition is essential for the dispersal of the spores. It may be held as probable that the circumstances of an amphibial life, to which the early vegetation of the land must naturally have been exposed, would tend to accentuate in their phylesis this characteristic alternation of biologically different generations. But a further consideration should also be taken into account. It is a fact illustrated in both of the kingdoms of living things that the highest developments have been attained by the diploid somatic phase. More especially is this so in the characteristic vegetation of the land, in which a ventilation-system with stomata, a conducting system of vascular tissues, and a high evolution of external form are the specific features of the sporophyte. These are all either absent from the haploid gametophyte, or are represented only in limited degree in some of the most highly elaborated Mosses. The facts both in plants and animals suggest that some higher potentiality lies in the diploid state, which may in itself account in some degree for the ascendancy of the sporophyte, and for its general dominance over the gametophyte in the vegetation of the land. We may hold that it is able by its higher adaptability to seize upon and make its own use of those sub-aerial conditions which are less easily met by the haploid gametophyte. Whether or not this is a true picture of the origin of the marked difference between the generations of the Archegoniatae, it at least accords with the facts, and makes intelligible that high degree of constancy, as well as of differentiation, which is seen in the alternate generations of the Archegoniatae. On the other hand, the more uniform circumstances of many Thallophytes would allow of such latitude as they actually show in the adjustment of their chromosome-cycle to their somatic development. Never having been exposed in the same way to the stress of sub-aerial conditions, we may recognise that their life-cycle has not been standardised to the same degree as that of the Archegoniatae.

The further step of attempting to trace some exact source from which this more highly standardised alternation of the Land-Plants may have sprung would lead, in the absence of specific fact, to mere speculation. In a treatise on the phylogeny of the Filicales such speculation can find no place. It must suffice to state the plain facts of their alternation and the modifications to which they are subject, and to compare them in general terms with what is seen in the Thallophytes. But this should be done without advancing those comparisons beyond the bounds of general Organography. They should not be presented in the light of any definite phyletic suggestion until the relations between certain Archegoniatae with certain definite Thallophytic types have been drawn much closer than the facts at present available would justify. Moreover homoplasy is written so large across the face of the Vegetable Kingdom, and in particular of the lower organisms, that it should give pause to rash speculation. It should suggest caution in recognising as truly phyletic such similarities, whether of somatic development or of the various life-histories, as have been based upon an obligatory chromosome-cycle. Whatever further interest the normal alternation or its abnormalities seen in the life-cycle of Ferns may ultimately present for comparison with that more lax alternation seen in the Thallophytes, such features cannot yet be profitably used in phyletic discussion. In point of fact, after all the intensive study which has been devoted to the chromosome-cycle of the Filicales, and of the Archegoniatae at large, the question of the origin of alternation in them stands at the moment very much as it stood in 1890, before the cytological difference of the alternating generations had been recognised.

Crested Varieties

There yet remain certain abnormalities frequently seen in Ferns of very various affinity, which are described as "crested." In these the leaves are in greater or less degree marked by extra branchings, chiefly at or near to the distal end of the phyllopodium, pinnae, or pinnules. Often the branchings appear as very exact bifurcations, and may even differ in this mode of branching from that seen in the normal parts of the same leaf. Though the distal regions of the leaves are most commonly affected, the forking may appear sometimes in the lower regions of the leaf. *Scolopendrium* often shows its blade forked into two almost equal shanks. The same is not uncommon in *Polypodium vulgare*, while in certain strains of *Nephrodium molle* the leaf frequently appears to fork near to its base, giving the semblance of two equal leaves borne upon a single stalk (Fig. 305, *B*). In cultures grown in Glasgow from spores this abnormality of *N. molle* recurred with some degree of persistence in successive generations. Such simple forkings as those named are frequently combined with extra pinnations beyond those normally present. The result may be complex structures which are difficult to refer back to the exact factors to which they owe their origin (Fig. 305, *A*). In extreme cases crested Ferns present a very full habit, and are often of great beauty. In particular, the crested forms of *Nephrolepis* are raised for the market in large numbers by nursery gardeners as decorative plants. But the extreme types are liable to revert under ordinary conditions of culture towards the normal and less decorative form of leaf.

A very considerable number of British Ferns have been found to show crested varieties. Many are recorded under such names as vars. *multifida laciniata*, *cristata*, *polydactyla*, etc. by Moore(300), Lowe(301), Druery, and others. Reference may also be made to Luerssen for a record of the continental examples(302). Such varieties bear, however, in so high a degree the stamp of individual aberrations from the normal that they suggest that they should properly be ranked as individual "sports." Nevertheless they present many legitimate problems. From the comparative, morphological

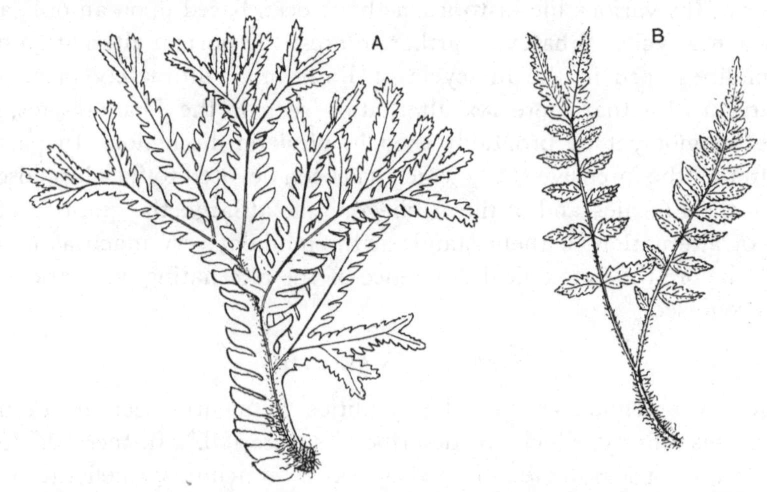

Fig. 305. *A. Nephrodium Filix-mas*, *multi-cristata*, Lowe. A single pinna. (After Lowe.) *B. Nephrodium Filix-mas*, *Schofieldii*, Sim. Leaf twice forked. Reduced. (After Lowe.)

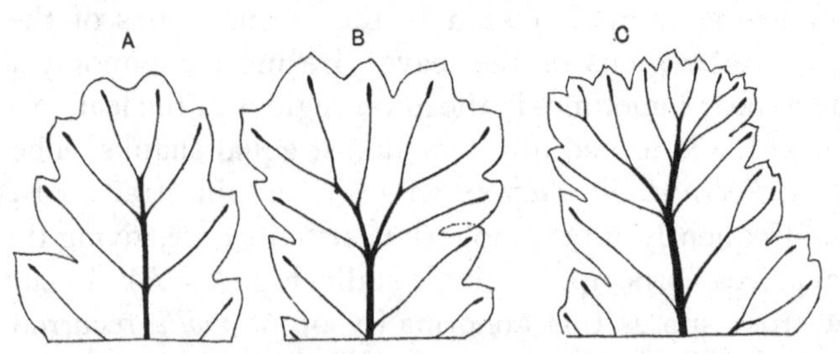

Fig. 306. Tips of pinnae from juvenile leaves of *Nephrodium Filix-mas*, var. *cristatum*, raised apogamously in the Glasgow Botanic Garden, 1889. They show three progressive examples of cresting. *A* is normal; *B* the distal dichotomy is almost equally expanded; in *C* the right-hand shank is more profusely crested than the left. Enlarged.

point of view the cresting is essentially a reversion. A leading feature in it certainly is a partial, or indeed often a very perfect, return to that equal forking which comparison shows to have been primitive (Chapter V). The very gradual steps leading to this have been observed in crested pinnules of *Osmunda regalis*. These give an opportunity of relating the forked development to the regular dichopodial architecture of the normal leaf.

Three examples are shown in Fig. 307. In (*A*) the forking is slightly unequal, the stronger shank being to the left. There is a regular alternation of the veins forming the dichopodial midrib, and the whole supply to the smaller right-hand lobe may be held to represent one of those veins which has developed more strongly, and branched more frequently to supply the marginal outgrowth. On this view it would represent a partial reversion from the inequality of the forking in the sympodium towards the primitive equality seen in the juvenile leaves. (Compare Chapter V, Fig. 80.) A similar interpretation would apply to (*B*). Such conditions harmonise readily with that in (*C*), where the veins of the two lobes are very equally developed, so that it is difficult to say whether the one lobe or the other is the apex of the dichopodium. In other words, the dichopodium has reverted very perfectly to the primitive, equal dichotomy.

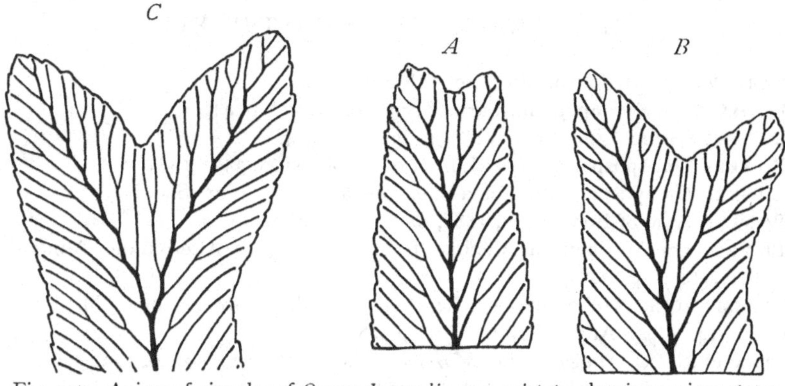

Fig. 307. Apices of pinnules of *Osmunda regalis*, var. *cristata*, showing various states of furcation of the apex. (× 4.)

It cannot escape notice that two of the Ferns quoted earlier in this Chapter as showing irregularities of the chromosome-cycle are "crested" varieties. *Scolopendrium vulgare*, var. *crispum Drummondiae* has been seen to be diploid throughout its life-cycle, while there is strong probability that the converse holds for the Fern styled *Lastraea pseudo-mas*, var. *cristata* (= *Nephrodium Filix-mas*, var. *cristatum*), which is presumably haploid. The cresting appears in the juvenile leaves of the apogamously-produced plants of the latter, and examples of their pinnae may be seriated so as to show progressively gradual steps from the normal (Fig. 306, *A*) to pronouncedly crested tips (Fig. 306, *C*). Such facts suggest that some relation exists between the crested state and sexual irregularity. A crested state may be, like apospory and apogamy, in some way the expression of an incompatibility of chromosome-number in synaptic pairing, consequent on the promiscuous hybridisation which is always possible in Ferns. The necessary facts are not yet available for establishing the suggested relation: but so far as they go they appear to indicate that the question deserves to be tested by further cytological enquiry. Such enquiry would not only bear interest in relation to the chromosome-cycle itself: but it might also help to show what degree of relation exists in plants between form and nuclear constitution. Yet still the fact remains that crested varieties commonly show branching by equal dichotomy. This may be held as a reversion to that state recognised in Chapter v as the more primitive. The fact of its occurrence in crested forms of very various affinity may be held as supporting once more the general thesis, that leaf-architecture in Ferns is based upon sympodial progression from a primitive dichotomy.

BIBLIOGRAPHY FOR CHAPTER XVI

268. FARLOW. Quart. Journ. Micr. Sci. 1874, p. 266.
269. DE BARY. Ueber apogame Farne. Bot. Zeit. 1878, p. 449.
270. DRUERY. Proc. Linn. Soc. Vol. xxi, p. 254. 1884.
271. BOWER. On Apospory. Proc. Linn. Soc. Vol. xxi, p. 355. 1884.
272. BOWER. On apospory and allied phenomena. Trans. Linn. Soc. N. S. Vol. ii, 1887.
273. BOWER. Ann. of Bot. Vol. i, 1888, p. 269, Vol. iv, p. 168.
274. BOWER. On Antithetic as distinct from Homologous Alternation. Ann. of Bot. Vol. iv, 1890, p. 347.
275. STRASBURGER. Ueber periodische Reduktion der Chromosomenzahl. Biol. Centralbl. 14, 1894. Ann. of Bot. 1894, p. 281.
276. STRASBURGER. Apogamie bei *Marsilia*. Flora. Bd. 97, 1907, p. 123.
277. LANG. Phil. Trans. Vol. 190, 1898, p. 187.
278. FARMER & DIGBY. Ann. of Bot. Vol. xxi, 1907, p. 161.
279. WINKLER. Ueber Parthenogenesis u. Apogamie. Progressus Rei Bot. Vol. ii, p. 293. Here the literature is fully quoted.
280. VON GOEBEL. Experimentelle Morphologie. 1908, p. 187. Organographie. II Auflage. Part I, p. 413, Part II, p. 996.
281. YAMANOUCHI. Apogamy in *Nephrodium*. Bot. Gaz. May 1908. Vol. 45, p. 289.
282. STEIL. Ann. of Bot. 1919, p. 109.
283. SVEDELIUS. *Scinaia*. Nova acta Reg. Soc. Ups. Ser. IV. Vol. 4, No. 4. 1915.
284. SVEDELIUS. Das Problem des Generationswechsels bei den Florideen. Naturwiss. Wochenschrift, xv, 1916.
285. TANSLEY. Meiosis and Alternation. New Phyt. 1912, p. 145.
286. DAVIS. Life histories of Red Algae. Amer. Naturalist. 1916, p. 502.
287. KLEBS. Beding. d. Fortpfl. bei einigen Algen u. Pilzen. Jena. 1896.
288. OLTMANNS. Morph. u. Biol. der Algen. Jena. 1904.
289. LLOYD WILLIAMS. Studies in Dictyotaceae. Ann. of Bot. 1904, p. 141.
290. LEWIS. *Griffithsia*. Ann. of Bot. 1909, p. 639.
291. STRASBURGER. *Fucus*. Pringsh. Jahrb. 1897.
292. FARMER. *Fucus*. Phil. Trans. B. 1898.
293. ALLEN. *Coleochaete*. Ber. d. D. Bot. Ges. 1905, p. 285.

294. YAMANOUCHI. *Polysiphonia.* Bot. Gaz. 1906, p. 401.
295. STEVENS. *Albugo.* Bot. Gaz. xxviii, p. 149.
296. TROW. *Achlya.* Ann. of Bot. xviii, p. 541.
297. BLACKMAN. *Phragmidium.* Ann. of Bot. xviii, p. 323.
298. KLEBAHN. Desmids and Diatoms. Pringsh. Jahrb. xxii, xxix.
299. SCOTT. Address to Sec. K. Brit. Assn. Report. Liverpool. 1896.
300. MOORE. Ferns of Great Britain and Ireland; Nature-printed. 2 vols. 1855.
301. LOWE. Our Native Ferns. London. 1865.
302. LUERSSEN. Die Farnpflanzen. Rab. Krypt. Flora. iii, 1889.

CHAPTER XVII

ORGANOGRAPHIC COMPARISON OF THE FILICALES WITH OTHER PLANTS

IN the foregoing Chapters an analysis has been made of the several outstanding features, somatic and propagative, which may be used as a basis for the seriation of Ferns according to the probable history of their evolution. In each of them, partly on comparison of living forms, partly by reference to the fossil history, the characteristics held to be relatively primitive have been distinguished from those held to be relatively recent in Descent. In these determinations each criterion of comparison has been regarded as standing upon its own footing, and has been judged as far as possible independently of others. But naturally if the conclusions arrived at are sound, it may be expected that the sequences in respect of the various criteria of comparison will run parallel; and in the measure in which they do so they will gain mutual support. By collecting the conclusions from comparison in respect of the different criteria, and putting them together, it will be possible to construct a composite picture of an *archetype* which shall embody all the features which are recognised as the most primitive for Ferns. The visualisation of such an archetype would thus be *based upon actual knowledge of Ferns living or fossil.* It can then be compared with the earliest fossil land-plants of which we have detailed knowledge. It is believed that this method is more trustworthy and scientific than any general comparisons of land-plants with Algae, and will be more helpful than these in elucidating the probable evolution of the Class. The reason for holding this opinion is that a comparison of Archegoniate with Archegoniate is nearer than that of an Archegoniate with any Alga, and it is more reliable because it leaves less scope for the slack joint of homoplasy.

In Chapter III *the simple shoot* composed of an axis and an acropetal succession of leaves has been recognised as the unit of construction of the sporophyte for the Ferns at large. In the primitive condition the shoot as a whole is *unbranched*, as it is at the start of every ontogeny of living Ferns. This condition is often maintained in the adult of the Ophioglossaceae, Osmundaceae, and Marattiaceae, in which the axis is as a rule upright, a position held to be primitive. Such shoots are radially constructed. The prone position is probably derivative, though it certainly was acquired early, and it follows naturally from the heavy leafage, and the prevailing absence of secondary thickening of the stem. The symmetry of the prone shoots is dorsiventral, which is held to be a derivative state in Descent, as it is actually seen to be in the individual life of *Helminthostachys*, *Danaea*, or *Christensenia*.

Terminal branching of the shoot is not uncommon (Frontispiece). As a rule it is *dichotomous* in primitive types: but the shanks are frequently developed unequally, so that a sympodial shoot-system results. If the inequality is great it may lead to the weaker shank appearing as an appendage, either in the axil of the nearest leaf (Ophioglossaceae, Hymenophyllaceae, *Ankyropteris*), or in various other positions, most frequently as an abaxial bud on the leaf-base (*Lophosoria*, *Metaxya*, etc.). Various grades may be seen between equal dichotomy and monopodial branching, with axillary branching as a special case. But all may be referred to dichotomy, which appears to have been the primitive mode of branching; that is, if branching occurred at all in the most primitive forms. In many no such branching is seen.

The shoot is fixed in the soil by numerous roots, of which all the later are clearly adventitious. The nature and origin of the first root may be open to question. That origin is not constant in time or place for all Ferns. The root may actually be absent in *Salvinia*. Its emergence where there is a suspensor present is lateral. These facts indicate that it is accessory to the shoot, as are all those which follow accessory to the shoots which bear them. *Accordingly it is possible to contemplate a primitive Fern-sporophyte as a simple, upright, radial, rootless shoot, either unbranched or showing dichotomy.*

The leaves of living Ferns are bifacial, and most of them have two rows of pinnae, one on either side of the rachis: but in certain Zygopterids the higher appendages were in alternating pairs, forming four longitudinal rows, and giving a radial construction. Thus the bifacial leaf was not universal. In Chapter IV it is shown how the whole leaf of Ferns is traceable in origin to *an elongated rachis with dichotomous distal region*, and frequently with stipular growths at the base. All manner of steps of sympodial development can be traced in the distal branch-system, so as to establish a *dichopodium*, which being continuous with the rachis constitutes the *phyllopodium*. Upon this in advanced cases the earlier pinnae may arise monopodially, the later being dichotomous. In many primitive types the segments are all separate, each containing a single vein. This is held to be the original state. Webbed leaf-expansions are held as derivative, but they may still retain the dichotomous venation, with free vein-endings. A further advance was the looping of the veins to form a reticulum. Thus *a primitive type of Fern-leaf may be figured as long-stalked, with a distal dichotomy of narrow, separate, single-veined segments, arranged either radially or bifacially.*

Such Ferns as existed in the Primary Rocks were mostly, and perhaps exclusively Eusporangiate. It is shown in Chapter V that corresponding

Ferns of the present day have not only their sporangia complex in segmentation—which is the feature of Eusporangiate types—but all their apical meristems are also relatively complex. In the primary segmentation of each of their parts more than a single initial cell is the rule. This indicates a general robustness of organisation of the whole plant. But in the Leptosporangiate Ferns there is a relative simplicity and regularity of all the apical regions, with a single initial cell in each, a state which runs parallel with their precise sporangial segmentation. This precision is regarded as secondary and derivative. Naturally the apical meristems of Palaeozoic Plants can only occasionally be examined in detail, and cannot be expected to give exact results. But such comparison as is possible suggests that *the primitive Fern-types would probably have been relatively robust in the primary organisation of all their parts, as are the living types to which they are related.*

The vascular system gives a better foundation than any other structural features for comparison, and its features are specially applicable in the phyletic treatment of the later and derivative types of Ferns, in which more complex primary structure is seen in the adult shoot than in any other class of plants. Passing backwards in Descent from these elaborate and derivative states, the vascular system is found as a rule to be of simpler structure, until in some of those which in other respects are held as primitive, a condition is seen where the axis is traversed by a solid cylindrical column, or protostele. A like structure is found also in very early fossil Ferns, and a similarly simple structure appears in all Fern sporelings, however complex the adult structure may be. Thus the evidence from comparison of adult living Ferns held as primitive, whether the facts be derived from early fossils, or from the first steps of the ontogeny of living Ferns, all points to the same conclusion; that *the protostele is the primitive vascular construction for the stem of the Filicales.* Similarly the leaf-stalk is traversed by strands of the leaf-trace: and comparison along similar lines shows that *the primitive vascular supply to the leaf was a single undivided strand, and that in the earliest types it was oval or almost cylindrical in transverse section, thus resembling that of the axis.*

The non-vascular tissues give less useful material for phyletic treatment. The most constant features about them are the facts that they serve for nutrition and storage, and that they embed the conducting tracts. But it is otherwise with the dermal appendages. All living Ferns have them of one sort or another, though sometimes only sparingly: and they are frequently absent from the adult parts. Simple hairs are found in primitive types such as the Osmundaceae, and they are recorded for *Botryopteris.* Flattened scales are an indication of advance, and they are found in most of the later Leptosporangiate Ferns. *The most archaic Filical types would therefore be expected either to bear simple hairs, or possibly to be glabrous.*

Comparison of living Ferns has led definitely to the conclusion that the superficial position of the sorus, so prevalent in modern Ferns, is a later and derivative state, and that *a distal or marginal position of the sorus was primitive.* The latter is found in the Botryopterideae among the fossils, and in the Ophioglossaceae, in *Osmunda*, and the Schizaeaceae, and in many other primitive types of living Ferns. The sporangia of the Simplices are all *formed simultaneously*, and they are usually *without indusial protections.* These are also indications of a primitive state. In many of the Simplices the sporangia are grouped in sori: but sometimes the sporangia are *solitary*, as in the Schizaeaceae, and each is seated on its own vein-ending, as in *Botrychium.* This is the condition styled by Prantl the "monangial sorus," and it is probably the most general, primitive soral condition of all. It is well represented by the fossil *Stauropteris*, which is considered to be a very early type of the Filicales. The primitive sporangium was a relatively large one, of the Eusporangiate type, with a thick wall and a simple opening mechanism, and with a large spore-output. All these characters are seen in the Ophioglossaceae, Marattiaceae, and Osmundaceae, as well as in the Botryopterideae, and in *Stauropteris.*

Using the archaic characters thus recognised by comparison of the sporophyte in living and fossil Ferns, it is possible to visualise a primitive type which would represent them all, and so to sketch an *archetype* for the Class. It would consist of a simple upright shoot of radial symmetry, possibly rootless, dichotomising if it branched at all, and with the distinction between axis and leaf ill-defined. The leaf, where recognisable as such, long-stalked with distal dichotomy, tending in advanced forms towards the sympodial origin of a dichopodium. All the limbs of the dichotomy would be narrow, and distinct from one another. The whole plant would be relatively robust as regards cellular construction, and traversed by conducting strands with a solid xylem-core. The surface might be glabrous, or invested with simple hairs. The solitary sporangia would be relatively large, and distal in position, with thick walls, and a simple method of dehiscence: and each would contain numerous homosporous spores. Though all the chief primitive features are concentrated into this description, it is not to be assumed that they will all be represented in the same individual or type: but it is quite conceivable that they should be.

The specification thus given is entirely based upon comparisons of plants recognised as belonging to the Filicales, living or fossil. But if it be checked by reference to the form and structure of the fossil Psilophytales of the Devonian Period, it is at once apparent that a real similarity exists between the verbal specification and plants which have actually lived (Fig. 308). No attempt is made to link this archetype of the Filicales definitely with any one of the described fossils, or to point to any one of them as their actual ancestor.

It is not even suggested that the Psilophytales include the direct ancestry of the Filicales. The intention is only to indicate the general similarity which the verbal sketch bears to these ancient examples of land-vegetation. The chief points of difference lie in the higher differentiation of the shoot shown by Ferns, and the establishment of a root-system. The possible origin of the Fern-shoot from a simpler source was forecasted from a comparative study of Ferns and Cycads in 1884 (Bower, *Phil. Trans.* 1884, Part II, p. 605). It was asked, "May we not with good reason think that, just as the phyllopodium gradually asserts itself as a supporting organ among parts originally of similar origin and structure to itself, so also the stem may have gradually acquired its characters by differentiation of itself as a supporting organ from other members originally similar to itself in origin and development? Thus the stem and leaf would have originated simultaneously by differentiation of a uniform branch-system into members of two categories, and this is what is actually illustrated in the case of the phyllopodium and pinnae in the series of plants above discussed." This suggestion was "left in the air," owing to the entire absence at the time when it was written of any direct evidence to support it, such as might be supplied by more primitive vascular plants. Now with the Psilophytales before us, and with the striking similarity of vascular structure between the leaf and the axis of *Botryopteris cylindrica*, together with other anatomical evidence now available, the idea of the differentiation of axis and leaf as seen in the Filicales from an indifferent dichotomising system emerges as a reasonably tenable hypothesis.

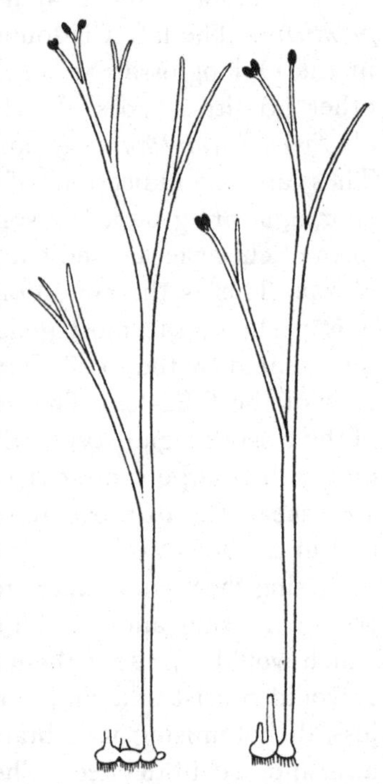

Fig. 308. *Hornea Lignieri*, reconstruction by Kidston and Lang; quoted here as an example of a very early type of land-living sporophyte.

It has been suggested in the final paragraph of Chapter XV, p. 315, on Embryology, that as the first leaf and the axis originate jointly from the epibasal hemisphere in all Ferns, the primitive spindle may have had the power of branching at its apex, and that these two parts were the result. It has lately been shown by Holloway that the embryo of *Tmesipteris* actually does this occasionally, and that the branches of the indifferent "shoot-region" may be equal, or in various degrees unequal. The former is shown in Fig. 309. Here the epibasal region has divided into two equal parts, each endowed with apical growth. The fact thus brought into the comparison is not quoted as any indication that *Tmesipteris* represents an ancestral form for Ferns. It is cited to show that the condition required by our induction from the developmental facts has now been actually observed in a primitive type of living sporophyte. Thus for the first time the suggestion made in 1884 takes its foundation upon a basis of observed fact. The hypothesis may now be accepted as passing from the region of mere surmise to the arena of scientific theory. Once the distinction between axis and leaf is attained, the further derivation of a Fern-leaf, as seen in its simpler forms, will follow readily by reduction of the radial to a bifacial type, and by sympodial development and webbing of the dichotomous twigs to form the pinnae, along lines traced in Chapter V. Finally, if such a distal position of the sporangia as is seen in the early Devonian forms were retained throughout the progress of these changes, they would then appear terminal on the vascular strands, as they actually are in *Stauropteris*, and in *Botrychium* (compare Lignier, 305, 308).

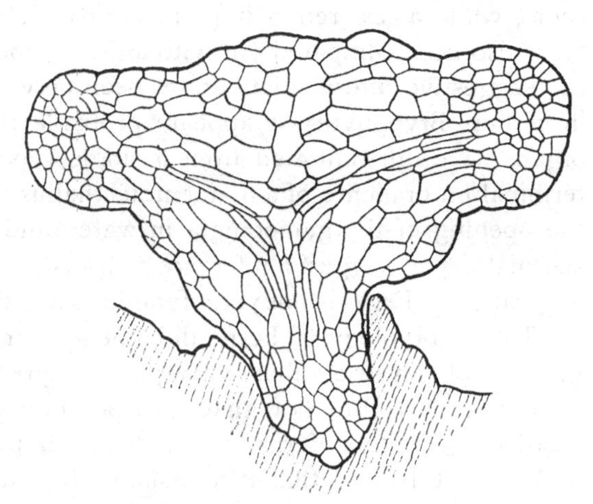

Fig. 309. Young sporophyte of *Tmesipteris* still attached to the prothallus, cut in longitudinal section, and showing two equally developed shoot-regions. (After Holloway.)

The origin of the root-system is not clear. The distended protocorm, so prominently seen in some of the Devonian fossils, does not appear in living Ferns. It may never have been a feature in them. It may be suggested that the prototype of the root is to be seen in the branched rhizome of such a type as *Asteroxylon*. But of this there is no direct evidence, and the question of its origin for the Filicales must be left undecided. There is, however, some reason to think that the most primitive Ferns may have been rootless, as the Psilophytales are, a view which readily accords with the adventitious character of the later roots, and the appearance of the first root in the embryos of some primitive Ferns in a manner that clearly suggests an accessory organ.

Passing from these comparisons in respect of the sporophyte to those of the gametophyte the ground is less satisfactory, partly because the features of the prothallus are less pronounced and more variable, partly because fossil prothalli are almost unknown, and comparison is thus thrown back solely upon those of modern Ferns. In Chapter XV it was concluded that a filamentous structure probably underlies all gametophyte-development in Ferns.

It is possible to derive all types of their prothalli from the simple filament, and many of them show this structure in their first development from the spore, while a few retain it permanently. The antheridia and archegonia have also been compared one with another, and partly from this, partly from the suggestive intermediate states seen between antheridia and archegonia in certain Bryophytes, it appears probable that these apparently distinct organs were differentiated from a single type of gametangium, which was terminal on branches of a filamentous thallus. These features, together with the opening of the gametangia in water, and the motility in water of the spermatozoids, suggest a reference to an Algal origin. But the gametophyte generation of Ferns is not yet referable to any definite type of known Algae.

The embryology of Ferns did not appear to yield any intelligible suggestions of phyletic value so long as the types of embryo with a suspensor were unknown. But now that a suspensor has been found in a number of relatively primitive Ferns, it would appear probable that this organ is an archaic feature, and that it was shared by early Filicales with the Lycopodiales. It is, however, apt to be eliminated in the more advanced types. This seems to be a natural conclusion from the known facts. Professor Lang has extended the comparison on the one hand to the Seed-Plants, in which also a suspensor is the rule, and on the other to the Bryophytes (310, 311). Many of these suggest by the first segmentations of their sporogonia a derivation from the filamentous construction. The basal cell of the sporogonium of thc Jungermanniaceae is specially significant. Such a widespread reference of the young sporophyte to a filamentous origin as is indicated by the first steps of the ontogeny cannot be overlooked. Especially does it claim attention where, as not uncommonly happens, the filamentous suspensor makes an awkward curve of the embryo necessary before the shoot can take the usual upright position. The facts appear to point to a filamentous construction underlying the first steps of the ontogeny of the sporophyte in large sections of the plant-kingdom, a feature which certain of the most primitive of the Filicales share(311). Thus in both generations of Ferns, the facts suggest a reference of the initial structure back to a filamentous source, such as is seen in the general development of the soma not of any special types of Algae alone, but of Algae in general. Moreover this comparison receives further support in that certain Algal types show a filamentous stage preceding a more massive development, but still of the same "generation." For instance the *Chantransia* filaments lead to the more complex *Lemanea* stage: or the filamentous proembryo leads to the more complex segmented shoot of *Chara*. It would be an error to drive such comparisons into detail in the present state of our knowledge. All that is suggested here is that a precedent filamentous stage, leading to a more complex somatic development, is seen in both the generations of the Filicales: and that a like progression can be matched in Algal types, and often in such as are systematically quite apart from one another. They probably illustrate a wide polyphylesis in somatic development of organisms at large. But it is noteworthy that a departure from the filamentous state, and an early substitution of the massive soma, is particularly likely to occur under the biological conditions attending life upon land. In fact the elimination of the filamentous stage would be almost a biological necessity in terrestrial plants: and this is probably the reason why evidence of its existence is so often absent, or effectively disguised, in the Archegoniatae.

In all such comparisons between large divisions of the Vegetable Kingdom the probability of far-reaching polyphylesis should be constantly kept in view. The prevalence of homoplasy within the nearer circles of affinity is becoming clearer every day. The Filicales show it among themselves in very high degree in external form, in anatomical structure, and in soral and sporangial characters. It is evident also in their gametophyte-generation. It appears on comparison of the Filicales with other Archegoniatae. The Algae also show homoplasy in high degree between their own Red, Brown, and Green Series. All sexually produced organisms start from the egg, and each must attain its adult form by progressive development from that simple source. Is it not then probable that homoplasy should produce forms and structures in Algae and Archegoniatae which, though similar, are analogous and nothing more? The tracing of analogies between the form, structure, and propagative organs of plants is an easy and a seductive occupation. But to put any more direct interpretation than analogy upon the similarities which they, and especially their higher forms, show could only be admitted after close comparison of sequences of proved relationship. It is only in this way that analogies can be raised from the atmosphere of vague surmise to the level of scientific hypothesis(315). Hitherto such sequences have not been demonstrated between the Algae and the Archegoniatae. It still remains an open question whether points of contact are to be found between the Archegoniatae and the Red, the Brown, or the Green Algae, though analogies may be traced with representatives of each of these distinct series. This in itself should give pause to unrestrained theorising. But whatever real relationship there may be between Algae and Archegoniatae, indications of it should be sought among lowly and generalised forms rather than among the more specialised Algal types. The comparative analysis of both the generations of the Filicales makes it seem probable that their Algal progenitors may have been still filamentous at the time when they essayed the transition from water to life on land.

That transition has involved, in one way or another, the encapsulation of the embryo(306). Its protection within such an organ as the archegonium was a matter of the highest importance for successful sub-aerial life. Hence an archegonium is present in all the known primitive plants of the land. How the archegonium originated, and the consequent encapsulation came about, are still matters of surmise rather than of demonstration. An internal embryology is a necessary consequence. But this would itself restrict any filamentous development of the embryo. Consequently, though the evidences of this primitive form of construction of the sporophyte are held to be a sufficient indication of its having been general, those evidences have in many cases been almost eliminated. This is the interpretation here given of the state of the embryo seen in all the more advanced Ferns, where no suspensor is present. They are all held to have lost in the course of their Descent that most obvious indication of their ultimate filamentous origin (310, 311).

General Conclusion, and Morphological Comparison with other Plants

A theory of the "Primitive Spindle" has recently been formulated for plants at large(317). It brings into prominence that filamentous structure which, though frequently disguised, is shown by a wide comparison to be inherent in their embryology. Its application to the sporophyte of Ferns, as based upon the comparisons and arguments contained in this volume, may be stated thus: The structure of the sporophyte is essentially based upon the primitive or, better perhaps, the primordial spindle, the polarity of which is defined by the first segmentation of the zygote. The apex of the spindle lies as near as possible to the centre of the epibasal hemisphere: its base, as seen in certain primitive Ferns still living, is the tip of the suspensor. The basal pole of it, known as the suspensor, is filamentous, and is regarded as vestigial. It forms no permanent organ of the plant, and it is present only as an embryonic structure in the more primitive types: in those which are more advanced it may be entirely absent, presumably by abortion. It is believed that in the course of Descent the spindle has developed dichopodially at its apex to form the leafy shoot as it is seen in Ferns. Theoretically the whole shoot of Ferns may be regarded as a specialised dichopodium, which originated from equal dichotomy of the apex of the primitive spindle. The shoot is developed primarily as a radial structure, but it is liable to dorsiventral modification. One shank of the first distal forking of the spindle, recognised in the embryo of a Fern as the *axis*, is endowed with slow but continuous apical growth: the other, which grows more strongly, is the first leaf, or *cotyledon*. The successive later leaves may also have originated in Descent as shanks of such forkings of the apex repeated, and developed dichopodially. But all the post-cotyledonary leaves are actually seen to arise laterally below the apex of the axis, that is monopodially. This position may, however, have been attained secondarily in Descent, just as the monopodial origin of pinnae laterally upon the phyllopodium has been a secondary modification of dichotomy of the leaf-apex (Chapter V, p. 91). In the latter case the correctness of the view is upheld by the gradual transition to equal dichotomy in the distal branchings of the leaves of many Ferns(303). It is true that in the relations of the leaves to the axis no similar evidence of such a transition reveals the actual origin of leaf and axis. But the view expressed is based on analogy with what is seen in the leaves of Ferns. In point of demonstration the relation of leaf to axis in Ferns is as open a question as that of the pinnae to the phyllopodium of Flowering Plants would have been if no Ferns had survived to give the explanation.

The simple shoot thus constructed may grow indefinitely at its apex. But frequently it is further amplified by distal branching. This may be either by equal dichotomy, or it may be dichopodial. The former would give a forked shoot as seen in the Frontispiece: the latter would account for the origin of axillary and other buds related to the leaf-base (Chapter IV). Adventitious buds are, however, often formed in addition to these at points not directly related to the apical region. The origin of the whole shoot-system of Ferns may thus be accounted for. On the other hand, the embryonic spindle as above defined normally gives rise at points apart from its apex to the haustorium or foot, and to the first root. These are held to be accessory organs formed laterally upon the spindle. The later roots are also accessory, and may be borne on the axis or on the leaf-bases. Hairs and emergences may appear indiscriminately on the free surfaces.

This hypothesis accounts for the primary origin of all the vegetative parts of the sporophyte. It is stated here in its special application to the Filicales. But the same hypothesis is also applicable—perhaps with some modification as to the mode of origin of the leaves—to other Pteridophytes, and even to vascular plants at large. Those Eusporangiate Ferns which retain the suspensor are regarded in this and in other features as relatively primitive in their embryology. They form a natural link between the Class of the Filicales and other great Divisions of the Vegetable Kingdom. The Leptosporangiate Ferns have been shown to be derivative among the Filicales, and they are less clearly in accord with the hypothesis. The facts that their suspensorless embryos were the first to be examined, and that they are so easily obtained for class purposes, have given them an entirely false importance in general morphology, and obscured the recognition of the suspensor as an important vestige. The Leptosporangiatae constitute in fact a line of specialisation peculiar to themselves, which reached its highest development in the Ferns of the present day. Turning to the Lycopodiales their embryology accords with the hypothesis of the primitive spindle, since with the exception of *Isoetes* they have a suspensor, and endoscopic orientation (see p. 309). But difficulties of embryological comparison certainly exist in such facts as the absence of a suspensor and the exoscopic orientation of the embryo in *Equisetum* and *Tmesipteris* (see p. 309). It is possible that in these primitive plants the abortion of the basal pole of the spindle happened early, thus giving the recognised advantage of freedom from the inconvenient tie of the suspensor: and that this was followed by exoscopic orientation of the spindle in accordance with their horizontal or upward-turned archegonia. But on this point the facts do not suffice for a decided opinion. On the other hand the embryology of Seed-Plants, both Gymnosperms and Angiosperms, accords well with that of the primitive Filicales and Lycopodiales in having the filamentous suspensor as a very constant feature, which serves only an embryonic function and develops no further. Passing to the Bryophytes, the hypothesis is in harmony with their embryology, for the sporogonium itself is a simple primitive spindle without appendicular organs. But the embryos of Bryophytes differ in the fact that their orientation is constantly exoscopic, while in the Filicales and Lycopodiales it is primarily endoscopic. Finally, the primitive spindle with its single row of cells at the base is believed to find its true correlative in that filamentous structure which commonly results from the germination of the spores of benthic Algae. Such comparisons as these appear to place the morphology of the sporophyte of Ferns in its natural relation to that of other great sections of the Vegetable Kingdom. They throw some welcome light upon the general architecture of the Plant-Body, and point in the direction of a filamentous origin for even the most complex sporophytes.

BIBLIOGRAPHY FOR CHAPTER XVII

303. BOWER. Comp. Morph. of the Leaf in Vascular Cryptogams and Gymnosperms. Phil. Trans. 1884. Part II, p. 565.
304. POTONIÉ. Lehrbuch der Pflanzen-palaeontologie. Berlin. 1899.
305. LIGNIER. Bull. Soc. Linn. de Normandie. Serie 5. Vol. 7. Caen. 1903, p. 93.
306. SCHENCK. Ueber die Phyl. der Archegoniaten. Engler's Bot. Jahrb. xlii, Leipzig. 1908.
307. TANSLEY. Lectures on the Evolution of the Filicinean Vascular System. Lecture I. New Phyt. Reprints, No. 2. 1908.
308. LIGNIER. Bull. Soc. Bot. de France. 1911, p. 87.
309. POTONIÉ. Eine neue Pflanzenmorphologie. Naturwiss. Wochenblatt. Juni. 1912.
310. LANG. Embryo of *Helminthostachys*. Ann. of Bot. xxviii, 1914, p. 19.
311. LANG. Presidential Address to Sec. K. Report Brit. Assn. Manchester. 1915.
312. LAWSON. The prothallus of *Tmesipteris*. Trans. Roy. Soc. Edin. Vol. 51, Part III, p. 785. 1917.
313. HOLLOWAY. The prothallus and young plant of *Tmesipteris*. Trans. New Zeal. Inst. Vol. 50, 1918.
314. HOLLOWAY. Further Studies on *Tmesipteris*. Trans. New Zeal. Inst. Vol. 53. 1921.
315. CHURCH. Thalassiophyta. Oxford Press. 1919.

316. CHURCH. Somatic organization of the Phaeophyceae. Oxford Press. 1920.
317. BOWER. Presidential Address. Roy. Soc. Edin. Anniversary Meeting. 1922.
318. KIDSTON and LANG. Old Red Sandstone Plants from the Rhynie Chert. Parts I—V. Trans. Roy. Soc. Edin. Vols. li—lii, 1917—1921.
319. CHAUVEAUD. La constitution des Plantes Vasculaires révélée par leur Ontogénie. Payot. Paris. 1921.
320. PAUL BECQUEREL. La découverte de la Phyllorhize. Rev. Gén. des Sciences. 28 Feb. 1922.
321. M. P. BUGNON. Quelques critiques à la théorie de la Phyllorhize. Bull. Soc. Bot. de France. Vol. 68, p. 495. 1921.
322. M. P. BUGNON. L'origine phylogénétique des Plantes Vasculaires d'après Lignier. Bull. Soc. Linn. de Normandie. Caen. 1921, p. 196.

POSTSCRIPT TO VOLUME I

This treatise on the Filicales is divided into two parts. The object of the first has been to recognise, to describe in detail, and to evaluate the several criteria which are most suitable for use in the comparative treatment of the Class: by comparison of the fossil remains to check the several features and their details according to such evidence of their time of appearance as the successive geological ages will afford: and so, partly by comparison of living forms and partly by the more direct record of the Earth's crust, to arrive at a sound basis for the phyletic grouping of the Class.

This having been carried out in the above pages with such detail as the limits of a single volume will allow, it remains to apply the method in reconstructing the phylesis of the Filicales from the ample, but nevertheless fragmentary, facts available. This will be the subject of the Second Volume, which will thus deal constructively with the material. The First Volume has been analytical: the Second Volume will be mainly synthetic in its method.

INDEX

Abnormalities, 318
Abortion of basal pole, 346
Accessory strands, 156
Acrostichoid derivatives, 232; species, 232; state, 231, 233, 238
Acrostichum aureum, 146, 239
Adiantum, embryo of, 20 (Fig. 25)
Adult plant of *Dryopteris Filix-mas*, Rich., 26 (Fig. 31)
Adventitious buds, 72, 79; buds of *Cystopteris bulbifera*, 72 (Fig. 66); roots, 159
Alsophila blechnoides, pinna of, 227 (Fig. 223)
A. crinita, lenticel-like pneumathodes of, 203
Alternating generations, 21
Amphicosmia Walkerae, 68 (Fig. 60)
Amphiphloic solenostely, 141, 145
Anadromic helicoid branching, 88
Analysis, morphological, 66
Anatomical memoirs, list of, 193
Anatomy, vascular, 192
Anemia adiantifolia, juvenile leaves of, 85 (Fig. 77)
A. phyllitidis, young plant of, 125 (Fig. 118)
Angiopteris, stipules of, 100 (Fig. 96); wings of leaves of, 112 (Fig. 109)
A. evecta, apex of stem of, 113 (Fig. 112); sporangium of, 250 (Fig. 244)
Angular leaf, 73
Ankyropteris Grayi, 68 (Fig. 59); stele of, 123 (Fig. 117), 182
Annulus, 13, 14 (Fig. 17), 242, 249, 251, 255 (Fig. 250); many-rowed, 252; of *Loxsoma*, 255; position of, 254 (Fig. 250); single-rowed, 252; vestigial, 255
Anogramme chaerophylla, perennation in, 276
A. leptophylla, 33
Antheridia (male organs), 15, 17 (Fig. 20); of various Ferns, differences in number of spermatocytes in, 291 (Fig. 281); of various Ferns seen in surface view, 293 (Fig. 283); of *Woodsia ilvensis*, 287 (Fig. 277); parallel between, and sporangia, 296; position of, 278; segmentation of, 293 (Fig. 282)
Antheridium of *Marattia Douglasii*, 288 (Fig. 278); of *Matteuccia Struthiopteris*, 285 (Fig. 276); of *Ophioglossum pendulum*, 288 (Fig. 278)
Apex of leaf of *Osmunda*, 111 (Fig. 108); of root of *Pteridium*, 10 (Fig. 12); of stem of *Angiopteris evecta*, 113 (Fig. 112); of stem of *Osmunda regalis*, 9 (Fig. 10); of stem of *Trichomanes radicans*, 107 (Fig. 101)
Aphlebiae, 99 (Fig. 95)
Apical cell, 106, 107; cell, segments of, 107, 111 (Fig. 107); meristem, 9; meristem of leaf of *Zygopteris corrugata*, 109 (Fig. 104); segmentation, 115
Apices of pinnules of *Osmunda regalis*, var. *cristata*, 333 (Fig. 307)
Apogamous growth in *Nephrodium pseudo-mas*, var. *polydactylum*, 325 (Fig. 302)
Apogamy (apomixis), 24, 320, 323; in *Nephrodium pseudo-mas*, var. *cristatum*, 322 (Fig. 297); in *Pteris cretica*, 319 (Fig. 295); in *Scolopendrium vulgare*, 322 (Fig. 298)
Apomixis (apogamy), 24, 320
Apospory, 24, 320 (Fig. 296), 321 (Fig. 296 *cont.*), 323; in *Nephrodium pseudo-mas*, var. *cristatum*, 322 (Fig. 297)
Appendages, dermal, 61
Archegonia (female organs), 15; bisexual, of *Mnium cuspidatum*, 290 (Fig. 280); of *Marattia Douglasii*, 287, 288 (Fig. 278); of *Mnium*, 289 (Fig. 280); of *Nephrodium dilatatum*, 323 (Fig. 299); of *Ophioglossum pendulum*, 287, 288 (Fig. 278); of *Polypodium vulgare*, 18 (Fig. 21); position of, 278
Archegoniophore of *Trichomanes pyxidiferum*, 282 (Fig. 272)
Archegonium of *Matteuccia Struthiopteris*, 288 (Fig. 279); standardised, 296
Arches, commissural, 228 (Fig. 225)
Archesporium, 242, 257
Archetype for Class, 336, 339
Architecture of leaf, 61, 81
Armature, 202
Aspidium anomalum, sori of, 225
A trifoliatum, spores of, 261 (Fig. 257)
Asplenium nidus, epiphytic Nest Fern, 45 (Fig. 52)
A. Trichomanes, sporangia of, 245 (Fig. 239)
Asterocalamites scrobiculatus, 103 (Fig. 98)
Asterochlaena laxa, stele of, 183
Athyrium (Lady Fern), 26
A. Filix-foemina, var. *clarissima*, apospory of, 320 (Fig. 296); diploid throughout, 326

Axial system, polycyclic, of *Saccoloma elegans*, 153 (Fig. 146)
Axillary bud, 76; bud of *Trichomanes radicans*, 70 (Fig. 63); position, 70
Axis, 344; arrangement of leaves on, 67; relation of leaf to, 315; vascular system of, 140
Azolla, 48, 261; massula of, in young state, showing glochidia, 269 (Fig. 264)
A. filiculoides, megasporangium of. 268 (Fig. 263)

Basal indusium, 237; pole, abortion of, 346; wall, 19, 301
Basipetal sorus, 212
Bifurcated trunk of *Cyathea medullaris*, 158 (Fig. 152)
Bisexual archegonia of *Mnium cuspidatum*, 290 (Fig. 280)
Blechnoid fusion-sorus, disintegration of, 230 (Fig. 228)
Blechnum, flange in species of, 239 (Fig. 237); indusium of, 238; mucilaginous hairs of, 197 (Fig. 185)
B. longifolium, fusion-sorus of, 229 (Fig. 227)
B. punctulatum, Sw. var. *Krebsii*, 230 (Figs. 228, 229), 231 (Fig. 229)
B. tabulare, 29 (Fig. 35)
Blepharoplast, 286
Botrychioxylon, 137
Botrychium, 131; young plant of, 184
B. daucifolium, sporangium of, 247 (Fig. 242)
B. Lunaria, embryo of, 306 (Fig. 292); medullation of, 125 (Fig. 119); old embryo of, 307 (Fig. 293); reconstruction of stelar structure of, 131 (Fig. 124); seedling of, 307 (Fig. 293); sporangia of, 208 (Fig. 198)
B. obliquum, 306; suspensor of, 305; young sporophyte of, 305 (Fig. 289)
B. virginianum, stele of, 137 (Fig. 129)
Botryopteris cylindrica, 121 (Fig. 114), 161 (Fig. 153); transverse section of stem of, 182 (Fig. 174)
B. forensis, equisetoid hairs of, 199 (Fig. 187); solid xylem-core of, 127
Bracken (*Pteridium aquilinum*), meristele of, 5 (Fig. 4), 6 (Fig. 5); rhizome of, 4 (Fig. 3)
Bramble Ferns, 39 (Fig. 44), 171
Branching, 69, 79; anadromic helicoid, 88; catadromic helicoid, 88; dichopodial, 89; helicoid, 88; monopodial, 91; terminal, dichotomous, 337
Bud, axillary, 76
Budding, gametophytic, 15, 320; sporophytic, 206, 320
Buds, 69, 70; adventitious, 72, 79; adventitious, of *Cystopteris bulbifera*, 72 (Fig. 66); extra-axillary, 70
Bulk, proportion of surface to, 177, 181

Cambial increase, 178
Canal-cell, 17, 18 (Fig. 21)
Cap-cells, 293
Capsule of sporangium, 242
Caspary-band, 185
Catadromic helicoid branching, 88
Cauline stele, 139; vascular system, 139
Cell, apical, 106, 107; apical, segments of, 107, 111 (Fig. 107); embryonic, 300; opercular, 287 (Fig. 277); opercular, single, 293
Cell-cleavages, embryology based upon, 299
Cells, initial, 105, 107, 109; marginal, 109; of endodermis, 186 (Fig. 177); sporogenous, 257
Cellular construction, 105; segmentation of embryo, 314
Ceratopteris, young leaf of, 9 (Fig. 11)
C. thalictroides, 324
Chantransia filaments, 342
Cheiropleuria, 138, 233 (Fig. 231); four-rowed stalk of, 254 (Fig. 251); protostele of, 122 (Fig. 115); sporophyll of, 233 (Fig. 231); two-rowed segmentation of, 246
C. bicuspis, 71 (fig. 65)
Christensenia, scales of, 201; synangia of, 249
Chromosome-cycle, 22; normal, 324
Chromosome-distinction between generations, 324
Chromosomes, 21
Cibotium Barometz, solenostelic rhizome of, 151 (Fig. 142); vascular system of, 170 (Fig. 163)
Cilia, 286
Classification, early methods of, 57; phyletic, 53
Cleavages, segmental, in embryo, 299
Climbing leaf, 38 (Fig. 43)
Coaxial scheme, 114 (Fig. 113)
Coenopterideae, 162
Coenosori, 228
Commissural arches, 228 (Fig. 225)
Commissures, 229
Common vascular system, 139
Comparison, criteria of, 59; organographic, 336
Compensation strand, 152
Cone, 179
Conical form of stele, 180; form of young stem, 179; stem of *Polypodium vulgare*, 179 (Fig. 171); stem of *Pteris podophylla*, 187 (Fig. 178)
Conjunctive parenchyma, 5
Cordate type of prothallus, 273 (Fig. 265)
Corrugation, 151; of solenostele of *Histiopteris incisa*, 151 (Fig. 143)

Corynepteris Essenghi, sorus of, 250 (Fig. 245)
Cotyledon, 311, 344
Cover-cell, 286
Covering of scales, 42 (Figs. 47, 48)
Creeping habit, 32, 33
Crested state related to sexual irregularity, 333; varieties, 331
Criteria of comparison, 59
Cushion, 273
Cutting of leaf-blade, 83
Cyathea dealbata, double-headed specimen of, Frontispiece
C. Imrayana, 156 (Fig. 150)
C. medullaris, bifurcated trunk of, 158 (Fig. 152)
C. Schansin, 34 (Fig. 40)
Cyatheaceae, 54
Cylindrical processes, 284 (Fig. 275)
Cystopteris bulbifera, adventitious buds of, 72 (Fig. 66)
C. fragilis, scale or ramentum of, 201 (Fig. 189)

Danaea, scales of, 201; synangia of, 249
D. alata, 226 (Fig. 221)
Davallia Griffithiana, 215 (Fig. 208)
D. speluncae, petiole of, 166 (Fig. 159)
Dehiscence, slit of, shifting of, 256
Delay of certain parts, 314
Dendroid habit, 30; types, 29
Dennstaedtia, rhizomes of, 152 (Fig. 144)
D. apiifolia, sorus of, 255 (Fig. 252)
D. dissecta, 221 (Fig. 215)
D. rubiginosa, sorus of, 215 (Fig. 208)
Deparia Moorei, sori of, 225; starved, 179 (Fig. 172)
Dermal appendages, 61; tissues, 195
Dichopodial branching, 89
Dichopodium, 76, 337; scorpioid, 88
Dichotomous terminal branching, 337
Dichotomy, 69, 79, 86, 158 (Fig. 152); equal, 90; of meristele of *Schizaea dichotoma*, 172 (Fig. 166); of rhizome, 69 (Fig. 61); sympodial, 91
Dicksonia Scheidei, young sorus of, 218 (Fig. 211), 219 (Fig. 212)
Dictyostelic stem of Ferns, 155 (Fig. 149)
Dictyostely, 147
Dictyo-xylic ring, 133
Differentiation of leaves, 49
Diffusing surface, area of, 178
Dimorphic leaves, 46 (Fig. 53)
Diplodesmic condition, 233
Diploid generation, 323; sporophyte, 21
Diplolabis Römeri, 82 (Fig. 73)
Dipterid-derivatives, 232
Dipteris, two-rowed segmentation of, 246
D. conjugata, young sori of, 216 (Fig. 209)
D. Lobbiana, fertile pinna of, 227 (Fig. 222)
Disintegration of fusion-sorus, 230 (Fig. 228); of stele, 135, 157, 189 (Fig. 180), 191
Distal region of wall, 252
Divergence, two-fifths, 68
Divergent unit, 168
Dorsiventral and dichotomous rhizome of *Lygodium scandens*, 67 (Fig. 58); rhizome of *Polypodium vulgare*, 67 (Fig. 57 *bis*)
Dorsiventrality of shoot, 67
Drymoglossum subcordatum, 42 (Fig. 49)
Drynaria, 46 (Fig. 53)
D. Filix-mas, Rich. (Male Shield Fern), 1, 26; adult plant of, 26 (Fig. 31)
D. Linnaeana, rhizome of, 27 (Fig. 32)
D. vivipara, extra-marginal type of pinna-supply in, 172 (Fig. 168)

Early methods of classification, 57
Ectophloic condition, 141
Embryo, 19 (Fig. 24), 298; as a living whole, 300; cellular segmentation of, 314; encapsulation of, 343; of *Adiantum*, 20 (Fig. 25); of *Botrychium Lunaria*, 306 (Fig. 292); of *Helminthostachys*, 304 (Fig. 288); of *Macroglossum*, 303 (Fig. 287); of Marattiaceae, 303 (Fig. 286); old, of *Botrychium Lunaria*, 307 (Fig. 293); parts of, origin of, 311; parthenogenetic, of *Marsilia Drummondii*, 325 (Fig. 303); polarity of, 300, 313; rotation of, 312; segmental cleavages in, 299; segmentation of, 301 (Fig. 284)
Embryological method, 299
Embryology based upon cell-cleavages, 299; endoscopic, 309; exoscopic, 309; of sporophyte, 63; theoretical problem in, 298
Embryonic cell, 300
Encapsulation of embryo, 343
Endodermis, 5, 138, 180, 185; as physiological barrier, 181; cells of, 186 (Fig. 177); continuity of, 143; internal, 130, 131, 133; involution of, 145; of *Helminthostachys zeylanica*, 181 (Fig. 173)
Endoscopic embryology, 309
Epibasal hemisphere, 19, 311; tier, 301
Epiphytic Ferns, 43, 44; habit, 35, 45, 46 (Fig. 53); Nest Fern, 45 (Fig. 52)
Equal dichotomy, 90
Equisetoid hairs of *Botryopteris forensis*, 199 (Fig. 187)
Eusporangiate Ferns, 242, 244; Ferns, segmentation in, 111; sporangium, 270
Exoscopic embryology, 309
Exospore, 261 (Fig. 257)
Extra-axillary buds, 70
Extra-marginal pinna-trace, 173; type of pinna-supply, in *Dryopteris vivipara*, 172 (Fig. 168)
Extrastelar pith, 188

Factor, limiting, 181
Female organs (archegonia), 15
Fern, life-cycle of, 20, 21 (Fig. 28); starved by unfavourable culture, 179 (Fig. 172)
Fern-leaf, primitive type of, 337
Fern-Plant, 1
Ferns, branching in, 79; epiphytic, 43, 44; Eusporangiate, 242, 244; Filmy, 41, 54; Leptosporangiate, 244; segmentation in Eusporangiate, 111; solenostelic stem of, 155 (Fig. 149); stems of, 189 (Fig. 180); differences of number of spermatocytes in, 291 (Fig. 281); antheridia of, seen in surface view, 293 (Fig. 283)
Fern-sporophyte, primitive, 337
Fertile leaf of *Leptochilus tricuspis*, 234 (Fig. 233); pinna of *Dipteris Lobbiana*, 227 (Fig. 222); pinnule of *Klukia exilis*, 252 (Fig. 248)
Fertilisation in *Onoclea sensibilis*, 18 (Fig. 22)
Filament, simple, 342; transversely septate, 301
Filamentous construction, 342; origin, 346; type of prothallus, 275 (Fig. 266), 278, 280 (Fig. 270)
Filmy Ferns, 41, 54
First leaf, 311; root, 312, 345
Fission of sorus, 227
Flange, 239 (Fig. 237); indusial, 238
Floating habit, 48
Foliar gap, 142, 145
Foot, 19, 310, 345
Fossil Osmundaceae, 129; actual size of steles of, 184 (Fig. 176)
Frond, 81
Fructification of *Ptychocarpus unitus*, 210 (Fig. 201)
Fungus, symbiotic, 284 (Fig. 274)
Fusion of sori, 227; of veins, 64
Fusion-sori of *Lindsaya lancea*, 228 (Fig. 225)
Fusion-sorus, 228, 229 (Fig. 227); disintegration of, 230 (Fig. 228); of *Blechnum longifolium*, 229 (Fig. 227)

Gametangia, 273, 285, 295; analogies between, and sporangia, 285, 289; arrangement of, 278; homology of, 289; male and female, 285
Gametophyte, 21, 273; comparison based on, 295; haploid, 21
Gametophytic budding, 15, 320
Gaps, foliar, 142, 145
Gemmae, 15; of *Trichomanes*, 280; of *Trichomanes alatum*, 281 (Fig. 271)
Generations, alternating, 21
Germination of spores, 15 (Fig. 18)
Glands, water, of *Polypodium vulgare*, 205 (Fig. 196)
Glandular hairs, 198; of *Notholaena trichomanoides*, 199 (Fig. 186)
Gleichenia, sorus of, 208
G. dicarpa, base of petiole of, 170 (Fig. 164)
G. flabellata, 142 (Fig. 132); protostele of rhizome of, 141 (Fig. 131); sori of, 209 (Fig. 200)
G. pectinata, juvenile plant of, 188 (Fig. 179); plan of stelar construction, 144 (Fig. 134); solenostelic structure of, 143 (Fig. 133); sorus of, 212 (Fig. 203); stiff hairs of, 202
Gleicheniaceae, 54
Glochidia, 261, 268, 269 (Fig. 264)
Gradatae, 213
Gradate sorus, 212
Gymnogramme japonica, vascular system of, 167 (Fig. 160)

Habit, 26; creeping, 32, 33; dendroid, 30; epiphytic, 35, 45, 46 (Fig. 53); floating, 48; monophyllous, 31; nest, 44; scandent, 35; upright, 29; xerophytic, 41
Habitat, 26
Hairs, 195, 197; equisetoid, of *Botryopteris forensis*, 199 (Fig. 187); glandular, 198; glandular, of *Notholaena trichomanoides*, 199 (Fig. 186); indusial, 239; mucilaginous, of *Blechnum*, 197 (Fig. 185); on prothalli, 277 (Fig. 268); peltate, 239; simple linear, 198; stellate, of *Polypodium lingua*, 200 (Fig. 188); stiff, of *Gleichenia pectinata*, 202
Haploid gametophyte, 21; generation, 323; sporophyte of *Lastraea pseudo-mas*, var. *cristata*, 326
Haustorium, 310, 345
Helicoid branching, 88
Helminthostachys, 131; embryo of, 304 (Fig. 288); root of, 175; saprophytic prothallus of, 283; suspensor of, 304; young plant of, 184
H. zeylanica, 31 (Fig. 37); endodermis of, 181 (Fig. 173); prothalli and sexual organs of, 283 (Fig 274)
Hemisphere, epibasal, 19, 311; hypobasal, 312
Hemitelia capensis, aphlebiae of, 99 (Fig. 95)
H. grandifolia, scales of, 202 (Fig. 191)
Heterospory, 267
Histiopteris incisa, corrugation of solenostele of, 151 (Fig. 143); young sori of, 224 (Fig. 219)
Homology of gametangia, 289
Homoplasy, 343
Hornea Lignieri, reconstruction of, by Kidston and Lang, 340 (Fig. 308)
Hymenophyllaceae, 54, 115
Hymenophyllum dilatatum, 282 (Fig. 273)
H. Wilsoni, sorus of, 212 (Fig. 204)
Hypobasal hemisphere, 312; tier, 19, 301
Hypolepis repens, pinnule of, 221 (Fig. 216); young sorus of, 222 (Fig. 217)

Increase by spores, 206; cambial, 178; in size, 188; primary, 178
Indusial flange, 238; hairs, 239; protections, 62
Indusium, 234, 236 (Fig. 236), 237, 238, 240; basal, 237; of *Blechnum*, 238; of *Pteris*, 238; peltate, 238; prototype of, 235; reniform, 238; types of, in Polypodiaceae (Fig. 236); vestigial lower, 222
Initial cells, 105, 107; cells of leaves, 109
Intercellular spaces, 180, 184
Internal endodermis, 130, 131, 133; phloem, 145
Intrastelar pith, 127, 188
Inversicatenales, 162
Involution of endodermis, 145
Isolated tracheides, 126

Juvenile leaf of *Todea superba*, 85 (Fig. 76); leaves of *Anemia adiantifolia*, 85 (Fig. 77); leaves of *Osmunda regalis*, 87 (Fig. 80); plant of *Gleichenia pectinata*, 188 (Fig 179)

Klukia exilis, fertile pinnule of, 252 (Fig. 248)

Lady Fern (*Athyrium*), 26
Lastraea pseudo-mas, var. *cristata apospora*, haploid sporophyte of, 326
Lattice-work, 149
Leaf, 7; angular, 73; apex of, of *Osmunda*, 111 (Fig. 108); apical meristem of, 109 (Fig. 104); architecture of, 61, 81; climbing, 38 (Fig. 43); definition of, 81; fertile, of *Leptochilus tricuspis*, 234 (Fig. 233); first, 311; juvenile, of *Todea superba*, 85 (Fig. 76); relation of, to axis, 315; vascular tissue of, 61; venation of, 61; young, of *Ceratopteris*, 9 (Fig. 11)
Leaf-architecture, 81
Leaf-blade, cutting of, 83; venation of, 83
Leaf-fall, 33
Leaf-margin, overlapping of, in *Matteuccia intermedia*, 238
Leaf-trace, 160, 161, 162, 163 (Fig. 156); in *Thamnopteris Schlechtendalii*, 162 (Fig. 155); origin of, *Lophosoria quadripinnata*, 163 (Fig. 156)
Leaves, arrangement of, on axis, 67; differentiation of, 49; dimorphic, 46 (Fig. 53); initial cells of, 109; juvenile, of *Anemia adiantifolia*, 85 (Fig. 77); juvenile, of *Osmunda regalis*, 87 (Fig. 80); segmentation at margin of, 110 (Fig. 105); of Zygopterideae, 81; wings of, of *Angiopteris*, 112 (Fig. 109); young, of *Schizaea rupestris*, 220 (Fig. 213)
Lecanopteris carnosa, 43 (Fig. 50)
Leptochilus tricuspis, fertile leaf of, 234 (Fig. 233); soral region of, 233 (Fig. 232)
Leptosporangiate Ferns, 244; sporangium, 270
Life-history of a Fern, 1
Limiting factor, 181; surfaces, 178
Lindsaya lancea, fusion-sori of, 228 (Fig. 225)
L. linearis, stele of, 146 (Fig. 136)
L. scandens, sections through node of, 147 (Fig. 137)
Lindsaya-condition, 146
Linear hairs, simple, 198
Liverworts, sporogonia of, 117
Lonchitis pubescens, 173 (Fig. 169)
Lophosoria, pinna-trace in, 174 (Fig. 170)
L. quadripinnata, origin of leaf-trace in, 163 (Fig. 156)
Loxsoma, 145; annulus of, 255; meristele of, 165 (Fig. 158)
L. Cunninghamii, mature sorus of, 214 (Fig. 206); plan of stelar construction of, 145 (Fig. 135); vascular system at node of (Fig. 130)
Lygodium, monangial sorus of, 235 (Fig. 235)
L. circinatum, sporangium of, 265 (Fig. 260)
L. scandens, dorsiventral and dichotomous rhizome of, 67 (Fig. 58)

Macroglossum, embryo of, 303 (Fig. 287)
Male organs (antheridia), 15
Male Shield Fern (*Dryopteris Filix-mas*, Rich), 1
Marattia, synangia of, 249
M. Douglasii, antheridium of, 288 (Fig. 278); archegonia of, 287, 288 (Fig. 278)
M. fraxinea, sporogenous tissue of, 258 (Fig. 254)
Marattiaceae, 55, 135; embryos of, 303 (Fig. 286); sori of, 211 (Fig. 202); suspensors of, 303
Margin of leaves, segmentation at, 110 (Fig. 105)
Marginal cells, 109; origin of sorus of *Pteridium*, 222; pinna-trace, 173; position of sorus, 217, 219; type of pinna-supply in *Pteris umbrosa*, 172 (Fig. 167)
Marginales, 140
Marsilia, 48
M. Drummondii, parthenogenetic embryo of, 325 (Fig. 303)
Marsiliaceae, 55
Massulae, 261, 268, 269 (Fig. 264)
Matonia, 154
M. pectinata, stelar system of, 154 (Fig. 148)
Matoniaceae, 54
Matteuccia intermedia, overlapping of leaf-margin in, 238; sorus of, 229 (Fig. 226)
M. Struthiopteris, antheridium of, 285 (Fig. 276); archegonium of, 288 (Fig. 279)

Mature sorus of *Loxsoma Cunninghamii*, 214 (Fig. 206)
Mechanisms, opening, 253
Medulla, 124
Medullation, 124, 183; of *Botrychium Lunaria*, 125 (Fig. 119)
Mega-gametangia, 290
Megasporangium, 268 (Fig. 263), 290
Megaspores, 267, 268 (Fig. 263)
Memoirs, anatomical, list of, 193
Meristele, 4; dichotomy of, of *Schizaea dichotoma*, 172 (Fig. 166); of Bracken, 5 (Fig. 4), 6 (Fig. 5); of *Loxsoma*, 165 (Fig. 158)
Meristeles in petioles, 171 (Fig. 165)
Meristem, apical, 9; of leaf of *Zygopteris corrugata*, 109 (Fig. 104)
Metaclepsydropsis duplex, mixed pith of, 127; stele of, 128 (Fig. 122)
Metaxya rostrata = *Alsophila blechnoides*
Micro-gametangia, 290
Microsporangia, 290; of Salviniaceae, 267 (Fig. 262)
Microspores, 267
Mitosis, vegetative, 22 (Fig. 29)
Mixed pith, 126, 184; of *Metaclepsydropsis duplex*, 127; of *Osmunda regalis*, 127 (Fig. 120)
Mixtae, 214, 216
Mnium, archegonia of, 289 (Fig. 280)
M. cuspidatum, bisexual archegonia of, 290 (Fig. 280)
Mohria, scales of, 201
M. caffrorum, position of sporangium of, 220 (Fig. 214)
Monangial sorus, 208, 241; of *Lygodium*, 235 (Fig. 235)
Monophyllous habit, 31
Monopodial branching, 91
Morphological analysis, 66
Morphology, stelar, 177
Mosaics, theory of, 299
Mucilaginous hairs of *Blechnum*, 197 (Fig. 185)
Mycorhizic habit, 47; prothalli of primitive Ophioglossaceae, 295; type, 283

Neck of archegonium, 295
Nectary of *Pteridium aquilinum*, 205 (Fig. 195)
Nephrodium dilatatum, prothalloid cylindrical process of, bearing archegonia, 323 (Fig. 299)
N. Filix-mas, multi-cristatum, 332 (Fig. 305)
N. Filix-mas, var. *cristatum*, raised apogamously, 332 (Fig. 306)
N. pseudo-mas, var. *cristatum*, apogamy and apospory of, 322 (Fig. 297)
N. pseudo-mas, var. *polydactylum*, apogamous growth of, 325 (Fig. 302)
Nephrolepis, stolons of, 190 (Fig. 182); tubers of, 43
Nervatio Anaxeti, 98 (Fig. 94); *Caenopteridis*, 97 (Fig. 93); *Cyrtophlebii*, 98 (Fig. 94); *Doodyae*, 98 (Fig. 94); *Drynariae*, 98 (Fig. 94); *Eupteridis*, 97 (Fig. 93); *Goniophlebii*, 97 (Fig. 93); *Goniopteridis*, 97 (Fig. 93); *Marginariae*, 98 (Fig. 94); *Neuropteridis*, 97 (Fig. 93); *Pecopteridis*, 97 (Fig. 93); *Sageniae*, 98 (Fig. 94); *Sphenopteridis*, 97 (Fig. 93); *Taeniopteridis*, 97 (Fig. 93)
Nest Ferns, 45 (Fig. 52), 49
Nest-habit, 44
Node of *Lindsaya scandens*, sections through, 147 (Fig. 137); of *Loxsoma Cunninghamii*, vascular system at, 141 (Fig. 130)
Notholaena trichomanoides, glandular hairs of, 198, 199 (Fig. 186)

Octant-division, 300
Onoclea sensibilis, fertilisation in, 18 (Fig. 22)
Ontogeny, 120
Opening mechanisms, 253
Opercular cell, 287 (Fig. 277); single, 293
Ophioglossaceae, 55; primitive, mycorhizic prothalli of, 295
Ophioglossum Bergianum, 136 (Fig. 128)
O. palmatum, storage-stock of, 30 (Fig. 36)
O. pedunculosum, view of Campbell on, 316
O. pendulum, antheridium of, 288 (Fig. 278); archegonia of, 287, 288 (Fig. 278)
O. simplex, 48 (Fig. 55)
O. vulgatum, sporogenous tissue of, 260 (Fig. 255)
Organographic comparisons, 336
Osmunda, apex of leaf of, 111 (Fig. 108)
O. cinnamomea, 133 (Fig. 126)
O. regalis, apex of stem of, 9 (Fig. 10); juvenile leaves of, 87 (Fig. 80); mixed pith of, 127 (Fig. 120); root-tip of, 112 (Fig. 110), 113 (Fig. 111); sporangium of, 265 (Fig. 259); spore of, 260 (Fig. 256)
O. regalis, var. *cristata*, apices of pinnules of, 333 (Fig. 307)
Osmundaceae, 55; fossil, 129; living and fossil, actual size of steles of, 184 (Fig. 176)
Osmundaceous series, 128
Osmundites Carnieri, 134, 135 (Fig. 127)
O. Kidstoni, 129, 137
O. skiegatensis, 134
Output per sorus, 266
Ouvirandra, 149
Ovum, 17; uniformity of, 298

Parenchyma, 195; conjunctive, 5
Parkeriaceae, 54

Parthenogenetic embryo of *Marsilia Drummondii*, 325 (Fig. 303)
Pellaea, root of, 175
P. rotundifolia, vascular system of, 148 (Fig. 139)
Peltate hairs, 239; indusium, 238; scales of *Polypodium incanum*, 200 (Fig. 188)
Perennation in *Anogramme chaerophylla*, 276
Perennials, 33
Perforation, 149, 169
Pericycle, 5
Perispore, 259, 261 (Fig. 257)
Petiole, base of, of *Gleichenia dicarpa*, 170 (Fig. 164); of *Davallia speluncae*, 166 (Fig. 159); of *Saccoloma elegans*, 169 (Fig. 162)
Petioles, meristeles in, 171 (Fig. 165); sections of, 168 (Fig. 161); transverse sections of, 189 (Fig. 181)
Phlebodium aureum, 42 (Fig. 48); scales of, 202 (Fig. 190); sporangium of, 245 (Fig. 240)
Phloem, 5; internal, 145
Phloeoterma, 138
Phyletic classification, 53; drift, 270; slide, 221, 222
Phyllopodium, 82, 160, 165, 337
Phyllorhize, 315
Pilularia, 48
P. americana, sporangium of, 256 (Fig. 253)
P. globulifera, 78 (Fig. 72)
Pinna, fertile, of *Dipteris Lobbiana*, 227 (Fig. 222); of *Alsophila blechnoides*, 227 (Fig. 223)
Pinna-supply, extra-marginal type of, in *Dryopteris vivipara*, 172 (Fig. 168); marginal type of, in *Pteris umbrosa*, 172 (Fig. 167)
Pinna-trace, 160, 172; extra-marginal, 173; in *Lophosoria*, 174 (Fig. 170); marginal, 173
Pinnule, fertile, of *Klukia exilis*, 252 (Fig. 248); of *Hypolepis repens*, 221 (Fig. 216)
Pinnules, apices of, of *Osmunda regalis*, var. *cristata*, 333 (Fig. 307)
Pith, 124; extrastelar, 188; intrastelar, 127, 188; mixed, 126, 184; mixed, of *Metaclepsydropsis duplex*, 127; mixed, of *Osmunda regalis*, 127 (Fig. 120)
Plagiogyria, 150; pneumatophores of, 204 (Fig. 194)
P. pycnophylla, rhizomes of, 148 (Fig. 138)
Plant-body, constitution of, 60
Plasmodium, 259
Platycerium, sporangium of, 254 (Fig. 251); three-rowed segmentation of, 247
P. angolense, sporophyll of, 233 (Fig. 231)
Platyzoma, 131; spores of, 265 (Fig. 258); stele of small plant of, 132 (Fig. 125)
Pneumathode-areas, 169
Pneumathodes, 169; lenticel-like, of *Alsophila crinita*, 203
Pneumatophores, 203; of *Plagiogyria*, 204 (Fig. 194)
Polarity of embryo, 300, 313; of spindle, 313
Pole, basal, abortion of, 346
Polybotrya, sori of, 234 (Fig. 234)
Polycyclic axial system of *Saccoloma elegans*, 153 (Fig. 146)
Polycycly, 151, 155
Polyphylesis, 343
Polypodiaceae, 54; types of indusium in, 236 (Fig. 236)
Polypodium bifrons, 44 (Fig. 51)
P. incanum, peltate scales of, 200 (Fig. 188)
P. lingua, stellate hairs of, 200 (Fig. 188)
P. vulgare, 33 (Fig. 39); archegonia of, 18 (Fig. 21); conical stem of, 179 (Fig. 171); dorsiventral rhizome of, 67 (Fig. 57 *bis*); spores of, 260 (Fig. 256); water-glands of, 205 (Fig. 196)
Polystichum angulare, var. *pulcherrimum*, 321 (Fig. 296 *cont.*)
Primary increase, 178
Primitive Fern-sporophyte, 337; Ophioglossaceae, mycorhizic prothalli of, 295; spindle, 313, 344; type of fern-leaf, 337; vascular construction, 338
Primordial organs, relations of, 314
Principal walls, 107, 112
Principle of similar structures, 177, 191
Procambial destination, 144, 145
Progressive simplification of structure, 117
Propagation, vegetative, 10
Protections, indusial, 62
Prothalli, hairs on, 277 (Fig. 268); mycorhizic, of primitive Ophioglossaceae, 295; of *Helminthostachys zeylanica*, 283 (Fig. 274); of *Pteris longifolia*, 276 (Fig. 267); of *Scolopendrium*, 284 (Fig. 275)
Prothalloid cylindrical process of *Nephrodium dilatatum*, bearing archegonia, 323 (Fig. 299)
Prothallus, 1, 14, 16 (Fig. 19); cordate type of, 273 (Fig. 265); filamentous type of, 278, 280 (Fig. 270); form and physiology of, 62; of *Schizaea pusilla*, 279 (Fig. 269); of *Trichomanes alatum*, 281 (Fig. 271); saprophytic, of *Helminthostachys*, 283
Protocorm, 341
Protostele, 121, 181; of *Cheiropleuria*, 122 (Fig. 115); of rhizome of *Gleichenia flabellata*, 141 (Fig. 131); primitive vascular construction, 338; upgrade development from, 133; varying form of, 183
Protoxylem, 6
Protoxylem-groups, 164

Proximal region of wall, 253
Pteridium, apex of root of, 10 (Fig. 12); marginal origin of sorus of, 222
P. aquilinum (Bracken), nectary of, 205 (Fig. 195); origin of sorus of, 223 (Fig. 218); rhizome of, 4 (Fig. 3); young plant of, 75 (Fig. 69)
Pteris, indusium of, 238
P. cretica, apogamy in, 319 (Fig. 295)
P. elata, vascular system of, 152 (Fig. 145)
P. longifolia, prothalli of, 276 (Fig. 267)
P. podophylla, 153 (Fig. 147); conical stem of, 187 (Fig. 178)
P. serrulata, young sori of, 224 (Fig. 220)
P. umbrosa, marginal type of pinna-supply in, 172 (Fig. 167)
Ptychocarpus unitus, 208 (Fig. 201); fructification of, 210 (Fig. 201)

Radial symmetry, 66
Ramentum, 200; of *Cystopteris fragilis*, 201 (Fig. 189)
Receptacle, origin of, 218
Reniform indusium, 238
Reserve materials, 35
Rhizome, dichotomy of, 69 (Fig. 61); dorsiventral, of *Polypodium vulgare*, 67 (Fig. 57 *bis*); dorsiventral and dichotomous, of *Lygodium scandens*, 67 (Fig. 58); of Bracken (*Pteridium aquilinum*), 4 (Fig. 3), 34; of *Dennstaedtia*, 152 (Fig. 144); of *Dryopteris Linnaeana*, 27 (Fig. 32); of *Plagiogyria pycnophylla*, 148 (Fig. 138); of *Pteridium aquilinum*, 4 (Fig. 3); protostele of, in *Gleichenia flabellata*, 141 (Fig. 131); solenostelic, in *Cibotium Barometz*, 151 (Fig. 142)
Ring, dictyo-xylic, 133
Ring-cells, 286, 287 (Fig. 277)
Root, 175; apex of, in *Pteridium*, 10 (Fig. 12); first, 312, 345; in *Helminthostachys*, 175; of *Pellaea*, 175
Roots, adventitious, 159; of Ferns, 8; of *Pellaea*, 8 (Fig. 9)
Root-tip, 108 (Figs. 102, 103); of *Osmunda regalis*, 112 (Fig. 110), 113 (Fig. 111)
Rotation of embryo, 312

Saccoloma elegans, petiole of, 169 (Fig. 162); polycyclic axial system of, 153 (Fig. 146)
Sac-shaped urns, 44 (Fig. 51)
Salvinia, 48
Salviniaceae, 55; microsporangia of, 267 (Fig. 262)
Saprophytic prothallus of *Helminthostachys*, 283
Scalariform tracheides, 7
Scale, flattened, 199; or ramentum of *Cystopteris fragilis*, 201 (Fig. 189)
Scales, 195, 197; covering of, 42 (Figs. 47, 48); of *Christensenia*, 201; of *Danaea*, 201; of *Hemitelia grandiflora*, 202 (Fig. 191); of *Mohria*, 201; of *Phlebodium aureum*, 202 (Fig. 190); peltate, of *Polypodium incanum*, 200 (Fig. 188)
Scandent habit, 35
Schizaea dichotoma, dichotomy of meristele of, 172 (Fig. 166)
S. pusilla, prothallus of, 279 (Fig. 269)
S. rupestris, young leaves of, 220 (Fig. 213)
Schizaeaceae, 55
Sclerenchyma, 195, 196
Scolopendrium, 230 (Fig. 228); prothalli of, 284 (Fig. 275)
S. vulgare, 323 (Fig. 300); apogamy of, 322 (Fig. 298)
Scorpioid dichopodium, 88
Seedling of *Botrychium Lunaria*, 307 (Fig. 293)
Segmental cleavages in embryo, 299
Segmentation, apical, 115; at margin of leaves, 110 (Fig. 105); cellular, of embryo, 314; in Eusporangiate Ferns, 111; of antheridia, 293 (Fig. 282); of embryo, 301 (Fig. 284), 314; of sporangia, 110 (Fig. 106), 246, 247; primary, 301; seriation according to, 115; sporangial, 115; three-rowed, of *Platycerium*, 247; two-rowed, of *Cheiropleuria*, 246
Segments of apical cell, 111 (Fig. 107)
Selaginella, 117
Senftenbergia elegans, sporangium of, 252 (Fig. 247)
Seriation according to segmentation, 115
Sextant walls, 107, 112
Sexual irregularity, crested state related to, 333; organs, 15, 273; organs of *Helminthostachys zeylanica*, 283 (Fig. 274); organs, structure and position of, 63
Shade-form, 38
Shoot, dorsiventrality of, 67; form of, 27, 60; proportions of, 60; simple, 336; symmetry of, 67
Sieve-tubes, 5, 7; of *Pteridium*, 7 (Fig. 7)
Similar structures, principle of, 177, 191
Simple shoot, 336; sorus, 208, 209, 210, 211
Simplices, 212
Simplification of structure, progressive, 117
Siphonostele, 140
Size, 177; increase in, 188
Slide, phyletic, 221, 222
Slit of dehiscence, shifting of, 256
Solenostele, 140; corrugation of, in *Histiopteris incisa*, 151 (Fig. 143)
Solenostelic rhizome of *Cibotium Barometz*, 151 (Fig. 142); stem of Ferns, 155 (Fig. 149); structure of *Gleichenia pectinata*, 143 (Fig. 133)
Solenostely, 140, 141; amphiphloic, 141, 145
Soleno-xylic stele, 124

Solitary sporangia, 207
Soral region of *Leptochilus tricuspis*, 233 (Fig. 232)
Sori, fusion of, 227; of *Aspidium anomalum*, 225; of *Deparia Moorei*, 225; of *Gleichenia flabellata*, 209 (Fig. 200); of Marattiaceae, 211 (Fig. 202); of *Polybotrya*, 234 (Fig. 234); uniseriate, 210; young, of *Dipteris conjugata*, 216 (Fig. 209); young, of *Histiopteris incisa*, 224 (Fig. 219); young, of *Pteris serrulata*, 224 (Fig. 220)
Sorus, 12 (Fig. 14), 206, 207; basipetal, 212; constitution of, 240; fission of, 227; gradate, 212; individuality of, 240; loss of individuality of, 225, 227; marginal, 219; marginal origin of, in *Pteridium*, 222; marginal position of, 217; mature, of *Loxsoma Cunninghamii*, 214 (Fig. 206); mixed, 214; monangial, 208, 241; monangial, of *Lygodium*, 235 (Fig. 235); of *Corynepteris Essenghi*, 250 (Fig. 245); of *Dennstaedtia apiifolia*, 255 (Fig. 252); of *Dennstaedtia rubiginosa*, 215 (Fig. 208); of *Gleichenia*, 208; of *Gleichenia pectinata*, 212 (Fig. 203); of *Hymenophyllum Wilsoni*, 212 (Fig. 204); of *Matteuccia intermedia*, 229 (Fig. 226); of *Thyrsopteris elegans*, 213 (Fig. 205); origin of, in *Pteridium aquilinum*, 223 (Fig. 218); output per, 266; position of, 62, 216, 240; primitive marginal position of, 339; protection of, 234, 237; relations of, 62; simple, 208, 209, 210, 211; superficial position of, 217; young, of *Dicksonia Scheidei*, 218 (Fig. 211), 219 (Fig. 212); young, of *Hypolepis repens*, 222 (Fig. 217)
Spaces, intercellular, 180, 184
Spermatocytes, 16, 18 (Fig. 21), 286; differences in number of, in antheridia of various Ferns, 291 (Fig. 281); diminishing output of, 292; number of, 292
Spermatozoid, 16, 17 (Fig. 17), 286, 298
Sperm-cells, 286
Sperm-numbers, 296
Sphenophylleae, 103 (Fig. 99)
Sphenopteris crenata, aphlebiae of, 99 (Fig. 95)
Spindle, polarity of, 313; primitive, 313; primitive, theory of, 344
Sporangia, 209 (Fig. 199); of *Botrychium Lunaria*, 208 (Fig. 198); analogies between, and gametangia, 285, 289; brevicidal, 255; formed simultaneously, 339; longicidal, 255; of *Asplenium Trichomanes*, 245 (Fig. 239); of *Thyrsopteris elegans*, 213 (Fig. 205); of *Zygopteris*, 209 (Fig. 199); parallel between, and antheridia, 296; protection of, 240; segmentation of, 110 (Fig. 106), 244 (Fig. 238); solitary, 207; two-sided initial cell of, 246; two-sided segmentation of, 246 (Fig. 241)
Sporangial segmentation, 115; stalks, sections of, 248 (Fig. 243)
Sporangium, 207 (Fig. 197), 214 (Fig. 207), 242; capsule of, 242; definition of, 243; development of, 13 (Fig. 16); Eusporangiate, 270; Leptosporangiate, 270; of *Angiopteris evecta*, 250 (Fig. 244); of *Botrychium daucifolium*, 247 (Fig. 242); of *Lygodium circinatum*, 265 (Fig. 260); of *Osmunda regalis*, 265 (Fig. 259); of *Phlebodium aureum*, 245 (Fig. 240); of *Pilularia americana*, 256 (Fig. 253); of *Platycerium*, 254 (Fig. 251); of *Senftenbergia elegans*, 252 (Fig. 247); of *Trichomanes speciosum*, 214 (Fig. 207); ontogenetic origin of, 243; origin of, 62; position of, in *Mohria caffrorum*, 220 (Fig. 214); segmentation of, 12 (Fig. 15); stalk of, 247; wall of, 249
Spore of *Osmunda regalis*, 260 (Fig. 256)
Sporeling, stele of, 180
Spore-mother-cells, 13, 258; number of, 292; of sporangium, 242
Spore-output, 261; actual, 270; large, importance of, 271; numerical, 62; per sporangium, 264
Spore-producing organs, 206
Spores, 11, 13, 20; germination of, 15 (Fig. 18); increase by, 206; of *Aspidium trifoliatum*, 261 (Fig. 257); of *Platyzoma*, 265 (Fig. 258); of *Polypodium vulgare*, 260 (Fig. 256); of sporangium, 242
Spore-tetrad, 14
Sporogenous cells, 257; tissue of *Marattia fraxinea*, 258 (Fig. 254); tissue of *Ophioglossum vulgatum*, 260 (Fig. 255)
Sporogonia of Liverworts, 117
Sporophyll of *Cheiropleuria*, 233 (Fig. 231); of *Platycerium angolense*, 233 (Fig. 231)
Sporophylls, 11
Sporophyte, 21; adaptation of, 49; diploid, 21; embryology of, 63; haploid, of *Lastraea pseudo-mas*, var. *cristata*, 326; primitive, 337; young, of *Botrychium obliquum*, 305 (Fig. 289); young, of *Tmesipteris*, 341 (Fig. 309)
Sporophytic budding, 206, 320
Stalk, four-rowed, of *Cheiropleuria*, 254 (Fig. 251); of sporangium, 242
Stauropteris, sporangia of, 249
S. oldhamia, sporangium of, 207 (Fig. 197)
Stelar construction, plan of, of *Gleichenia pectinata*, 144 (Fig. 134); construction, plan of, of *Loxsoma Cunninghamii*, 145 (Fig. 135); morphology, 177; structure,

reconstruction of, of *Botrychium Lunaria*, 131 (Fig. 124); system of *Matonia pectinata*, 154 (Fig. 148); system of *Stenochlaena tenuifolia*, showing perforations, 150 (Fig. 141); theory, 138
Stele, 57, 139; cauline, 139; conical form of, 180; disintegration of, 135, 157, 189 (Fig. 180), 191; of *Ankyropteris Grayi*, 123 (Fig. 117), 182; of *Asterochlaena laxa*, 183; of *Botrychium virginianum*, 137 (Fig. 129); of *Lindsaya linearis*, 146 (Fig. 136); of *Metaclepsydropsis duplex*, 128 (Fig. 122); of small plant of *Platyzoma*, 132 (Fig. 125); of sporeling, 180; of *Trichomanes scandens*, 123 (Fig. 116); soleno-xylic, 124
Steles, actual size of, in living and fossil Osmundaceae, 184 (Fig. 176); xylem of, outlines of, 183 (Fig. 175)
Stellate hairs of *Polypodium lingua*, 200 (Fig. 188)
Stem, apex of, of *Angiopteris evecta*, 113 (Fig. 112); apex of, in *Osmunda regalis*, 9 (Fig. 10); apex of, in *Trichomanes radicans*, 107 (Fig. 101); conical, of *Polypodium vulgare*, 179 (Fig. 171); conical, of *Pteris podophylla*, 187 (Fig. 178); dictyostelic, 155 (Fig. 149); solenostelic, 155 (Fig. 149); transverse section of, in *Botryopteris cylindrica*, 182 (Fig. 174); vascular tissue of, 61; young, conical form of, 179
Stem-apices, 106 (Fig. 100)
Stems of Ferns, 189 (Fig. 180)
Stenochlaena sorbifolia, fertile pinnae of, 232 (Fig. 230)
S. tenuifolia, stelar system of, showing perforations, 150 (Fig. 141)
Stiff hairs of *Gleichenia pectinata*, 202
Stipular growths, 100
Stipules of *Angiopteris*, 100 (Fig. 96)
Stock of *Dryopteris*, 149 (Fig. 140)
Stolons of *Nephrolepis*, 190 (Fig. 182)
Stomium, 13, 251, 256; of sporangium, 242; position of, 254 (Fig. 250)
Storage and climbing, 35
Storage-pith, 130
Storage-stock of *Ophioglossum palmatum*, 30 (Fig. 36)
Strand, compensation, 152
Strands, accessory, 156
Structure, progressive simplification of, 117
Structures, similar, principle of, 177, 191
Sun-form, 38
Superficial position of sorus, 217
Superficiales, 140
Suppression of certain parts, 314
Surface, diffusing, area of, 178; proportion of, to bulk of organ, 177, 181
Surfaces, limiting, 178
Suspensor, 300, 342; elimination of, 302; liable to be suppressed, 308; of *Botrychium obliquum*, 305; of *Helminthostachys*, 304; of Marattiaceae, 303
Symbiotic fungus, 284 (Fig. 274)
Symmetry of shoot, 67; radial, 66
Sympodial dichotomy, 91
Synangia, 234; of *Christensenia*, 249; of *Danaea*, 249; of *Marattia*, 249
Synangium, 208, 242
Synapsis, 22
Syngamy, 298

Tapetum, 13, 258; of sporangium, 242
Terminal branching, dichotomous, 337
Tetrad-division, 23 (Fig. 30), 258
Thamnopteris Schlechtendalii, 130; leaf-trace in, 162 (Fig. 155)
Thyrsopteris elegans, 153; sorus of, 213 (Fig. 205); sporangia of, 213 (Fig. 205)
Tier, epibasal, 301; hypobasal, 19, 301
Tissue, sporogenous, of *Ophioglossum vulgatum*, 260 (Fig. 255)
Tissues, dermal, 195
Tmesipteris, young sporophyte of, 341 (Fig. 309)
Todea barbara, Frontispiece; sporangium of, 251 (Fig. 246)
T. superba, juvenile leaf of, 85 (Fig. 76)
Tracheides, 6; isolated, 126; of *Pteridium*, 7 (Fig. 6); scalariform, 7
Tree-Ferns, 54
Trichomanes, gemmae of, 280
T. alatum, prothallus and gemmae of, 281 (Fig. 271)
T. pyxidiferum, archegoniophore of, 282 (Fig. 272)
T. radicans, apex of, 107 (Fig. 101); axillary bud of, 70 (Fig. 63)
T. reniforme, 86 (Fig. 78)
T. rigidum, filamentous prothallus of, 280 (Fig. 270)
T. scandens, stele of, 123 (Fig. 116)
T. speciosum, sporangium of, 214 (Fig. 207)
T. venosum, Frontispiece
Trunk, bifurcated, of *Cyathea medullaris*, 158 (Fig. 152)
Tubers of *Nephrolepis*, 43
Tubicaulis solenites, 128 (Fig. 121), 161 (Fig. 154)
Two-fifths divergence, 68
Typical number, 262

Uniseriate sori, 210
Upgrade development from protostele, 133
Urns, sac-shaped, 44 (Fig. 51)

Variability, 271
Vascular anatomy, 192; construction, primitive, 338; skeleton (Fig. 1), 2; system, 120; system, cauline, 139; system,

common, 139; system at node of *Loxsoma Cunninghamii*, 141 (Fig. 130); system of *Cibotium* (*Dicksonia*) *Barometz*, 170 (Fig. 163); system of *Gymnogramme japonica*, 167 (Fig. 160); system of axis, 140; system of *Pellaea rotundifolia*, 148 (Fig. 139); system of *Pteris elata*, 152 (Fig. 145); tissue of leaf, 61; tissue of stem, 61
Vegetative mitosis, 22 (Fig. 29); propagation, 10
Veins, fusion of, 64
Venation, 160, 174; of leaf, 61; of leaf-blade, 83
Venter, 295
Ventilating areas, 203; system, 184
Ventilation, 130
Ventral-canal-cell, 17
Vestigial annulus, 255; lower indusium, 222
Vestiture, 57

Wall, basal, 19, 301; distal region of, 252; proximal region of, 253
Walls, principal, 107, 112; sextant, 107, 112
Water-glands of *Polypodium vulgare*, 205 (Fig. 196)
Water-storage, 43
Webbing, 83, 102; progressive, 85
Wings of leaves of *Angiopteris*, 112 (Fig. 109)
Woodsia ilvensis, antheridia of, 287 (Fig. 277); filamentous type of, 275 (Fig. 266)
Woodwardia, 11 (Fig. 13)

Xerophytic habit, 41
Xylem, 6; of steles, outlines of, 183 (Fig. 175)
Xylem-core, solid, of *Botryopteris forensis*, 127
Xylic-gaps, 143
Xylic-perforations, 149

Young leaf of *Ceratopteris*, 9 (Fig. 11); leaves of *Schizaea rupestris*, 220 (Fig. 213); plant of *Anemia phyllitidis*, 125 (Fig. 118); plant of *Botrychium*, 184; plant of *Helminthostachys*, 184; plant of *Pteridium aquilinum*, 75 (Fig. 69); sori of *Dipteris conjugata*, 216 (Fig. 209); sori of *Histiopteris incisa*, 224 (Fig. 219); sori of *Pteris serrulata*, 224 (Fig. 220); sorus of *Dicksonia Scheidei*, 218 (Fig. 211), 219 (Fig. 212); sorus of *Hypolepis repens*, 222 (Fig. 217); sporophyte of *Botrychium obliquum*, 305 (Fig. 289); sporophyte of *Tmesipteris*, 341 (Fig. 309); stem, conical form of, 179

Zalesskia, 129
Zygopterideae, leaves of, 81
Zygopteris, sporangia of, 209 (Fig. 199)
Z. corrugata, apical meristem of leaf of, 109 (Fig. 104)
Zygote, 19, 20, 298

CAMBRIDGE: PRINTED BY THE SYNDICS OF THE PRESS
AT THE UNIVERSITY PRESS

Printed in the United States
By Bookmasters